Applied Environmental Materials Science for Sustainability

Takaomi Kobayashi
Nagaoka University of Technology, Japan

A volume in the Advances in Environmental Engineering and Green Technologies (AEEGT) Book Series

IGI Global
DISSEMINATOR OF KNOWLEDGE
www.igi-global.com

Published in the United States of America by
IGI Global
Information Science Reference (an imprint of IGI Global)
701 E. Chocolate Avenue
Hershey PA, USA 17033
Tel: 717-533-8845
Fax: 717-533-8661
E-mail: cust@igi-global.com
Web site: http://www.igi-global.com

Library of Congress Cataloging-in-Publication Data

Names: Kobayashi, Takaomi, 1958- editor.
Title: Applied environmental materials science for sustainability / Takaomi Kobayashi, editor.
Description: Hershey PA : Information Science Reference, [2017] | Includes bibliographical references.
Identifiers: LCCN 2016050384| ISBN 9781522519713 (hardcover) | ISBN 9781522519720 (ebook)
Subjects: LCSH: Materials science. | Renewable natural resources. | Green technology.
Classification: LCC TA418.5 .A67 2017 | DDC 620.1/1--dc23 LC record available at https://lccn.loc.gov/2016050384

This book is published in the IGI Global book series Advances in Environmental Engineering and Green Technologies (AEEGT) (ISSN: 2326-9162; eISSN: 2326-9170)

British Cataloguing in Publication Data
A Cataloguing in Publication record for this book is available from the British Library.

Advances in Environmental Engineering and Green Technologies (AEEGT) Book Series

ISSN:2326-9162
EISSN:2326-9170

Mission

Growing awareness and an increased focus on environmental issues such as climate change, energy use, and loss of non-renewable resources have brought about a greater need for research that provides potential solutions to these problems. Research in environmental science and engineering continues to play a vital role in uncovering new opportunities for a "green" future.

The **Advances in Environmental Engineering and Green Technologies (AEEGT)** book series is a mouthpiece for research in all aspects of environmental science, earth science, and green initiatives. This series supports the ongoing research in this field through publishing books that discuss topics within environmental engineering or that deal with the interdisciplinary field of green technologies.

Coverage

- Green Transportation
- Policies Involving Green Technologies and Environmental Engineering
- Sustainable Communities
- Radioactive Waste Treatment
- Green Technology
- Biofilters and Biofiltration
- Water Supply and Treatment
- Renewable Energy
- Contaminated Site Remediation
- Air Quality

IGI Global is currently accepting manuscripts for publication within this series. To submit a proposal for a volume in this series, please contact our Acquisition Editors at Acquisitions@igi-global.com or visit: http://www.igi-global.com/publish/.

The Advances in Environmental Engineering and Green Technologies (AEEGT) Book Series (ISSN 2326-9162) is published by IGI Global, 701 E. Chocolate Avenue, Hershey, PA 17033-1240, USA, www.igi-global.com. This series is composed of titles available for purchase individually; each title is edited to be contextually exclusive from any other title within the series. For pricing and ordering information please visit http://www.igi-global.com/book-series/advances-environmental-engineering-green-technologies/73679. Postmaster: Send all address changes to above address.

Titles in this Series

For a list of additional titles in this series, please visit: www.igi-global.com

Reconsidering the Impact of Climate Change on Global Water Supply, Use, and Management
Prakash Rao (Symbiosis International University, India) and Yogesh Patil (Symbiosis International University, India)
Information Science Reference • copyright 2017 • 430pp • H/C (ISBN: 9781522510468) • US $215.00 (our price)

Environmental Sustainability and Climate Change Adaptation Strategies
Wayne Ganpat (The University of the West Indies, Trinidad and Tobago) and Wendy-Ann Isaac (The University of the West Indies, Trinidad and Tobago)
Information Science Reference • copyright 2017 • 406pp • H/C (ISBN: 9781522516071) • US $200.00 (our price)

Technologies for the Treatment and Recovery of Nutrients from Industrial Wastewater
Ángeles Val del Río (University of Santiago de Compostela, Spain) José Luis Campos Gómez (University Adolfo Ibáñez, Chile) and Anuska Mosquera Corral (University of Santiago de Compostela, Spain)
Information Science Reference • copyright 2017 • 391pp • H/C (ISBN: 9781522510376) • US $200.00 (our price)

Agricultural Development and Food Security in Developing Nations
Wayne G. Ganpat (The University of the West Indies, Trinidad and Tobago) Ronald Dyer (Grenoble School of Management, France) and Wendy-Ann P. Isaac (The University of the West Indies, Trinidad and Tobago)
Information Science Reference • copyright 2017 • 370pp • H/C (ISBN: 9781522509424) • US $200.00 (our price)

Oceanographic and Marine Cross-Domain Data Management for Sustainable Development
Paolo Diviacco (Istituto Nazionale di Oceanografia e di Geofisica Sperimentale (OGS), Italy) Adam Leadbetter (Marine Institute, Ireland) and Helen Glaves (British Geological Survey, UK)
Information Science Reference • copyright 2017 • 425pp • H/C (ISBN: 9781522507000) • US $205.00 (our price)

Examining the Development, Regulation, and Consumption of Functional Foods
Sailas Benjamin (University of Calicut, India)
Medical Information Science Reference • copyright 2017 • 344pp • H/C (ISBN: 9781522506072) • US $185.00 (our price)

Applying Nanotechnology for Environmental Sustainability
Sung Hee Joo (University of Miami, USA)
Information Science Reference • copyright 2017 • 558pp • H/C (ISBN: 9781522505853) • US $225.00 (our price)

Exploring the Nutrition and Health Benefits of Functional Foods
Hossain Uddin Shekhar (University of Dhaka, Bangladesh) Zakir Hossain Howlader (University of Dhaka, Bangladesh) and Yearul Kabir (University of Dhaka, Bangladesh)
Medical Information Science Reference • copyright 2017 • 523pp • H/C (ISBN: 9781522505914) • US $200.00 (our price)

IGI GLOBAL
DISSEMINATOR OF KNOWLEDGE
www.igi-global.com
701 E. Chocolate Ave., Hershey, PA 17033
Order online at www.igi-global.com or call 717-533-8845 x100
To place a standing order for titles released in this series, contact: cust@igi-global.com
Mon-Fri 8:00 am - 5:00 pm (est) or fax 24 hours a day 717-533-8661

Editorial Advisory Board

Table of Contents

Detailed Table of Contents

Advances in environmental materials are described especially concerning about utilization of natural source and their waste, for example biomass and their re-uses. They strongly support development to sustainable society. So, these classify and demonstrate environmental technologies, which strongly contribute to overcome pollution environment for the purpose of sustainable society. In view point of green chemistry and the related technologies to representative biomasses and inorganic wastes are applying for advanced functional materials. Since biomass and inorganic wastes are very interesting candidates for regenerated materials, it is very meaningful to understand and attention to environmentally and friendly materials and technologies. This is because that an increasing environmental concerns to sustainability.

Among all biomass sources, cellulose is the most diverse material sourced from plants and is abundant in nature. It possesses some promising properties such as biodegradability, biocompatibility and hydrophilicity. In several industrial biomasses, the waste products especially for food manufacture have high content of cellulose and their disposals are serious in reuse problem. Therefore, people have given attention with regard to their regeneration to the useful materials. In this chapter, it is worth noting to describe that natural biomass cellulose is usefully applied with the regeneration to functional hydrogel films. The treatment processes of the bagasse waste used as a main source of cellulose and the fabrication of hydrogel films are described and their potential characteristics, especially on the medical field are reviewed.

Chapter 3

Although natural rubber is a challenging elastomeric material for both dynamic and static engineering applications, there are some inherent drawbacks such as the poor oxygen, ozone, oil and heat resistance and also the low compatibility with the polar materials. To loosen these inferior properties and also to transform natural rubber into new polymeric materials, natural rubber needs to be either chemically or physically modified. Moreover, to make natural rubber become a promising elastomeric material for the advanced application, the development of natural rubber with the superior mechanical strength, the low gas permeability or the high conductivity is recently focused. And these can be done by reinforcing natural rubber with nanofillers. This chapter presents a broad review on the recent research and development of natural rubber including the modification of natural rubber and the preparation of various natural rubber nanocomposites for advanced application.

Chapter 4

Natural rubber (NR) is representative biomass polymer and the effective uses are strongly contributed to sustainable society. This chapter presents the innovative and advanced rubber nanocomposites with polystyrene-encapsulated silica nanohybrids (PS-nSiO_2) subsequently used as a nanofiller for NR and NR/styrene butadiene rubber (NR/SBR). The PS-nSiO_2 were prepared via 'in situ' differential microemulsion polymerization. The core-shell nanohybrids of PS-nSiO_2 were achieved with an average diameter of 40 nm using a smaller amount of surfactant, compared to microemulsion polymerization method. Moreover, the effects of the NR and NR/SBR filled with PS-nSiO_2 nanohybrids on the mechanical properties, thermal stability, flammability and morphology are also discussed. The results indicated that the encapsulation of nSiO_2 with PS can provide not only the well-dispersion of nanoparticles in the rubber matrix but also the synergistic properties of two components from the polymer and the inorganic nanoparticles by improving mechanical properties, thermal stability and flammability of rubber nanocomposites.

Chapter 5

The consumption of plastic products from petrochemical feedstock has increased sharply resulting in plastic waste problems while raw materials from fossil fuels tend to decrease rapidly. Researchers and the plastic industry have since proposed a sustainable solution through the development of bioplastics. Ideally, bioplastics which are synthesized from renewable bioresources normally render biodegradability in appropriate conditions. Polyester, one of the most diversely used synthetic polymers today, is an ideal choice for biodegradable polymers due to the relative ease of breaking ester linkages. Poly(lactic acid) or polylactide or PLA which is a thermoplastic polyester with many advantageous properties, for instance, environmentally friendly, biocompatibility, processability, and high chemical resistance is now available in the plastic market as a promising bioplastics. However, the cost of PLA is still much higher than that of general commodity plastics. In order to make PLA commercially competitive, advanced and innovative applications should thus be explored. In this chapter, technological background of PLA production

Geopolymer is alumino-silicate material which can apply for many applications due to that geopolymers have several attractive properties of high strength, low permeability, high acid resistance, hazardous materials, and immobilization of toxic materials. This chapter presents synthesis of the dense and the porous structures of geopolymer by using fly ash and industrial waste as the raw materials. Aluminum hydroxide waste (Al-waste) and fly ash (FA) were used to synthesis the dense geopolymers for cement materials. The Al-waste based geopolymer influenced the geopolymer strength, when sodium hydroxide (NaOH) concentration was changed at different curing temperatures. As preliminary microwave oven was exposed in the lower NaOH paste of geopolymer, the successful synthesis of geopolymer cement was obtained. In addition, porous fly ash geopolymers was achieved using a household microwave oven. The geopolymer paste was cured within 1 min by using a microwave oven at different output power. Porous geopolymers were formed immediately as ued at 850 W power of the microwave oven.

Conventional vapour compression system is one of the most utilized cycles in refrigeration and air conditioning systems, due to its compact size, the relatively low running cost, the high coefficient of performance and the wide range of the operating temperatures. Nevertheless, the system suffers from the high initial cost and the high-energy consumption. Unlike the vapour compression cycle, adsorption heat pumps have the advantage of utilizing waste heat reducing the energy consumption and the carbon emissions. Activated carbon is a porous adsorbent material that can be efficiently used in low temperature adsorption refrigeration systems. This chapter reviews the recent developments in the compact adsorption cooling systems using activated carbon regarding the enhancement of the material properties, the design of the sustainable adsorption systems and their environmental and cost perspectives.

Petroleum reserves have been decreasing in recent years and microalgae are attractive as a potential source of new biomass petroleum. Microalgae are unicellar microscopic algae and most species microalgae produce lipids. In particular, Botryococcus braunii produces large amount of lipids found with nearly 70% on the basis of the dry weight. This chapter reviews high lipid-producing microalgae found from

Okinawa area around National Institute of Technology, Okinawa College (NIT, Okinawa). The microalgae collected were isolated on an AF-6 agar plates, and incubated in AF-6 medium. The fatty acids were extracted from the algae, converted into fatty acid methyl esters, and analysed by GC/MS. As a result, two microalgae strains were identified that the produced fatty acids was loaded in the algae with nearly 20% in the dry weight base. In addition, these two microalgae strains produced palmitic acid as nearly 40% of the total produced lipids. Therefore, the two microalga strains isolated are potentially and highly efficient for the organisms applied for the production of biodiesel fuel.

This chapter presents an overview of the study of cellulose sheets used as a bio-membrane in water/alcohol pervaporation for a variety of purposes in order to offer an environmentally sustainable, renewable, low-energy use and economical alternative. Particular interest is the utilization of readily available biological materials such as cellulose and chitosan to create biomass polymer membranes for the environmentally responsible and sustainable process of reclaiming alcohol from alcohol/water mixtures through the use of cellulose in the pervaporation process. This process is a promising and innovative alternative to the more energy intensive, environmentally destructive methods and materials currently used.

Bacterial cellulose, synthesized by bacteria, is one of the most highly pure cellulose sources. It has gained a great attention due to its unique properties. With molecular similarity to cellulose derived from plants, bacterial cellulose can be modified based on diverse techniques established for the plant-derived cellulose. Modification of cellulose has become one of the major areas of cellulose research to provide cellulose-based materials with novel properties. The progress in cellulose science has also opened up more potentials for bacterial cellulose. This chapter describes an overview of biosyntheses, modifications, and applications of bacterial cellulose.

The amounts of "Electronic wastes" including heavy metals are increasing day by day. Such waste is in the rich resource of various metals having the precious metals. Therefore, these wastes are considered as urban mine, if people successfully can separate them to each source. In sustainable viewpoints, separation technologies applied for such electronics waste are essential and important to efficiently recover various metals at a low concentration from these sources. This chapter reviews functional adsorbents made of polymers, ionic liquid, and dendrimer. Also, membrane technology is introduced as separation toll for heavy metals. Among them, topics of phytoremediation are made as an effective sustainable method, utilizing certain plants to clean up the environmental contaminants. Here, plants are able to remove harmful chemicals such as metals, which are present in the soil, when their roots absorb water and nutrients from the contaminated soil, sediment and surface, or ground water. The contaminants are removed by trapping them into harvestable plant biomass. Furthermore, cleanup methods of environments and recovery of precious and rare metals are mentioned for sources of urban and submarine mines with low cost and high recovery efficiency.

Governments in developed countries specify water quality standards to be applied in their own countries. Recently the allowance values of harmful metals are getting tighter for environmental water and industrial wastewater throughout the world. For example, World Health Organization (WHO) recommends the criterion value of 3 ppb for cadmium for protection of human health, furthermore, governments of some countries provide the wastewater standard for zinc for conservation of aquatic life. High-performance analytical instruments such as AAS, ICP, and ICP-MS have been employed as official analytical methods of ultra-trace elements. Besides the costly initial and running expenses and the material and energy consuming of the instruments, specific technical skills and long analytical time are required for sample pre-treatment. Therefore, simple yet highly sensitive and eco-friendly on-site methods have been demanded for quick judgment of industrial and environmental water in developed countries as well as evaluation of drinking water in developing countries.

Ionic liquids (ILs) have been known as a class of green solvent and played an important role in green chemistry in recently years. Materials based on ILs, such as poly(ionic liquid) and polymer/ILs composites, are also widely investigated and possess many unique properties and applications. In this chapter, green solvent IL is reviewed and the responsive behavior of ILs based materials to external stimulus, like ultrasound (US), is introduced. Especially, it is described that FT-IR spectroscopy was effective for analyzing hydrogen bonding in ILs, polymeric ILs and PVA/IL composites. In addition, shear viscosity and viscoelasticity measurement and techniques of two-dimensional correlation and deconvolution are applied to analyze the change of interaction between IL to different substances and US responded behavior.

Preface

Since editor decides to ask each author who contributes each book chapter for sustainable matters, we have firsthand experience of how the materials are important and made the description understanding advanced topics concerning with environmental materials. This should broaden to students of engineering in universities to be known in well reviewing about the importance of such material and their applications. So, this book describes in details about advance in environmental materials, especially for utilization of biomass and waste for advanced materials and classifies environmental materials, which strongly contribute to sustainable society in coming future and about importance in one book for next generation people.

This book surveys advanced materials prepared from of biomass and waste in Chapter 1. Regeneration or direct utilization of waste plant sources can provide a new generation of recyclable-engineered materials. Understanding the structures and functional characteristics of these novel biomass materials can expedite the design and fabrication for eco-friendly, recyclable to advanced materials. In Chapter 2, firstly, natural plant polymer, cellulose, originated from bagasse wastes is utilized for the regeneration process to cellulose and is converted by a new preparation technique for pure cellulose hydrogel films. Among biomass polymers, this book focuses functional polymers sourced from biomass natural rubber (NR) and their reinforcement technique, in Chapters 3 and 4. These biomass NR plants are economic agricultural products that widely utilize in many applications. Although NR is known to exhibit numerous outstanding properties, reinforcing fillers are necessarily added into NR in order to enhance the appropriate properties for a variety of rubber applications. Especially, NRs in Thailand are introduced by advanced topics of NR modification. In Chapter 5, poly(lactic Acid) (PLA) generated for advanced materials is introduced with blending PLA and nanohybridization. Innovative techniques are described in PLA blends and composites with technological processing. Chapter 6 describes inorganic-organic composite materials from liquid NR with its unique properties. The sol-gel technique is explained as excellent technique for the composite materials. Clay minerals can be converted to porous materials and their application, challenge and perspective are shown in Chapter 7. Among inorganic sustainable materials for Chapter 8, geopolymers is one of representative sustainable materials sourced with fly ash and industrial aluminum wastes. Chapter 9 explains sustainable cooling research and application with using activated carbon adsorbents in energy saving systems and their environmental impact. Chapter 10 contains analysis of lipids produced by microalgae isolated from the area around Okinawa, Japan. As new potential energy source in petroleum reserves, attention is focused to biomass petroleum with microalgae. Chitosan and their hybrid materials are introduced for environment treatment materials in Chapter 11. The use of biomass polymer sheets in water/alcohol pervaporation can provide energy saving alcohol production, which is candidate for sustainable energy source. Chapter 12 describes bacterial cellulose for biosyntheses and applications. Bacterial cellulose can be modified based on diverse techniques as established for the plant-derived cellulose. Plant sustainable materials are applied for metal ion separation with functional adsorbents in Chapter 13. Phytoremediation used as decontamination technologies for heavy metal pollution is also described. There are several techniques for eco-friendly on-site water analyses for ultra-trace harmful ions. Chapter 14 mentions powerful analytical tools for the environmental monitoring of metal ions with water quality standard and current analytical methods. In the final chapter, Chapter 15, ionic liquids and poly(ionic liquid)s is introduced for uses of green solvent.

The editor is grateful to the many contributors who have aided the production of this edition. I am confident that the readers will find the present book of great interest. Therefore, it was very meaningful to edit the articles for this book.

With cooperation of editor's friends, I am grateful to faculty members of Chularlongkorn university in Thai land, Hanoi University of Science and Technology and National University of Ho Chiminh City Science in Vietnam and Science Malaysia University in Malaysia for their participations and to be strong supporters with discussion to the contents of this book. We believe that this will make significant contribution to improving knowledge and skill for young generation engineering students.

I would like to thank Ms. Naomi Tominaga and students in Nagaoka University of Technology for their assistance in the preparation of this book. Finally, the editor thanks for cooperative support on Ms. Nobuko Kobayashi.

Takaomi Kobayashi
Nagaoka University of Technology, Japan

Introduction

This book describes advances in environmental materials, especially concerning the utilization of biomass and waste for advanced materials. So, we classify and describe Environmental materials, which strongly contribute to sustainable society in view point of representative biomasses and inorganic wastes for advanced functional materials. Since biomass and inorganic wastes are very interesting candidates for regenerated materials, it is very meaningful to publish such articles by this book in Environmental Materials. Also, recently, attention has been environmentally and friendly paid to bio-based materials, because of an increasing environmental concern to sustainability in addition with their bioactive properties.

This book surveys advanced materials prepared from of biomass and waste. Regeneration or direct utilization of waste plant sources can provide a new generation of recyclable-engineered materials. Understanding the structures and functional characteristics of these novel biomass materials will expedite the design and fabrication for eco-friendly, recyclable, advanced materials, especially for cellulose. Firstly, natural plant polymer, cellulose, originated from bagasse wastes is utilized for the regeneration process to cellulose and is converted by a new preparation technique for pure cellulose hydrogel films. Among biomass polymers, this book will focus functional polymers sourced from biomass cellulose, natural rubber (NR) and their reinforcement technique. These biomass plants including NR are economic agricultural products that widely utilize in many applications. It is widely known to possess many attractive properties including high green strength, outstanding tensile and tear strength, high resilience, good abrasion resistance, excellent dynamic properties, high tack and low cost. Among them, although NR is known to exhibit numerous outstanding properties, reinforcing fillers are necessarily added into NR in order to enhance the appropriate properties for a variety of rubber applications. Especially, NRs in Thailand are introduced by topics of NR modification, blending with polylactic acid (PLA) and nanohybridization. In addition, since consumption of plastic products from petrochemical feedstock has increased sharply resulting in plastic waste problems, the development of bioplastics ideally has been proposed as a sustainable solution. The detail in the applications as biobased edible packaging materials, encapsulation matrix and inorganic composites is mentioned as advanced materials.

Inorganic nanomaterials based on metallic or metal oxide composition are extensively used in a variety of products. It will be introduced that they are transported to wastewater treatment plants and deposited in biosolids, which eventually end in soils. When nanomaterials are used in remediation process, it is described that they are directly released into the soil or groundwater. Geopolymerization is an effective manner to process and reuse silicates, aluminates or aluminosilicates in mineral waste. It is known that geopolymers are alkaline activated aluminosilicates consisting of reactive species of SiO_2 and Al_2O_3 to produce high-strength materials. In addition, the reactive species consisting of a solid reactive component that contains fly ashes, active clays, pozzolanas and slags as re-uses of their wastes for eco-friendly materials. By using active carbon adsorbents, environmental concerns of adsorption cooling system was mentioned for the aspects of energy saving. The activated carbon materials and recent research are classified for the refrigerant adsorption as eco-energy sustainable system. Study on the use of biomass polymer sheets in water/alcohol pervaporation is introduced as a sustainable source of alcohol energy. The biomass polymers of cellulose and chitosan are used for the pervaporation membranes with energy intensive, environmentally destructive methods. As a new potential source of energy in petroleum reserves, attention has been focused to microalgae as a new biomass petroleum. Microalgae are unicellar microscopic algae and most microalgae species produce lipids, *Botryococcus braunii*, which produces large amount of lipids that are harvested from the Okinawa island. In addition, bacterial cellulose for their biosyntheses, modifications, and applications is reviewed in the point of view of the cellulose modification and applications.

For decontamination process of pollutants, separation process becomes important technologies. There are functional adsorbents made of polymers, ionic liquid, and dendrimer targeted to heavy metal pollutants. The utilization of their separation technologies are mentioned in filter-membrane and adsorbents. Also, phytoremediation is made to utilizing certain plants to clean up the environmental contaminants in several heavy metal ions. Analyses for ultra-trace harmful ions are applied to the determination of several alternative techniques for water quality assessment. Eco-friendly on-site water analyses for ultra-trace techniques are presented as powerful analytical tools for the environmental monitoring of metal ions. It is attractive recently for ionic liquids and poly(ionic liquid)s that people use as green solvent, which is played an important role in green chemistry. These subjects above mentioned will classify and review in detail.

OBJECTIVE OF THE BOOK

This comprehensive and timely publication aims to be an essential reference source, building on the available literature in the field of electronic government in developing countries while providing for further research opportunities in this dynamic field. It is hoped that this text will provide the resources necessary for policy makers, technology developers and managers to adopt and implement electronic government platforms in developing nations across the globe.

Chapter 1 Introduction of Environmental Materials

Takaomi Kobayashi
Nagaoka University of Technology, Japan

ABSTRACT

Advances in environmental materials are described especially concerning about utilization of natural source and their waste, for example biomass and their re-uses. They strongly support development to sustainable society. So, these classify and demonstrate environmental technologies, which strongly contribute to overcome pollution environment for the purpose of sustainable society. In view point of green chemistry and the related technologies to representative biomasses and inorganic wastes are applying for advanced functional materials. Since biomass and inorganic wastes are very interesting candidates for regenerated materials, it is very meaningful to understand and attention to environmentally and friendly materials and technologies. This is because that an increasing environmental concerns to sustainability.

INTRODUCTION

Chemistry brings comfort in modern life. For example, clothing is made with a synthetic fiber, and dye of various clothing is a product of the chemical industry. Computers have each outside of home electric appliances made with plastic and metal. These are mostly made from oil and a mineral through many chemical steps. So, chemistry greatly contributes to securing of richness of the human and supports to construct the high civilization. On the other hand, pollution of air, water and soil increases as industry becomes active. The pollution has influenced human health and been injured. However, many developed countries including Japan have overcome the pollution now, and the global environment has been improved by green technologies. However, the problem of the global environment was gradually actualized. Resources and the acceleration of the energy consumption lead to anxiety of earth resources. Furthermore, the consumption of a large quantity of fossil fuel causes the sharp increases of carbon dioxide and soot particles such as particle pollution (PM2.5) in atmosphere (Liu, Hsu, Lee, Ye, Zheng, Liu, Li, & Cui, 2015). When the growth of the human reaches the limit by the resources drying up and environmental aggravation, the report "The limit grow" of 1972, in particular, Club of Rome, sounds an alarm bell

DOI: 10.4018/978-1-5225-1971-3.ch001

(Meadows, Meadows, Randers, & Behrens,1972). Therefore, this has reviewed the development of the economy that the earth is infinite. This also comments that to keep world balance is important. In addition, the toxic substance has an influence on the people and gives uneasiness in the society. For sustainable development, the planethood has various actions and global environment improvement has been directed to deeply consider it, for example, by United Nations Environment Programme (UNEP) (UNEP, 2014) In late 2014 world climate change was already affecting many contents in global warming. This gives strong impact to food security and therefore, emphasizes the chemistry and the related technology with the global community. It is desirable to develop the consideration for environment conservation, since green chemistry can solve problems in energy, a material, an environmental pollution to pull sustained society. In other words, because the green chemistry strongly relates to eco-friendly manufacturing, the harmony with environment and the process adapted to technical development leads to development of clean technology (Williams, 2001; Andrews, Brecombe, Jikel, & Brianreid, 2004). Recently, therefore, attention has been increasingly devoted to bio-based environmentally and friendly materials because of increasing environmental concerns related to sustainability.

BACKGROUND

As already mentioned, people should recognize a limit of the growth of the global environment. The coexistence of global environmental conservation and the industrial activation becomes extremely important. This is because that, as one of example, the danger signals such as the frequent abnormal weather strongly suggest an importance in sustainable society. From such a background, the practice of the concept called "green" becomes required in a chemical technique and materials development for the sustainable society. In the minimum waste and the recycling, people should bring highly efficient environmental symbiosis in energy technology, reuse of the biomass waste and the switch to a re-used material and to the remediation technology, which is applied to the earth pollution for the conservation of our environment. Therefore, the substantial use is urgent in this century. In this chapter, an action to the sustainable society and green chemistry is introduced in the points of view of the green technology using about biomass and waste re-utilization for environment materials.

Green Chemistry for Sustainably Renewable Sources

Sustainable means operating with a long-term mindset that includes responsibly maximizing our use of renewable resources. So one of the goal of sustainability is for renewable resource management (Rosenberg, Fogarty, Sissenwine, Beddington, & Shepherd, 1993). To provide both positive examples of sustainable resource use and lessons for future improvements in materials, a renewable resource is mainly an organic natural resource and can replenish to overcome usage and consumption. In such a background, United States Environmental Protection Agency (EPA) proposed green chemistry in 1994. In green chemistry, people design chemical products and processes for reduction or elimination of the use or generation of hazardous substances. Now, green chemistry strongly relates with the product life cycle by approaching its design, manufacture, use, and ultimate disposal. So it is also known as sustainable chemistry. The rapid development of new chemical technologies and the vast number of new chemical products in the last decades turned the attention of environmentalists to remedial actions for the negative impacts. Two decades after the implementation of EPA, the Pollution Prevention Act (1990) was created to enforce eco-friendly strategies. The EPA defines green chemistry as following principles (Figure 1) (EPA,1990); Prevention of pollution at the molecular level, application to all areas of chemistry, not a single discipline of chemistry, innovation of scientific solutions to real-world environmental problems, results in source reduction because it prevents the generation of pollution, reduction of the negative impacts of chemical products for better reflection on human health and the environment, sometimes elimination of hazard from existing products and processes, designs of chemical products and processes to reduce their intrinsic hazards.

What is different point in green chemistry and general remediation process unlike cleaning up pollution? The remediation processes involve treating waste streams or cleanup of environmental spills and other releases. So remediation mainly relates to removal process for hazardous materials from the environment. Green chemistry, on the other hand, keeps the hazardous materials out of the environment for purpose of clean up environmental contaminants. Here, this needs technology to remove pollutant. The green chemistry technology also includes remediation technology, but also relates to non-disposable technology. This contains that processes prevent waste, high energy efficiency and effective uses of renewable feedstocks and then can produce less hazardous chemicals with no toxicity to either humans or the environment. Also this minimizes potential for accidents with explosions, fires, and releases chemicals to the environment or less byproducts in the reaction. In addition these processes should have that the final product contains the maximum proportion of the renewable starting materials. To degrade after use to innocuous substances and to minimize waste in reaction, catalytic reactions without stoichiometric reagents are preferred for the prevention of chemical accumulation in the environment. Green chemistry may change all these negative impacts and through design, innovation and green processes to restore the planet's sustainable development.

To implement sustainable practices, green chemistry technologies can modify their practices in ways that provide better and cleaner tomorrow. The technology aims to be innovative to restore, protect and ensure the sustainability of water resources and also identifies the renewable energy potential of current and formerly contaminated lands. The next bre1akout to a solution of environmental problems is beneficial to improve environmental protection and human health.

Figure 1. Green chemistry

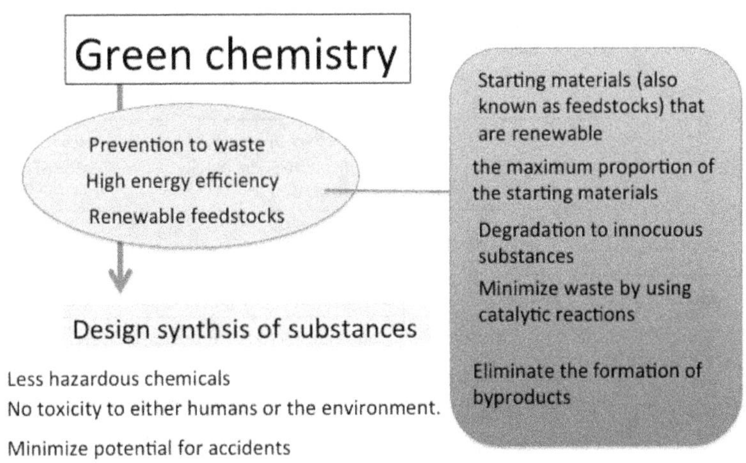

ENVIRONMENTAL MATERIALS IN SUSTAINABILITY

In such situation, now, materials science and technology intersects with the fields of environmental science and sustainability in green chemistry ways. Environmental materials are related with challenging a serious environmental problem such as an energy problem and the water pollution in a various materials technology. Both developing new materials and conducting basic science research to pollution and hazardous chemicals are included as environmental prevention technologies (Table 1). To date, most sustainability or green technologies have attempted to quantify sustainable materials and focused primarily on energy, material, and water conservation. The sustainable materials also cover lower regeneration cost and energy and long term uses for environmentally friendly materials.

Table 1. Prevention technologies for earth environment

Classification	Development Theme	Specific Examples	Materials & Technologies
Energy	Liquid fuel	Biofuel Biodiesel	Alcohol fermentation Vegetable oil Fats Peterorium
		Bioalchol	Methanol Ethanol Buthanol
	Gaseous fuel	Hydrogen	Photocatalysis Electro-decomposition of water
		Methane	Fermentation
	Solid fuel	Biomass	Tree Bagasse Coal
		Batteries Fuel cell	Electricity Hydrogen Methane Alchole
Regeneration of biomass and mine resources	Biomass	Biopolymer	Polysaccharide Protein Nucleic acid Natural rubber
		Bioplastics	Polyglycolic acid Polylactic acid
	Mine	Heavy metals	Cadmium Lead, Zinc
		Ash	Geopolymer
Environmental remediation technology	Air	Filter Adsorbents	Gas adsorbents
	Water	Chemical precipitation,	Precipitation induced by the addition of an alkali, sulfide, coagulant, or other reagent.
		Chemical stabilization	Using phosphate for treatment of solid mining wastes and reducing divalent heavy metals.
		Electrocoagulation	Technologies using an electrical current that coagulates organic constituents and suspended solids in water.
		Biochemical reactors	Microorganisms to transform contaminants in the treated water to another chemical and to increase pH; Sulfate reduction to sulfide and bicarbonate and heavy metal precipitation in high pH.
		Pressure-driven membrane separation	Thin barrier that permits selective mass transport of solutes and solvents across the barrier; Reverse osmosis, ultrafiltration and microfiltration.
		Ion exchange	Interchange or exchange of ions between a solid media and water containing ion; Various resins and peat or zeolites.
	Soil	Biosolids	Using organic phosphate to eliminate heavy metals
		Bioremediation	Controlling biological processes to reduce the contaminants of concern to target levels of heavy metals or fuel oils.
		Soil chemical oxidation	Treatment of chemical oxidation to reduction of contaminants.
		Thermal desorption	Heating certain contaminants to reduce their concentrations

There is a way of thinking called the carbon neutral. This is based on the biomass being generated by photosynthesis with CO_2 as raw materials. So, there is an idea that the atmospheric CO_2 is not to increase even if biomass is used to burn fuel derived from biomass. This is expressed for the Carbon neutral; the CO_2 income = the expenditure zero in the atmosphere, as seen in Figure 2 for carbon cycle of biomass. However, the problem is not easy. The biomass fuel uses fossil fuel from seeding into crops cultivation, crop work, fermentation, manufacturing process with the chemical processing. Therefore, when additional injection of fossil fuel is performed and then the amount of energy is more than that of the last fuel, the process cannot be said as carbon neutral. However, although the burned carbon is returned to the atmosphere as CO_2, these processes have happened for plants on earth as the carbon cycle.

The representative trial is biofuel. The production of the biofuel is provided by converting biomass to fuel and the biofuel is available as a substitute of gasoline and the diesel. Generally, the bioethanol is made more by the fermentation of the saccharide (Bajiai, 2013; Goldemberg, 2007; Yaman, 2004; Chavez-Guerrero,& Hinojosa, 2010). The alcohol fermentation is processed, for example, by yeast to the sugarcane bagasse containing saccharide. Through this, mono-saccharides is able to be converted into ethanol and carbon dioxide; $C_6H_{12}O_6 \rightarrow 2C_2H_5OH + CO_2$

On the other hand, oils and fats are the matters synthesized with the long chain fatty acid of 2-3 molecules to form ester combination of three values glycerin. The oils and fats are used generally by edible oil or butter. The oil and fat itself has high viscosity and is not suitable for an internal combustion engine, but, it becomes the fuel of the diesel engine with ester glyceride of mono-substitute of long chain molecules or the methanol ester. Therefore, the organic matter provided with exchange reaction of methanol with rapeseed oil or used edible oil is used for biodiesel fuel, which has low enough viscosity to allow proper atomization of the fuel. These biofuels have a characteristic not to include sulfur. However, the fossil fuel includes sulfur derived from creatures like an amino acid such as methionine and cysteine

Figure 2. Carbon cycle scheme of biomass

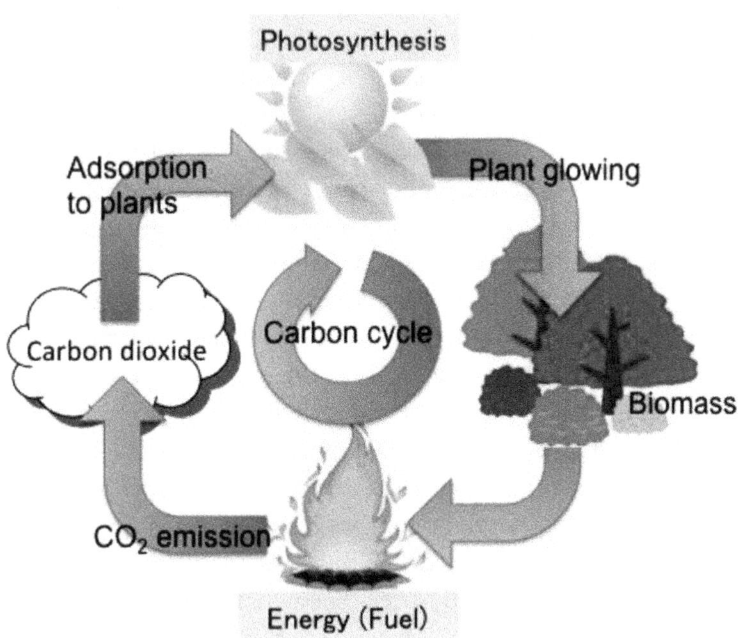

constituting protein. When the fuel is burned, sulfur of the fuel becomes the sulphur dioxide, SO_2. The SO_2 released in the atmosphere finally becomes the sulfuric acid and causes acid rain in earth environment. Therefore, the biofuel without the problem attracted attention from the side of the environmental conservation. After oil crisis of the 1970s, Brazil was considered to have the world's first sustainable biofuels economy. As the biofuel industry leader, ethanol fuel program is based on the most efficient agricultural technology in this country (Hofsetz & Silva, 2012), especially for sugarcane cultivation. Its sugarcane ethanol is the most successful alternative fuel. Since 1976 the government made it mandatory to blend anhydrous ethanol with gasoline, fluctuating is between 10% to 22%. Also, alternative fuels are known to including biodiesel, bioalcohol (methanol, ethanol, and butanol), chemically stored electricity (batteries and fuel cells), hydrogen, non-fossil methane, non-fossil natural gas, vegetable oil, and other biomass sources.

In addition to the biofuel, another representative group is biopolymer and bio-plastic (Pilla, 2011). Actually, there are various kinds of nature macromolecules in the natural world. A representative macromolecule is polysaccharide, protein and nucleic acid (Kalia, 2011). For instance, glucose forms α-1,4 bond of the starch which is kind of polysaccharides. Starch is divided into amylose and amylopectin by the structure. These are soluble in warm water and are decomposed by an enzyme in the body. Also, glucose performs α-1,4 bond of cellulose and the chief ingredient constitutes the cell wall of the plant (Brett & Waldron,1990). The polymeric chains interact with hydrogen bond and form fibrous fibers. Cellulose is relatively rigid and insoluble in water. The human cannot digest cellulose due to the absence of hydrolysis enzyme in our body. Furthermore, natural rubber (NR) is the candidate in gum tree growing in the tropics and 1,4-isoprene, $CH_2=C(CH_3)CH=CH_2$, polymerizes, as NR hardens the sap of the tree. Latex is existed as water dispersed organic component in the sap. This is a material composed primarily of cis-polyisoprene $[(C_5H_8)n]$. After the latex is refined and dried, the aggregated product is NR which has elasticity property and has been used for a tire, elastic string and ball. Protein is polymer condensed between the carboxylic acid and amino groups in amino acid molecule. The amide linkage to form here is called peptide bond in particular and belongs to polypeptide;-C(=O)- N(H)-. The peptide structure of protein molecules forms hydrogen bonding in the same molecules nearby relatively peptide and takes spiral structure (α helix) (Figure 3). Also, hydrogen bonding between the molecules next to each other can take plane structure (β sheet). The former spiral structure body is fully elastic and the latter plane structure can give superior mechanical strength.

On the other hand, there is biomass plastic based on idea of the carbon neutral. The representative plastic is polyester-based material of polyglycolic acid or polylactic acid. These have already came to be industrialized and are plastics made from creature resources (biomass) made from corn or sugarcane with much content of starch and the sugar mainly. A definition of these materials is to include a polymer derived from the organic resources which can reproduce as raw material and provided by synthesizing it chemically or biologically. In other words, the renewable resources of corn and sugarcane are used for raw materials of the plastic. Therefore, there is not influence for a rise in density of the atmospheric greenhouse gas and is less cause of the global warming, even if it returns to carbon dioxide by combustion after use. For these carbon neutral materials, following future is found; Even if burn and discard it, the income and expenditure of carbon dioxide is zero, When it burns up, heat of combustion is low, but no dioxin occurs. As seen in Figure 4, polylactic acid shows hydrolysis to lactic acid and then biodegradates to H_2O and CO_2 (Steavens, 2002; Ebnesajjsd, 2013). Litter is a problem with a very negative social and environmental impact. As an environmentally-friendly solution, one way to tackle this problem is to use biodegradable plastics.

Figure 3. Chemical structure of representative polysaccharides and biodegradable polymers

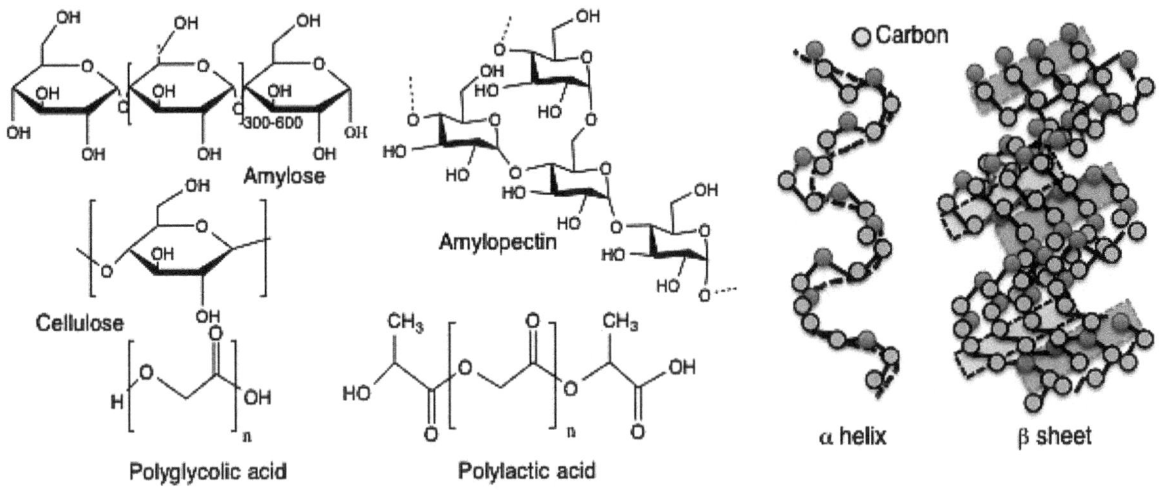

Figure 4. Polymerization and degradation scheme of polylactic acid

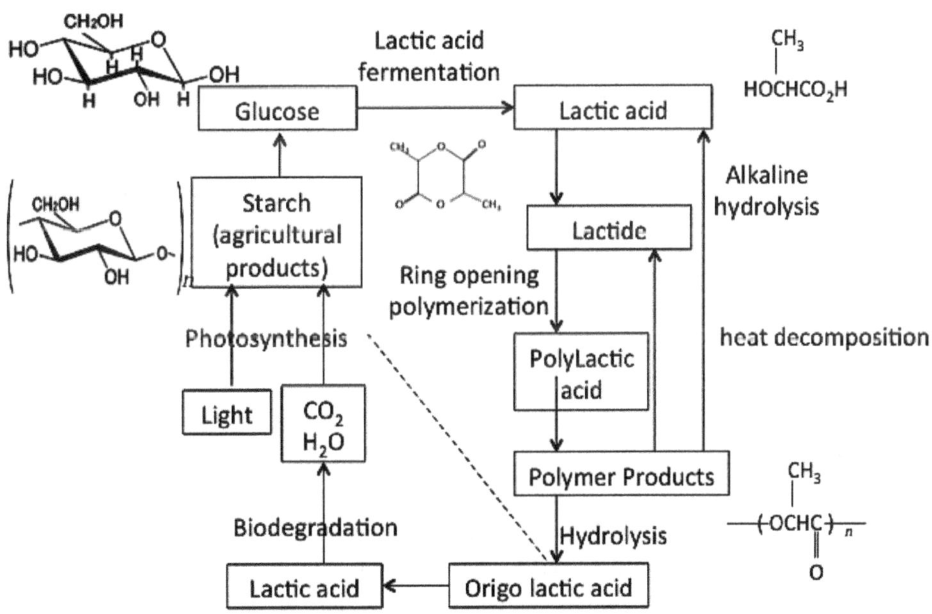

Environmental Green Technology

Environmental technologies, which have been developed in pollutant decontamination for the earth, incorporates environmental materials. The contribution of environmental materials to quality of the water purification and the detection of the pollutant is in particular large. The water is a resource circulating universally because of supporting our ecosystem. Therefore, it has deep relation to environment, industry and life. As shown in Figure 5, the water circulates in various forms through earth environment, as the evaporation, condensation, raining or snowing, running off, infiltration and flowing occur in the natural

Figure 5. Water cycle with contaminants through hydrosphere-lithosphere-atmosphere

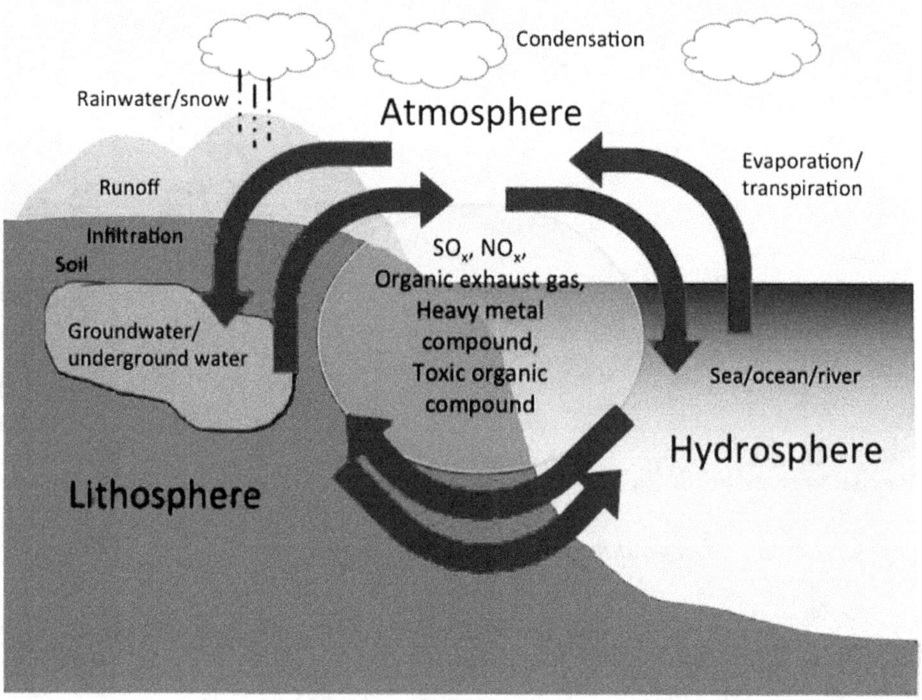

world. Chemicals (SO_x, NO_x, organic exhaust gas, heavy metal compound, and toxic organic compound) are carried through the medium of water and fixed to the soil and hydrosphere. In the water cycle, the animals, fishes and plants adsorb and immobilize the chemicals in their body. Furthermore, heavy metal may dissolve by acid rain. In addition, the acid rain acidifies a river and wetlands, affects the habitation of the creature and causes deterioration of the cultural heritage. In the present when industry developed, it becomes the problem source of water pollution such as organic matter or the heavy metal.

Therefore, recovery process for decontamination to their pollutant is needed by introduction of functional adsorbents. For example, after the Great East Japan Earthquake and Tsunami damaged the Fukushima Daiichi Nuclear Power Plant during and after March 11, 2011, long-term radiation contamination ensued, presenting severe difficulties to human life and the local environment (Morino, Ohara, Watanabe, Hayashi, & Nishizawa, 2013). The neighboring inhabitants within a radius of 30km are evacuated forcibly in now when 5 years pass for the accident. It is felt uneasy about that diffusion of the radiocesium through the medium of the water from the soil and the forest polluted by emission. Figure 6 shows adsorption materials that adsorb radiocesium efficiently are developed to prevent this in Tomioka town between 2013 and 2014 (Kobayashi, Ohshiro, Nakamoto & Uchida, 2016). The porous composite fibers showed decontamination to low-level Cs in the flowing river, preventing its flow further downstream by the diffusion of radioactive Cs through water contaminant in natural environments.

Figure 6. Pictures of zeolite composite fibers for decontamination of radioactive cesium in Tomioka town, Fukushima, Japan
Source: Kobayashi, Ohshiro, & Uchida, 2014

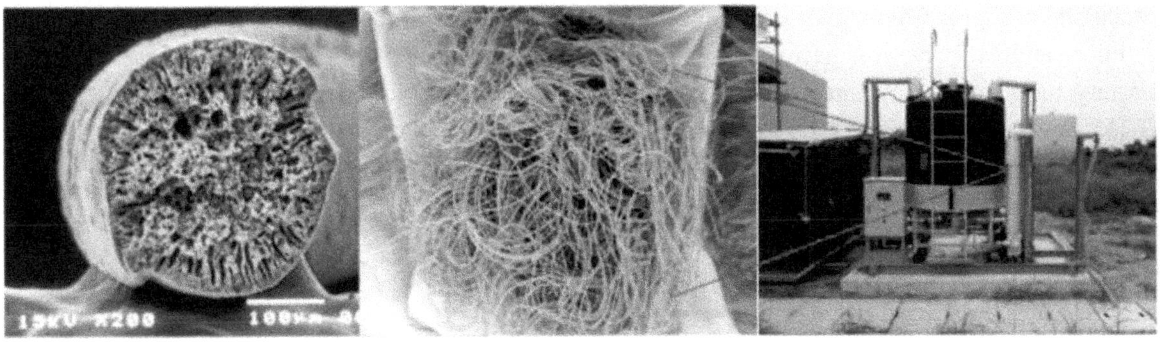

ENVIRONMENTAL MATERIALS IN SUSTAINABILITY

Agro-Industrial Biomass Wastes and Biopolymers

As many global issues start to influent our daily life, people begin to focus attention to renewable resources mainly included in agricultural production. In recent years, there have been many studies on the effective utilization of agro-industrial residues including sugarcane bagasse, cassava bagasse, rice husk, wheat straw and coconut fiber, etc (Pandey, Soccol, Nigam, & Soccol, 2000; Ashori & Nourbakhsh, 2010). It is currently used mainly as an biofuel for power and heat generation (Goldemberg, 2007;Yaman, 2004; Jorapur, & Rajvanshi, 1997) and in the manufacture of pulp for paper. However, eco-profile studies measure the re-generation of the waste bagasse and other inputs require to make a product obtained from the extraction process or harvesting of raw plant materials for eco-friendly, recyclable, advanced materials, especially for cellulose. The natural plant polymer, cellulose, originated from bagasse wastes is utilized for the regeneration process to cellulose (Kobayashi, 2015; Tovar-Carrillo, Tagaya, & Kobayashi, 2013). The cellulose fibrous fibers are sustainable materials in plants and have numerous functional possibilities, as expected to be developed into a broad range of applications, especially with the demand for environmentally friendly and biocompatible products.

Among them, sugarcane bagasse is generated as a biomass fibrous residue in extraction process of sugar juice from the sugarcane (Chandel, daSilva, Carvalho, & Singh, 2012). The sugarcane was produced about 1.69×10^3 million tons in the world in 2010 and generally 280 kg of humid bagasse is generated from 1 ton of sugarcane. Therefore, this is one of the largest cellulosic agro-industrial by-products. At the present time, the most of sugarcane bagasse is used as manure, feedstock and fuel of the boiler in the sugar industry (Hofsetz, & Silva, 2012). Several attempts for potentially renewable and sustainable have been made to utilize sugarcane bagasse as a low material. These contain ethanol production by hydrolysis and fermentation, synthesis of cellulose derivatives (Jorapur, & Rajvanshi, 1997) and also are converted to functional materials (Nakasone, & Kobayashi, 2016; Saucedo-luna, Castro-Montoya, Campos-Garcia, & Rico, 2010; Nakasone, Ikematsu, & Kobayashi, 2016). Also, agave tequilana weber azul, which is an economically important plant cultivated in central Mexico for the production of tequila (Goldemberg, 2007;Yaman 2004) and the bagasse is a byproduct material consisting of the cellulose tissue of the blue agave after extraction of the sap. More than 800,000 tons of agave bagasse has been disposed as a waste product per year (Figure 7). The disposal becomes similar problem and gives attention with regard to its regeneration. It is well known that cellulose is the most abundant natural polymer consisting of β-glucose in the nature and forms about half to one-third of plant tissues (Reid,1997). The cellulose and its derivatives have attracted many researchers due to their unique properties such as low toxicity, biocompatibility and biodegradability (Kobayashi, 2015; Kobayashi, Tovar-Carrillo & Tagaya, 2015). Several research studies have been performed to offer an alternative use for their bagasse waste, but, until now, none has offered us a final solution, due to the necessity for development of new technologies to solve this problem.

Mine Wastes for Environmental Materials

Mine reuse often brings economic, environmental, and social benefits to our communities. It also may serve as a catalyst for expediting environmental risk reduction, as EPA recognizes (EPA,1990). The economical matter strongly relates with industry and the environmental one with pollution decontaminant. For example, Interstate technology & regulatory council in USA repots that heavy metal impacts in the world, primary including cadmium, lead, and zinc (ITRC, 2010; Broadbent, 1998). The specific criteria for the reuse are for lead (15 μg/L) and cadmium (5 μg/L) in potential drinking water sources by the National Primary Drinking water standards and freshwater chronic National Recommended Water Quality Criterion for zinc of 120 μg/L. The re-uses and reprocessing technologies are effective for remediation of solid mining wastes. Also, there is attractive technology to turning mining waste into beneficial products. This can decrease human and ecological exposure to contaminated materials. The toxic heavy metal contamination, for example, can be remediated by one or more of the technologies. Their remediation technologies generally refer to chemical precipitation, chemical stabilization, electrocoagulation, biochemical reactors, pressure-driven membrane separation, permeable reactive barrier. Among them, efficient removal and recovery of toxic heavy metals from industrial wastewater is a major challenge (Gaballah & Kilbertus,1998; Saeed, Akhter, & Iqbal, 2005). Furthermore, reutilization of solid wastes of coal combustion ashes is reported (White & Stegemann, 2001; Chindaprasirt, 2009) and reused to adsorbents of gaseous waste (Suchiai, Shizaka, Nakamura, Ueno, & Hattori,1996) and heavy metals (Sekhar & Chakravarthy, 2001) and also to glass ceramic (Strad, 1986) and concreat materials

Figure 7. Pictures in Mexico, Guanajuato, for the production of tequila and agave bagasse. The agave pinier (left)is steamed (middle) and bagasse (right) is obtained after extraction process of sugar juice

(Onutai, Sirithan, Thavorniti, & Kobayashi, 2015). Geopolymers are also sustainable materials, which can be prepared from mine and ash (Davidovits, 2011), as mentioned below.

For example, the presence of the toxic metals generated by mineral processing in industries causes a major hazard to the water environment (Matis, Zouboulis, & Lazaridis,1998). Thus, serious regulations are required to establish in many countries to remove effectively the toxic metal ions from the waste waters prior to discharge into natural environment. So, the adsorption techniques present excellent qualities for treating industrial waste waters containing metal ions, when solid adsorbents are employed for recovery of metal ions as lead, copper, cadmium, nickel and zinc (Brauckmann, 1990: Cheng, Lee, Ko, Ueng, & Yang, 2012: Lopez, Sugita, & Kobayashi, 2014). Geopolymers are alkali activated aluminosilicates, consisting of a solid reactive component that contains SiO_2 and Al_2O_3 (Komnitsas & Zaharaki, 2007; Ahmaruzzaman, 2010). By using rice husk, fly ashes, active clays, pozzolanas and slags, decontamination of heavy metal ions were reported as geopolymer adsorbents (Li, Wang, & Zhu, 2006).

GREEN TECHNOLOGIES LEADING FOR SUSTAINABLE SOCIETY

Scientific and technological change brought the human a great benefit until now. By various kinds of harmful chemical substances released abundantly to environment, the various issue of environmental pollution is taking place. To raise the possibility that humanity can continue soundly and to succeed a global environment in a future generation, conservation technologies are necessary. It is essential to purify and have maintenance of environments of air, water and soil. To this end, green technology can prevent the diffusion circulation of the pollutant, causing the environmental damage. The goal is the

Figure 8. Protocol of geopolymerization synthesis with rice husk and metakaolin
Source: Lopez, Sugita, Tagaya, & Kobayashi, 2014

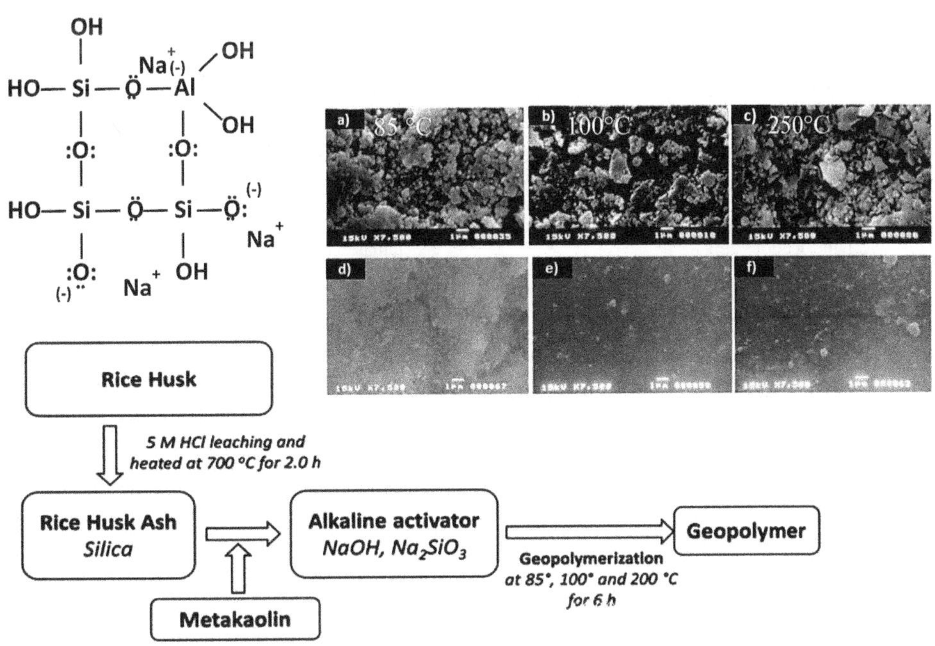

global environment construction, where people can live under sustainable energy society. Also, the cyclically used technology conserves resources of the earth. The green technique can prevent the cycle. As seen in Figure 5 diffusion circulation of the pollutants, SOx, NOx, hazardous organic chemicals and heavy metal ion takes part in the environmental purification especially in the developed country. The green technology has the following evaluation guidelines:

1. Technology to reduce consumption resources and energy,
2. To produce energy,
3. To make use of unused resources in,
4. To reduce environmental load,
5. To make quality of life and
6. To make economic competitiveness.

The critical technology of the materials enables energy saving. New energy, resource saving recycling, and living environment risk reduction is summarized in Table 2. These technical examples are followed; biodegradable polymers as an environmental ecad chemistry product, a compact detergent as a life-related article, a energy saving chemical products, a quickness diagnosis technology with a small size sensor, the high-performance material in eco-tire, health and medical care with the field of diagnosis. Among others, carbon fibers made of polyacrylonitrile is mainstream and are light weight and high strength in the rainforced materials. These are used for a aeroplane, a car, a golf club, engineering works, construction materials or an inquiry material widely.

Table 2. Lists of critical technologies of the field of materials

Classification	Development Theme	Specific Examples
Energy saving	High strength, lightweight resin composite materials	CFRP, carbon / cellulose nano fiber
	Polyolefin having super heat-resistant, and high strength	Post-metallocene catalysts
	Metal substitute super lightweight materials by the organic inorganic hybrid material	Metal-replaced boad
	Materials having a low friction surface and lubrication	Carbon materials, surface coating, nanolubrication materials
	semiconductor for light source materials	Next-generation LED, organic electroluminescence, inorganic EL
	Heat train conversion materials for the low temperature waste	Peltier element, polymer materials, organic no plane based materials
	high-performance decicant air conditioning by high temperature adsorbent materials	Energy saving de-humidification air conditioner

continued on next page

Table 2. Continued

Classification	Development Theme	Specific Examples
New energy	Polymer actuator	Micromachine, sensor for the biomaterial
	Super high accumulation of electricity type rechargeable battery materials	Original cathode active substance, carbon electrode, electrolyte, separator (film)
	New battery materials including the lithium substitute	Lightweight small batteries
	Organic radical battery, redox polymer materials	Thin and light devices
	Electrolyte for the high accumulation of electricity	High performance ionic liquid
	High accumulation of electricity energy materials for the movement / fixation power supply	New solid electrolyte, electrolyte membrane, ionic liquid
	New materials for the fuel cell	Pt substitute electrode, fluorine substitute film, CO reduction catalyst, H_2 production device
	Dye sensitization type solar battery materials	Titania dye sensitization type, plastic board type
	Thin film silicon type solar battery materials	Amorphous; crystalline
	Organic semiconduction type solar battery materials	P-type semiconductor polymer, n type fullerene
	High performance organic semiconductor materials	Fullerene type new materials
Resource saving recycling	Substitute materials for the fluorine-based functional chemicals	Refrigerant, blowing agent, jet agent, insulating material, glass for the semiconductor / liquid crystal
	Substitute materials of the laminated film, easy recycling materials	Single-like layer film
	Easy recycling and decomposable adheisive, self-detachment material	Non-epoxy adheisives
	Rare metal, noble metal substitute materials	Ni,Co,Mo,W,Li,In, rare earth elements, Ta, platinum group, boron
	Non-organic solvent-based lithography and patterning technology	Liquid hardening resin nano imprint methods
Life environment risk reduction	Halogen, phosphorus free new incombustible materials	Nanoparticles composite, carbonization layer formation, high density filler composites
	Complete water-based paints	De-VOC (volatile organic compounds)
	Non-VOC component building materials for the residence environment	Material-related industry for building and the residence materials
	Non-VOC component adhesive, ink	Public welfare, assembling processing-related industry
	Safe insect killing, insect repellent, mold agent-proof, sanitization	Photocatalyst, nanocatalytic model antibiotics
	Security deodorization, deodorant materials	Nanocatalytic model deodorant, nanopore type deodorant
	Easy biodegradability surfactant	Natural sourced soap and detergent
	Future model cosmetics materials using a nanotechnique and an interface technique	Anti-aging function, safety
	High functional foods wrapping materials	Materials for detection of cold insulation, oxidization prevention, dehydration, humidity retention, shading and, abnormality
	Self-repairing materials	Organic-inorganic hybrid, plastic
	Insulation materials for sound, heat and electromagnetic wave	New glass, new building materials, insulation paint

Source: Ministry of Economy, Japan, 2010

CONCLUSION

This chapter accounts environmental materials in the point of view of green chemistry and technology. Sustainability in eco-friendly materials and the related technologies for biomass resource is described especially for biomass fuel, biomass waste and inorganic waste. Also, trials of pollution problems in earth are reviewed in this chapter. As prevention techniques applied, it is described that green technology is suitable for the future sustainable society.

REFERENCES

Ahmaruzzaman, M. (2010). A review on the utilization of fly ash. *Progress in Energy and Combustion Science*, *36*(3), 327–363. doi:10.1016/j.pecs.2009.11.003

Andrews, J. E., Brecombe, P. B., Jikel, T. D., & Brianreid, P. L. (2004). *An introduction to environmental chemistry.* Blackwell Sieme, LTD.

Ashori, A., & Nourbakhsh, A. (2010). Bio-based composites from waste agricultural residues. *Waste Management (New York, N.Y.)*, *30*(4), 680–684. doi:10.1016/j.wasman.2009.08.003 PMID:19740643

Bajiai, P. (2013). *Advances in bioethanol.* New York: Springer. doi:10.1007/978-81-322-1584-4

Brett, C., & Waldron, K. (1990). *Physiology and biochemistry of plant cell walls.* London, UK: Unwin Hyman. doi:10.1007/978-94-010-9641-6

Broadbent, J. (1998). *Environmental Politics in Japan.* Cambric, UK: Cambridge University press. doi:10.1017/CBO9780511571060

Chandel, A. K., daSilva, S. S., Carvalho, W., & Singh, O. V. (2012). Sugarcane Bagasse and Leaves:Foreseeable Biomass of Biofuel and Bio-products. *Journal of Chemical Technology and Biotechnology (Oxford, Oxfordshire)*, *87*(1), 11–20. doi:10.1002/jctb.2742

Chavez-Guerrero, L., & Hinojosa, M. (2010). Bagasse from the mescal industry. *Fuel*, *89*, 4049–4052.

Cheng, T. W., Lee, M. L., Ko, M. S., Ueng, T. H., & Yang, S. F. (2012). The Heavy Metal Adsorption Characteristics on Metakaolin-Based Geopolymer. *Applied Clay Science*, *56*, 90–96. doi:10.1016/j.clay.2011.11.027

Chindaprasirt, P., Jaturapitakkul, C., Chalee, W., & Rattanasak, U. (2009). Comparative study on the characteristics of fly ash and bottom ash geopolymers. *Waste Management (New York, N.Y.)*, *29*(2), 539–543. doi:10.1016/j.wasman.2008.06.023 PMID:18715775

Ebnesajjsd, S. (2013). *Handbook of Biopolymers and Biodegradable Plastics: Properties, Processing and Applications.* MA, USA: Elsevier Inc.

Environmental Protection Agency (EPA). (1990). *Green chemistry.* Retrieved from http://www.epa.gov/greenchemistry/basics-green-chemistry

Gaballah, I., & Kilbertus, G. (1998). Recovery of heavy metal ions through decontamination of synthetic solutions and industrial effluents using modified barks. *Journal of Geochemical Exploration*, *62*(1-3), 241–286. doi:10.1016/S0375-6742(97)00068-X

Goldemberg, J. (2007). Ethanol for sustainable energy future. *Science*, *315*(5813), 808–810. doi:10.1126/science.1137013 PMID:17289989

Hofsetz, K., & Silva, M. A. (2012). Brazilian sugarcane bagasse:Energy and non-energy consumption. *Biomass and Bioenergy*, *46*, 564–573. doi:10.1016/j.biombioe.2012.06.038

Interstate Technology & Regulatory Council (ITRC). Mining Waste Team. (2010). *Technology overview Re-use and reprocess (r2) technologies.* Retrieved August, 2010, from http://www.itrcweb.org/miningwaste-guidance/to_reuse.htm

Jorapur, R., & Rajvanshi, A. K. (1997). Sugarcan leaf-bagasse gasifiers for industrial heating applications. *Biomass and Bioenergy*, *13*(3), 141–146. doi:10.1016/S0961-9534(97)00014-7

Kalia, S., & Averous, L. (2011). Biopolymers: Biomedical and environmental applications. Scrivener Publishing. doi:10.1002/9781118164792

Kobayashi, K., Tovar-Carrillo, K. L., & Tagaya, M. (2015). Bagasse sustainable polymers for cellulose hydrogel sheets showing tissue regeneration. In Handbook of Sustainable Polymers: Structure and Chemistry. Pan Stanford Publishing Pte Ltd.

Kobayashi, T. (2015). *Fabrication of cellulose hydrogels and characterization of their biocompatible films. In Studies in Natural Products Chemistry, 45* (pp. 1–15). Amsterdam, Netherlands: Elsevier B.V.

Kobayashi, T., Ohshiro, M., Nakamoto, K., & Syuji Uchida, S. (2016). Decontamination of Extra-Diluted Radioactive Cesium in Fukushima Water Using Zeolite–Polymer Composite Fibers. *Industrial & Engineering Chemistry Research*, *55*(25), 6996–7002. doi:10.1021/acs.iecr.6b00903

Kobayashi, T., Ohshiro, M., & Uchida, S. (2014). *Zeolite fukugouka seni niyoru Cesium zyosen gizyutu (Japanese).* Tokyo, Japan: NTS.

Komnitsas, K., & Zaharaki, D. (2007). Geopolymerisation: A Review and Prospects for the Minerals Industry. *Minerals Engineering*, *20*(14), 1261–1277. doi:10.1016/j.mineng.2007.07.011

Li, L., Wang, S. B., & Zhu, Z. H. (2006). Geopolymeric adsorbents from fly ash for dye removal from aqueous solution. *Journal of Colloid and Interface Science*, *300*(1), 52–59. doi:10.1016/j.jcis.2006.03.062 PMID:16626729

Liu, C., Hsu, P., Lee, H., Ye, M., Zheng, G., Liu, N., & Cui, Y. et al. (2015). Transparent air filter for high-efficiency PM2.5 capture. *Nature Communications*, *6*, 1–9. doi:10.1038/ncomms7205 PMID:25683688

Lopez, F. J., Sugita, S., & Kobayashi, T. (2014). Cesium-adsorbent geopolymer foams based on silica from rice husk and metakaolin. *Chemistry Letters*, *43*(1), 128–130. doi:10.1246/cl.130851

Lopez, F. J., Sugita, S., Tagaya, M., & Kobayashi, T. (2014). Geopolymers using rice husk silica and metakaolin derivatives; Preparation and their characteristics. *J. Mater. Sci. Chem. Eng.*, *2*, 35–43.

Matis, K. A., Zouboulis, A. I., & Lazaridis, N. K. (1998). Removal and recovery of Metals from Dilute Solutions: Applications of Flotation Techniques. *Mineral Processing and the Environment*, *43*, 165–196. doi:10.1007/978-94-017-2284-1_9

Meadows, D. H., Meadows, D. L., Randers, J., & Behrens, W. W. (1972). *The limits to glow*. New York: Universe Books.

Ministry of Economy. Japan. (2010). *Trade and Industry technology strategy map 2010*. Retrieved from http://www.meti.go.jp/policy/economy/gijutsu_kakushin/kenkyu_kaihatu/str2010.html

Morino, Y., Ohara, T., Watanabe, M., Hayashi, S., & Nishizawa, M. (2013). Episode analysis of deposition of radiocesium from Fukushima daiich neuclear power plant accident. *Environmental Science & Technology*, *47*, 2314–2322.

Nakasone, K., Ikematsu, S., & Kobayashi, T. (2016). Biocompatibility evaluation of cellulose hydrogel film regenerated from sugarcane bagasse waste and its in vivo behavior in mice. *Industrial & Engineering Chemistry Research*, *55*(1), 30–37. doi:10.1021/acs.iecr.5b03926

Nakasone, K., & Kobayashi, T. (2016). Effect of pre-treatment of sugarcane bagasse on the cellulose solution and application for the cellulose hydrogel films. *Polymers for Advanced Technologies*, *27*(7), 973–980. doi:10.1002/pat.3757

Onutai, S., Sirithan, J., Thavorniti, P., & Kobayashi, T. (2015). Aluminium hydroxide waste based geopolymer composed of fly ash for sustainable cement materials. *Construction & Building Materials*, *101*, 298–308. doi:10.1016/j.conbuildmat.2015.10.097

Pandey, A., Soccol, C. R., Nigam, P., & Soccol, V. T. (2000). Biotechnological potential of agro-industrial residues I, Sugar Bagasse. *Bioresource Technology*, *74*(1), 69–80. doi:10.1016/S0960-8524(99)00142-X

Pilla, S. (2011). Handbook of bioplastics and biocomposites Engineering applications. Scrivener Publishing. doi:10.1002/9781118203699

Reid, J. S. G. (1997). Carbohydrate methbilism: Structural carbohydrates. In Plant Biochemistry. Academic Press.

Rosenberg, A. A., Fogarty, M. J., Sissenwine, M. P., Beddington, J. R., & Shepherd, J. G. (1993). Achieving sustainable use of renewable resources. *Science*, *262*(5), 828–829. doi:10.1126/science.262.5135.828 PMID:17757341

Saeed, A., Akhter, M. E., & Iqbal, M. (2005). Removal and recovery of heavy metals from aqueous solution using papaya wood as a new biosorbent. *Separation and Purification Technology*, *45*(1), 25–31. doi:10.1016/j.seppur.2005.02.004

Saucedo-luna, J., Castro-Montoya, A. J., Campos-Garcia, J., & Rico, J. L. (2010). Optimization of acid hydrolysis bagasse from agave tequilana weber. *Revista Mexicana de Ingenieia Quimica*, *9*(1), 91–97.

Sekhar, M., & Chakravarthy, C. (2001). Coal and clay minerals removal of zinc by adsorption. In Advances in Environmental Materials, Volume II Environmentally preferred materials. MRS (Singapore).

Steavens, E. S. (2002). *Green Plastics: An Introduction to the New Science of Biodegradable Plastics.* United Kingdom: Princeton University Press.

Strnad, Z. (1986). *Glass-ceramic materials.* Amsterdam: Elsevier.

Suchiai, H. T., Shizaka, T. I., Nakamura, H., Ueno, T., & Hattori, H. (1996). Study of flue gas desulfurizeation absorbent prepared from coal fly ash. *Industrial & Engineering Chemistry Research, 35*, 322–2326.

Tovar-Carrillo, K. L., Tagaya, M., & Kobayashi, T. (2013). Bamboo fibers elaborating cellulose hydrogel films for medical applications. *J. Mater. Sci. Chemical Engineering (Albany, N.Y.), 1*(7), 7–12.

United Nations Environment Programme (UNEP). (2014). *Climate change Minimizing the scale and impact of climate change,* Retrieved 2014, from http://www.unep.org/annualreport/2014/en/pdf/climate_change.pdf

White, T., & Stegemann, J. A. (2001). Advances in Environmental Materials, Volume II Environmentally preferred materials. MRS (Singapore).

Williams, L. (2001). Environmental chemistry a modular approach. John Wiley & Sons, LTD.

Yaman, S. (2004). Pyrolisis of biomass to produce fuels and chemical feedstock. *Energy Conversion and Management, 45*(5), 651–671. doi:10.1016/S0196-8904(03)00177-8

KEY TERMS AND DEFINITIONS

Biomass: In general, this means plant as the source of supply of the alternative energy. In this chapter, supply to alternative energy and material source of plants.

Biofuel: Fuels derived from biomass; any matter derived from plants or animals.

Biopolymer: Polymers produced by living organisms.

Carbon Neutral: Achievement of a product, service or event being equal to zero, as net greenhouse gas emissions of an organization to earth.

Environmental Material: Materials applied to a serious environmental problem such as an energy problem and the water pollution in a various materials technology.

Geopolymer: Typically inorganic ceramic materials having non-crystalline amorphous networks of silico-oxide (-Si-O-Si-O-), silico-aluminate (-Si-O-Al-O-), ferro-silico-aluminate (-Fe-O-Si-O-Al-O-) and alumino-phosphate (-Al-O-P-O-) by geopolymerization.

Green Chemistry: Chemistry related science to solve problems in energy, a material, an environmental pollution and for development of sustained society.

Green Technology: Technology for clean and sustainable communities.

Remediation: The following technologies, isolation, immobilization, toxicity reduction, separation and extraction remediate contaminant level in soil, water and air.

Sustainable Society: Ensuring the health and vitality of human life and culture and of nature's capital, for present and future generations.

Toxic Heavy Metal: Heavy metals, particularly cadmium, mercury, lead and arsenic has potential toxicity, especially in environmental contexts. Additionally, the major public concern of other examples includes manganese, chromium, cobalt, nickel, copper, zinc, selenium, silver, antimony and thallium.

Water Cycle: Hydrological cycle occurs continuously in the movement of water in earth through atmosphere, hydrosphere and lithosphere. When water moves, it goes through different phases of liquid, ice and vapor phases.

Chapter 2
Natural Material Source of Bagasse Cellulose and Their Application to Hydrogel Films

Karla L. Tovar Carrillo
Universidad Autónoma de Cd. Juárez, Mexico

Takaomi Kobayashi
Nagaoka University of Technology, Japan

ABSTRACT

Among all biomass sources, cellulose is the most diverse material sourced from plants and is abundant in nature. It possesses some promising properties such as biodegradability, biocompatibility and hydrophilicity. In several industrial biomasses, the waste products especially for food manufacture have high content of cellulose and their disposals are serious in reuse problem. Therefore, people have given attention with regard to their regeneration to the useful materials. In this chapter, it is worth noting to describe that natural biomass cellulose is usefully applied with the regeneration to functional hydrogel films. The treatment processes of the bagasse waste used as a main source of cellulose and the fabrication of hydrogel films are described and their potential characteristics, especially on the medical field are reviewed.

INTRODUCTION

Cellulose is one of structural component in green plants and is the most common organic compound on earth. Now in our days, several approaches mainly undergo the effective uses of biomass as alternative energy source as biofuel (Edgar, Buchanan, Debenham, Rundquist, Seiler, & Shelton, 2001). But, the elaboration of cellulose material is very limited to cover society necessities. It is recognized that the importance is to protect earth environment by promoting the reuse of the waste products. The potential uses of the bagasse cellulose are important. For example, in the food industries using sugarcane for fermentation and agave to production of tequila, the bagasse is fibrous after sugarcane or agave is crushed to extract their juice. In the cellulose regeneration processes, due to mainly obtaining cellulose fibers from

DOI: 10.4018/978-1-5225-1971-3.ch002

bagasse, the chemical treatment of the cellulosic fibers is a crucial point. The treatments diminish the structure of the fibers helping to extract non-cellulosic compounds as lignin and others. In this process, cellulose leaving high cellulose content is necessary for the regeneration process.

Among their applications, cellulosic hydrogel is one of candidate material having a three dimensional network of crosslinked swellable of water liquid in the polymer chains (Liu, Ma, Mao, & Gao, 2011). This is because that cellulose exhibits hydrophilic nature, the hydrogel application is widely potential in several fields of food, agriculture, environmental remediation and medical using as biomaterials. However, several approaches failed due to their poor properties of less solubility and the self-standing scaffold. In biocompatibility, biodegradability and hydrophilicity, it is much interesting to focus on their characteristic properties of the regenerated cellulose. Therefore, to re-use for the development of functional materials is advance. Since the chains crosslinked chemically or physically each other, hydrogels are insoluble in water. In contact therewith swelling considerably increasing their volume is caused, but maintaining its shape is led to be elastic. Its hydrophilic characteristic is because of the presence of OH-groups in the chain of cellulose. In fact, hydrogels have wide potential applications in the fields (Sionkowske, 2011; Chen, Yuan, Song, Wu, & Li, 2008). Rapid progress has been made recently in the field of biomedical materials, which utilize both natural and synthetic polymers. Thus, cellulose-based materials can be used in a variety of applications, including wound closure (Lloyd, Kennedy, Methacanon, Paterson, & Knill, 1998), drug delivery systems (Nair & Laurencin, 2007), novel vascular grafts (Chandy & Sharma, 1998), or scaffolds for in vitro or in vivo tissue engineering (Cai, Sharma, Lui, Mu, Lui, Zhang, & Deng, 2014). As mentioned, biopolymers are important in all aspects of medicine, surgery and healthcare. Because of the increasing demand for environmentally friendly and biocompatible products, such application of cellulose hydrogels involves casting membranes and sponges for food companies, flexible gels and weak gels for industry, films and membranes for medical applications.

In recent years, much energy was devoted to the development of novel hydrogels as scaffolds for medical applications (Edgar., Buchanan, Debenham, Rundquist, Seiler, & Shelton, 2001). Several approaches were worked in a scaffold studies for more suitable for tissue regeneration of biomass polymers. In the past, synthetic polymers were the first option for the development of scaffolds, but, these could not provide an ideal environment to support cell adhesion and tissue formation due to their bioinert nature (Nair & Laurencin, 2007). For this reason, researchers focus their attention on cellulose biopolymers having excellent nature of biopolymers with the biocompatibility. However, most of biopolymer film shows less mechanical property, especially in hydrogels. This is due to much water retained. Therefore, very few works have been reported about for self-standing biopolymer hydrogels. In addition, several researches have shown problems in the development of scaffolds with the capacity, not only to offer a non-toxic and biocompatible surface, but also a material capable to promote cell adhesion, proliferation and tissue formation (Gibson, 2005; Tovar-Carrillo, Nakasone, Sugita, Tagaya, & Kobayashi, 2014). In these points, most of the approaches have failed in biopolymer films. More recently, authors reported regeneration process of bagasse wastes for cellulosic hydrogel films (Tovar-Carrillo, Tagaya, & Takaomi, 2013). The hydrogels are self-standing films having high mechanical strength, when these have loading about 1000-3000% water to the cellulosic fibrous chains formed physically crosslinked network.

Background

It is well known that cellulose is considered as renewable organic polymeric source produced in the biosphere (Wooding, 2001). Although cellulose is the main raw material used in the industry widely, a high percentage of the important raw material is ended as industrial waste product. Unfortunately, a mere 2% is recovered industrially. Agave bagasse is waste in Mexico and composed of the residual fibers after the cooked agave heads are shredded, milled and the sugars water-extracted. Recently, the tequila market has grown and gained international recognition. Since increasing the disposal problems is caused for tequila companies and the serious economic and environmental problems have been caused by the disposal of these resources, the re-use of agro-industrial waste is indispensable (Iñuguez-Covarrubias, Lang, & Rowel, 2001). But, it becomes only to biofuels and still less work with regard to its regeneration. Several classes of biopolymers already are proved to be most useful in biomedical and cytocompatibility (Sionkowske, 2011).

For one of advanced example of regeneration, this chapter accounts cellulose regenerated from waste product to fabricate the biopolymer hydrogel films. In addition with the fabrication process, the future of cellulosic hydrogel films is introduced.

CELLULOSIC BAGASSE FOR SUSTAINABLE MATERIAL SOURCE

Agave Weber is economically important in Mexico for the production of tequila. Through the process, more than 800,000 tons of agave bagasse is disposed as waste product per year. The disposal has been given attention with regard to its regeneration as a useful material. Several researchers are performed to offer an alternative use for the bagasse waste product (Kobayashi, Tovar-Carrillo & Tagaya 2015; Tovar-Carrillo, Sugita, Tagaya, & Kobayashi, 2013). As an advanced work, one of the possible solutions is of cellulose hydrogel films regenerated from the bagasse and for application to scaffolds of tissue engineering.

Cellulosic Fibers and Its Classification

Originally, natural fibers come from plants such as cotton, hemp, kenaf and flax, or from animal sources such as wool, hair and fur, or insects such as silk (Salem, Tedler, & Roberts, 2002). Such natural polymers basically fall into two origin categories: protein and cellulose. The origin of regenerated protein comes from plant source such as corn, soy, alginate, and peanut or from animal one such as casein from milk. On the other hand, the cellulose origin is from bamboo, rayon, lyocell / TENCEL®, Modal® and Viscose®. There are made of cellulose source harvested from tree wood and inner pith and leaves from bamboo plants using differing fiber manufacturing processes with common roots. It was appearance in France in the 1890s that “artificial silk” or “art silk” was produced as first textile. The textile industry adopted the term “rayon” in 1924. This was family of regenerated cellulose fibers for textiles called reconstructed fibers or natural synthetic fibers. The natural cellulosic fibers include cotton, flax, hemp, jute, and ramie.

In contrary, cellulose has a basic molecular unit in $C_6H_{10}O_5$ so called anhydroglucose unit (AGU) (Figure 1). The cellulose molecule is linked in the form of β-1, 4-glucan. The polymer chain length is expressed as the number of AGUs or degree of polymerization, DP. Generally, 20-30 repeating units give all the cellulose properties. For example, cotton composes 90% cellulose fibers and bamboo varies from 50% to 60% cellulose. The natural fibers are also sourced from vegetable, animal, or mineral in origin. Vegetable fibers, as the name implies, are derived from plants. Wood is composed of cellulosic fibers in the range of 40% to 50% cellulose, 15% to 25% hemicellulose and fortified with 15% to 30%

Figure 1. Chemical structure of cellulose including anhydroglucose unit (AGU)
Source: Tovar-Carrillo, 2014

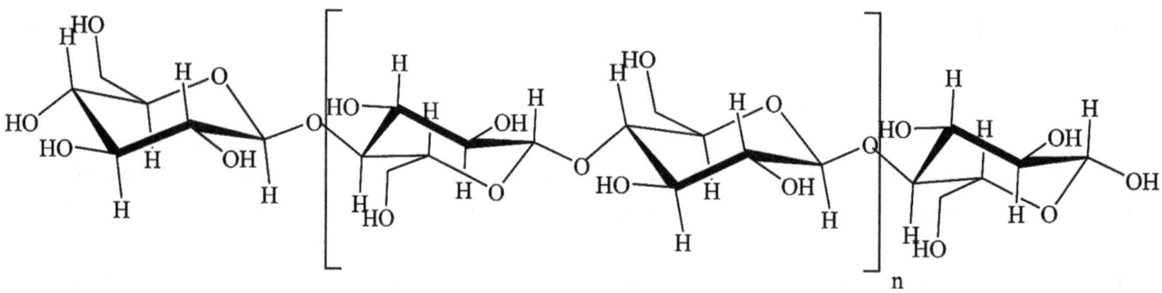

lignin. Here, hemicellulose is also a polysaccharide but it is composed of short and weak sugar chains that can be easily hydrolyzed and decomposed by dilute acids or alkalis in water and by some enzymes. Lignin is acted as glue which fills the spaces in plant cell walls between the cellulose, hemicellulose and other compounds found in the cell walls. The fibers are usually bound by a natural phenolic polymer of lignin, which also is frequently present in the cell wall of the fiber; thus vegetable fibers are also often referred to as lingo-cellulosic fibers, except for cotton, which does not contain lignin (Figure 2). The long 1inear chains of cellulose permit the hydroxyl functional groups on each AGU to interact with hydroxyl groups on adjacent chains through hydrogen bonding and van der Waal s forces. These strong intermolecular forces between chains, coupled with the high linearity of the cellulose molecule, account for the crystalline nature of cellulosic fibers (Wooding, 2001).

Table 1 lists the classification of cellulosic fibers for plants sourced. As classified according to their source in plants, it is followed that

1. The bast or stem fibers form the fibrous bundles in the inner bark of the plant stems and are often referred to as soft fibers for textile use,
2. The leaf fibers run lengthwise through the leaves of monocotyledonous plants and are also referred to as hard fibers, and
3. The seed-hair fibers for the source of cotton is the most important vegetable fiber.

There are >250,000 species of higher plants. However, only a very limited number of species is exploited for commercial uses (<0.1%, see in Table 1). For mostly general uses, the composite-type strand having long fiber is used directly and cpommonly. When such fiber strand is pulped by chemical treatments, the strand is broken down into much shorter and finer fibers, meaning the ultimate fibers (Wooding, 2001; Tovar-Carrillo, 2014; Nakasone & Kobayashi, 2016).

Cellulose Fibrous Structure

Even part from the variety of natural cellulose fibers, wood pulp fibers are presently provided to a major area for several end-use applications. The fibers can be utilized as reinforcement in bio-degradable composites and as a source of raw materials for bio-energy (Kalia, Dufresne, Cherian, Kaith, Aveous, Njuguna, & Nassiopoulos, 2011) and biochemical products (Kalia, Kaith, & Kaur, 2009). The wood pulp materials are introduced and defined as microfibrillated cellulose (Wooding, 2001).

Table 1. Classification of fibers

Fiber Type	Description
Cotton	Is the most commonly used natural cellulosic fiber. Cotton fibers grow from the seeds in the boll (seedpod). Each boll contains seven or eight seeds, and each seed may have up to 20,000 fibers growing from it. Seeds hairs
Coconut	Is from the fibrous mass between the outer shell and husk of coconuts. It is a stiff fiber. It is usually used to make highly durable indoor and outdoor mats, rugs, and tiles.
Kapor	is from the seed of the Java or Indian kapok tree. The fiber is soft, lightweight, and hollow. It breaks down easily and it is difficult to spin into yarns. It is used as fiberfill and as the stuffing for pillows. It used to be used as a stuffing for lifejackets and the mattresses on cruise ships because it is very buoyant.
Kenaf	Is consider to be one of the highest producers of biomass among plants in terms of tons/land/time. One hectare of kenaf can yield 6500-9000 liters of ethanol. Can be used as raw material for many uses and applications, separating the fibers from core can be better alternative.
Ramie	Ramie fibers, are 4 to 6 inches long. The fibers are whiter and softer than flax. Ramie does not retain dyes well unless it is dry-cleaned. Ramie is strong for a natural fiber, but it lacks resiliency, elasticity, and elongation potential. It is resistant to mildew, insects, and shrinkage. It is used for apparel, window treatments, ropes, paper, and table and bed linens
Bamboo	Is extracted from the pulp of bamboo plants. It is usually not from the fibers of the plant. Bamboo can be cultivated quickly. Is extremely resilient and durable as a fiber. In studies comparing it to cotton and polyester, it is found to have a high breaking tenacity, better moisture-wicking properties, and better moisture absorption.
Sugarcane	Is a renewable, natural agricultural resource because it provides sugar, besides myriad of products/co-products with ecological sustainability. The solid fibrous part (bagasse) is the major product of sugarcane industry. Is left after the sugarcane is crushed in the factories of the sugar and alcohol industry, and is widely available renewable source.
Hemp	Is similar to flax. The fibers range in length from 3 to 15 feet. Hemp production is illegal in the U.S. Hemp has a low environmental impact; it does not require pesticides. It produces 250% more fiber than cotton and 600% more fiber than flax on the same amount of land. Hemp plants can be used to extract zinc and mercury pollutants from soil. Hemp is used for ropes, apparel, and paper. Potheads are willing to pay inflated prices for hemp apparel because it is related to the marijuana plant
Agave	Succulents with a large rosette of thick, fleshy leaves, each ending generally in a sharp point and with a spiny margin; the stout stem is usually short, the leaves apparently springing from the root. Major use as raw material for tequila manufacturing. Bagasse is left after agave cores are cooked and squeezed to extract the sugar juice for tequila preparation.
Pineapple	Are from the leaves of the pineapple plant. It is used to make lightweight, sheer, stiff fabrics for apparel, bags, and table linens. It is also used to make mats
Jute	Is one of the cheapest textile fibers, and one of the weakest cellulosic fibers. Jute has poor elasticity, elongation, sunlight resistance, mildew resistance, and colorfastness. It is used to produce sugar and coffee bagging, carpet backing, rope, and wall coverings. Burlap is made from jute.

Source: Calvin Wooding, 2001

As seen in Figure 2, the wood pulp fibers have multistate characteristics. Roughly, typical lengths of fibers are ranging from 1 to 3 mm and typical widths are 10-50 μm. The fiber wall thickness is roughly between 1 and 5 μm and the fiber wall is composed of defined layers, including the primary wall and several secondary wall layers. Each of these layers is characterized by a specific arrangement of fibrils. (Wooding, 2001; Kalia, Dufresne, Cherian, Kaith, Averous, Njuguna, & Nassiopoulos, 2011).

Chemical pulp fibers have a surface, which is characterized by a particular pattern having wrinkles and micriofibrils in the outer layers of the fiber wall structure. The surface of chemical pulp fibers corresponds mainly to the primary and secondary wall layers of the fiber wall. The secondary wall layer is characterized by having a structure of microfibrils organized in a helical manner (Figure 3). In the cellulose extraction of wood, which consists of chemical action with mechanical agitation, most of the lignin and other impurities are dissolved, and the fiber structure is disintegrated longitudinally

Figure 2. Components of primary cell wall
Source: Tovar-Carrillo, 2014

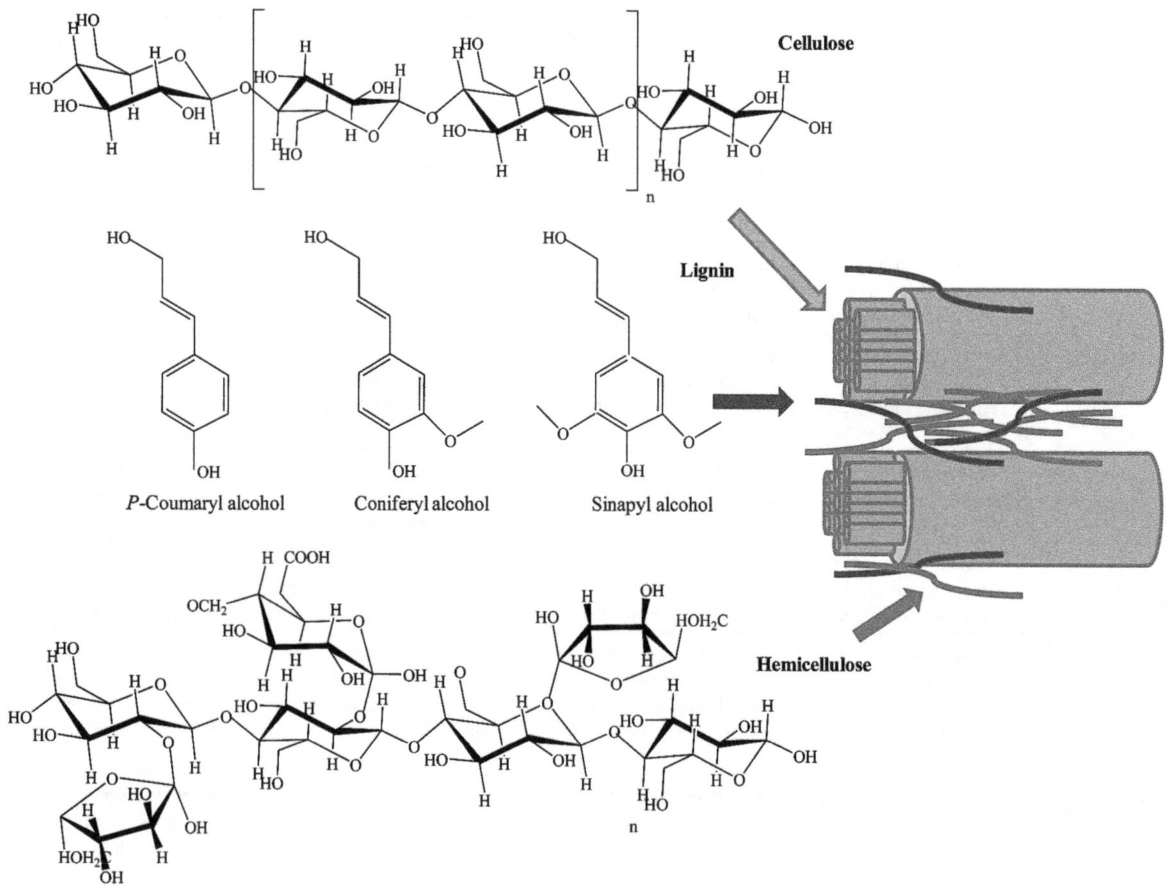

Figure 3. Model of basic cell wall structures of cotton and wood cells showing structures of cell walls
Source: Tovar-Carrillo, 2014

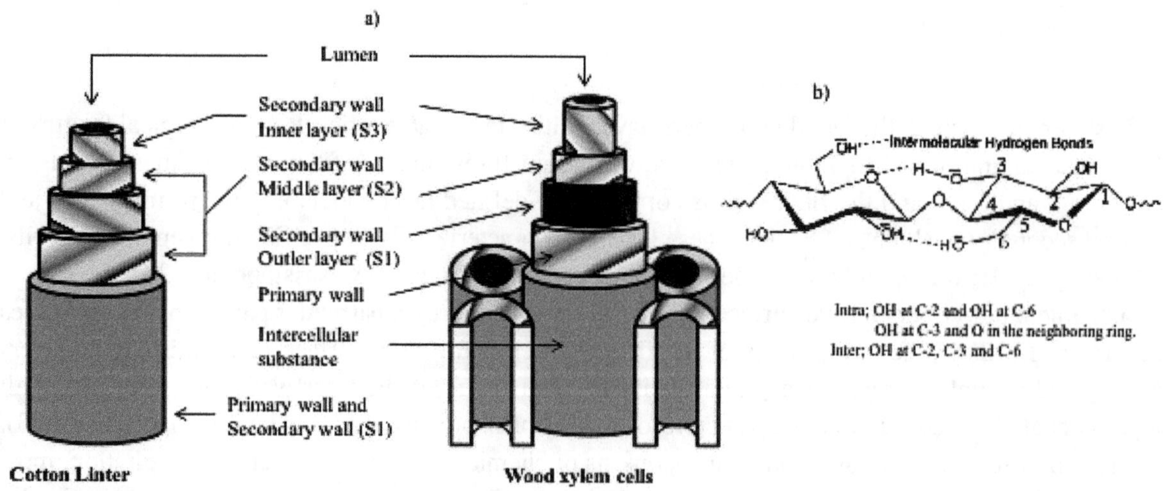

(Wooding, 2001). Cellulose fibers, depending on the part of the plant, can be classified as followed; grasses and reeds (bamboo and sugarcane), leaf fibers (as sisal and referred as hard fibers), bast fibers (as kenaf), seed and fruit hairs (as cotton and coconut) and wood fibers (as wooden pulp). These fibers consist of several fibrils. The cell wall in cellulose consists of highly crystalline and regular regions of pure cellulose interspaced with small regions of irregular or amorphous cellulose (Ramos, 2003). These amorphous regions usually contain small amounts of impurities. Chemical association in the plant cell wall is consisted of:

1. The cellulose backbone with its basic structural unit of cellobiose,
2. Framework of cellulose chains in the elementary fibril,
3. Cellulose crystallite, and
4. Microfibril cross-section, showing strands of cellulose molecules embedded in a matrix of hemicellulose and protolignin.

The bundles of these celluloses are called microfibrils, varying in width from 0.008 to 0.03 μm. The microfibril bundles are held together by hydrogen bonds formed between the protruding hydroxyl groups of the flat glucose molecules. The microfibrils are covalently bonded together by various polymeric sugars and proteins. The covalent bonded microfibrils are often found into fibrils (Wooding, 2001; Roberts, 2001). These cellulose fibrils are generally between 0.05-0.3 μm in diameter with an average diameter of approximately 0.15 μm and up to 20 cm or more in length. The cellulose fibers in the pulp fibers contain fibrous residue remaining after the mechanical chipping of pulp wood. Also, chemical treatment designs to remove lignin and other impurities and leaves the cellulose fibers in the pulp wood. The cell wall in cellulose consists of highly crystalline and regular regions of pure cellulose interspaced with small regions of irregular or amorphous cellulose (Kalia, Dufresne, Cherian, Kaith, Averous, Njuguna, & Nassiopoulos, 2011).

Bagasse's Wastes

Bagasse fibers are remaining pulping residue after crushing for extraction of juice. Bagasse is an extremely inhomogeneous material comprising around 30-14% of pith fiber, which is derived from the core of the plant and is mainly parenchyma material. (Kalia, Dufresne, Cherian, Kaith, Averous, Njuguna, & Nassiopoulos, 2011). For agave bagasse, a typical chemical analysis of bagasse reported the composition of bagasse fibers:

- Cellulose 45-55%,
- Hemicellulose 20-25%,
- Lignin 18-24%,
- Ash 1-4%, and
- Waxes ˂1% (Tovar-Carrillo, Sugita, Tagaya, & Kobayashi, 2013; Kalia, Kaith, & Kaur, 2009).

Agave bagasse waste is one of regeneration materials harvested in central Mexico (Inuguez-Covarrubias, Lang, & Rowel, 2001). In Tequila industry, as represented in Mexico for economic importance, the used blue agave plant (Figure 4) primarily distributed in the area surrounding Guanajuato and northwest of Guadalajara and also in the highlands (Los Altos) of the north western Mexican state of Jalisco. Mexican

Figure 4. (a) Mexico country map indicating the states cultivated for (b) agave azul plant

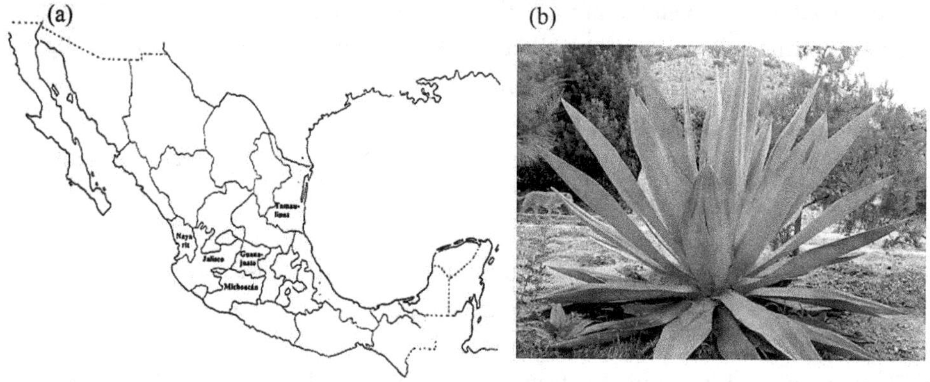

laws state that tequila can be produced only in the state of Jalisco (Iñuguez-Covarrubias, Diaz-Terez, Sanjuan-Duenas, Anzaldo-Hernandez & Rowel). In these areas, the disposal of the agro-industrial waste causes serious economic and environmental problems (Inuguez-Covarrubias, Lang, & Rowel, 2001). The possibility of utilizing at least part of agave bagasse as an efficient ruminant feed could have a major impact on livestock production practices in their areas. Much progress has been achieved lately as listed in Table 2 for uses and benefits of bagasses fibers. The bagasse fibers could also be used to produce a wide variety of other products such as filters, sorbents, geotextiles, fiberboards, packing, molded products or in combination with other resources (Kalia, Kaith, & Kaur, 2009; Inuguez-Covarrubias, Lang, & Rowel, 2001).

Table 2. Uses and benefits of bagasses fibers

Utilization	Benefits
Fuel: bagasse is often used as a primary fuel source for sugar mills; when burned in quantity, it produces sufficient heat energy to supply all the needs of a typical sugar mill.	The resulting CO_2 emissions are less than the amount of CO_2 that the sugarcane plant adsorbed from atmosphere during its growing phase, which makes the process of cogeneration greenhouse gas-neutral.
Paper: bagasse is commonly used as a substitute for wood in many tropical and subtropical countries for the production of pulp, paper and board.	Good quality wrapping and magazine paper can be produced. In South Africa, a company mixes 30% bagasse cellulose fibers in with recycled kraft paper fiber to make ultra-lightweight composite boards. These are sold as an environmentally-friendly, formaldehyde-free alternative to MDF and particleboard.
Electricity: The more straightforward solution is to produce electricity from a high pressure boiler and condensing turbo-alternator.	The generation of electricity from bagasse is undoubtedly the easiest and best utilization of this byproduct.
Biogas: Theoretically 1 kg of cellulose would produce 415 L of methane, but in practice the process is less efficient with a complex three-stage reaction operating in cascade.	Biogas has a caloric calue of about 22 000 kJ (which is equivalent to 27 500 kJ per m^3). This alternative use will reduce the cost per GJ.

Source: Kalia, Dufresne, Cherian, Kaith, Averous, Njuguna, & Nassiopoulos, 2011

Properties of Cellulose Fibers

Cellulose is important in our daily life. It constitutes around 1.5 x 10^{12} tons of the total annual biomass production and approximately 2% (3.2 million tons in 2003) are used for production of cellulose regenerate fibers and films by exploiting hierarchical structure design (Li, Mai, & Ye, 2000; Wooding, 2001; Nair, 2007). The properties of naturally sourced fibers strongly depend upon plant source and are influenced by many factors (Table 3), particularly for chemical composition and internal fiber structure. For example, cellulose fibers such as conifer, kraft fiber and cotton have the same tensile strength value due to their comparable density. Because of the hydrogen bonding in these microfibrils remains intact except under several processing conditions such as elevated temperatures, strong solvents, acids, and bases (Wooding, 2001). After this treatment, cellulose fibrils mainly constitute of the natural fiber full of protruding the hydroxyl groups.

CHEMICAL TREATMENT OF BAGASSE FIBERS

Cellulose is also the major chemical component in plant materials such as bamboo, bagasse, cotton, flax, hemp, jute, kenaf, and ramie, but has some limitations for many applications used in each areas (Roberts, 1997). In the chemical treatment process for agave waste (Tovar-Carrillo, Nakasone, Sugita, Tagaya, & Kobayashi, 2014) and sugarcane (Nakasone & Kobayashi 2016), lignin is dissolved from the bagasse under alkaline, neutral or acidic conditions (Figure 5). One treatment method is to use acid sulfite used with combinations of sulfur dioxide and water at high temperature (Roberts, 1997). Some degradation of the carbohydrate component occurs in the process. Cellulose, on the other hand, partially depolymerised via acid hydrolysis of glycosidic linkages through the process. The delignification produces soluble lignin in the aqueous sulfonic acid and retains a high molecular weight. Another method is

Table 3. Properties of some cellulose fibers harvested with different plant source (Calvin Wooding)

Fiber	Diameter (μm)	Density g/cm^3	Moisture Content (%)	UTS (MPa)	Modulus (GPa)
Cotton		1.5		500-880	0.05
Jute	200	1.45	12	460-533	2.5-13
Coir	100-450	1.15	10-12	131-175	4-6
Banana	80-250	1.35	10-12	529-754	7.7-20.8
Sisal	50-200	1.45	11	568-640	9.4-15.8
Conifer		1.50		1100	100
Kraft fiber		1.54		1000	40
Sunhemp	48	0.673		200-300	2.68
Pineapple	20-80	1.44		413-1627	34.5
Palmleaf	240			98.14	2.22
Kenaf	200	1.47		157.38	12.62
Kusha grass	390			150.59	5.69

Figure 5. Agave bagasse fiber chemical treatment

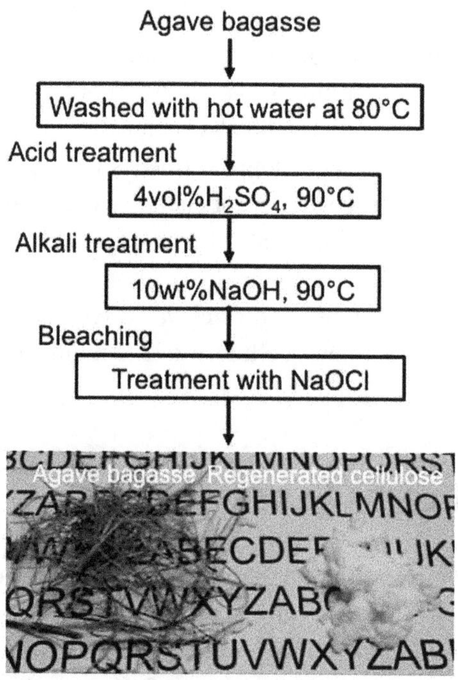

alkaline delignification with mixture of sodium hydroxide and sodium sulfide (Tovar-Carrillo, Tagaya, & Kobayashi, 2013). The presence of sodium sulfide improves the efficiency of the process. The effect of the alkaline method is to depolymerise the lignin molecule. At the end of the process the black liquor is rich in solubilised lignin. The fibers obtained are brown in the color and, whilst suitable for many applications. To obtain white fibers the process often requires bleaching for many end uses. The color of these fibers is mainly due to residual lignin (~3-6%) although products derived from carbohydrate may also make a contribution (Roberts, 1997; Pokhrel, 2004).

The bleaching of fibers involves the removal of residual lignin (Nakasone & Kobayashi, 2016). Until recently, chlorine and compounds of chlorine have been the most widely used bleaching agents (Tovar-Carrillo, Nakasone, Sugita, Tagaya, & Kobayashi, 2014). In the case of sugarcane fibers treated with different NaOCl condition, the higher temperature enhanced the bleaching of the sugarcane bagasse, but the molecular weight decline occurred in the 50 °C.

MATERIALS ELABORATED WITH REGENERATED CELLULOSE

Utilization of Regenerated Cellulose and Its Applications

Natural cotton composed of mainly cellulose is colorless, odorless and nontoxic solid polymer and possesses more additional properties of relative thermostabilization (Wooding, 2001), high sorption capacity (Roberts, 1997) and alterable optical appearance (Smook, 1992). These properties enable cellulose to be applied to a vast array of fields (Table 4). In the papermaking, cellulose fibers can be fabricated for

Table 4. Applications of cellulose in different forms

Material Forms	Applications
Fiber	Fiber, reinforcement material, biomaterial, magnetic paper.
Film/membrane	Drug delivery, separation, water treatment, package, optical media, biomembrane, and adsorption.
Nanocomposite	Biomaterials, drug delivery, reinforcement material, barrier film, membrane, conductive material, and adhesion.
Polymer	Drug delivery, biomaterial, water treatment, thickener, and stabilizer.

Source: Hon, 1997

paper structure like microporous materials, high efficiency binder fibers for other fibers/active particulates and coatings. Base fundamental reinforcement unit anables to strengthen all subsequent structures within trees, plants, some marine creatures, and algae. Alternatively, cellulose nanoparticles (CNs) are ideal materials on which to base a new biopolymer composites industry (Kalia, Dufresne, Cherian, Kaith, Averous, Njuguna, & Nassiopoulos, 2011). CN have characteristics with high aspect ratio, low density, and reactive surface of –OH side groups that facilitate grafting chemical species to achieve different surface properties (Roberts, 1997; Kalia, Kaith, & Kaur, 2009).

In biomedical field, cellulose and its derivatives have been extensively used for decades, for example for firstly proposing the use of regenerated cellulose hydrogels to orthopedic applications (Hoenich, 2006). Cellulose regenerated by the viscose process was patented and thoroughly investigated as usually found in refined wood pulp and converted into alkali cellulose by steeping in sodium hydroxide. Alkali cellulose is then converted into sodium cellulose xanthate with carbon disulfide, and finally the xanthate is dissolved in dilute alkali, and viscose is regenerated (Edgar, Buchanan, Debenham, Rundquist, Seiler, & Shelton, 2001; Kalia, Kaith, & Kaur, 2009; Filho, Ribeiro, Meireles, Da silva, Ruggiero, Junior, Cerqueira, Assuncao, Zeni, & Polleto, 2011). As implantable material in orthopedic surgery, as sealing material for the femoral component in hip prostheses, in place of the acrylic cement, as well as dyaphyseal obturator (Yamaoka, Ogasawara, Nishizawa, Nakatsuka, Koshima, Nakamura, Kawaguchi, Chung, Takato, & Hoshi, 2006). It was envisaged to take advantage not only of its biostability and good matching with mechanical properties of cortical bone but also of its hydroexpansivity, allowing therefore a satisfactory fixation to hard tissue (Sionkowske, 2011).

Cellulose phosphate (CP) as a biomaterial for orthopedic applications improved the osseo integrationand was used in the treatment of Ca^{2+} metabolism-related diseases, such as renal stones, due to its high Ca binding capacity. This associated with lack of toxicity and indigestibility (Langer & Vacanti, 1993). CP also is not cytotoxic, independently of the phosphate content, however, promotes poor rates of cell attachment, proliferation and differentiation. This is attributed to the negative charge as associated with the high hydrophilicity of the cellulose derivatives (Chen, Yuan, Song, Wu, & Li, 2008).

On the contrary, unmodified and slightly oxidized regeneration cellulose hydrogel promoted high rates of cell attachment, proliferation and differentiation. In this case, an apatite layer was observed between the cellulose surface and the cell layer, which was attributed to the synthesis of extracellular matrix by the osteoplastic cells (Tamada & Ikada, 1993; Tovar-Carrillo, Sugita, Tagaya, & Takaomi, 2013).

Fabrication of Regenerated Cellulose

Cellulose solubility is a crucial point for the obtaining of the cellulose films. In the fabrication of cellulose-based materials, there is a key process for solubilization of cellulose to solvent. So this it is necessary to be that cellulose can be prepared from a cellulose solution. Because cellulose has many hydroxyl groups which can form hydrogen bonding linked network easily. The basic requirement for cellulose dissolution is that the solvent is capable to interact with the hydroxyl groups of the AGU. It is needed that the solvent eliminate, at least partially, the strong inter-molecular hydrogen-bonding between the polymer chains (Wooding, 2001). However, cellulose is hard to dissolve due its stiffness with strong hydrogen bonds of cellulose. However, for preparing cellulose solution, lack of appropriate solvents delay the development for the preparation of cellulosic films (Striegel, 1997; Wooding, 2001; Lindma, Karlstrom & Stigsson, 2010).

Table 5 lists some conventional solvents for cellulose. Sulfuric acid hydrolysis is now widely used to yield cellulose whisker microcrystals (Lindma, Karlstrom, & Stigsson, 2010). It is believed that sulfuric acid destroys the amorphous regions of the cellulose fibers and allows the grafting of sulfate groups on the surface. This was useful for stabilizing the aqueous whisker suspensions. Alkali chemicals including LiOH and NaOH tended to be swelling cellulose. *N*-methylmorpholine-*N*-oxide (NMMO) is a solvent to be used in partially industrialized process. In general, alkali hydroxide solutions possess some of these requirements, e.g., low cost, ease of recycling, and high polarity. Indeed, a clear solution was obtained when microcrystalline cellulose was swollen by 8 to 9 wt% aqueous solution of NaOH. However, it is difficult to dissolve substantially all cellulose having a high degree of polymerization (DP>300) in an

Table 5. Conventional and new cellulose solvents

Solvents	Methods and Condition	Cellulose	References
8-10 wt% NaOH	Direct dissolution, 4°C	Treated cellulose, DP 330	Kamide et al. (1984)
7-9 wt% NaOH	Direct dissolution	Treated cellulose DP 330	Yamane et al. (1996)
8-9 wt% NaOH	Freeze-thaw	MCC, DP 200	Isogai and Atalla (1998)
6 wt% NaOH/4 wt% urea	Freeze-thaw	Cotton linter DP=690	Zhou and Zhang (2000)
7 wt% NaOH/12 wt% urea	Direct dissolution, -10°C	Cotton linter DP=700	Cai and Zhang (2005)
9.5 wt% NaOh/4.5 wt% thiourea	Direct dissolution, -4°to -5°C	Cotton linter DP=620	Ruan et al. (2004)
9 wt% NaOH/1 wt% PEG	Freeze-thaw	Cellulose powder DP 810	Yan and Gao (2008)
12-18 wt% NaOH	Two-step, -2 to 5°C	Cotton linter DP 570	Qi et al. (2011)
16-28 wt% urea			
12-18 wt% NaOH	Two-step, -2 to 5°C	Cotton linter DP 570	Qi et al. (2011)
4-6 wt% thiourea			
14-18 wt% thiourea	Two-step, -2 to 5°C	Avicel DP 570	Qi et al. (2011)
2-4 wt% PEG			
NMMO	Two-step, 90°C	Cottonlinter DP=690	Hon (1997)
DMAc	Two-step, room temp	Cottonlinter DP=700	Striegel (1997)

Source: Hon, 1997

alkali aqueous solution. Moreover, some limitations of NaOH-based aqueous systems have been observed on the preparation of solutions from wood pulps (Fukuzawa & Ohtani, 1990; Hon, 1997).

Due to that the intermolecular hydrogen bonds are present in cellulose, the effective destruction of intermolecular hydrogen bonding is essential for successful applications of cellulose. It was reported that intermolecular hydrogen bonding of polysaccharides could be broken by using urea (Tovar-Carrillo, 2014). Interestingly, NaOH and especially urea broke intermolecular hydrogen bonding of polysaccharides, allowing enhancement of water-solubility. The addition of organic compounds such as urea or thiourea to NaOH solution could showed substantial impact on cellulose solubility. However, this system could diminish cellulose molecular weight (Lindma, Karlstrom, & Stigsson, 2010). More extensive work has been carried out on binary and ternary mixtures, which will be designed as solvent "systems". The most investigated ones include inorganic or organic electrolyres in strong dipolar aprotic solvents. Examples are LiCl in DMAc (Striegel, 1997), in *N*-methyl-2-pyrrodinone (Lindma, Karlstrom, & Stigsson, 2010), in 1,3-dimethyl-2-imidazolidinone (Pokhrel & Visoraghavan, 2004), and tetra-n-butyl ammonium fluoride trihydrate in DMSO (TBAF/DMSO) (Wooding, 2001). Among them, the solvent system LiCl/DMAc has been most extensively employed because this has a capability of dissolving different celluloses including samples of high DP for example in cotton linters and bacterial cellulose. In the presence of aggregates in the LiCl/DMAc solvent system, the size depended on the pre-treatment employed, degree of polymerization, concentration of LiCl, and presence of water (Striegel, 1997; Tovar-Carrillo, Sugita, Tagaya, & Kobayashi, 2013). Molecularly dispersed cellulose solutions are obtained at low polymer- and high LiCl concentrations.

In the DMAc/LiCl system, the activation step is crucial for destroying the hydrogen bonded in the polymer chains into the most relaxed conformation. In order to enhance the solvation to the tightly packed crystalline regions, the step process spends mainly allowing sufficient time as an exchange solvent process. Water activation firstly is needed to be followed by DMAc exchange. At this time, water swells to loose cellulose structure. The inter-and-intramolecular hydrogen bonds are replaced by hydrogen bonds with water. Then, DMAc impedes the inter-and-intra-hydrogen bonds to be re-forming. The exchange works as solvation with water on the cellulose solution. After the activation phase, the cellulose substrate is ready to dissolve in DMAc/LiCl. It was apparently mentioned that the relative proportion of LiCl and cellulose was critical for optimal dissolution (Striegel, 1997; Tovar-Carrillo, Nakasone, Sugita, Tagaya, & Kobayashi, 2014) for cotton and wide variety of wood fibers such as softwood and hardwood in Kraft pulps. It was found that the concentration of LiCl to be critical in the formation of aggregates upon dissolution of wood pulp and cotton linter, independently of the sample concentration (Striegel, 1997). Over the years, various dissolution mechanisms for cellulose in DMAc/LiCl were proposed as following:

1. A $[DMAc_n{+}Li]^+$ macrocation must exist.
2. In the ion cluster with cellulose, the Cl^- anion is dissociated from the Li^+ cation by intercalating with one or more DMAc molecules.
3. The Li^+ cation interacts with the carbonyl group oxygens of the DMAc molecules.
4. The Cl^- anion disrupting the hydrogen bonds of cellulose can create hydrogen-bond-type interactions with the hydroxyl group hydrogens of cellulose.
5. The macrocation must have weak interactions with the cellulose oxygens.

But, these processes should be mentioned to be no conclusive evidence. Until today, the interactions between Li^+ cation and the glycosidic oxygen were indicated in the solution system described though semi-empirical MNDO computer models for interaction between cation and various disaccharides (Striegel, 1997).

HYDROGEL FILMS ELABORATED FROM REGENERATED CELLULOSE FOR MEDICAL APPLICATIONS

It is worth noting that hydrogels and films have wide potential applications in the field of food, biomaterials, agriculture, and water purification. Cellulose film could be applied to several fields such as food industry, pharmaceutics, and medical. In pharmaceutical, cellulose has a long history of use in the pharmaceutical industry. The material has excellent compaction properties when blended with other pharmaceutical excipients so that drug-loaded tablets form dense matrices suitable for the oral administration of drug. Cellulose membrane was fabricated to separate heavy metal ions and organic molecules (Sionkowske, 2011) and to detect them by sensors (Svensson, Nicklasson, Harrah, Panilatis, Kaplan, Brittgerg, & gatenholm, 2005).

Polysaccharides, natural polymers, fabricated into hydrophilic matrices remain popular biomaterials for controlled-release dosage forms and uses of a hydrophilic polymer matrix is one of the most popular approaches in formulating an extended release dosage forms (Table 6). Cellulosic hydrogels were applied for drug delivery (Chen, Yuan, Song, Wu, & Li, 2008; Jinag & Kobayashi, 2016) and for super absorbent hydrogels (Gibson, 2005). The stimuli on/off switch was made by changes of pH, temperature, ionic concentration, as used for smart materials based on cellulose (Lloyd, Kennedy, Methacanon, Paterson, & Knill, 1998). Cellulose film offers several potential advantages as a drug delivery excipient, in advanced pelleting systems, but the rate of tablet disintegration and drug release may be controlled by microparticle inclusion, excipient layering or tablet coating (Shen, Shamshina, Berton, Gurrau, & Rogers, 2016). In such drug delivery, carboxymethyl cellulose (CMC) in the hydrogel system contributes to enhance the ability due to the electronic repulsion of carboxyl groups, leading to a large swelling ratio as well as faster release of protein drugs.

Cellulose-based biocompartable hydrogels were reported for variety of applications, including wound closure (Lloyd, Kennedy, Methacanon, Peterson, & Knill, 1998) and drug delivery systems (Sionkowske, 2011) vascular grafts (Langer & Vacanti, 1993) or scaffolds for in vitro or in vivo tissue engineering (Nakasone, Ikematsu, & Kobayashi, 2016) (Table 7. Cellulose was used successfully as scaffold for cell adhesion, proving that it could become a high-value product in the field of tissue engineering (Langer & Vacanti, 1993).

Scaffolds and Cell Adhesion Materials for Tissue Engineering

Another interesting field in cellulose application is tissue engineering. The tissue engineering is a most recent application of hydrogels by using as scaffolds to mimic many roles of extracellular matrixes and to engineer new tissues. These scaffolds provide space and nutrients for new tissue formation, and potential control the structure and function of the engineered tissue in situ or in vitro. Currently, hydrogel scaffolds are at or near clinical uses in engineering many tissues in the body including cartilage, bone, muscle, skin, adipose, artery, ligament, tendon, liver, bladder, and neurons (Gibson, 2005). Recently,

Table 6. Medical applications of cellulose

	Hydrogel	The 2nd Component	Cross-Linker	Reference
Drug Delivery				
Controlled release by composition	Cellulose	CMC	ECM	Chang, Duan, Cai & Zhang, 2010.
	Cellulose	Lignin	ECM	Ciolacu, Oprea, Anghel, Cazacu & Cazacu, 2012.
	Cellulose	CNW	N/A	Tsutsumi, Koga, Qi, Saito & Isogai, 2014
Thermosensitive	Cellulose	PNIPAAm	MBAAm	Wang, Zhou & Xiao, 2013.
	Cellulose	PNIPAAm	MBAAm and GA	Sanna, Fourtunati, Alzari, Nuvoli, & Terenzi, 2013
pH-Sensitive	Cellulose	PAAm	MBAAm	Pandey, Amin, Ahmad & Abeer, 2013.
Magnetic	Cellulose	Fe_3O_4	GA	Luo, Liu, Zhou & Zhang, 2009.
Tissue Engineering				
	Cellulose	N/A	N/A	Gavillon & Budtova, 2008.
	Cellulose	Carbodimide	Collagen	Cai, Sharma, Liu, Mu & Liu, 2014
	Cellulose	$CaCl_2$	Pectin	Saska, Teixeira, de Oliveira & Gaspar, 2012.
	Cellulose	N/A	HA	Zadegan, Hosainalipour, Rezaire & Ghassai, 2011.
Wound Dressing				
	Cellulose	N/A	N/A	Gelin, Bodin, Gatenhol &, Mihranyan, 2007.
	Cellulose	AgNPs	N/A	Maneerung, Tokura & Rujiravanit, 2008.
	Cellulose	ZnO NPs	N/A	Katepetch, Rujiravanit, & Tanura, 2013.
	Cellulose	Alginate	Ca^{2+}	Chiaoprakobkij, Sanchavanakit, Subbalekha & Phisalaphong, 2011.

Carboxymethyl cellulose (CMC); cellulose nanowhisker (CNW); N,N′-methylenebisacrylamide (MBAAm); polyacrylamide (PAAm); poly(N-isopropylacrylamide) (PNIPAAm); poly(N-vinylcaprolactam) (PNVCL); hydroxyapatite (HA; nanoparticles (NPs);

novel hydrogels were developed and used for tissue engineering applications (Langer & Vacanti, 1993). This has a purpose in polymers occurring in new strategies for producing engineered tissue on synthetic or natural materials as a scaffold (Sionkowske, 2011; Chen, Yuan, Song, Wu, & Li, 2008; Liu, Ma, Mao, & Gao, 2011). In general, synthetic scaffolds could not provide an ideal environment to support cell adhesion and tissue formation due to their bio inert nature. Therefore, one strategy has been working with cellulosic polymers, in order to improve the properties to support cell adhesion and tissue formation on the surface (Kobayashi, 2015). In addition, because of their hydrophilicity, biocompatibility, and biodegradability, biopolymers were easily assimilated in the body by cellulosic hydrogel materials (Tamada & Ikada, 1993; Tovar-Carrillo, Tagaya, & Kobayashi, 2013).

Since cellulose itself has no antimicrobial activity to prevent wound infection, ZnO or Ag nanoparticles (NPs) can be impregnated into the cellulose gel system to achieve antimicrobial ability (Katepetch, Rujiravanit, & Tanura, 2013.). The mechanism of the bactericidal effects of ZnO and AG NPS are different: it is assume that water molecules can react with incorporated ZnO particles, leading to the generation of reactive oxy-radicals or hydroxyl-radicals that may cause oxidative injury inside bacteria or attach to the surface of the bacteria disturbing permeability and respiration functions.

Cellulose has advantage for similar structural material with plant cell wall found in nature. The environmentally friendly and renewable biomaterial is, therefore, reason for the suitable uses. The ultimate goal of tissue engineering is to replace, repair or enhance the biological function or damage, absent of dysfunctional elements of tissue or an organ. Engineered tissues are produced by using cells that are manipulated trough their extracellular environment to develop living biological substitutes for tissue that are lacking or malfunctioning (Svensson, Nicklasson, Harrah, Panilaitis, Kaplan, Brittberg, & Gatenholm, 2005). There are three ways in tissue engineering (Langer & Vacanti, 1993) by:

1. Inducing migration of tissue regeneration,
2. Using to encapsulate cells and acting as immuno-isolation barrier, and
3. Using as a matrix to support cell growth and cell organization.

In general, biodegradable biomaterials are desirable in tissue engineering to be replaced by newly formed tissue upon regeneration. For example, hydrogels provide a moist wound covering conductive to healing, and also protect the wound from infection. A disadvantage of many hydrogel networks, however, is their lack of strong mechanical properties. Several approaches have been done to elaborate cellulose hydrogel films using cross linkers or modifying cellulose structure to improve its solubility (Lloyd, Kennedy, Methacanon, Paterson & Knill, 1998). Since even fewer were known to deal with effective applications of cellulose for hydrogel materials (Sionkowske, 2011), most works could prepare biomaterials using cellulose derivatives or cellulose powder. Based on the characteristics of cellulose such as excellent biocompatibility, biodegradable and safety, wooden pulp was used to elaborate hydrogel films for tissue regeneration.

For the tissue cultivation, fibroblast cells were used to test the cytotoxicity of the obtained films (Kobayashi & Tovar-Carrillo, 2015). Cell morphology of adherent fibroblast on pulp hydrogels was analyzed and it was observed that fibroblasts adhered and proliferated well on the films surface. In addition, the obtained results showed that cell densities in the hydrogel films were higher than the observed on the commercial dishes used control. Cell morphology results showed that cell area, aspect ratio, and the

Table 7. Cellulose-based hydrogels prepared direct from natural celluloses

Gelation Methods with Solvent	Hydrogels		
LiCl/dimethylacetamide (DMAc) system	Natural cellulose films	Industry, medical applications	Tovar-Carrillo, Nakasone, Sugita, Tagaya, & Kobayashi, 2014 Kobayashi & Tovar-Carrillo, 2015 Nakasone, Ikematsu & Kobayashi, 2016
NMMO system	Natural cellulose films, fibers, food castings, membranes, sponges, beads.	Industry, food companies	Lindma, Karstrom & Stigsson, 2010
IL system	Flexible gels obtained	Industry	Pokhrel & Viroraghavan, 2004
Alkaly/urea or Thiourea aqueous	Weak gels can be obtained	Industry	Tovar-Carrillo, Tagaya & Kobayashi, 2013

length on the long axis were higher than the observed on the control dishes (Figure 6). Moreover, the boundaries of the adhered cells on cellulose films seemed to be tightly adhered on the hydrogel surface, showing a diffuse shape in the first 4 hours of cell culture. This morphologic signs were observed on the surface of the commercial dishes used as control at 24h of cell culture.

The chemical treatment process of the cellulose scaffold involves several solvents and compounds, traces of the compounds remain in the material, affecting the biocompatibility of the scaffold. For example, the type of treatment used on the bagasse fibers could affect the biocompatibility of the obtained material. In the case of NaOCl for cellulose bleaching, the cytocompatibility and biocompatibility of the obtained agave films were strongly influenced depending on the concentration of NaOCl used. It was observed an increment in the number of adherent cells on films as NaOCl treatment increased. In addition, the trace lignin in the cellulosic hydrogel effected on the fibroblast cultivation (Nakasone & Kobayashi, 2016).

One of the goals of tissue engineering is to manufacture scaffolds, suitable for transplantation, which are infiltrated with cells. Both viscose cellulose sponges and non-woven cellulose have been used experimentally to fabricate cartilage transplants. Cellulose derived from Acetobacter xylinum (Sionkowske, 2011), as discussed above in the context of wound healing, has also been explored as a potential scaffold material (Lloyd, Kennedy, Methacanon, Paterson & Knill, 1998), due to its unusual material properties and degradability. The cell type should be suitable for the implanted site, but the cultivated cells should be provided to the greater surface area in the scaffold (Salem, 2002; Klocking & Helbing, 2005). Thus, biomaterials in tissue-engineered substitutes serve as a structural component and provide the proper three-dimensional (3D) architecture of the construct (Gibson, 2005). The scaffold should prefer to provide

Figure 6. Phase-contrast light images of adherent fibroblast cells (a,b) commercial PS dish and (c,d) hydrogel films prepared with 6 wt% of LiCl, The cell culture time was (a,c) 4h and (b,d) 24h

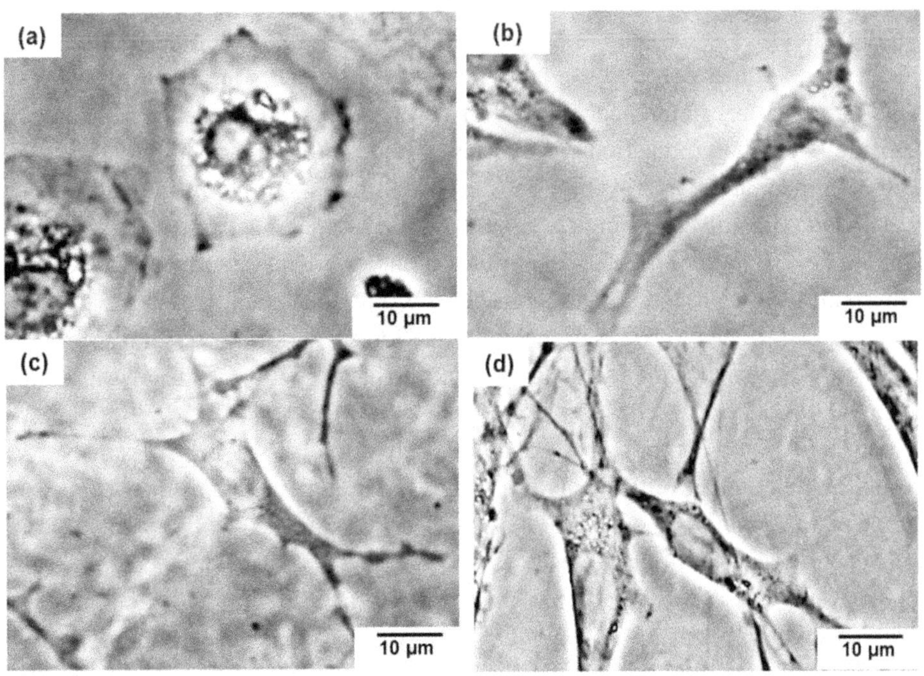

the 3D matrix for guided cell proliferation and control the shape of the bioartificial devise. It has been recently paid attention that the application of biomaterials in tissue engineering is a truly interdisciplinary field, (Klocking & Helbing, 2005; Williams, 2004; Rogers, 2001; Williams, 2004; Klocking & Helbing, 2005; Williams, 2004; Hollander & Hatton, 2004). In many cases scientists highly focused expertise in one discipline and cross boundaries into the new area.

The recent shift in their emphasis is way tissue engineering towards biomaterials. There was a steady growth over the last 40 years in the annual number of papers with biomaterials as a key word of title word, although the phrase tissue engineering was not cited in the literature until the mid-1980s and during the 1990s. Indeed, as dawn of the new millennium, tissue engineering undoubtedly revolutionizes the disease treatment in the near future (Langer & Vacanti, 1993).

APPLICATIONS DEVELOPMENT IN CURRENT AND FUTURE

For tissue-engineering scaffold, the material should be compatible, absorbable, and easily and reproducibly processability. In addition, it is necessary that the material can interact with cells and tissues at the surface. The material should not transfer antigens, and should be immunologically inert. Among on the criteria, biocompatibility becomes a phenomenological concept and is for the essential property. For instance, the inner surface of an implanted vascular graft or blood pump must be blood-compatible in artificial heart, while its outer surface has to be tissue-compatible. In other words, the material surfaces must not exert any adverse effects upon blood or tissue, or upon other biological elements at the interfaces.

Since activated charcoal is effective for the direct removal of various circulating toxic materials and waste metabolites from the blood (Chandy & Sharma, 1998), coating of the material results in an improved biocompatibility, and a variety of coatings, including cellulose nitrate has been used (Hoenich, 2006). There are materials as well as charcol for a variety of clinical applications in cellulose-based microspheres for the delivery of drugs (Gibson, 2005) and to bovine spermatozoa (Sionkowske, 2011).

Furthermore, cellulose is candidate material to implanting in body. Sutures are the largest group of devices implanted into humans (Lloyd, Kennedy, methacanon, Paterson, & Knill, 1998). Silk is a traditional fiber used as sutures in medical applications. Although biodegradable, its major disadvantage is its lower tensile strength and reaction with tissues. The problem of strength retention in the case of bio-absorbable commercial sutures is being tackled by different surface modification techniques. Although polypropylene and polyester filament coated with Teflon, for example, are employed for vascular surgery (Sionkowske, 2011), modification of cellulose filaments in sutures also have been produced and on carbon fiber surface (Svensson, Nicklasson, Harrah, Panilaitis, Kaplan, Brittberg, & Gatenholm, 2005) Haemorrhage continues to be a serious complication of surgical procedures. Surgical (oxidized regenerated cellulose) is used widely to control haemorrhage, as this material, when saturated with blood, swells rapidly (Gibson, 2005).

More recently in vivo implantation in rats of cellulose hydrogels (Nakasone, Ikematsu, & Kobayashi, 2016) was tested for biocompatibility evaluation of the film regenerated from sugar cane bagasse waste. The results showed that small influence of the implanted hydrogel films on the growth of mice and no inflammation reaction in the intraperitoneal was observed by post-mortem examination for 4weeks implantation. In vivo research was conducted in rats testing biocompatibility of cellulose films (Tovar-Carrillo & Kobayashi 2015), as shown in Figure 7. Surgery area was observed under microscope after 30 days of the material implantation with diminish of the material portion in no adverse body reaction.

Figure 7. Optical microscope images of the surgery area after 30 days. (a,c) cellulose sample at 10 x, (b) section of sample 1 at 40 x, and (d) typically inflammatory reaction to polyacrylamide

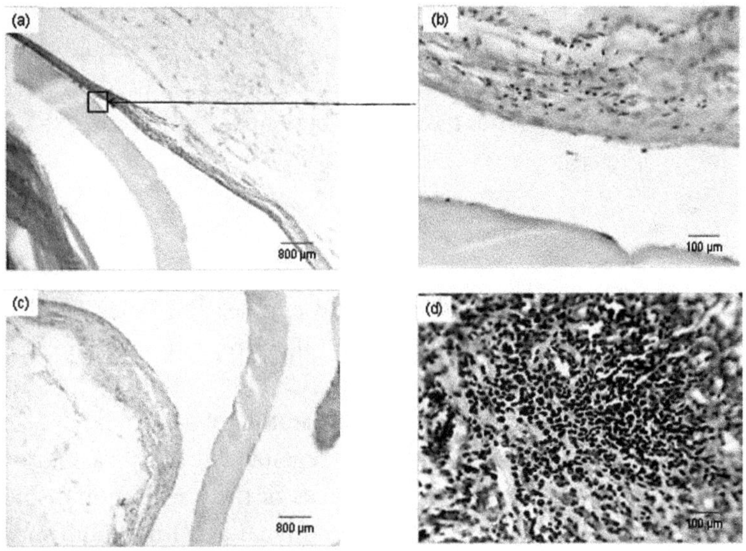

The optical microscope images showed thin fibrous tissue, (a and b). No cell agglomeration was observed near or far the hydrogel material. The presence of cells showed in dark purple, are inflammatory cells, but this cells are located far from de-implanted material. These cells in the images are not due to an inflammatory reaction against the implanted hydrogel. Moreover, the a and c presented fibrous section and remained traces cellulose film in the surgery area. The picture d showed a typical inflammatory adverse reaction in a tissue section after the placement of polyacrylamide material. After 30 days of the surgery a tissue section was observed. The results showed the typical inflammatory reaction in the area with the classic cell agglomeration reaction to the synthetic material. On the other hand, no cell agglomeration or inflammatory reaction was observed in the hydrogel samples. Consequently, we can suggest the cellulose hydrogel material samples are not toxic to the organism with biocompatibility. Base on the obtained results, we could suggest that the tested cellulose hydrogels could be applied for medical applications.

The Future

The most widely used application of cellulose in bio-medicine materials was used as a membrane in the treatment of renal failure (Svensson, Nicklasson, Harrah, Panilaitis, Kaplan, Brittberg, & Gatenholm, 2005). Worldwide trends in the use of membranes for the treatment of chronic renal failure indicated a move away from cellulose-based membranes in favour of synthetic membranes. Despite review in the European Best practice guidelines (Lloyd, Kennedy, Methacanon, Paterson, & Knill, 1998), dialyzer membranes would be the lowest degree of complement with leukocyte activation, inflammatory reactions, and/or a blunting of the response of leukocytes. Furthermore, such cost is important in favoring the continuing use in emerging economies. In the clinical application of cellulose in dialysis declines, the most promising appears to uses of microbial cellulose synthesized by Acetobacter xylinum as a novel would heal system.

Cellulose-based materials have vast applications in the sensing field as a result of their behavior changes in response stimulus (Chen. Yuan, Song, Wu, & Li, 2008) for pH sensors immobilizing dye on cellulose materials (Svensson, Nicklasson, Harrah, Panilaitis, Kaplan, Brittberg, & Gatenholm, 2005) ammonia sensors (Gibson, 2005), composited cellulose-conductive materials for humidity and temperature sensors and series of chemosensors (Sionkowske, 2011). It is quite fair to propose that the development of cellulose-based materials from biomass has a wide number of possibilities for cytocompatible materials.

CONCLUSION

Natural polymer, cellulose, has been applied in several engineering fields for decades. Especially in medical and bio-medicine materials, tissue engineering more recently involves cell seeding on a scaffold followed by cultivating in vitro prior implantation. Recently, cellulose is gaining interesting in the scientific community as a potential polymer for the preparation of scaffolds.

This chapter describes the importance of cellulose regenerated from waste products and cellulosic materials including hydrogel films. Moreover, advantages of the preparation of biomass cellulose and their hydrogels are described for point of view of economical and important impacts in environment sustainability As a frontier research, cellulose biomass can provide importance in the uses of cellulose fibers regenerated from waste products.

REFERENCES

Andersson, A. D. (2006). A study of the nanostructure of the cell wall of the tracheids of conifer xylem by C-ray scattering. *University of Helsinki, Reports Series in Physics. HU-P-D, 135*, 18–25.

Branco, A., Santos, J. D., Pimente, M., Osuno, J. T., Lima, L. S., & David, J. M. (2010). D-Mannitol from Agave sisalana biomass waste. *Industrial Crops and Products, 32*(3), 507–512. doi:10.1016/j.indcrop.2010.06.025

Cai, H. L., Sharma, S., Liu, W. Y., Mu, W., Liu, W., Zhang, X. D., & Deng, Y. L. (2014). Aerogel microspheres from natural cellulose nanofibrils and their applications as cell culture scaffolds. *Biomacromolecules, 15*(7), 2540–2547. doi:10.1021/bm5003976 PMID:24894125

Chandy, T., & Sharma, C. P. (1998). Activated Charcoal Microcapsules and their Applications. *Journal of Biomaterials Applications, 13*, 128–157. PMID:9777464

Chang, C. Y., Duan, B., Cai, J., & Zhang, L. N. (2010). Superabsorbent hydrogels based on cellulose for smart swelling and controllable delivery. *European Polymer Journal, 46*(1), 92–100. doi:10.1016/j.eurpolymj.2009.04.033

Chen, H., Yuan, L., Song, W., Wu, Z., & Li, D. (2008). Biocompatible polymer materials. Role of protein-surface interactions. *Progress in Polymer Science, 33*(11), 1059–1065. doi:10.1016/j.progpolymsci.2008.07.006

Chiaoprakobkij, N., Sanchavanakit, N., Subbalekha, K., Pavasant, P., & Phisalaphong, M. (2011). Characterization and biocompatibility of bacterial cellulose/alginate composite sponges with human keraticytes and gingival fibroblast. *Carbohydrate Polymers*, *85*(3), 548–553. doi:10.1016/j.carbpol.2011.03.011

Ciolacu, D., Oprea, A. M., Anghel, N., Cazacu, G., & Cazacu, M. (2012). New cellulose-lignin hydrogels and their application in controlled release of polyphenols. *Materials Science and Engineering C*, *32*(3), 452–463. doi:10.1016/j.msec.2011.11.018

Dupont, A. L., Egasse, C., Morin, A., & Vasseur, F. (2007). Comprehensive characterization of cellulose and lignocellulose degradation products in aged papers: Capillary zone electrophoresis of low molar mass organic acids, carbohydrates, and aromatic lignin derivatives. *Carbohydrate Polymers*, *68*(1), 1–8. doi:10.1016/j.carbpol.2006.07.005

Edgar, K. J., Buchanan, C. M., Debenham, J. S., Rundquist, P. A., Seiler, B. D., Shelton, D., & Tindall, D. (2001). Advances in cellulose ester performance and application. *Progress in Polymer Science*, *26*(9), 1605–1611. doi:10.1016/S0079-6700(01)00027-2

Filho, G. R., Ribeiro, S. D., & Meireles, C. S. (2011). Release of doxycycline through cellulose acetate symmetric and asymmetric membrane produced from recycled agroindustrial residue: Sugarcane bagasse. *Industrial Crops and Products*, *33*(3), 566–576. doi:10.1016/j.indcrop.2010.10.037

Gavillon, R., & Budtova, T. (2008). Aerocellulose: New highly porous cellulose prepared from cellulose-NaOH aqueous solutions. *Biomacromolecules*, *9*(1), 269–277. doi:10.1021/bm700972k PMID:18085745

Gelin, K., Bodin, A., Gatenholm, P., Mihranyan, A., Edwards, K., & Stromme, M. (2007). Characterization of water in bacterial cellulose using dielectric spectroscopy and electron microscopy. *Polymer*, *48*(26), 7623–7631. doi:10.1016/j.polymcr.2007.10.039

Gibson, L. J. (2005). Biomechanics of cellular solids. *Journal of Biomechanics*, *38*(3), 377–399. doi:10.1016/j.jbiomech.2004.09.027 PMID:15652536

Guimaraes, J. F., Frollini, E., da Silva, C. G., Wypych, F. K. G., & Satyanarayana, K. G. (2009). Characterization of banana, sugarcane bagasse and sponge gourd fibers of Brazil. *Industrial Crops and Products*, *30*(3), 407–415. doi:10.1016/j.indcrop.2009.07.013

Hamisso, A. M., Lodi, A., Seffen, M., Finocchio, E., Botter, R., & Converti, A. (2010). Asorption of Cd (II) and Pb (II) from aqueous solution onto Agave Americana fibers. *Chemical Engineering Journal*, *159*(1-3), 67–76. doi:10.1016/j.cej.2010.02.036

Hoenich, N. (2006). Cellulose for medical applications: Past, present, and future. *BioResources*, *2*, 270–280.

Hollander, A. P., & Hatton, P. V. (2004). *Biopolymer methods in tissue engineering*. Human press.

Hon, D. N., (1997). *Chemical modification of lignocelluosic materials.* New York: Marcel Dekker, Inc.

Iñuguez-Covarrubias G., Diaz-Terez R., Sanjuan-Duenas R., Anzaldo-Hernandez J., & Rowel R. M. (n.d.). Utilization of by-products from the tequila industry. Part 2: potential value of agave tequilana weber azul leaves. *Bioresources Technology, 77*, 101-108.

Iñuguez-Covarrubias, G., Lang, S. E., & Rowel, R. M. (2001). Utilization of by-products from the tequila industry. Part 1: Potential value of agave tequilana weber azul leaves. *Bioresource Technology*, *77*, 25–40. doi:10.1016/S0960-8524(00)00137-1 PMID:11211072

Jiang, H., Tovar-Carrillo, K., & Kobayashi, T. (2016). Ultrasound Release of Mimosa Medicine Included in Cellulose Hydrogel Films. *Ultrasonics Sonochemistry*, *32*, 398–406.

Kalia, S., Dufresne, A., Cherian, B. M., Kaith, B. S., Averous, L., Njuguna, J., & Nassiopoulos, E. (2011). Cellulose-based bio and nanocomposites: A review. *International Journal of Polymer Science*, *2011*, 1–35. doi:10.1155/2011/837875

Kalia, S., Kaith, B. S., & Kaur, I. (2009). Pretreatments of natural fibers and their applications as reinforcing materials in polymers composites- a review. *Polymer Engineering and Science*, *49*(7), 1253–1272. doi:10.1002/pen.21328

Katepetch, C., Rujiravanit, R., & Tamura, H. (2013). Formation of nanocrystalline ZnO particles into bacterial cellulose pellicle by ultrasonic-assisted in situ synthesis. *Cellulose (London, England)*, *20*(3), 1275–1292. doi:10.1007/s10570-013-9892-8

Klocking, R., & Helbing, B. (2005). *Medical aspects and human applications of humic substances*. Wiley-VCH.

Kobayashi, K., Tovar-Carrillo, K. L., & Tagaya, M. (2015). Bagasse sustainable polymers for cellulose hydrogel sheets showing tissue regeneration. In Handbook of Sustainable Polymers: Structure and Chemistry. Pan Stanford Publishing Pte Ltd.

Kobayashi, T. (2015). *Fabrication of cellulose hydrogels and characterization of their biocompatible films. In Studies in Natural Products Chemistry, 45* (pp. 1–15). Amsterdam: Elsevier B.V.

Kobayashi, T., & Tovar-Carrillo, K. (2015). Fibroblast Cell Cultivation on Wooden Pulp Cellulose hydrogels for cytocompatibility scaffold method. *Pharmaceutic Analylical Acta*, *6*, 10. doi:10.4172/2153-2435.1000423

Langer, R., & Vacanti, J. P. (1993). Tissue engineering. *Science*, *260*(5110), 920–926. doi:10.1126/science.8493529 PMID:8493529

Li, Y., Mai, Y.-W., & Ye, L. (2000). Sisal fibre and its composites: A review of recent developments. *Composites Science and Technology*, *60*(11), 2037–2048. doi:10.1016/S0266-3538(00)00101-9

Lindma, B., Karlstrom, G., & Stigsson, L. (2010). On the mechanism of dissolution of cellulose. *Journal of Molecular Liquids*, *156*(1), 76–81. doi:10.1016/j.molliq.2010.04.016

Liu, X., Ma, L., Mao, Z., & Gao, C. (2011). Chitosan-based biomaterials for tissue repair and regeneration. *Advances in Polymer Science*, *244*, 81–87. doi:10.1007/12_2011_118

Lloyd, L. L., Kennedy, J. F., Methacanon, P., Paterson, M., & Knill, C. J. (1998). Carbohydrates polymers as wound management aids. *Carbohydrate Polymers*, *37*(3), 315–319. doi:10.1016/S0144-8617(98)00077-0

Luo, X. G., Liu, S. L., Zhou, J. P., & Zhang, L. N. (2009). In situ synthesis of. Fe3O4/cellulose microspheres with magnetic-induced protein delivery. *Journal of Materials Chemistry*, *19*(21), 3538–3545. doi:10.1039/b900103d

Maneerung, T., Tokura, S., & Rujiravanit, R. (2008). Impregnation of silver nanoparticles into bacterial cellulose for antimicrobial wound dressing. *Carbohydrate Polymers*, *72*(1), 43–51. doi:10.1016/j.carbpol.2007.07.025

Martin, A. R., Martins, M. A., da Silva, O. R., & Motto, L. H. (2010). Studies on the thermal properties of sisal fiber and its constituents. *Thermochimica Acta*, *506*(1-2), 14–22. doi:10.1016/j.tca.2010.04.008

Nair, L. S., & Laurencin, C. T. (2007). Biodegradable polymers as biomaterials. *Progress in Polymer Science*, *32*(8-9), 762–767. doi:10.1016/j.progpolymsci.2007.05.017

Nakasone, K., Ikematsu, S., & Kobayashi, T. (2016). Biocompatibility evaluation of cellulose hydrogel film regenerated from sugarcane bagasse waste and its in vivo behavior in mice. *Industrial & Engineering Chemistry Research*, *55*(1), 30–37. doi:10.1021/acs.iecr.5b03926

Nakasone, K., & Kobayashi, T. (2016). Effect of pre-treatment of sugarcane bagasse on the cellulose solution and application for the cellulose hydrogel films. *Polymers for Advanced Technologies*, *27*(7), 973–980. doi:10.1002/pat.3757

Nakasone, K., & Kobayashi, T. (2016, July). Cytocompatible cellulose hydrogels containing trace lignin. *Materials Science and Engineering C*, *64*, 269–277. doi:10.1016/j.msec.2016.03.108 PMID:27127053

Pandey, M., Amin, M. C. I. M., Ahmad, N., & Abeer, M. M. (2013). Rapid synthesis of superabsorbent smart-swelling bacterial cellulose/acrylamide-based hydrogels for drug delivery. *International Journal of Polymer Science*.

Pokhrel, D. T., & Viroraghavan, T. (2004). Treatment of pulp and paper mill wastewater a review. *The Science of the Total Environment*, *333*(1-3), 37–43. doi:10.1016/j.scitotenv.2004.05.017 PMID:15364518

Ramos, L. P. (2003). The chemistry involved in the steam treatment of lignocellulosic materials. *Quimica Nova*, *26*(6), 23–31. doi:10.1590/S0100-40422003000600015

Roberts, J. C. (1997). *The chemistry of paper*. Cambridge, UK: RSC Paperbacks.

Salem, A. K., Tendler, S. J., & Roberts, C. J. (2002). Interactions of 3T3 fibroblasts and endothelial cells with defined pore features. *Journal of Biomedical Materials Research*, *61*(2), 212–218. doi:10.1002/jbm.10195 PMID:12007201

Sanna, R., Fortunati, E., Alzari, V., Nuvoli, D., Terenzi, A., Casula, M. F., & Mariani, A. et al. (2013). Poly (N-vinylcaprolactam) nanocomposites containing nanocrystalline cellulose: A green approach to thermoresponsive hydrogels. *Cellulose (London, England)*, *20*(5), 2393–2402. doi:10.1007/s10570-013-9988-1

Saska, S., Teixeira, L. N., de Oliveira, P. T., Gaspar, A. M. M., Ribeiro, S. J. L., Messaddeq, Y., & Marchetto, R. (2012). Bacterial Cellulose-collagen nanocomposite for bone tissue engineering. *Journal of Materials Chemistry*, *22*(41), 22102–22112. doi:10.1039/c2jm33762b

Shen, X., Shamshina, J. L., Berton, P., Gurrau, G., & Rogers, R. D. (2016). Hydrogels based on cellulose and chitin: Fabrication, properties, and applications. *Green Chemistry*, *18*(1), 53–75. doi:10.1039/C5GC02396C

Sionkowske, A. (2011). Current research on the blends of natural and synthetic polymers as new biomaterials. *Progress in Polymer Science*, *36*(9), 1254–1261. doi:10.1016/j.progpolymsci.2011.05.003

Smook, G. A. (1992). Handbook for pulp and paper technologists. Agus Wilde Publications.

Striegel, A. M. (1997). Theory and applications of DMAc/LiCl in the analysis of polysaccharides. *Carbohydrate Polymers*, *34*(4), 267–273. doi:10.1016/S0144-8617(97)00101-X

Svensson, A., Nicklasson, E., Harrah, T., Panilaitis, B., Kaplan, D. L., Brittberg, M., & Gatenholm, P. (2005). Bacterial cellulose as a potential scaffold for tissue engineering of cartilage. *Biomaterials*, *26*(4), 419–425. doi:10.1016/j.biomaterials.2004.02.049 PMID:15275816

Tamada, Y., & Ikada, Q. (1993). Effect of preadsorbed proteins on cell adhesion to polymer surface. *Journal of Colloid and Interface Science*, *155*(2), 334–341. doi:10.1006/jcis.1993.1044

Tovar-Carrillo, K., Nakasone, K., Sugita, S., Tagaya, M., & Kobayashi, T. (2014). Effects of sodium hypochlorite on Agave tequilana Weber bagasse fibers used to elaborate cyto and biocompatible hydrogel films. *Materials Science and Engineering C*, *42*, 808–815. doi:10.1016/j.msec.2014.06.023 PMID:25063183

Tovar-Carrillo, K., Tagaya, M., & Kobayashi, T. (2013). Bamboo Fibers Elaborating Cellulose Hydrogel Films for Medical Applications. *Journal of Material Science Chemical Engineering*, *1*(07), 7–12. doi:10.4236/msce.2013.17002

Tovar-Carrillo, K. L. (2014). *Study on Cellulose Hydrogel Films Regenerated From Natural Plant Bagasse and Their Bio and Cytocompatible Properties for Tissue Engineering* (Thesis). Nagaoka University of Technology.

Tovar-Carrillo, K. L., & Kobayashi, T. (2015). Fibroblast Cell Cultivation on Wooden Pulp Cellulose Hydrogels for Cytocompatibility Scaffold Method. Pharm Anal Acta, 6(10), 423.

Tovar-Carrillo, K. L., Tagaya, M., & Kobayashi, T. (2013). Fibroblast compatibility on scaffold hydrogels prepared from Agave tequilana weber bagasse for tissue regeneration. *Industrial & Engineering Chemistry Research*, *60*(33), 11607–11673. doi:10.1021/ie401793w

Tsutsumi, Y., Koga, H., Qi, Z. D., Saito, T., & Isogai, A. (2014). Nanofibrillar chitin aerogels as renewable nased catalysts. *Biomacromolecules*, *15*(11), 4314–4319. doi:10.1021/bm501320b PMID:25285573

Vazquez, M. I., De Lara, R., & Benavente, J. (2008). Chemical surface diffusional, electrical and elastic characterization of two different dense regenerated cellulose membranes. *Journal of Colloid and Interface Science*, *328*(2), 331–338. doi:10.1016/j.jcis.2008.09.019 PMID:18842272

Wang, J., Zhou, X., & Xiao, H. (2013). Structure and properties of cellulose/poly (N-isopropylacrylamide) hydrogels prepared by SIPN strategy. *Carbohydrate Polymers*, *94*(2), 749–754. doi:10.1016/j.carbpol.2013.01.036 PMID:23544629

Williams, D. F. (2004). *The Biomaterials Silver Jubilee Compendium*. Amsterdam: Elsevier B. V.

Wooding, C. (2001). *Regenerated cellulose fibers.* Cambridge, UK: Woodhead Publishing Limited. doi:10.1533/9781855737587

Wu, J., Fukuzawa, K., & Ohtani, J. (1990). Lignin analysis in some tropical hardwoods using ultraviolet microscopy, Res. Bull. Of College Exp. For. *Hokkaido University*, *47*, 353–361.

Yamaoka, H., Ogasawara, T., Nishizawa, H., Takahashi, T., Nakatsuka, T., Koshima, I., & Hoshi, K. et al. (2006). Cartilage tissue engineering using human auricular chondrocytes embedded in different hydrogel materials. *Journal of Biomedical Materials Research. Part A*, *78*(1), 1–8. doi:10.1002/jbm.a.30655 PMID:16596585

Zadegan, S., Hosainalipour, M., Rezaie, H. R., Ghassai, H., & Shokrgozar, M. A. (2011). Synthesis and biocompatibility evaluation of cellulose/ hydroxyapatite nanocomposite scaffold in 1-n-allyl-3-methylimidazolium chloride. *Materials Science and Engineering C*, *31*(5), 954–961. doi:10.1016/j.msec.2011.02.021

KEY TERMS AND DEFINITIONS

Baggase: The pulp residue left after the extraction of juice from plants as sisals, agave and sugarcane.

Biomass: Organic matter is derived from living and generally can be used as a source of energy. Commonly this refers to plant or plant-based materials which are not used for food or feed.

Cell Adhesion: The behavior means capability of cells to interact and attach to a surface, substrate or another cell, as mediated by interactions between molecules of the cell surface.

Hydrogel: In a structure network-type of polymers, crosslinkage fablicates water insoluble matrix showing highly hydrophilic nature to retain over 90% water.

Scaffold: Matrix for cell cultivation by attaching and growing. This matrix is capable of supporting three-dimensional tissuc formation.

Tissue Engineering: Research field which combines cell biology, engineering and materials methods, as well as biochemical, to improve or replace biological functions.

Tissue Regeneration: It is the process of renewal and growth of damage tissue or desire portion of tissue to improve biological functions.

Chapter 3
Development and Modification of Natural Rubber for Advanced Application

Kanoktip Boonkerd
Chulalongkorn University, Thailand

ABSTRACT

Although natural rubber is a challenging elastomeric material for both dynamic and static engineering applications, there are some inherent drawbacks such as the poor oxygen, ozone, oil and heat resistance and also the low compatibility with the polar materials. To loosen these inferior properties and also to transform natural rubber into new polymeric materials, natural rubber needs to be either chemically or physically modified. Moreover, to make natural rubber become a promising elastomeric material for the advanced application, the development of natural rubber with the superior mechanical strength, the low gas permeability or the high conductivity is recently focused. And these can be done by reinforcing natural rubber with nanofillers. This chapter presents a broad review on the recent research and development of natural rubber including the modification of natural rubber and the preparation of various natural rubber nanocomposites for advanced application.

INTRODUCTION

Natural rubber (NR) is a high molecular weight polymeric substance with viscoelastic properties. Structurally it is cis 1,4-polyisoprene. Isoprene is a diene and 1, 4 addition leaves a double bond in each of the isoprene unit in the polymer. Natural rubber is considered as a renewable resource and an environmentally friendly materials. Because of this, natural rubber shows all the reactions of an unsaturated polymer. It gives addition compounds with halogens, ozone, hydrogen chloride and several other reactants that react with olefins. An interesting reaction of natural rubber is its combination with sulfur. This is known as vulcanization. This reaction converts the plastic and viscous nature of raw rubber into elastic. Vulcanized rubber will have very high tensile strength and comparatively low elongation. Its hardness and abrasion

DOI: 10.4018/978-1-5225-1971-3.ch003

resistance also will be high when compared to raw rubber. Because of the unique combination of these properties, natural rubber finds application in the manufacture of a variety of products.

Although natural rubber is a challenging elastomeric material, there are some inherent drawbacks. These drawbacks are attributed to the high unsaturation and also the non-polar character of natural rubber chain. Moreover, to make natural rubber become a promising elastomeric material for advanced application, the development of natural rubber with the superior mechanical strength, the low gas permeability and the high conductivity is recently focused. This broad review is contributed to the research and development of NR including the overview about modification of natural rubber and the preparation of various natural rubber nanocomposites for advanced application.

MODIFIED FORMS OF NATURAL RUBBER

Modification of natural rubber can be carried out in many forms. Mainly, it can be divided into two categories including physical and chemical modification as shown in Figure 1.

Physical Modification of Natural Rubber

Oil Extended Natural Rubber (OENR)

Petroleum oils are used in rubber, either as a process aid or as an extender. When the oil added is below 10 parts per hundred rubber, it is termed as process aid and above that is termed extender, which cheapens the product. Natural rubber, which contains oil as an extender is termed oil extended natural rubber or OENR. In tires OENR can impart high skid resistance on wet road, and resistance to wear and groove cracking. Oil addition to rubber can be carried out in latex or dry stage.

Table 1. Typical properties of natural rubber

Property	Specified Values
Specific gravity	0.92
Refractive index	1.52
Coefficient of cubical expansion	0.00062/°C
Cohesive energy density	63.7 Cal/cc.
Heat of combustion	10547.084 Cal/g
Thermal conductivity	0.00032 Cal/sec/cm/°C
Dielectric constant	2.37
Power factor (at 1000 cycles)	0.15-0.2
Volume resistivity	1015 ohm.cm
Dielectric strength	3937 V/mm

Figure 1. Physical and chemical modification of natural rubber

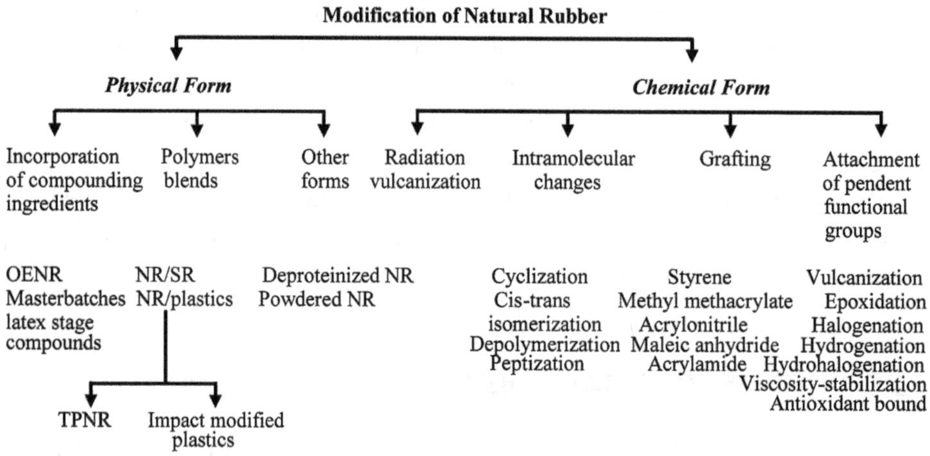

Latex Carbon Black Masterbatch

For latex carbon black masterbatch, it is made by mixing carbon black in the form of dispersion in latex and the resultant rubber-black mixture is coagulated, dried and baled. Latex stage mixing of carbon black in rubber can save energy.

Thermoplastic Natural Rubber (TPNR)

Thermoplastic elastomers (TPE) are a new class of materials which possess the processing characteristics of thermoplastics and properties of vulcanized rubbers at room temperature. TPEs can soften reversibly with heat like thermoplastics, but possess considerable resilience and flexibility, a characteristic of vulcanized rubbers. Thermoplastic natural rubber or TPNR made from a blend of NR and polypropylene (PP) with certain definite composition can function as a TPE. The elastic properties of TPNR could be improved, with partial crosslinking of the rubber phase, known as dynamic vulcanization. TPNR is generally resistant towards heat and ozone. TPNR may find application in footwear, sports goods, automobiles etc.

Superior Processing (SP) Natural Rubber

Superior processing (SP) rubber consists of an intimate mixture of vulcanized and unvulcanized rubber. It is made by mixing vulcanized latex with preserved field latex in different proportions, followed by coagulation and drying. Typical grades are SP 20, PA 80 and PA 57. SP 20 is a blend that contains 20 parts vulcanized and 80 parts unvulcanized NR. It is available in sheet and crepe forms. PA 80 is a blend of 80 parts vulcanized and 20 parts unvulcanized NR. PA 57 is an oil extended version of PA 80. Superior processing rubbers find application in extruded products as it can reduce die swell, especially in lightly filled components.

Deproteinized Natural Rubber (DPNR)

Deproteinized natural rubber (DPNR) is a highly purified form of natural rubber with very low nitrogen and ash contents. It is suitable for electrical and engineering applications. Removal of proteins from NR reduces its moisture sensitivity, thereby improving its engineering properties. Deproteinization is usually effected by an enzymic means. Enzymes hydrolyze the proteinaceous non-rubber materials into water soluble forms, which are then washed away during subsequent processing.

Chemical Modification of Natural Rubber

Chemical modification of natural rubber depends on the chemical reactivity of the rubber molecule. Different types of commercially relevant chemically modified forms of natural rubber are shown below.

Chlorinated Natural Rubber

It is one of the early forms of chemically modified NR which finds commercial application of chlorinated rubber. Chlorination of natural rubber is carried out in solution or in latex with gaseous chlorine. The chlorine content of the product is about 65 per cent. It is a pale cream colored thermoplastic powder, which is non-flammable and highly resistant to chemicals. Chlorinated rubber finds application in paints, adhesives, printing inks, paper coatings and textile finishes.

Epoxidised Natural Rubber (ENR)

Though the mechanical properties of natural rubber are superior to those of most synthetic rubbers, it cannot compete with specialty synthetic elastomers with regard to properties such as gas permeability and resistance to hydrocarbon oils. Latex stage epoxidation of natural rubber under controlled conditions gives a chemically-modified form of natural rubber, called epoxidised natural rubber (ENR) with improved resistance to hydrocarbon oils, low air permeability, increased damping and good bonding properties while retaining the high strength properties of natural rubber. Improvement in these properties depends on the degree of epoxidation (Roberts, 1998). Consequently two grades of ENR, ENR-25 and ENR-50 with 25 and 50 mole per cent of epoxidation respectively, have attained commercial importance. ENR can find application in products, which require hydrocarbon oil resistance and air impermeability such as in oil seals and tire inner tubes respectively. Another distinctive feature of ENR is the high degree of reinforcement achieved with silica fillers without the addition of a silane coupling agent. Silica reinforces ENR-25 and ENR-50 to an extent equivalent to that obtained with carbon black. Another interesting feature of ENR is its high wet grip and lower rolling resistance, which makes it a choice in tire tread compounds. High bondability of ENR with other materials finds application in adhesives and sealants. Other general applications envisaged for ENR are in milking inflation, pharmaceutical and food contact applications, conveyor belting, shoe soles, floor covering, engine mountings etc.

Constant Viscosity (CV) and Low Viscosity (LV) Natural Rubber

NR undergoes hardening or increases in viscosity during primary processing and subsequent storage under ambient conditions. This behavior is termed 'storage hardening' and is enhanced by low relative humidity. The hardening occurs via a crosslinking reaction between the carbonyl groups randomly distributed on the rubber chain and the amino acids present in the non-rubber constituents. Storage hardening causes processing problems like consumption of higher energy. The hardening problem can be avoided by pretreating the latex with certain chemicals viz., hydroxylamine hydrochloride, hydroxylamine neutral sulphate or semicarbazide. Low viscosity rubber (LV) is a viscosity stabilized rubber added with a small quantity of naphthenic oil.

Chemical Modification of Natural Rubber By Grafting

Grafted natural rubber is a chemically modified form of natural rubber prepared by polymerizing vinyl monomers onto natural rubber either in latex or in solution. Methyl methacrylate (MMA), styrene and acrylonitrile are the important monomers used for grafting on to natural rubber. Generally grafting can be achieved using free radical initiation by two methods: by use of chemicals and by irradiation with gamma rays. In the case of MMA graft natural rubber, it is produced by reacting methyl methacrylate monomer in NR latex with suitable initiators. A wide range of products is possible depending upon the amount of MMA monomer used for grafting. Malaysia have commercialized this product since 1950 under the trade name '*Heveaplus MG*'. Three grades viz. MG 30, MG 40 and MG 49 are popular where the number represent the percentage of MMA content in the grafted rubber. PMMA grafted-NR has higher modulus. It can provide self-reinforced vulcanizates. It finds use in adhesive and shoe industry. For polystyrene graft natural rubber (SGNR), polystyrene can be grafted onto natural rubber by polymerizing styrene in NR latex. Grafting can be initiated by chemical means or by gamma radiation. The SGNR could be used as a modifier in microcellular products.

Modification of Natural Rubber By Intramolecular Changes

For natural rubber modification by intramolecular changes, liquid natural rubber (LNR) and cyclized natural rubber are the two commercially important modifications under this group. For the former case, extensive size reduction of the molecular chains of natural rubber by depolymerization leads to the formation of LNR. LNR can be produced by thermal or chemical depolymerization of natural rubber. Thermal depolymerization involves heating of the masticated natural rubber at 220 to 240°C for a predetermined time under stirring, while chemical depolymerization involves an oxidation-reduction reaction using phenylhydrazine and air. LNR may find application in making flexible mold and printing industry and as binder for grinding wheels, reactive plasticizer in polymers etc. In latter case, the cyclized natural rubber is a hard brittle thermoplastic derivative of natural rubber obtained by treating natural rubber with acidic catalysts such as sulfuric acid, aromatic sulfonic acids and lewis acids such as stannic chloride, titanium tetrachloride and boron trifluoride. A simple method of preparation is by mixing 100 parts of natural rubber with 8 parts of para toluene sulfonic acid and heating the resulting mixture at 140°C. Cyclized rubber may find applications in adhesives, bonding agents, paints, shoe soles, hard moldings, industrial rollers etc. and as a reinforcing resin.

Radiation Vulcanization

Radiation vulcanization is a new method for vulcanization of rubber, especially latex with gamma rays, in the presence of a sensitizer namely normal butyl acrylate. The sensitizer is used to reduce the radiation dose. The advantage of radiation vulcanization over sulfur vulcanization are the absence of n-nitrosamines, very low cytotoxicity and protein allergy. Radiation vulcanized latex can be used for the commercial production of gloves, condoms, catheters etc. In India, radiation vulcanization plant with a capacity of 3 tones/day installed in Rubber Board is the only facility available for the production of vulcanized latex.

Natural Rubber Nanocomposites

Since the start of the twentieth, carbon black and amorphous silica have been utilized as conventional reinforcing fillers in various rubbers for a variety of applications. The incorporation of carbon black and silica into the rubber matrix tremendously increases mechanical properties of the rubber products. In general, carbon black provides an enhancement in tensile strength, tear strength, elastic modulus and abrasion resistance (Rattanasom *et al.,* 2007), whereas silica gives a unique combination of tear strength, abrasion resistance, aging resistance and adhesion (Hashim *et al.,* 1998). However, to achieve the acceptable reinforcing level, large amount of these conventional fillers are needed, giving a high density rubber product and leading to a filler network with different level of aggregation (Vilgis, 2005). Moreover, none of these two conventional fillers can provide the specific properties such as gas barrier property, electrical properties, etc. Therefore, in order to develop high-performance materials with a combinations of different properties at very small amount of filler, the incorporations of nanofillers have attracted the attention of many rubber researches.

Nanofillers offer enormous advantages over traditional macro-or micro-particles (e.g. talc, glass) due to their higher surface area and aspect ratio. To obtain the compound with desired properties only small amount of nanofiller is needed, thus giving product with lighter weight and lower cost. However, also due to their higher surface area and aspect ratio, nanofillers tend to form interaction between each other, thus leading to the filler aggregation. The most vital feature to obtain the useful nanocomposite is to establish well dispersion of nanofillers and good adhesion between matrix and incorporated nanofillers. To produce high performance nanocomposite materials and also to have uniform and thermodynamically controlled distribution of nanofillers in the phase of rubber matrix, two main obstructions have to solve. These are deaggregation of the nanofillers association which consists of millions of filler particulates, accumulated with a strongly active inter-particulate force (Israelachvili, 1991), and formation of strong rubber-filler interfacial adhesion. It is essential for better physical interconnecting between matrix and filler phases. The above mentioned two requisites are very much important in the rubber based nanocomposites. The dimensions of the various nanofillers have some additional obstruction which should be solved for the nanocomposites formation. Such as the entropic reactions between the two dimensional fillers and the rubbers in nanoscopic range, de-agglomeration of filler accumulation like CNTs; and conquering the extensively faster kinetics of the nanofiller's reaggregation. To overcome all of the above problems, many researches proposed various fabricating processes for preparing the desired useful nanocomposites.

Based on how many dimensions of the dispersed particles having the nanometer range, nanofillers can be classified into three categories. If all the dimensions are in the order of nanometers, as in spherical nanoparticles, they are called isodimensional nanoparticles such as silica and carbon black. When two dimensions are in the order of nanometers while third is large, it is considered to be nanotubes such as carbon nanotubes (CNTs), cellulose whiskers and carbon nanofibers (CNFs). The third type of nanofiller is characterized by only one dimension in the nanometer range such as layered silicates, expanded graphite (EG) sheets and layered double hydroxides (LDHs) flakes sheet.

Although there are tremendous researches in the field of polymer nanocomposites are reported during the last two decades, the development of rubber nanocomposites compared to the thermoplastic or thermoset, seems to be infancy. This is due to the fact that for rubber to exhibit real functionality such as elasticity or high aging resistance, several ingredients have to be incorporated into them. Hence, it is a great challenge. Here will focus on the key characteristics of applicable nanofillers for natural rubber, how to obtain the useful and desirable natural rubber nanocomposites and their roll on the properties of natural rubber. Only three types of natural rubber nanocomposites including natural rubber/clay nanocomposite, natural rubber/carbon nanotube and natural rubber/graphene nanocomposite are focused here.

Rubber Clay/Nanocomposites

Clays

Clay minerals are hydrous aluminum silicates and generally classified as layered silicates. This is their most important characteristic. The layered silicates of clay are generated by a combination of silicon tetrahedral sheet and aluminium octahedral sheets (Theng, 1974). They are formed as natural and synthetic clays such as mica, bentonite, magadite, laponite, fluorohectorite and so on. There are many kinds of the layered silicates. The commonly used layered silicates for preparing polymer nanocomposites belong to the smectile family with the structure consisting of nanometer thick layers (platelets) of aluminium octahedral sheet sandwiched in between two silicon tetrahedral sheets as called 2:1 layered silicates. These sheet-like nanofillers possess a high aspect ratio with 1 nm thick and 100-1000 nanometers long. Stacking of the layers leads to a Van der waals gap between the layers. This implies that layers can be separated from each other by applying a small amount of energy. Isomorphic substitution of Al with Mg, Fe, Li in the octahedron sheets and/or Si with Al in tetrahedron sheets gives each three-sheet layer an overall negative charge, which is counterbalanced by exchangeable metal cations residing in the interlayer space, such as Na, Ca, Mg, Fe, and Li. After the inclusion of these nanofillers, the properties such as stiffness, strength and barrier properties of nanocomposites are improved compared to the pure matrix. The major concern when using these nanofillers is how to enhance the compatibility between these inorganic nanofiller with the organic polymer. Among the smectile clays, montmorillonite (MMT) is the most commonly used layered silicate for preparing nanocomposites due to its high cation exchange capacities, large surface area, good surface reactivity and high surface adsorptive properties (Alexandre & Dubois, 2000).

Polymer/Clay Nanocomposite Structure

Generally, the structure of polymer/clay nanocomposites are classified on the basis of delamination of the clay sheets as seen in the drawing (Figure 2). Phase separation, intercalation and exfoliation of polymeric chains into the interlayer spaces or clay galleries are effected by clay nature, organic modifier, polymer matrices and preparation method. On the basis of structure of clay layers in the polymer matrix the polymer matrix nanocomposites (PCN) can be divided into three types: a. Phase separated or conventional composite b. Intercalated nanocomposites and c. Exfoliated nanocomposites (Galimberti, 2001). Details in this regards are given in Figure 2.

Figure 2. Schematic drawing of a) phase separated clay composite b) intercalated clay nanocomposite and c) exfoliated clay nanocomposite

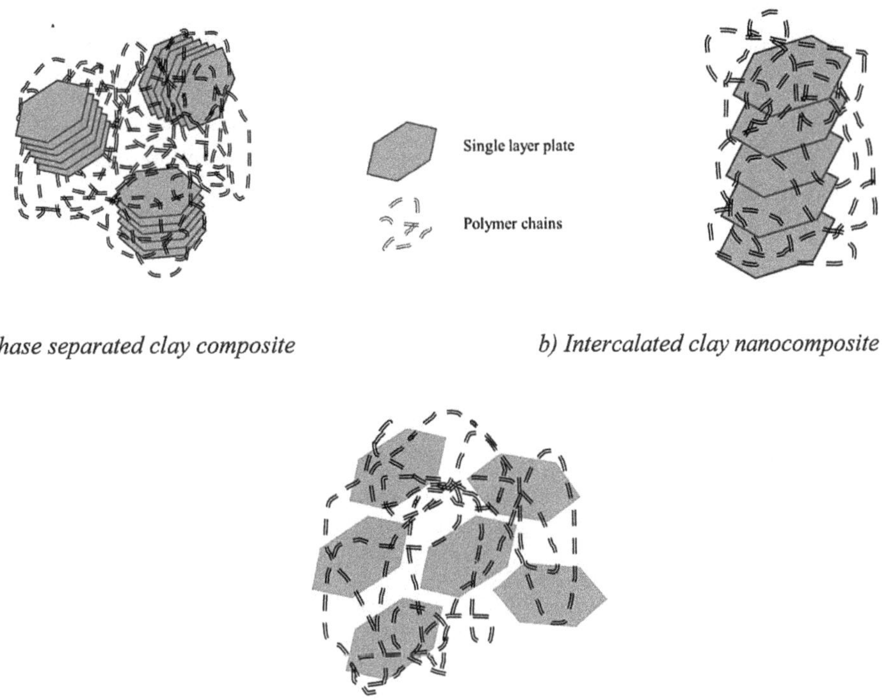

a) Phase separated clay composite

b) Intercalated clay nanocomposite

c) Exfoliated clay nanocomposite

In phase separated composites, when a polymer is incapable to intercalate (or penetrate) between the filler/silicate sheets, a morphologically phase-separated composite is obtained, and the properties remain in the same range as those for the traditional microcomposites. The clay layers remain unseparated in the matrix referred as tactoids. The distance between the silicate layers is not affected by mixing procedures. The phase separation results in microcomposites structure without achieving the desired benefits. This is the most probable but least desirable morphologies for the microcomposites. Here the clay acts as an "inactive" constituent and scattered all over the polymer matrix with very little or no interactions between the two the filler and polymer.

As shown for the intercalated nanocomposite (Figure 2), the nanocomposite is the state in which the extended polymer chains are between the clay layers and these results in a multilayered structure with alternating polymers and silicate. Polymers diffuse into clay lamellae but the pristine clay structure is still evident. Intercalations of polymer chains cause less than 2-3 nm separations of silicate layers. This type of nanocomposites shows improvement in properties but not substantial. Intercalated structure results in a well ordered multilayered morphology built up with an alternating polymeric and silicate layers.

It is seen for the foliated nanocomposite that the exfoliated or delaminated type of nanocomposite structure originates when the silicate layers having dimensions in the nanometer range are dispersed completely and uniformly in the continuous polymer matrix. In exfoliated nanocomposites the polymer separates the clay platelets e.g. by 8-10 nm or more. It shows the greater phase homogeneity than the intercalated configuration is of particular interest because it maximizes with the polymer that leads to the most effective changes in mechanical and physicals properties. Yet these terms define only ideal cases, it is further likely that the dispersion of a material morphology drop somewhere between these extremes. The polymer chain insertion is the cause behind the swelling of interlayer region, thus spacing between successive clay layers increases. Exfoliated nanocomposite formed when the clay becomes delaminated and exfoliated once the Van der waals forces are no longer capable to grip the clay layers together. This will be the reason for clays to change its structure that resembles to a "deck-of-deck" or to a "house-of-cards". Afterwards, the exfoliated clay will be held together mainly by the positive and negative charged surface of each other.

Organoclay

Although clay are is one of the most promising nanofillers due to their richest intercalation chemistry, high strength and stiffness, and high aspect ratio of individual platelets, abundance in nature and low cost, there is the incompatibility between hydrophilic clay and hydrophobic polymer, which often causes agglomeration of clay in the polymer matrix. To modify the surface property of clay, the most popular clay modification is to exchange the interlayer inorganics cations (e.g Na^{+}, Ca^{2+}) with organophilic cations. The modified clay is usually called as "*organoclay*". The modification of clay with organophilic cations does not only improve hydrophobicity of clay, buy also swell the interlayer space up to a certain extent (normally over 20 A), thus reducing layer-layer attraction which allow more polymer chains to diffuse into the interlayer. Generally, alkylammonium ions are mostly used. The structure of alkylammonium ions plays a role on the properties of nanocomposite. Sometimes, a mixture of alkylammonium clay modifier with different alkyl chains is used as intercalating agent referred to as "*tallow*" ammonium.

Methods for Preparation of Polymer/Clay Nanocomposite

Although the followings are mostly based on methods for preparation of polymer/clay nanocomposite but it can also be applicable for other nanofillers. Basically, there are three major methods that can be used to prepare polymer/clay nanocomposites, including in-situ polymerization, solution blending and melt blending. For in situ intercalative polymerization technique, the layers of clay get dispersed and distributed in monomer followed by the polymerization. Sheet like silicates are inflated in the liquid monomer or the monomer solution which give raise the polymer chains in between space of interlayer of layered silicates. Polymerization takes place either by radiation or heat and also occurs by diffusion of suitable initiator or catalyst attached by the cation exchange inside the inter layer gallery before swelling step (Ray & Okamoto, 2003). This process for synthesis of nanocomposites has been employed at early stages (Theng, 1979). Figure 3 shows the schematic of in-situ polymerization techniques. The nanocomposite "nylon 6/MMT" had been synthesized by Toyota research group via in situ polymerization method and noticed that at also very low loading of nanoclay giving results on the better service life and enhanced properties of the synthesized materials (Okada, Kawasumi, Kurauchi & Kamigaito, 1987). In 2002, Zeng *et al.*, successfully developed some nanocomposites based on polystyrene-layered silicate and polyaniline-layered silicate by this method and they had observed a high ordered poly-aniline chain, alternatively attached with the natural clay layers. There are disadvantages of in-situ polymerization technique as followed"

1. The rate of reaction is very slow.
2. Exfoliation of clay depends on the extent of swelling of clay and diffusion rate of monomers in the clay layer gallery.
3. Oligomer may be formed upon incomplete polymerization.

In case of solution intercalation method, this process involves with mixing of a polymer with the clay in solution medium. The properties of the resultant product from this method depend strongly on the solvent system. The selected solvent must dissolve polymer very well, at the same time it must swell the layered silicates. Thus, the solution contains polymer and layered silicate are mixed together and then the polymeric chains are interacted and forcefully bring the solvent out from the interlayer gallery space of the silicates as shown in Figure 4. The formed intercalated configuration persists after removal of solvent resulting in polymer-layered silicate based nanocomposites. Here the solvent is important in this scheme to start the inclusion of the polymer chains into inner layer gallery space of silicates and polarity of a solvent is also being a powerful factor for the intercalation (Ganter, Gronski, Reichart, & Mulhaupt, 2001). Surface modified polyamide (PA), polyethylene (PE), polyurethanes (PU), polyethylene oxide (PEO), polyvinyl alcohol (PVA) are commonly used in solution intercalation (Aranda & Ruiz-Hitzky, 1992; Xu *et al.*, 2003). The approach of disadvantages of solution blending are as followed:

Figure 3. In-situ polymerization technique

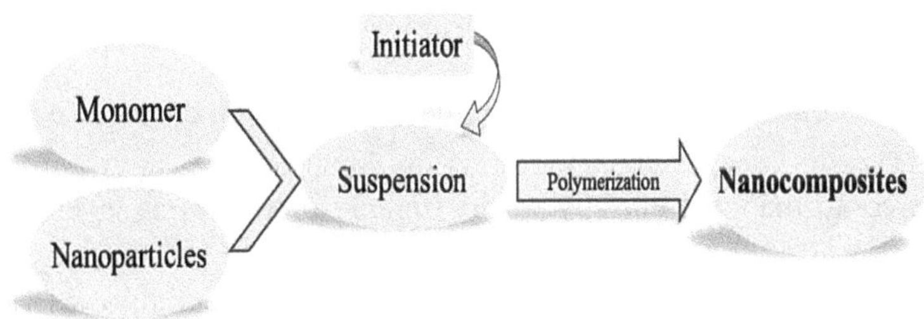

Figure 4. Solution blending technique

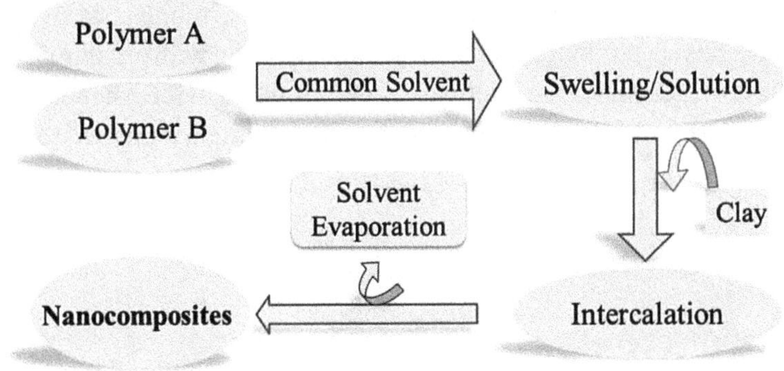

1. It is not acceptable for mass production, as large amount of solvent is required.
2. It is difficult to find a solvent that is compatible with both clay and polymer.
3. The organic solvent used in this process is dangerous and environment friendly. Thus, this method is hazardous regarding industrial or commercial point of view.

As known with melt intercalation method (as shown in Figure 5), melt intercalation, a form of melt blending, is another promising method that has been developed for the preparation of organic-inorganic nanocomposites by utilizing a conventional thermoplastic mixer such as extruder or internal mixer. This method involves the annealing, statically or under shear, of mixture of polymer and the clay above the glass transition temperature (Tg) during the processing time, the polymer chains can move into the pre-treated clay galleries beyond of its softening point and lead to either intercalation or exfoliation type of nanocomposites. Melt intercalation method is better among other two methods such as in situ intercalative polymerization and solution intercalation. This method is very much environment friendly because there is no requirement of organic solvent and effective with the industrial processes like extrusion and injection molding. If polymers are not suitable for the in situ interactive polymerization and solution intercalation process, melt intercalation method is employed in that cases for the development of nanocomposites. Rai and Okamoto had noticed that a suitable intergallery structure for the organoclays regarding the quantity per unit area and also the length of surfactant chain are most promising for nanocomposites formation. Also they had observed that the existence of polar interaction between the modified clay and the matrix polymer stimulates for the polymer intercalation. The proposed driving force of this mechanism can be rationalized based on three competing effects as followed:

1. The decrease in the conformational entropy of the flexible polymer chains when confined between the silicate layers.
2. An increase in conformational entropy of the organic modifiers as layers to accommodate the polymers.
3. A net energy gain associated with the establishment of more favorable secondary interaction between the polymer and the organically modified layered silicates than organically present in the unintercalated silicate.

Furthermore, in melt blending technique, there is no requirement of solvent and at the molten state layered silicates are mixed within the polymer matrix. Generally thermoplastics are mechanically blended via conventional approaches such as extrusion and injection molding with organophilic clay at a prominent temperature. Then the polymer chains are intercalated or exfoliated to form nanocomposites. This is very popular way to process thermoplastic nanocomposites. This technique is applied to those polymers which

Figure 5. Melt blending technique

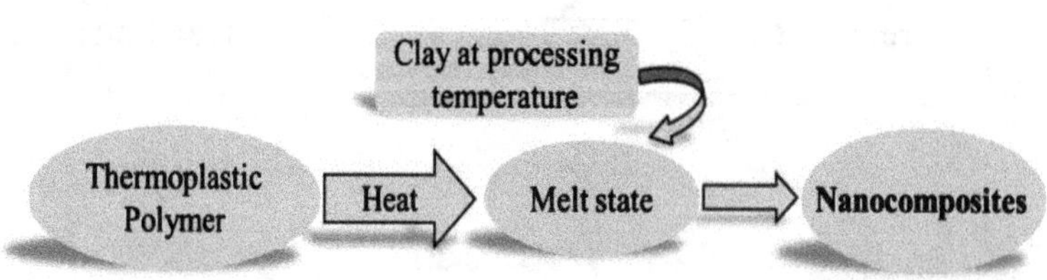

are neither suitable for adsorption nor for in-situ polymerization. In case of melt intercalation layered clays are directly mixed with the polymer matrix at the molten state. Melt intercalation method takes less than 10 minutes to get the nanocomposites. From an industrial standpoint, melt intercalation is the most direct, most cost effective, and environment friendly method to prepare polymer/clay nanocomposites.

Many rubbers are available in the latex form which is nothing but aqueous dispersion of rubber particles in the submicron-micron range. Therefore, in addition to three major methods described above, preparation of polymer/clay nanocomposites can also be carried out using latex compounding method. Moreover, clays are easily dispersed in water as water acts as a swelling agent due to the hydration of the intergallery cations, hence mixing of the latex with the clays is a promising method for preparing rubber nanocomposites. During the late 1990s, Zhang *et al.* and Wang *et al.* conducted a series of pioneering studies on the preparation of rubber/clay nanocomposites by mixing the rubber latex with a pristine clay water suspension, followed by coagulation.

Beyond the above basic methods, various techniques have been additionally applied in order to obtain the exfoliated polymer/clay nanocomposites. Exfoliated nanocomposites are desired due to the fact that the exfoliated layers exhibit the greatest reinforcement as mentioned above. Exfoliation depends on the establishment of favorable interaction between the polymer and clay surface and subsequent energy reduction. Exfoliated structure can be achieved by the following consideration given below:

- Organic modification of the clay by choosing proper modifier. The modifiers are generally quaternary ammonium ions. Following organic modification the inter layer space between the layered silicates increases as a result of long chain alkyl ammonium cation, thereby weakening the interaction between successive layers. Moreover these small chain molecules electrostatically bonded to silicates layers provide the active site for polymer interaction so as to attach them to the clay layer indirectly through the modifiers.
- Exfoliation can be obtained straight forwardly when the polymer contains functional group for example amides or imides groups.
- Using compatibilizer reduce the interfacial tension hence increase interphase adhesion.
- Exfoliation can also be achieved during polymer compounding. The driving force is the intense shear stresses experienced by the nanoclays during melt processing of the composites and the diffusion of polymers between the silicate lamellae. Successful melt exfoliation requires presence of strong interaction between the clay and polymer molecule, proper temperature, a stress field and an adequate residence time of the stress field while still preventing organic modifier volatilization/decomposition.

Natural Rubber/Clay Nanocomposite

The interest in natural rubber/clay nanocomposites is mainly due to the reinforcement and barrier properties that can be obtained by exfoliated silicate layers having nanoscale thickness. The followings review some research approaches focused on the preparation, characterization and properties of natural rubber/clay nanocomposites.

Madhusoodanan and Varghese prepared sulfur cured natural rubber/clay nanocomposites by melt-mixing of natural rubber with organically modified clays including octadecylamine-modified montmorillonite (MMT-ODA) and methyltallow bis-2-hydroxyethyl ammonium-modified montmorillonite (MMT-TMDA) in a two-roll mixing mill. Cure characteristics and mechanical properties of the obtained nanocomposites filled with modified silicates were compared with those of the ones filled with a pristine-layered silicate (sodium bentonite, BNT) and a nonlayered version (English Indian clay or EIC). This study showed that the addition of MMT-ODA and MMT-TMDA reduced the cure time and scorch time of the respective compounds, while the addition of unmodified clays (BNT and EIC) had no effect on those. It was believed that the clay modifiers, being amines, took part in amine complexation reaction (like an accelerator), which will in effect accelerate the curing reaction. The results from Mooney viscometer showed that the compounds filled with the modified silicates had higher viscosity than the ones filled unmodified clay. The increased viscosity indicated that the rubber molecules underwent intercalation during mixing. Because of this interaction process, the surface area of the layered silicate increased, which might be the reason for the high compound viscosity. The compound filled with BNT showed only one X-ray diffraction peak, which corresponds to an original interlayer distance of BNT at 1.24 nm, while the compound filled with amorphous EIC showed no X-ray diffraction peak. XRD diffractograms of the compounds filled with MMT-ODA and MMT-TMDA showed a shoulder at lower diffraction values, which corresponded to an interlayer distance of 3 nm (the initial layer distance of MMT-ODA and MMT-TMDA are 2.10 and 1.85 nm, respectively). This indicated that natural rubber chains were intercalated between the gallery spaces of the silicate. However, apart from the shoulder peak, the compounds filled with the modified clays also showed diffraction peaks corresponding to interlayer distances that were less than their original interlayer distance. This indicated that some layers were closed during the preparation process. It was proposed that this confinement of layers may be due to the coordination complexing of accelerated sulfur curing in the presence of zinc. Zinc is prone to coordination complexing with amine groups of the intercalants. Thus the intercalant amines may be extracted from the intergallery space and used for sulfur vulcanization and hence the reduced layer distance. Although TEM images showed that the modified clays were partially exfoliated, partially intercalated, and partially confined in the composites, the addition of them even at low loading (5-10 phr) gave the composites with better mechanical properties such as tensile strength, modulus, hardness, abrasion resistance, tear strength, compression set resistance than the gum and also the composites filled unmodified clay. Moreover, it was reported that under dynamic testing, the composites filled with the modified silicate showed lower heat build up values. Moreover, it was seen from TEM images that a portion of the MMT-ODA was exfoliated homogeneously throughout the composites, while the exfoliation extent of MMT-TMDA was somewhat less than that of the MMT-ODA. Some agglomeration of the silicate particles can be seen in the case of MMT-TMDA composites. The reason for this is because the initial layer distance of MMT-ODA is larger than that of and MMT-TMDA. The improvement in the mechanical properties of the modified silicates is mainly due to the exfoliation of the silicates, which causes several fold increase in surface area. This caused the MMT-ODA be more efficient reinforcing filler than MMT-TMDA.

Arroyo *et al.* compared properties of natural rubber filled with 10 phr of organo-montmorillonite with those of the one filled with 40 phr of conventional reinforcing carbon black. Organoclay was obtained by modifying Na^+-activated montmorillonite (MMT) with octadecylamine. Natural rubber/organoclay nanocomposites were simply prepared by melt blending using a two roll mill. The cure curves showed that the presence of organoclay even at 10 phr significantly accelerated the cure reaction. This behavior was comparable to the work of Madhusoodanan and Varghese. Furthermore, the natural rubber/organoclay nanocomposites showed higher torque value than the carbon black filled natural rubber, thereby indicting a higher degree of crosslinking as was also confirmed by swelling measurement and differential scanning calorimetry. From tensile testing, the result showed that besides being a more effective reinforcing filler than carbon black, the organoclay improves the strength of the natural rubber without hardly any reduction in the elasticity of the material. The unmodified clay showed very low reinforcing effect. This was due to poor compatibility between unmodified clay and hydrophobic rubber.

Varghese and Karger-Kocsis also compared the properties of the composites based on natural rubber containing organophilic and pristine layered silicates of natural and synthetic origin with the one reinforced with a conventional filler, silica. The compounds were prepared by melt compounding using a two-roll mill. At the same filler loading, the property improvements caused by the fillers were ranked as follows: organophilic clays > pristine synthetic layered silicate (sodium fluorohectorite) > pristine natural clay (purified sodium bentonite) > precipitated nonlayered silica.

The effect of the structure and length of alkylammonium ions on the structure and properties of natural rubber/clay nanocomposites prepared by melt intercalation method was investigated by Hrachova *et al.* Table 2 lists the structure of organo-cation used in this study. The base of C10 and C12 modifiers is the benzene ring, the only difference between the two being is the presence of a vinyl group in C12. The other set of modifiers was selected considering the differences in size of the cation and the length of the alkyl chains. The highest values of modulus 100, stress, and strain at break were obtained for nanocomposites with organo-clays 4C8 and 2C16 with the largest interlayer distances. It was found that organo-cations with the same number of carbon atoms but different structure affect the mechanical behavior of clay/rubber nanocomposites differently. Cations with carbon atoms divided into more alkyl chains are more effective. No improvement of mechanical properties was observed for small organic cations. Although it was expected that the double bond of vinyl group in C12 may increase interactions between the rubber matrix and the filler on the phase boundaries, the mechanical properties observed for the nanocomposite with the filler C12 was close to those of the system with C10, the filler of similar structure but without the highly-reactive vinyl group. This indicates that the double bond in the C12

Table 2. Structures of organo-cations used for preparation of organo-clays from Na-MMT

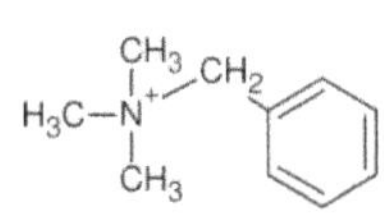 C10: Benzyltrimethylammonium	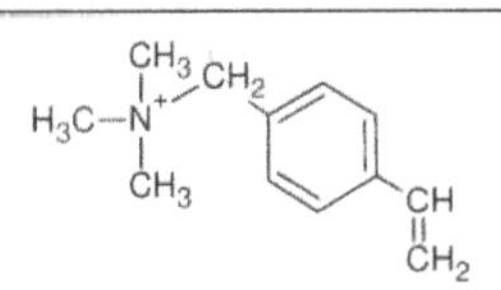 C12: 4-vinylbenzyl-trimethylammonium
$H_3C-(CH_2)_6-CH_2-NH_3^+$ 1C8: Octylammonium	$H_3C-(CH_2)_6-CH_2-NH_2^+-CH_2-(CH_2)_6-CH_3$ 2C8: Dioctylammonium
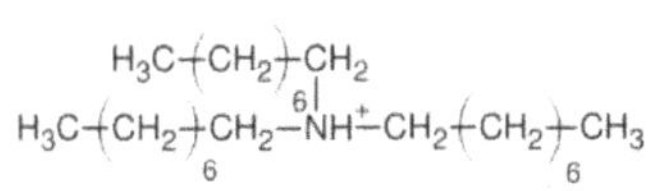 3C8: Trioctylammonium	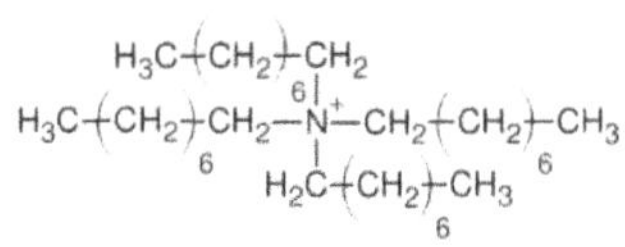 4C8: Tetraoctylammonium
$H_3C-(CH_2)_{14}-CH_2-NH_3^+$ 1C16: Hexadecylammonium	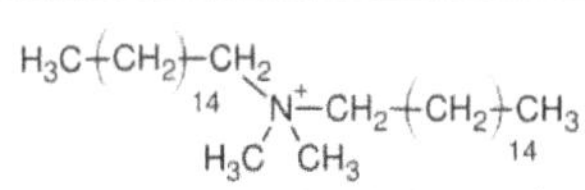 2C16: Dihexadecyldimethylammonium

filler, expected to possibly affect the vulcanization process, was indeed ineffective in curing reaction in natural rubber and did not affect the stress and strain values.

In addition to the use of origin natural rubber in preparing natural rubber based clay nanocomposites, many researches also pay attention to prepare natural rubber based clay nanocomposite from epoxidized natural rubber. Due to its high polarity, it should be beneficial when compounding with polar fillers, such as organoclays. Hakim and Ismail compared curing characteristics, tensile properties, thermal stability and morphology between epoxidised natural rubber with 50 mol% epoxidation (ENR 50) and natural rubber filled with octadecylamine modified montmorillonite. The content of organoclay was varied from 0 to 2, 4, 6, 8, and 10 phr. The rubber was mixed with organoclay and other ingredients using laboratory-sized two roll mills then the obtained compounds were cured at 150°C. The results indicated that the tensile strength and tensile modulus reached a maximum at 8 phr of organoclay, but elongation at break and thermal stability increased with increasing organoclay loading. Overall results show that organoclay-filled ENR nanocomposites exhibited shorter processing time and higher tensile properties than organoclay-filled natural rubber nanocomposites. This was related to the higher compatibility due to the interaction between the epoxy groups of ENR and the amine functionality of the organoclay. TEM and XRD results confirmed the better dispersion of organoclay in ENR compared to NR. Moreover, comparing the natural rubber and ENR nanocomposite SEM micrographs, the ENR nanocomposites exhibitd relatively rougher tensile fracture surfaces. The rougher the fracture surface the better the tensile properties of the related nanocomposites are. A smooth fracture surface usually indicates low compatibility accompanied with premature, rather brittle-type fracture. This observation validates the tensile results reported above. It was expected that the hydroxyl groups of the filler surface are able to react with the epoxy groups of ENR, giving rise to a better interaction between the organoclay and ENR matrix. Therefore, both the interaction between the epoxy groups of ENR with the amine functionality of the organoclay and with hydroxyl groups of the filler surface contributed to the better filler-rubber interaction in organoclay filled ENR compounds.

To further enhance the properties of natural rubber clay nanocomposites prepared by melting mixing, it is necessary to use compatibilizing agents being compatible and more polar than the rubber matrix. As mentioned early, polar rubber such as ENR can form strong interaction with polar clay. Therefore, it can be used as a compatibilizing agent when preparing natural rubber clay nanocomposites. Karger-Kocsis *et al.* investigated the effect of two different grades of ENR with 25 and 50 mol % epoxidation used as a compatibilizer in natural rubber matrix on the properties of natural rubber based nanocomposites containing 2 phr of onium ion modified montmorillonite. The result showed that ENR with 50 mol % epoxidation (ENR 50) gave better results than the 25 mol % epoxidation with regards to clay dispersion and mechanical properties. This was due to its higher polarity, thus leading the ENR 50 containing nano-composites in having better dispersion of the clay (i.e. pronounced intercalation without agglomeration) accompanied with strong interfacial adhesion between matrix and filler.

Besides a melt mixing method, literature reported some researches that successfully prepared natural rubber clay nanocomposite by mixing natural rubber latex with clay aqueous dispersion. Considering that the latexes are aqueous polymer dispersions and water is an excellent exfoliating agent for clays. Therefore, it should be possible to mix clay and latex within an aqueous dispersion. Zhang et al. first introduced this technique applied for blending MMT in rubber latex. After mixing, the nanocomposite can be obtained by two different techniques including casting and co-coagulating.

Varghese and Karger-Kocsis prepared natural rubber based nanocomposites with 10 wt% natural and synthetic layered silicates via latex compounding method. The latex was mixed with the aqueous dispersion of layered clay and compounding ingredients. Then, the obtained compound latex was kept overnight to allow the diffusion of added chemicals, then casted on glass plates. The casting was then allowed to dry and vulcanize. As layered silicates, sodium bentonite (natural) and sodium fluorohectorite (synthetic) were selected in addition to a non-layered inert filler (English India clay or commercial clay) as reference material. In this work, the layered silicates were used without modifying. They reported that layered clays filled natural rubber vulcanizates exhibited a great increase in modulus and tensile strength especially in the case of fluorohectorite. Dynamic mechanical analysis revealed that the layered clay filled composites possessed very high stiffness and low damping characteristics. TEM images showed that for the bentonite filled vulcanizate, there is some level of intercalation and the silicate layers form clusters with the formation of a secondary filler network, while for the fluorohectorite filled nanocomposites, individual layers of the silicate are well discernible, thus giving a relatively large amount of exfoliated layers structure. On the other hand, for the commercial clay filled vulcanizates, the filler existed as large particles. Thermogravimetric analysis showed that both layered clay filled composites retained higher percentage of weight at high temperature than the commercial clay filled composites. This increase in thermal stability of the layered clay filled composites may result from the dispersion of the clay and from a strong interaction between the clay platelets and rubber molecules. The results from this study showed that when the latex compounding technique was used to prepare natural rubber nanocomposites, the enhanced mechanical nanocomposites can be obtained without the necessity of the organophilic clay modification.

In 2005, Valadares *et al.*, investigated morphologies and mechanical properties of the unvulcanized natural rubber nanocomposites prepared by mixing the latex with the NaMMT aqueous dispersion under a mild shearing. The water of the latex compound was first partially removed using a spinning evaporator, thus giving a thick slurry. Then, the rubber nanocomposites films with various clay contents including 5, 10, 20 and phr were obtained using a casting method. XRD diffractogram of pure clay showed that NaMMT has a basal spacing of 1.17 nm. The incorporation of rubber into the clay dispersion led to the enhanced spacing distance. The degree of intercalation was strongly dependent of clay content. XRD diffractograms showed that the 30, 20 and 10 phr clay filled samples has inter-lamellar distance of 1.59 1.69, 172 nm, respectively, while the 5 phr clay filled sample showed a low-angle broad halo, indicating the extensive layer separation associating to the clay lamellae intercalation and exfoliation. TEM images of the 5 phr clay filled sample also confirmed that the clay was well exfoliated in the natural rubber matrix. In the case of the 30 phr clay filled sample, it can be clearly seen that there were many lamellae stacks but it was also possible to observe exfoliated structure. It is noteworthy that there was no void formation in the rubber-clay interface, providing a strong evidence for good adhesion between the clay lamellae and the rubber. Micrographs taken by a transmission electron microscopy associated to electron spectroscopy imaging (ESI-TEM) of very dilute clay-rubber latex dispersions that were allowed to dry on carbon films over the microscope holder grids showed that many clay particles were in contact with the rubber particle. The result here indicated that clay particles was a strong evidence in favour of clay-rubber compatibility, that is all but unexpected due to the great differences between the properties of the two phases. The reason that may explain why the latex mixing can be a useful method to prepare rubber/clay nanocomposite is first, water is an excellent exfoliating agent for montmorillonite clay and both clay and rubber particles carry excess of negative charges, thus they form stable dispersions. Upon drying, the clay and rubber particles undergo capillary adhesion that is sufficiently strong to produce the rubber particle–clay platelet clusters. After drying, counter-ions are clustered in between rubber and clay particles thus making an electrostatic contribution to cohesion in this system. Demixing between clay and rubber is expected on thermodynamic grounds but it is countered by adhesion as described in the previous paragraph and it is also kinetically limited by the slow diffusion of the clay platelets and rubber chains so that the intercalated or exfoliated patterns persist and they can be observed in the dry nanocomposites. Interesting, the results from tensile testing showed that the mechanical behavior of the 20-30 phr unvulcanized nanocomposites was similar to that of a semi-crystalline thermoplastic like LDPE or MDPE, considering the high initial modulus, the significant flow at higher stresses and the presence of a yield point.

In work of Wu *et al.*, several rubber-clay nanocomposites, including styrene butadiene rubber (SBR)-clay, natural rubber (NR)-clay, nitrile butadiene rubber (NBR)-clay, carboxylated acrylonitrile butadiene rubber (CNBR)-clay nanocomposites were prepared by directly co-coagulating the rubber latex and clay aqueous suspension. About 3% clay aqueous suspension and the rubber latex were mixed and vigorously stirred. After that, the mixture was co-coagulated in the electrolyte solution (2% dilute triethylenetetrammonium chloride solution for the NR and SBR systems and 1% calcium chloride aqueous solution for the NBR and CNBR systems), then washed with water and dried in an oven at 80 °C for 18 h, giving the rubber-clay nanocompound (uncured nanocomposite). The uncured nanocomposite was mixed with curing agents and then compression molded. One objective of this study was to investigate the mechanism of intercalation when using co-coagulating. From XRD results, it was revealed that both NR-clay and SBR-clay nanocomposites showed a peak corresponding to the layer spacing of 1.34 nm while the NBR–clay and CNBR-clay nanocomposites showed a peak corresponding to the layer spacing of 1.50 and 1.51 nm, respectively. Although all of these intergallery distances was larger than the initial value of NaMMT (1.25 nm), which seemed to indicate that rubber molecules were intercalated into the clay interlayer. However, the basal spacing of the four nanocomposites (1.34-1.51 nm) is smaller than that of organoclay in the literature (larger than 1.7 nm) and thus it is unreasonable to conclude that the intercalation of rubber macromolecules into the clay interlayer. However, the basal spacing 1.34 nm of NR-clay and SBR-clay nanocomposites using triethylenetetrammonium chloride as the flocculant was close to 1.31 nm of RNH_3^+-MMT, and in the cases of NBR-clay and CNBR-clay nanocomposites using calcium chloride as the flocculant, it was 1.50 and 1.51 nm, respectively, almost the same as 1.52 nm of Ca^2+-MMT. Therefore, the diffraction peaks of rubber-clay nanocomposites by co-coagulation should originate from the cations of flocculant in the intergallery, and no rubber molecules intercalate into clay galleries. On the basis of these results, it was assumed here that a cation exchange reaction occurred during the process of co-coagulating. This study also proposed that at the stage of mixing, the rubber latex particles were mixed with the clay aqueous suspension, in which clay was dispersed into individual silicate layers. After adding a flocculant, the flocculant coagulated the rubber latex and the silicate layers simultaneously, but the rubber macromolecules did not exactly intercalate into the galleries of clay. This mainly resulted from the competition between separation of rubber latex particles and reaggregation of single silicate layers upon addition of flocculant. Since rubber latex particles are composed of several molecules, the existence of latex particles between the galleries of silicate layers in the water medium should result in a completely separated (exfoliated) silicate layers. However, cations of flocculant cause separated silicate layers to re-aggregate so that the rubber latex particles between the silicate layers may be expelled. As a result, there are some non-exfoliated layers in the nanocomposites. In the meantime, due to the fact that the amount of latex is more than that of silicate layers and the latex particles agglomerate rapidly, the reaggregation of silicate layers is evidently obstructed to some extent by the agglomerated latex particles around the silicate layers. Consequently, the size of aggregates of silicate layers is at the nano-meter level, and thus the obtained nanocomposites contain both the exfoliated silicate layers and non-exfoliated (not intercalated) aggregates of nanometer thickness in the rubber matrix. Comparable to other works, the result from this also showed that the clay is potential reinforcing filler. The gas permeabilities of the rubber/clay nanocomposites were compared with those of the gum rubber. It was found that the gas permeabilities of the rubber/clay nanocomposites containing 20 phr of clay decreased 50%.

From above, it can be seen that in the past decade natural rubber/clay nanocomposites were successfully prepared using either melt mixing or latex mixing. The improved mechanical properties both static and dynamic mode, thermal property, gas barrier and solvent resistance can be achieved even at low loading clay. This will create natural rubber with unique properties that cannot be obtain from the use of convention filler, broadening the application of natural rubbers.

Rubber/Carbon Nanotube Nanocomposites

Carbon Nanotube

Among different types of nanotubes, carbon nanotube has gained tremendous momentum from the scientific community in the present decade because of their wide range of potential applications in several areas since its invention by Iijima in the year of 1991.

With the intensive studies on CNT, it is seen that CNT has unique properties like low mass density, high flexibility, high mechanical properties, enhanced electrical conductivity, high aspect ratio, good magnetic and improved thermal properties (Wang & Dang, 2005; Xu, Zhang, & Suhir, 2006; Zhao, Bagayoko, & Yang, 2006). Mainly, there are two types of CNTs named multi walled CNTs (MWCNTs) and single walled CNTs (SWCNTs) which have been prepared through different techniques like chemical vapor deposition (CVD) method, and electrochemical method (Sumio, 1991; Ebbesen & Ajayan, 1992; Ting, 1995; Ting, 1995). MWCNT consists of two or more concentric cylindrical shells of graphene sheets with inter planar distance of 0.36 nm, whereas SWCNT is made up of a single graphene cylindrical layer. The diameter of CNTs ranges from 1-2 nm with few micron lengths. CNTs possess stronger mechanical properties due to the presence of sp^2 hybridized carbon-carbon (C-C) chemical bond in CNTs compared to diamond in which sp^3 types bonding structure is located. It shows unique electric transport behavior because of the presence of π-π electronic conjugation between the carbon atoms. Carbon nanotubes exhibit electrical conductivity of as high as 5000 S/cm, thermal conductivity of 6000 W/mK, high thermal stability (stable up to 2800 °C in vacuum) and better mechanical properties (Dresselhaus, 1996; Berber, Kwon & Tománek, 2000).

Carbon nanotube-reinforced composites are of great interest to develop significantly light weight strong materials. It is very difficult to achieve a homogeneous dispersion of nanotubes in the polymer matrix due to the aggregation of these tubes. To reach optimum dispersion of nanotubes in the polymer matrix, scientists have utilized different techniques including:

1. Solution mixing (Pham *et al.*, 2003),
2. Sonication (Xie *et al.*, 2004),
3. Coagulation (Du, Fischer, & Winey, 2003),
4. Melt compounding (Dondero & Gorga, 2006),
5. In-situ emulsion polymerisation (Yu *et al.*, 2005),

6. The use of surfactants (Gong, Liu, & Baskaran, 2000), and
7. Chemical functionalization of nanotubes (Ramanathan, Liu, & Brinson, 2005; Wu, Yang, Ma, & Kuan, 2005).

Carbon nanotubes can be homogeneously dispersed by chemical modification of carbon nanotubes through covalent and noncovalent attachments of functional groups in nanotubes with the matrix. Because of these unique properties, it is applied in various applications like microwave absorbing applications, supercapacitors, sensors, hydrogen storage devices, photovoltaic cells, field-effect transistors, diodes etc. (Dai, Soundarrajan & Kim, 2002; Du, Yeh & Pan, 2005; Emmenegger *et al.*, 2003) and so on.

Natural Rubber/Carbon Nanotube Nanocomposites

Although the incorporation of the CNTs into the natural rubber matrix is expected to provide rubber product with a superior mechanical properties. But the dispersion of the CNTs in the natural rubber is still problematic due to the high aspect ratio of the CNTs and high viscosity of natural rubber. A fair amount of researches have carried out on the preparation of natural rubber/CNTs nanocomposites. Several methods have been developed for preparation of natural rubber /CNTs composites, such as direct melt mixing, solution mixing and latex compounding process. Direct melt mixing by conventional twin roll is difficult to prepare well dispersed natural rubber/CNTs composites. In the solution mixing technique, CNTs are dispersed in a polymer solution and then solvents are evaporated. Latex compounding technology involves natural rubber latex mixing with aqueous dispersion of CNTs and co-coagulated process. The content in this section will review various methods used to prepare the natural rubber natural rubber/CNTs nanocomposites from the literatures. Moreover, the change in properties of natural rubber when mixing with the CNTs is also reported here.

Fakhrul-Razi *et al.* prepared natural rubber/MWCNTs nanocomposites by incorporating MWCNTs in a rubber solution and subsequently evaporating the solvent. Firstly, in order to disentangle the MWCNTs that typically tend to cling together and form lumps, making it difficult to disperse, the MWCNTs were added into toluene used as solvent, then the solution of MWCNTs was sonicated using a mechanical probe sonicator. An efficient dispersion of nanotubes in solvent was now induced. The well dispersed MWCNTs was then mixed with the natural rubber solution, resulting in a solution that consisted of a good blend of nanotubes in the rubber. The solution was evaporated, then the dried compound was pressed using a hot press and cut into standard shape for tensile testing. The obtained nanocomposites contained various MWCNTs loadings including 1, 3, 5, 7, 10 wt%. From the stress and strain curves, it was found that as the MWCNTs content in the rubber increases, the stress level gradually increased but at the same time the strain of the nanocomposites decreased. The increment of stress level was due to the interaction between the CNTs and the rubber. A good interface between the CNTs and the rubber is very important for a material to stand the stress. At 10 wt% of MWCNTs, the stress value obtained reached 2.55 MPa which was 9 times that of pure natural rubber. TEM images showed the methods used to prepare the nanocomposites gave the well dispersed MWCNTs nanocomposites. TEM images also showed that at the low MWCNTs loading, the MWCNTs well oriented. However, at higher MWCNTs loading, the well MWCNTs became less oriented and more randomized.

Similar to the work of Fakhrul-Razi *et al.*, Bokobza also used a solution blending accompanying with sonication to prepare rubber/MWCNTs nanocomposites. Besides natural rubber, two different rubbers including SBR and EPDM were also used to prepare the elastomeric nanocomposites in this study. However, only natural rubber/MWCNTs nanocomposites was focused here. The key difference between these two researches is the latter research investigated the properties of the cured rubber/MW-CNTs nanocomposites, while the previous research investigated the properties of the uncured natural rubber /MWCNTs nanocomposites. The amount of /MWCNTs loading was increased from 0.5 to 1, 3, 5, and 10 phr. In general, in order to conduct electricity the certain amount of conducting filler is needed. Above a certain amount of conductive particles, called the percolation threshold, an interconnecting filler network is formed, which results in the sharp drop of the electrical resistance of the composites. The percolation threshold of composites containing conventional microscale conducting fillers like carbon black or graphite is as high as 10-50 wt%. In this study, the result showed that the percolation threshold of composites containing MWCNTs was around 0.5 phr being much lower than that of composites containing conventional microscale conducting fillers. This major attribute of CNTs to form a conducting network at a very low loading content was not only due to their intrinsically high conductivity and high aspect ratio, but also due to the use of solution method accompanied with sonication, giving a better nanotube dispersion within the elastomeric matrix. From tensile testing, it showed that the modulus at 100% strain of the natural rubber/MWCNT nanocomposites significantly increased with increasing filler loading. The pronounced enhancement of modulus is clearly due to the incorporation of rigid filler into the soft matrix. Moreover, Shanmugaraj *et al.*, proposed that the nanoparticle fillers have large aspect ratio that could provide an additional source of entanglement or physical crosslinks in the rubber matrix, increasing stiffness more. However, it was found that the strengths at break of the natural rubber/MWCNTs nanocomposites were negatively affected by the nanotubes. Perhaps it might be due to the lack of adhesion between filler and rubber matrix. The MWCNTs presented three Raman bands located at 1348, 1577, and 2684 cm^{-1} and assigned to the disordered graphite structure (D band), the tangential stretching mode of carbon-carbon bonds (G band), and the overtone of the D band (G′ band), respectively. Changes in these vibrational characteristics have been observed when nanotubes were incorporated in a host matrix or submitted to a strain. A shift to a higher-frequency especially for the G band has been reported in MWCNTs composites and has been attributed to a disentanglement and dispersion of the CNT bundles in the polymer matrix. The Raman spectra of the natural rubber composites filled with 4 phr MWCNTs showed that the peak of MWCNTs shift to a higher wave number when embedded in the elastomeric matrix. This indicated that the rubber chains penetrated into filler agglomerates, thus decreasing nanotube-nanotube interactions. This is comparable to TEM images that revealed good levels of dispersion with well-isolated nanotubes for composites on the basis of natural rubber.

To elucidate the effect of interfacial bonding between the rubber matrix and the CNTs on the properties of the nanocomposite, Sui *et al.*, pretreated the acid purified CNTs (or p-CNT) prepared by chemical vapor deposition of propylene gas with a blend acid and HRH (hydrated silica, resorcinol and hexamethylene tetramine with a weight of 15:10:6) bonding system during milling in a ball-milling machine. The obtained product was named as b-CNTs. The incorporation of p-CNTs and b-CNTs into NR was carried out a solvent mixing method. FT-IR spectrum showed that both CTNs had OH functional groups. After treatment, the b-CNTs showed a new peak appears at 1720 cm^{-1} which was attributed to the >C=O stretching vibration. It indicated that the carbonylic groups were loaded onto the b-CNT surface due to the oxidation of blended acid. Furthermore, the appearance of peaks at 1110 and 1050 cm^{-1} for the b-milled CNTs that corresponded to Si-O-Si and Si-O-C stretching vibration revealed the interaction between the silica-containing bonding system and CNTs. After adding the CNTs into natural rubber, the FT-IR spectrums of natural rubber/CNTs nanocomposites showed that the vibrational peak position of OH functional groups in natural rubber/CNTs nanocomposites shifts from 3450 to 3300

cm^{-1} attributing to hydrogen-bond action or other physical adsorptions between OH functional group and α-H on the macromolecular chain of rubber. The NR composites containing b-CNTs presented two peaks at 1670 and 1700 cm^{-1}. By contrasting with the neat natural rubber, the first peak can be assigned to the stretching vibration of C=C bonds in the NR chains. The second peak was the stretching vibration of >C=O groups, which was shifted from 1720 cm^{-1} in the FT-IR spectrums of CNTs. Therefore, this indicated that the carbonylic groups on the b-CNTs surface can form physical interaction with NR macromolecular chain. After loading the CNTs, it was reported that the hardness, tensile modulus and tensile strength of NR exhibited remarkable increases, while the elongation at breaking decreased, but to a relatively smaller degree. It can be deduced that the reinforcing effect of CNTs in NR is noticeable. By comparing with the p-CNTs, it showed that the b-CNTs had more pronounced effect on the improvement of mechanical properties of the rubber. Clearly, this was because many functional groups were introduced onto the ball-milled CNTs surface, physical interaction between the CNTs and natural rubber macromolecular chain as discussed earlier can be produced, which effectively restricted the changes in polymer molecular configuration, and then enhanced the stiffness of the rubber materials. Moreover, due to the strong interfacial reaction, the CNTs can transfer stress effectively throughout the rubber matrix, providing an obvious reinforcement role in the natural rubber. It can also be explained from the fracture surface of the nanocomposites. Lower reinforcing efficiency of the p-CNTs was corresponding to the existence of some aggregation observed from SEM image.

Ismail *et al.* studied the effect of MWCNTs loading and preparing methods including mechanical and solution mixing on the cure characteristics, rubber-filler interaction and mechanical properties of the NR nanocomposites. For solution method, MWCNT was dispersed in toluene and stired using magnetic stirrer for 20 minutes. The dispersion of MWCNT was poured into natural rubber/toluene solution and maintained under vigorous stirring for 2 hour. The solvents were then evaporated. The dried compound was mixed with other rubber additives on a laboratory-sized two roll mill. For mechanical method, natural rubber was mixed with various rubber additives prior to the addition of MWCNTs, and finally, the sulfur was added using a two roll mill. It was found that scorch time and cure time of natural rubber /MWCNTs nanocomposites prepared from both methods decreased with increasing MWCNTs loading but the maximum torque increased linearly with filler loading. This indicated that the addition of CNTs could absorb the basic accelerator species and delay the onset of vulcanization. At the similar loading, there were not much different value in cure characteristics using both methods. These indicated that different mixing methods did not affect the curing characteristics of natural rubber/MWCNTs nanocomposites. From a swelling measurement, the results indicated that the rubber filler interaction decreased with increasing filler loading. Moreover, it also showed that the nanocomposites prepared via solution method showed a better interaction between MWCNT and natural rubber than the nanocomposites prepared via mechanical method. Tensile test showed that the modulus of nanocomposite increased proportionally with filler loading. The nanocomposites prepared via solution method gave higher stiffness. This observation indicated that the presence of CNTs resulted in the increase stiffness of natural rubber macromolecular chain. For the natural rubber/MWCNTs nanocomposites prepared from solution method, tensile strength passed through the maximum with increasing MWCNTs loading. The natural rubber/MWCNTs nanocomposites contained of 0.5 and 1 phr MWCNTs had higher tensile strength than the neat natural rubber. However, regardless of filler loading, it was found that tensile strengths of the natural rubber/MWCNTs nanocomposites prepared from mechanical method were lower than that of the neat natural rubber and the nanocomposites prepared via solution method. The similar trend was also observed when considering the fatigue life. The reason for this was due to the better dispersion of

MWCNTs in the natural rubber matrix when using a solution method. Besides the results from swelling measurement, this can also be confirmed by SEM images.

Unlike the layered silicates, the CNTs do not disperse well in aqueous. Therefore in order to prepare the useful natural rubber/CNTs nanocomposites via a latex method, the CNTs must be well dispersed in aqueous prior to the adding into the latex. The non-covalent functionalization of CNTs by incorporating amphiphilic molecules known as surfactants onto the CNTs surface offers the desired technique. The stabilities of colloidal nanotube systems are promoted by electrostatic (ionic surfactant) or steric (non-ionic surfactant) repulsion in which their molecular chemical structures are largely responsible for their role as a third components, stabilizing the colloidal system through adsorption onto the interfaces of the immiscible bulk components.

Anoop-Anand *et al.* and *Mohamed et al.* investigated the effect of surfactant on the dispersing efficiency of the CNTs in aqueous and also on the properties of the obtained nanocomposites. Table 3 lists all surfactants used in these two researches. In the work of Anoop-Anand *et al.*, the natural rubber/SWCNTs nanocomposites were prepared through a latex stage mixing method. SWCNTs were dispersed in water through ultrasonication and stabilized by the subsequent addition of a surfactant. There were five different surfactants including sodium dodecyl benzene sulfonate (NaDDBS), sodium dodecyl sulfate (SDS), polyvinyl alcohol (PVA), isooctyl phenoxy polyethoxy ethanol (Criton X-100) and sodium benzoate used here. The stabilized aqueous dispersion of SWCNTs was mixed with the compounded natural rubber latex at various compositions so as to result in SWNT concentrations of 0.5-2.0 phr in natural rubber. The samples were cast onto flat glass trays, dried and subsequently cured. Firstly, the suitable surfactant was concerned. The result showed that NaDDBS was the best surfactant to prepare the SWCNTs dispersion. This was because the NaDDBS-SWCNTs dispersion was homogenous and remained dispersed for at least six weeks. It was believed that the introduction of the phenyl ring in the surfactant chains makes NaDDBS most effective. The existence of π–π stacking is proposed as a main factor promoting the exceptional performance of NaDDBS in dispersing nanotubes in polymer matrices. This paper reported a significant improvement in tensile strength and modulus of natural rubber. The tensile strength of natural rubber with a nanotube concentration as low as 2.0 phr increased by around 56%. The tensile modulus (at 300% strain) of the nanocomposite at this concentration is roughly 63% higher than that of pure rubber. However, the elongation at break of the matrix reduced with increasing nanotube loading due to stiffening of the elastomer matrix. Tear strength was also higher by around 30% as compared with pure rubber at SWCNTs fractions as low as 2.0 phr. It was proposed here that the improved reinforcement observed in natural rubber-SWCNTs nanocomposites was due to the interaction between SWCNTs and the rubber matrix. An essential condition for this interaction was an efficient dispersion of the nanofillers in the matrix, which is accomplished through surfactant assisted sonication and subsequent latex stage mixing. A good interface between SWCNTs and the rubber matrix supports the material to withstand stress. From the room temperature D.C. electrical conductivity results, it showed that SWCNTs were effective in imparting electrical conductivity to the insulating rubber matrix. At concentrations exceeding 1.0 phr, the conductivity surpassed the antistatic criterion of thin films. From thermogravimetric curves the results indicated that the presence of SWCNTs at low weight fractions had no significant effect on the thermal stability of natural rubber.

According to the above research, Mohamed *et al.* evaluated the importance of the phenyl ring in surfactant chains. A series of single-, double-, and triple-sulfosuccinate anionic surfactants containing phenyl ring moieties, representing by SDBS, AOTPh, and TCPh were used here. The preparing method used this study was only slightly different from the work of Anoop-Anand *et al.*, The major difference was

Table 3. List of surfactants used

Anoop-Anand *et al.*, 2010		Mohamed *et al.*, 2015	
Sodium dodecyl sulfate (SDS)	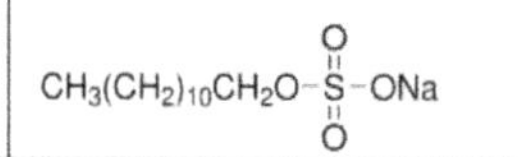	Sodium dodecyl sulfate (SDS)	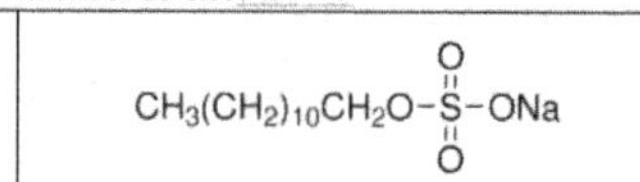
Sodium dodecyl benzene sulphonate (NaDDBS)	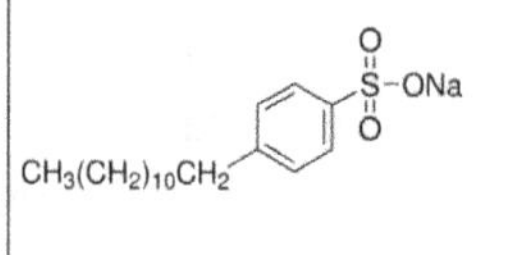	Sodium dodecyl Benzene sulfonate (NaDDBS or SDBS)	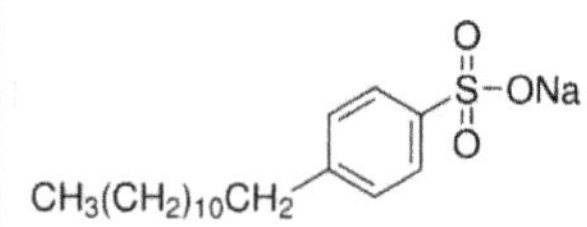
Polyvinyl alcohol (PVA)	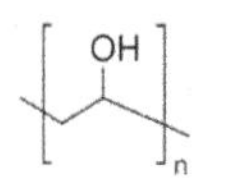	Sodium 1,4-dioxo-1,4-bis(3-phenylpropoxy) butane-2-sulfonate (AOTPh)	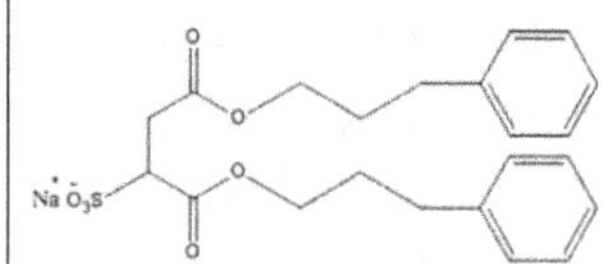
Isooctyl phenoxy polyethoxy ethanol (Criton X-100)	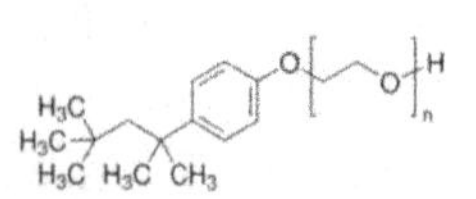	Sodium 1,5-dioxo-1,5-bis (3-phenylpropoxy)-3-((3-phenylpropoxy) carbonyl) pentane-2-sulfonate (TCPh)	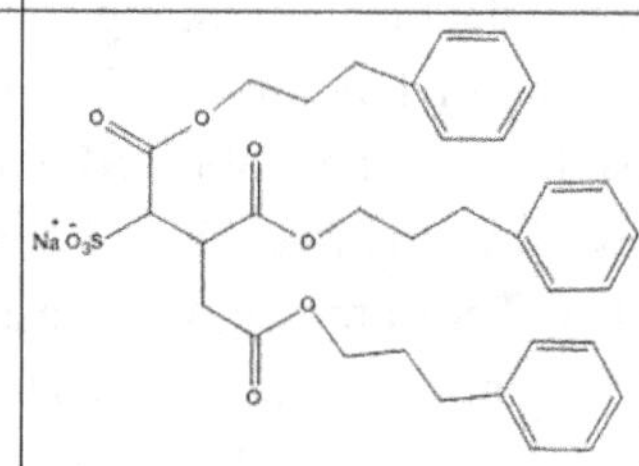
Sodium benzoate	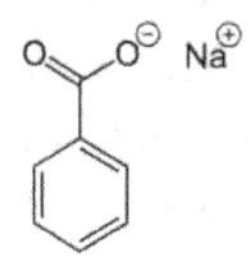		

Source: Anoop-Anand et al., 2010;Mohamed et al., 2015

the type of CNTs. This study used MWCNTs as a nanotube filler. From TEM observation, remarkably it was found that the addition of tri-chain phenyl-functionalized surfactant (TCPh) exhibited the most notable improvement in the level of dispersion of MWCNTs in the natural rubber matrix. The results from electrical conductivity measurement also confirmed that TCPh was the most efficient surfactant for the stabilization of colloidal nanotube systems because it gave a pronounced nanocomposite electrical conductivity enhancement. At low conductive filler loading, the enhanced electrical conductivity is achieved only when the conductive fillers are well dispersed and form an interconnecting filler network.

As known due to very high aspect ratios and also the lack of polar functional groups on the surface, the CNT fibers form strong Van der waals attractions between each other and poor filler-matrix interactions, causing very poor dispersion of the filler. Nakaramontri *et al.* investigated the alternative ways to prepare well dispersed CNTs in the natural rubber matrix nanocomposites. As in rubber-silica composites, silane is added directly during the mixing of rubber and silica. The silanization of silica particles can take place during mixing if the mixing conditions are suitable. Therefore, in this study the authors used two alternative mixing methods including melt mixing and latex mixing CNTs with natural rubber followed an additional mixing step. The natural rubber/CNT compounds from both methods were further mixed with silane coupling agent (bis(triethoxysilylpropyl) tetrasulfide (TESPT) using an internal mixer at high temperature to induce in-situ silanization. The presence of peaks at 1077 and 1126 cm^{-1} associated to Si-O-Si and Si-O-C asymmetric stretching vibrations in the ATR-FTIR spectrum confirmed that the grafting of silane molecules onto the sidewalls of the CNT structures by in-situ silanization was successfully taken place. Moreover, due to the presence of tetrasulfide in a silane molecule, this silanized CNTs can further react with the natural rubber during curing process. Therefore, the TESPT silane facilitated chemical linkages of CNT surfaces and natural rubber molecules. From rheocurves, the torque difference of compounds increased with the addition of CNTs, due to filler reinforcement effects. In addition, the torque difference further increased with the combined addition of CNTs and TESPT. This is due to the improved filler-rubber interactions between CNT surfaces and rubber molecules with added TESPT. From tensile testing, it was found that natural rubber/CNTs vulcanizates prepared from both melt mixing and latex mixing without adding silane showed lower than tensile strengths than the gum natural rubber vulcanizates. Nevertheless, this situation was improved by incorporating silane coupling agent and adding CNTs above 4 phr. The tensile strength increased with the addition of TESPT, done by in situ silanization, relative to the gum rubbers and the composites without TESPT. At the same CNT loading, the natural rubber/CNTs compounds prepared by latex mixing gave higher tensile strengths than by melt mixing. The highest tensile strength was achieved by the 5 phr CNTs composite prepared from a latex mixing with adding silane. From electrical conductivity measurement, it was found that percolation thresholds at approximately 2 and 1 phr of CNTs were observed for the natural rubber/CNT composites prepared via melt mixing and latex mixing with in situ silanization, respective. Electrical conductivity of the CNT composites is related to the dispersion and alignment of CNT particles in the rubber matrix. If CNTs agglomerates are broken down by the shear forces during mixing, an infinite diameter CNT network could be formed, and this would sharply increase the electrical conductivity. The natural rubber/CNT composites prepared via latex mixing with in situ silanization showed percolation thresholds at lower loading than melt mixing. This indicated that mixing method and in situ silanization strongly influenced on CNTs dispersion. Therefore, it is concluded that improved CNT dispersion and enhanced properties of natural rubber-CNT composites were achieved with latex mixing in combination with in situ functionalization of CNTs with silane coupling agent.

From above, it can be seen that the incorporation of CNTs into natural rubber gives the rubber nanocomposites with the excellent electrical properties. However, the mechanical properties is not clearly improved. The major obstacle is the poor dispersion and also low interaction between natural rubber chain and CNTs. To overcome this, the mixing method must be further studied.

Rubber/Graphene Nanocomposites

Graphene

Recently, graphene is one of the most interesting nanofiller in nanocomposites due to its extraordinary high specific surface area, mechanical properties, thermal conductivity, mobility of charge carries and gas impermeability. Therefore, graphene can be a good candidate to improve mechanical, thermal and electrical properties of composite materials. Graphene is the basic building block of all graphitic forms of carbon. Consequently, graphene is a two-dimensional carbon nanofiller with a single atomic layer of sp^2 hybridized carbon atoms arranged in a honeycomb crystal structure. It had previously been thought that the preparation of a single atomic layer would be impossible due to the unstable thermodynamic and also possibly rolling into scrolls. In 2004, Novoselov *et al.* reported about the preparation and isolation of single graphene layer. This led to a rapid rise of interest in the study of the structure and properties of graphene. The different methods have been used to prepare graphene. Mechanical cleavage or liquid phase exfoliation is commonly used to explode graphite agglomerate down into graphene layers. The other methods is to synthesize graphene using chemical vapour deposition (CVD), epitaxial growth on silicon carbide, etc.

In order to prepare large amount of graphene for preparing nanocomposites, one of the useful methods is the exfoliation and dispersion of graphite in organic or water-surfactant solutions. Another method to prepare graphene-based materials for composites and is reduction of graphite oxide (GO). GO is firstly produced by the treatment of graphite using strong mineral acids and oxidizing agents, typically via treatment with $KMnO_4$ and H_2SO_4, as in the Hummers method. These reactions achieve similar levels of oxidation (C:O ratios of approximately 2:1) (Dreyer, Park, Bielawski, & Ruoff, 2010) which ultimately disrupts the delocalized electronic structure of graphite and imparts a variety of oxygen-based chemical functionalities to the surface. It is thought that hydroxyl and epoxy groups are present in highest concentration on the basal plane, with carboxylic acid groups around the periphery of the sheets. Although the oxygen functionalities wearing the attractive forces between layers, the graphite oxide retains stacked similar to the graphite but with much wider spacing, enabling facile exfoliation. The sonication in polar solvents is often used for the exfoliation of GO. After exfoliation or during the exfoliation the obtained GO shall be reduced by chemical reduction or thermal reduction in order to restore the sp^2-hybridized carbon sheets of graphene. Therefore, the obtained product usually is termed as reduced graphite oxide (rGO). GO has an expanded interlayer spacing relative to graphite which depends on humidity due to intercalation of water molecules (Buchsteiner, Lerf, & Pieper, 2006). GO can be exfoliated using a variety of methods (most commonly by thermal shocking (Ruess & Vogt, 1948) or chemical reduction in appropriate media (Park & Ruoff, 2009; Park *et al.*, 2009), yielding a material reported to be structurally similar to that of pristine graphene on a local scale.

Natural Rubber/Graphene Nanocomposites

To prepare natural rubber/graphene nanocomposites, it is necessary to disperse graphene sheets homogenously and individually into the rubber matrix. However, due to mainly strong metallic bonding force (Rozploch *et al.*, 2007) and Van der waals between graphene platelets make of it much more difficult to disperse in natural rubber in comparison to carbon nanotube, where the contact area between two nanotubes is relatively much smaller. Various additional techniques have been adopted to prepare natural rubber/graphene nanocomposites. The context below describes the preparing methods that have been reported in the literatures about the preparing methods and also the influence of graphene on the cure behavior and properties of natural rubber composites.

Li *et al.* proved the reinforcing mechanism of GE and GO in natural rubber. The natural rubber/GE and natural rubber/GO composites were prepared through the previous route called ultrasonically assisted latex mixing and in-situ. For natural rubber/GO composites, the aqueous solution of GO prepared from natural flake graphite according to the Hummer's method was sonicated by bath sonication. After the coagulation and drying process, natural rubber/GO master batches were obtained. For natural rubber /GE composites, prior to the coagulation of the natural rubber/GO suspension, hydrazine hydrate was added into the natural rubber/GO latex and then the mixture was subjected for sonication. Both masterbatches with filler loadings of 0.3, 0.7 and 1 phr were mixed rubber chemical agents using two roll mill. From tensile tests, a noticeable enhancement in tensile strength was achieved for both GO- and GE-filled natural rubber systems, while GE has a better reinforcing effect. Moreover, the tensile strength increased with increasing filler loading. From Synchrotron WAXD patterns, it was found that the incorporation of GE produces a faster strain-induced crystallization rate and a higher crystallinity compared with the GO. From Mooney–Rivlin plots, it was found that elastic modulus (G_e) corresponding to the crosslink constraint and entanglement modulus and (G_c) corresponding to the topological tubelike constraint had an increase trend with the addition of filler. The crosslink density related to Gc and equivalent to an increase in the formation of polymer-filler couples and short bridging chains between of the GE/ natural rubber compounds was higher than the gum and the natural rubber/GO. Accompanied with the higher Ge and G_c for the natural rubber/GE compound, it was concluded that there was stronger interfacial interactions between GE and rubber. This was because GO consisting the hydrophilic polar oxygen functional groups does not exhibit good compatibility with the nonpolar rubber matrix. Therefore, GO tend to aggregate which will reduce the specific surface area of GO fillers, and thus lower the interaction extent between GO and rubber matrix. The entanglement-bound tube model was used to characterize the chain network structure of composites. It was reported that the contribution of entanglement to the conformational constraint increased and the network molecular parameters changed with the addition of GE and GO, while GE had a more evident effect than GO.

Jinrong *et al.* investigated the effect of graphene on the sulfur vulcanization kinetics of natural rubber nanocomposites. In this work, the preparation of nanocomposites were prepared by a latex mixing method without having to stabilize graphene using additional surfactant or ultrasonic field. In the method, aqueous dispersion of GO was mixed with natural rubber latex, and the mixture was co-coagulated with saturated solution of sodium chloride to form a suspension of micro-sized particles, in which GO was isolated by the nanospheres of natural rubber latex to achieve molecular level dispersion. Subsequently, hydrazine hydrate was added to reduce GO in situ. The solids were filtrated and washed with deionized water and then vacuum dried in an oven at 60 °C for 48 h. Then rubber and curing agents were compounded on two-roll mill. Raman spectroscopy was used to investigate the reduction of graphite oxide into graphene. Raman spectrum of graphite oxide contained a G-band at 1588 cm^{-1} and a D-band at 1347 cm^{-1} assigned to the in-phase vibration of the graphite lattice and the structural defects, respectively. After reduction, the intensity ratio of D-band to G-band increased from 0.89 for GO to 1.38 for graphene. The increase of this ratio after reduction has been proposed to be attributed the increase of I_D/I_G in the chemically reduced graphene to the smaller size of numerous newly-formed graphitic domains. This ratio of the natural rubber/GE nanocomposite with 5 phr GE was 1.37 which was basically equal to that of graphene prepared by chemical reduction in aqueous suspension. Therefore, it was concluded that the preparing method developed by the authors effectively in-situ reduced GO. From XPS measurement, it showed that the peak area corresponded to the nonoxygenated carbon increased from 54.7% to 75.4% to the total area for GE. This can also confirm that the GO was effectively reduced. However, GE still consisted of the same oxygen functionalities as that of GO. Moreover, reacting with hydrazine hydrate introduced an additional peak corresponding to carbon bounded to nitrogen. Moreover, the advantage of this preparing method was also confirmed by TEM and WAXD measurement. The TEM image of the natural rubber/GE nanocomposite with1 phr of graphene loading showed that GE nanosheets were exfoliated into single or a-few-stacked layers, which are finely dispersed throughout the natural rubber matrix. The XRD pattern of the natural rubber/GE nanocomposites displayed no detectable characteristic peaks of graphene stacks or graphite oxide in the nanocomposites. Some studies (Shanmugharaj *et al.*) indicate that adding carbon based fillers can delay the onset of vulcanization, due to the presence of functional groups such as carboxylic acid groups and oxygen groups on the surface of the nanofillers that absorb the basic accelerator additives. However, in this study the results from curemeter measurements indicated that incorporating graphene lower than 1 phr remarkably accelerated the crosslinking reactions indicting by the reduced scorch time and optimum cure time. However, when adding graphene more than 3 phr although the scorch time still further decreased, the optimum cure time inversely increased. As a result, the difference between the optimum cure time and the scorch time related to the vulcanization rate decreases at first, but then inversely increases. It can be inferred from that the vulcanization rate was promoted at low graphene loading (<1 phr), whereas it was suppressed subsequently. The enhanced vulcanization rate by incorporating graphene less than 1 phr was also reported by Hernandez *et al.* Furthermore, the increased crosslinking density indicating by the higher difference between minimum and maximum torque and the lower swelling ratio was observed when increasing GE loadings. From DSC results, it was proposed here that GE depresses the melting point of sulfur and plays as a role of accelerator. Both two factors result in lower curing temperature or shorter induction period of the vulcanization of natural rubber. Because GE takes part in the vulcanization process, GE provides additional crosslinks, leading to higher crosslinking density of the natural rubber/GE nanocomposites.

Because GO contains abundant hydrophilic groups that are not compatible with the hydrophobic natural rubber. Wu *et al.* treated surface of GO with bis(triethoxysilylpropyl)tetrasulfide or BTESPT. The interaction of BTESPT with the oxide surfaces included the hydrolysis of the ethoxy groups with a trace of water followed by the condensation of the silanol groups with the OH and COOH groups of GO. The obtained SGOs were mixed natural rubber by a solution mixing assisted with ultrasonication. The properties of natural rubber filled with pristine GO were compared with those filled with surface functionalized GO (or SGO). Due to the presence of absorption peaks in FT-IR spectrum at 1100, 1270 and 790 cm^{-1} assigned to Si-O stretching, Si-C stretching and Si-C vibration of SGO, it was confirmed that BTESPT are grafted onto the surfaces of GO. The XRD patterns showed that the introducing BTESPT molecules onto GO surface disturbed the layer-by-layer packing of GO sheets and increased interlayer spacing of GO. The tensile strength of the nanocomposites increased dramatically with increasing filler loading, but the effect subsequently leveled off when the SGO and GO loadings were higher than 0.3 wt% and 0.1 wt%, respectively. With the addition of only 0.3 wt% of SGO, the tensile strength of the SGO/NR nanocomposite was up to 14.7 MPa, a two-fold enhancement compared with that of unfilled NR. This prominent reinforcement effect has never been reported for NR filled with other nanofillers at such low loading. Compared with SGO, the reinforcement effect of GO was not so prominent. The tensile strength of the GO/NR nanocomoposite containing 0.1 wt% of GO was 9.7 MPa, which is enhanced by 33% compared with that of NR, much lower than that of the of double bonds in the rubber molecules, forming covalent linkages between the SGO nano-sheets and the rubber molecules via BTESPT molecules. The occurrence of these covalent linkages was confirmed by the lower swelling ratio. Compared with the gum NR, air permeability of the SGO/NR nanocomposites decreased dramatically to a value of 52% when the SGO content was 0.3 wt%. In the case of NR filled with pristine clay, the NR must be loaded with 16.7 wt% of clay in order to have the same ability to decrease air permeability as the SGO/NR nanocomposites (Wu *et al.*, 2005). Results from this study suggests that surface functionalized GO is a powerful method to increase the reinforcing efficiency and gas barrier properties of the rubber-based nanocomposites.

Besides latex mixing and solution mixing, a fair amount of researches reported the use of a conventional solid-state mixing to fabricate the natural rubber/graphene nanocomposites. Yaragallaa *et al.* incorporated thermally reduced graphene oxide (TRG) into dry NR using an internal mixer. The natural rubber/TRG composites with 0.5, 2, 3 wt% of TRG were obtained. The TEM micrographs of cured natural rubber/TRG composites with 0.5% wt of TRG showed relatively good uniform dispersion, however, in some regions (even at low concentrations) stacked and agglomerated layers were observed. The segregated and agglomerated TRG layers inside the natural rubber matrix could also be observed at 3% w/w of TRG. The result here indicates that it is difficult to achieve homogeneous dispersion of TRG platelets inside dry natural rubber using conventional mechanical mixing method. FT-IR technique was used to trace the chemical interactions between TRG and natural rubber phase. Neat natural rubber showed characteristic intense peaks at 2916 cm^{-1} and 1596 cm^{-1} which correspond to =CH, C=C stretching frequencies of cis 1,4 polyisoprene units. When 3 wt% of TRG was added into the natural rubber matrix, the intensity and peak area of the =CH peak decreased dramatically. Moreover, the =CH peak showed a slight shift towards lower frequency as compared to neat natural rubber. The C=C peak of natural rubber completely diminished, indicating existence of strong π-π interactions between the π electrons of TRG and C=C bonds of isoprene units. The stress-strain curves showed that there was a dramatic improvement in modulus with the addition of TRG platelets. This was due to the high modulus and reinforcement imparted by the rigid TRGs on to the natural rubber matrix. Besides, the elongation at break decreases with increase in TRG content. The presence of 0.5 wt% of TRG enhanced tensile strength of the neat NR. However, further increasing TRG loading tensile strength became inversely decreased. The NR/TRG composites with 3 wt% of TRG showed lower tensile strength than the neat natural rubber. Compared to the natural rubber/ GE nanocomposites prepared by latex mixing, at this graphene loading the obtained nanocomposites still had higher tensile strength. This might be due to the presence of TRG agglomeration when using a melt mixing. DMA results showed that with the addition of TRG into neat natural rubber the storage modulus increased by 25%. Moreover, the glass transition temperature was shifted to a large extent (15 °C) due to strong π–π interactions associated with π bonds of TRG and double bonds of natural rubber. This strongly indicated that a large number of natural rubber chains have been immobilized by the TRG filler surface leading to the formation of a rigid amorphous phase. A dramatic decrease in gas permeability values with the addition of TRG in natural rubber was observed. The arrangement of TRG layers led to a tortuous path in the natural rubber matrix. This study also compared the gas barrier properties of the TRG filled natural rubber with the one filled with GO and graphite. The TRG filled natural rubber composites exhibited lower permeability compared to that of those two composites. The highest O_2 gas barrier property was observed for the natural rubber/TRG composites with 3 wt% of TRG. The main reason for this difference is the nano-dimension of TRG platelets compared to GRT and GO and the relatively good dispersion of TRG platelets in the natural rubber matrix.

From the above, it can be seen that graphene can well disperse in the rubber matrix once the appropriate method is carried out. Although graphene is pricy filler, the enhanced properties can be obtained at very low loading. Therefore, graphene nowadays seems to be one of the most promising nanofillers.

CONCLUSION

On account of the recent emergence described above, it can be seen that the improvement of natural rubber can be done by modification and incorporation of nanofillers. Many forms of modified natural rubber nowadays are commercially available. The key of modification is to improve undesirable properties such as low hydrocarbon resistance and low heat tolerance and also to transform natural rubber into new class of materials such as thermoplastic elastomer. Nanotechnology also plays an important roll on the development of natural rubber. The enhancement of properties of natural rubber such as mechanical strength, gas impermeability and electrical conductivity can be obtained at much lower nanofiller loading. However, the well dispersion of nanofiller in the natural rubber must be developed. To achieve this, the suitable preparing method and filler loading must be carefully concerned. So far the latex mixing seems to be the most promising method to prepare natural rubber nanocomposites.

REFERENCES

Alexandre, M., & Dubois, P. (2000). Polymer-layered silicate nanocomposites: Preparation, properties and uses of a new class of materials. *Materials Science and Engineering*, *28*(1-2), 1–63. doi:10.1016/S0927-796X(00)00012-7

Anoop-Anand, K. (2010). Natural rubber-carbon nanotube composites through latex compounding. *International Journal of Polymeric Materials*, *59*(1), 33–44. doi:10.1080/00914030903172916

Aranda, P., & Ruiz-Hitzky, E. (1992). Poly (ethylene oxide)-silicate intercalation materials. *Chemistry of Materials*, *4*(6), 1395–1403. doi:10.1021/cm00024a048

Arroyo, M., Manchado, M. A., & Herero, B. (2003). Organo-montmorillonite as substitute of carbon black in natural rubber compounds. *Polymer*, *44*(8), 2447–2453. doi:10.1016/S0032-3861(03)00090-9

Berber, S., Kwon, Y. K., & Tománek, D. (2000). Unusually high thermal conductivity of carbon nanotubes. *Physical Review Letters*, *84*(20), 4613–4616. doi:10.1103/PhysRevLett.84.4613 PMID:10990753

Bokobza, L. (2012). Enhanced electrical and mechanical properties of multiwall carbon nanotube rubber composites. *Polymers for Advanced Technologies*, *23*(12), 1543–1549. doi:10.1002/pat.3027

Buchsteiner, A., Lerf, A., & Pieper, A. Jr. (2006). Water dynamics in graphite oxide investigated with neutron scattering. *The Journal of Physical Chemistry B*, *110*(45), 22328–223238. doi:10.1021/jp0641132 PMID:17091972

Dai, L., Soundarrajan, P., & Kim, T. (2002). Sensors and sensor arrays based on conjugated polymers and carbon nanotubes. *Pure and Applied Chemistry*, *74*(9), 1753–1772. doi:10.1351/pac200274091753

Dresselhaus, M. (1996). Carbon connections promise nanoelectronics. *Physics World*, *9*(5), 18–19. doi:10.1088/2058-7058/9/5/18

Dreyer, D. R., Park, S., Bielawski, C. W., & Ruoff, R. S. (2010). The chemistry of graphene oxide. *Chemical Society Reviews*, *39*(1), 228–240. doi:10.1039/B917103G PMID:20023850

Du, C., Yeh, J., & Pan, N. (2005). High power density supercapacitors using locally aligned carbon nanotube electrodes. *Nanotechnology*, *16*(4), 350–353. doi:10.1088/0957-4484/16/4/003

Du, F., Fischer, J. E., & Winey, K. I. (2003). Coagulation method for preparing single-walled carbon nanotube/poly(methyl methacrylate) composites and their modulus, electrical conductivity, and thermal stability. *Journal of Polymer Science. Part B, Polymer Physics*, *41*(17), 3333–3338. doi:10.1002/polb.10701

Ebbesen, T. W., & Ajayan, P. M. (1992). Large-scale synthesis of carbon nanotubes. *Nature, 358*(6383), 220–222. doi:10.1038/358220a0

Emmenegger, Ch., Mauron, Ph., Sudan, P., Wenger, P., Hermann, V., Gallay, R., & Züttel, A. (2003). Investigation of electrochemical double-layer (ECDL) capacitors electrodes based on carbon nanotubes and activated carbon materials. *Journal of Power Sources, 124*(1), 321–329. doi:10.1016/S0378-7753(03)00590-1

Fakhrul-Razi, A., Atieh, M. A., Girun, N., Chuah, T. G., El-Sadig, M., & Biak, D. R. A. (2006). Effect of multi-wall carbon nanotubes on the mechanical properties of natural rubber. *Composite Structures, 75*(1-4), 496–500. doi:10.1016/j.compstruct.2006.04.035

Galimberti, M. (2001). *Rubber-clay nanocomposites*. Hoboken, NJ: John Wiley & Sons, Inc.

Ganter, M., Gronski, W., Reichart, P., & Mulhaupt, R. (2001). Rubber Nanocomposites: Morphology and mechanical properties of BR and SBR vulcanizates reinforced by organophilic layered silicates. *Rubber Chemistry and Technology, 74*(2), 221–235. doi:10.5254/1.3544946

Gong, X., Liu, J., Baskaran, S., Voise, R. D., & Young, J. S. (2000). Surfactant-assisted processing of carbon nanotube/polymer composites. *Chemistry of Materials, 12*(4), 1049–1052. doi:10.1021/cm9906396

Hakim, R. N., & Ismail, H. (2009). Comparison of the effects of organoclay loadings on the curing and mechanical properties of organoclay-filled epoxidixed natural rubber nanocomposites and organoclay-filled natural rubber nanocomposites. *Journal of Applied Polymer Science, 20*(2), 37–59.

Hashim, A. S., Azahari, B., Ikeda, Y., & Kohjiya, S. (1998). The effect of Bis(3-triethoxysilylpropyl) tetrasulfide on silica reinforcement of styrene-butadiene rubber. *Rubber Chemistry and Technology, 71*(2), 289–299. doi:10.5254/1.3538485

Hernandez, M., Bernal, M., Verdejo, R., Ezquerra, T., & Lopez-Manchado, M. (2012). Overall performance of natural rubber/graphene nanocomposites. *Composites Science and Technology, 73*, 40–46. doi:10.1016/j.compscitech.2012.08.012

Hrachova, J., Komadel, P., Jochec-Moskova, D., Krajcl, J., Janigova, I., Slouf, M., & Chodak, I. (2013). Properties of organo-clay/natural rubber nanocomposites: Effect of orgonophilic modifiers. *Journal of Applied Polymer Science, 127*(5), 3447–3455. doi:10.1002/app.37602

Ismail, H., Ramly, F., & Othman, N. (2010). Multiwall carbon nanotube-filled natural rubber: The effects of filler loading and mixing method. *Polymer-Plastic Technology and Engineering, 49*(3), 260–266. doi:10.1080/03602550903413888

Israelachvili, J. (1991). *Intermolecular and surface forces*. San Diego, CA: Academic Press.

Jinrong, W., & Huang, G. (2013). Enhanced mechanical and gas barrier properties of rubber nanocomposite with surface functionalized graphene oxide at low content. *Polymer, 54*(7), 1930–1937. doi:10.1016/j.polymer.2013.01.049

Li, F., Yan, N., Zhan, Y., Fei, G., & Xia, H. (2013). Probing the reinforcing mechanism of graphene and graphene Oxide in natural Rubber. *Journal of Applied Physics, 129*(4), 2342–2351.

Madhusoodana, K. N., & Varghese, S. (2006). Technological and processing properties of natural rubber layered silicate-nanocomposites by melt intercalation Process. *Journal of Applied Polymer Science*, *102*(3), 2537–2543. doi:10.1002/app.24680

Mohamed, A., Anas, A. K., Bakar, S. A., Ardyani, T., Zin, W. M. W., Ibrahim, S., & Eastoe, J. et al. (2015). Enhanced dispersion of multiwall carbon nanotubes in natural rubber latex nanocomposites by surfactants bearing phenyl. *Journal of Colloid and Interface Science*, *455*, 179–187. doi:10.1016/j.jcis.2015.05.054 PMID:26070188

Nakaramontri, Y., Nakasan, C., & Kummerlowe, C. (2016). Enhancement of electrical conductivity and filler dispersion of carbon nanotube filled natural rubber composites latex mixing and in situ silanization. *Rubber Chemistry and Technology*, *89*(2), 272–291. doi:10.5254/rct.15.84848

Novoselov, K. S., Geim, A. K., Morozov, S. V., Jiang, D., Dubonos, S. V., Grigorieva, I. V., & Firsoc, A. (2004). Electric Field Effect in Atomically Thin Carbon Films. *Science*, *306*(5696), 666–669. doi:10.1126/science.1102896 PMID:15499015

Okada, A., Kawasumi, M., Kurauchi, T., & Kamigaito, O. (1987). Synthesis and characterization of a nylon 6-clay hybrid. *Papers Presented at the ... Meeting. American Chemical Society. Division of Polymer Chemistry*, *28*, 447.

Park, S., An, J. H., Jung, I. W., Piner, R. D., An, S. J., Li, X. S., & Ruoff, R. S. et al. (2009). Colloidal suspensions of highly reduced graphene oxide in a wide variety of organic solvents. *Nano Letters*, *9*(4), 1593–1597. doi:10.1021/nl803798y PMID:19265429

Park, S., & Ruoff, R. S. (2009). Chemical methods for the production of graphenes. *Nature Nanotechnology*, *5*(4), 217–224. doi:10.1038/nnano.2009.58 PMID:19350030

Pham, J. Q., Mitchell, C. A., Bahr, J. L., Tour, J. M., Krishanamoorti, R., & Green, P. F. (2003). Glass transition of polymer/single-walled carbon nanotube composite films. *Journal of Polymer Science. Part B, Polymer Physics*, *41*(17), 3339–4920. doi:10.1002/polb.10702

Ramanathan, T., Liu, H., & Brinson, L. C. (2005). Functionalized SWNT/polymer nanocomposites for dramatic property improvement. *Journal of Polymer Science. Part B, Polymer Physics*, *43*(17), 2269–2279. doi:10.1002/polb.20510

Rattanasom, N., Saowapark, T., & Deeprasertkul, C. (2007). Reinforcement of natural rubber with silica/carbon black hybrid filler. *Polymer Testing*, *26*(3), 369–377. doi:10.1016/j.polymertesting.2006.12.003

Ray, S. S., & Okamoto, M. (2003). Polymer/layered silicate nanocomposites: A review from the preparation to processing. *Progress in Polymer Science*, *28*(11), 1539–1641. doi:10.1016/j.progpolymsci.2003.08.002

Roberts, A. D. (1998). *Natural Rubber Science and Technology*. Oxford, UK: Oxford University Press.

Rozploch, F., Patyk, J., & Stankowski, J. (2007). Graphenes bonding forces in vulcanization kinetics of graphene/natural rubber nanocomposites. *Acta Phisica Polanica A*, *112*(3), 557.

Ruess, V. G., & Vogt, F. (1948). Hochstlamellarer Kohlenstoff aus Graphitoxyhydroxyd. *Monatshefte fur Chemie*, *78*(3-4), 222–242. doi:10.1007/BF01141527

Shanmughara, A. M., Bae, J. H., Lee, K. Y., Noh, W. H., Lee, S. H., & Ryu, S. H. (2007). Physical and chemical characteristics of multiwalled carbon nanotubes functionalized with aminosilane and its influence on the properties of natural rubber composites. *Composites Science and Technology*, *67*(9), 1813–1822. doi:10.1016/j.compscitech.2006.10.021

Sui, G., Zhong, W. H., Yang, X. P., & Yu, Y. H. (2008). Curing kinetics and mechanical behavior of natural rubber reinforced with pretreated carbon nanotubes. *Materials Science and Engineering A*, *485*(1-2), 524–531. doi:10.1016/j.msea.2007.09.007

Sumio, I. (1991). Helical microtubules of graphitic carbon. *Nature*, *354*(6348), 56–58. doi:10.1038/354056a0

Theng, B. K. G. (1974). *The chemistry of clay-organic reactions*. Academic Press.

Theng, B. K. G. (1979). *Formation and properties of clay-polymer complexes*. Amsterdam: Elsevier Scientific Publishing Company.

Ting, G. (1995). Self-assembly of tubular fullerenes. *Journal of Physical Chemistry*, *99*(27), 10694–10697. doi:10.1021/j100027a002

Ting, G. (1995). Electrical conductors comprising single-wall carbon nanotubes. *Chemical Physics Letters*, *243*, 49.

Vaia, R. A., Ishii, H., & Giannelis, E. P. (1993). Synthesis and properties of two-dimensional nanostructures by direct intercalation of polymer melts in layered silicates. *Chemistry of Materials*, *5*(12), 1694–1696. doi:10.1021/cm00036a004

Valadares, L. F., Letle, C. A. P., & Galembeck, F. (2006). Preparation of natural rubber-montmorillonite nanocomposites in aqueous medium: Evidence for polymer-platelet adhesion. *Polymer*, *47*(2), 672–678. doi:10.1016/j.polymer.2005.11.062

Varghese, S., & Karger-Kocsis, J. (2003, August). (2203). Natural rubber-based nanocomposites by latex compounding with layered silicates. *Polymer*, *44*(17), 4921–4927. doi:10.1016/S0032-3861(03)00480-4

Varghese, S., & Karger-Kocsis, J. (2004). Melt-compounded natural rubber nanocomposites with pristine and organophilic layered silicates of natural and synthetic origin. *Journal of Applied Polymer Science*, *91*(2), 813–5819. doi:10.1002/app.13173

Vilgis, T. A. (2005). Time scales in the reinforcement of elastomers. *Polymer*, *46*(12), 4223–4229. doi:10.1016/j.polymer.2005.02.060

Wang, L., & Dang, Z. M. (2005). Carbon nanotube composites with high dielectric constant at low percolation threshold. *Applied Physics Letters*, *87*, 0429031–0429033.

Wang, Y., Zhang, L., Tang, C., & Yu, D. (2000). Preparation and characterization of rubber-clay nanocomposites. *Journal of Applied Polymer Science*, *78*(11), 1879–1883. doi:10.1002/1097-4628(20001209)78:11<1879::AID-APP50>3.0.CO;2-1

Wu, Y. P., Wang, Y. Q., Zhang, H. F., Wang, Y. Z., Yu, D. S., Zhang, L. Q., & Yang, J. (2005). Rubber–pristine clay nanocomposites prepared by co-coagulating rubber latex and clay aqueous suspension. *Composites Science and Technology*, *65*, 1195–1202.

Xie, H., Liu, B., Yuan, Z., Shen, J., & Cheng, R. (2004). Cure kinetics of carbon nanotube/tetrafunctional epoxy nanocomposites by isothermal differential scanning calorimetry. *Journal of Polymer Science. Part B, Polymer Physics*, *42*(20), 3701–3712. doi:10.1002/polb.20220

Xu, G. C., Li, A. Y., Zhang, L. D., Wu, G. S., Yuan, X. Y., & Xie, T. (2003). Synthesis and characterization of silica nanocomposite in situ photopolymerization. *Journal of Applied Polymer Science*, *90*(8), 837–840. doi:10.1002/app.12650

Xu, Y., Zhang, Y., & Suhir, E. (2006). Thermal properties of carbon nanotube array used for integrated circuit cooling. *Journal of Applied Physics*, *100*, 074306.

Yaragallaa, S., Meera, A. P., Kalarikkal, N., & Thomas, S. (2015). Chemistry associated with natural rubber–graphene nanocomposites and its effect on physical and structural properties. *Industrial Crops and Products*, *74*, 792–802. doi:10.1016/j.indcrop.2015.05.079

Yu, Y., Ouyang, C., Gao, Y., Si, Z., Chen, W., Wang, Z., & Xue, G. (2005). Synthesis and characterization of carbon nanotube/polypyrrole core-shell nanocomposites via in situ inverse micro emulsion. *Journal of Polymer Science. Part A, Polymer Chemistry*, *43*(23), 6105–6115. doi:10.1002/pola.21114

Zeng, Q. H., Wang, D. Z., Yu, A. B., & Lu, G. Q. (2002). Synthesis of polymer montmorillonite nanocomposites by in situ intercalative polymerization. *Nanotechnology*, *13*(13), 549–553. doi:10.1088/0957-4484/13/5/301

Zhang, L., Wang, Y., Wang, Y., Sui, Y., & Yu, D. (2000). Morphology and mechanical properties of clay/styrene-butadiene rubber nanocomposites. *Journal of Applied Polymer Science*, *78*(11), 1873–1878. doi:10.1002/1097-4628(20001209)78:11<1873::AID-APP40>3.0.CO;2-8

Zhao, G. L., Bagayoko, D., & Yang, L. (2006). Optical properties of aligned carbon nanotube mats for photonic applications. *Journal of Applied Physics*, *99*(11), 114311–114315. doi:10.1063/1.2201738

KEY TERMS AND DEFINITIONS

Cure Property: A property that indicates the vulcanization rate.
Electircal Properties: A property of a substance which determines its response to an ekectric field.
Gas Permeability: An ability of a barrier material to allow gases to permeate.
Physical Property: Any property that is measurable, whose value describes a state of a physical system.
Reinforcement: A process that enhances properties of materials by adding reinforcing filler.
Vulcanization: A chemical reaction that crosslinks rubber chains together.

Chapter 4
Natural Rubber and Rubber Blend Nanocomposites:
Reinforcement of Natural Rubber with Polymer-Encapsulated Inorganic Nanohybrid Particles

Anyaporn Boonmahitthisud
Chulalongkorn University, Thailand

ABSTRACT

Natural rubber (NR) is representative biomass polymer and the effective uses are strongly contributed to sustainable society. This chapter presents the innovative and advanced rubber nanocomposites with polystyrene-encapsulated silica nanohybrids (PS-nSiO$_2$) subsequently used as a nanofiller for NR and NR/styrene butadiene rubber (NR/SBR). The PS-nSiO$_2$ were prepared via 'in situ' differential microemulsion polymerization. The core-shell nanohybrids of PS-nSiO$_2$ were achieved with an average diameter of 40 nm using a smaller amount of surfactant, compared to microemulsion polymerization method. Moreover, the effects of the NR and NR/SBR filled with PS-nSiO$_2$ nanohybrids on the mechanical properties, thermal stability, flammability and morphology are also discussed. The results indicated that the encapsulation of nSiO$_2$ with PS can provide not only the well-dispersion of nanoparticles in the rubber matrix but also the synergistic properties of two components from the polymer and the inorganic nanoparticles by improving mechanical properties, thermal stability and flammability of rubber nanocomposites.

INTRODUCTION

Natural rubber (NR) is an outstanding natural bio-based polymer that widely utilize in many applications. NR is known to exhibit numerous excellent properties; however, reinforcing fillers are necessarily added into NR in order to enhance the appropriate characteristics for a variety of the applications. Generally, the activity of reinforcing fillers depends on their primary particle size (Hewitt, 2007). Therefore, reinforcing nanoparticles play an important role to improve the NR properties for example, the large

DOI: 10.4018/978-1-5225-1971-3.ch004

surface area, which leads to the high interfacial adhesion between NR and nanoparticles. However, Most of reinforcing fillers contains a large number of hydroxyl groups (-OH groups) on their surface and causes the strong filler-filler interactions, leading to the reinforcing particle aggregation. Moreover, the acidicity (–OH groups) of the fillers results in the adverse effect on the NR curing system. This is due to the interaction between –OH groups on the surface of nanoparticles and basic accelerator in sulphur curing system (Choi, 2001). In order to enhance dispersability of nanoparticles and interaction between nanoparticles and rubber matrix, inorganic nanoparticles are usually encapsulated by polymer molecules such as polystyrene (PS) (Chuayjuljit & Luecha, 2011) and poly(methyl methacrylate) (PMMA) (Chuayjuljit, Sukasem, & Boonmahitthisud, 2014). Besides the enhanced dispersibility of nanoparticles, the polymer-encapsulated inorganic nanohybrid materials may exhibit the synergistic effects of polymer encapsulating onto nanoparticle and give the potential for preparing advanced engineering materials with specific and enhanced properties. Among various polymerization methods, *'in situ'* differential microemulsion polymerization method is an advanced technology to prepare nanosized encapsulation fillers.

In this chapter, the author presents the novel technique to synthesize polymer-encapsulated silica nanohybrid particles with small average diameter below 50 nm by means of *'in situ'* differential microemulsion polymerization for applying as a reinforcing nanofiller in rubber nanocomposites.

BACKGROUND

The rapidly increasing human population growth caused an increasing demand on the limited resources of petroleum. While the demand for world crude oil is growing at 2 percentages each year, the volume of a newly discovered oil has decreased dramatically every five years since the early 2000s (Figure 1). This has caused the impending world oil shortage and facing the rising oil price (Lindgren, 2003).

Figure 1. The volume of oil discovered worldwide during 1940-2000
Source: Lindgren, 2003

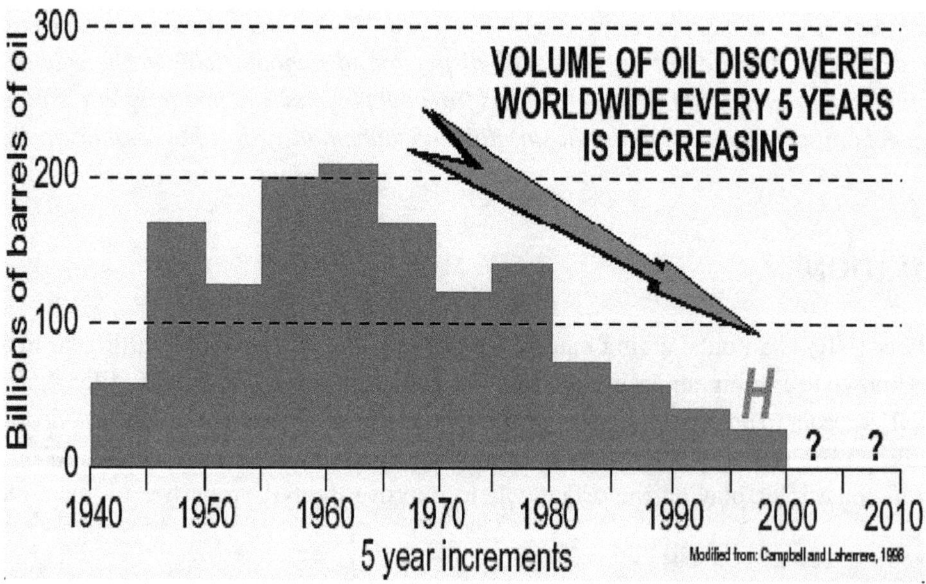

Currently, the environmental sustainability issues have received considerable attention in the scientific world. Utilization of bio-based resources including trees (Nair & Joseph, 2014), food crops (Chumeka, Pasetto, Pilard, & Tanrattanakul, 2014), shells of animals (Shahzadi et al., 2016), algae and agricultural byproducts (Noreen, Zia, Zuber, Ali, & Mujahid) is becoming an attractive alternative resource for the production of new materials because they can be naturally replenished themselves over short periods of time, which means they will never run out.

Natural rubber (NR) is a bio-based polymeric material harvested with the dispersed form as seen as a milky colloidal or latex in an aqueous medium. In the chemical structure, NR is known as polyisoprene consisting almost completely of cis-1,4 polyisoprene with the methyl groups (CH_2) located on the same side of the ethylene bond (C=C) (Figure 2), which is responsible for elasticity. Moreover, the high structural regularity of NR tends to low temperature crystallization and strain-induced crystallization resulting in possessing numerous outstanding properties as compared to any synthetic rubbers including high green (uncured) strength, excellence mechanical strength, good abrasion resistance and dynamic mechanical properties, high resilience, high tack and low cost (Huneau, 2011), (González, Custal, Lalaouna, Riba, & Armelin, 2016). These properties make NR useful a wide variety of applications ranging from ordinary household to aerospace products such as boots, toys, erasers, balloon, elastic bands, household gloves, surgical gloves, medical and dental devices, food-contact items, sports equipment, tire, automotive parts and aerospace equipment. NR is naturally harvested by tapping the rubber tree *Hevea brasiliensis* (Figure 3) (MREPC, 2016), which originated in Brazil and now mainly grows in South-east Asia. From the first introduction of rubber tree to Thailand in Trang province by Phraya Ratsadanupradit Mahison Phakdi (Khaw Sim Bee Na Ranong) in the year 1899 (Killmann, 2001), (Kaiyoorawong & Yangdee, 2006), NR has become the economic agricultural product of Thailand. Nowadays, Thailand is one of the biggest global share of NR production (30.6%), followed by Indonesia (26.6%), Malaysia (8.5%), Vietnam (7.5%) and India (7.0%) as seen in Figure 4 (Ono, 2014). The current production of NR in Thailand is 4.42 million tonnes per year and export is 3.4 million tonnes per year whereas the remnants are local consumption used (TRA, 2015a). A majority of NR products are exported in the raw form to the top export destination such as China, Malaysia, Japan and United State (TRA, 2015b). However, the current NR price in Thailand is about \$1.20 to \$1.50 per kilogram, decreased 70% from the peak of \$5.4 four years ago. Moreover, the NR is exported accounted for 1.6% of entire local product in 2014, falling down from 3.6% in 2011 and 2.4% in 2012 (Gallucci, 2015). Consequently, the Thai government has realized the need to increase the potential and performance of NR products manufactured in Thailand and changed to enhance the technological innovations in the value addition of NR in the country.

With the last few decades, the term "nanotechnology" has received more and more attention among scientist, engineer and technologists due to its application in various areas including chemistry, physics, materials science and biotechnology to produce new materials with unique properties such as:

- Textile (Wong, 2006),
- Cosmetics (Raj, Jose, Sumod, & Sabitha, 2012),
- Food (Chellaram et al., 2014),
- Electronic (Matsui, 2003),
- Fuel (Lee & Kjeang, 2013),
- Sensors (Comini et al., 2013), and
- Packaging (Duncan, 2011).

Figure 2. Chemical structure of cis-1,4 polyisoprene from natural rubber used in this research

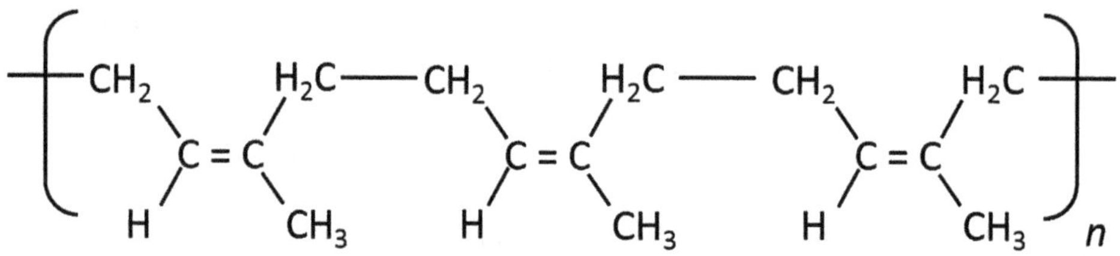

Figure 3. Natural rubber latex harvested by tapping the rubber tree Hevea brasiliensis
Source: MREPC, 2016

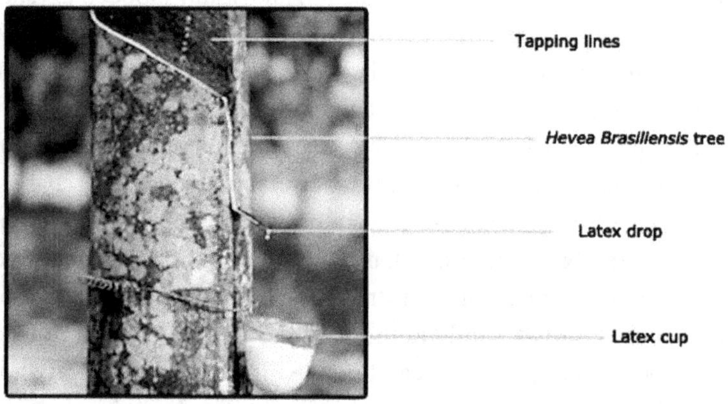

Figure 4. Natural rubber market share in the global
Source: Ono, 2014

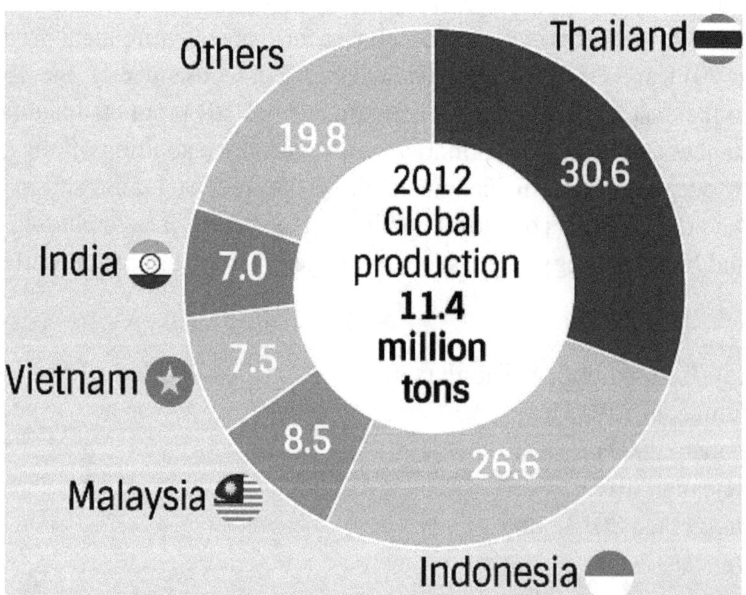

Nanocomposites made up of the combination of organic and inorganic components at the nanoscopic scale which can be classified into four major categories based on the type of matrix material; metal-matrix, ceramic-matrix, polymer-matrix and carbon-matrix (Masoud & G., 2011). The presence of inorganic nanoparticles can provide specific functionalities with magnetic, optical, catalytic activity, optoelectronic and biological activity in their properties and enhance mechanical and thermal properties of the matrix (Camargo, Satyanarayana, & Wypych, 2009). Compared to conventional micron-sized inorganic fillers, inorganic nanoparticles exhibit major advantages as follows: (Ma, Zhang, & Geng, 2010)

1. The applied stress transferred from matrix onto inorganic nanoparticles is more efficient due to their higher surface area.
2. The longer crack propagation length at the interface resulting in improvement in strength with maintaining elastic properties.
3. Lower density and as a result lighter weight composites with lower and easier processability owing to requiring very low loading levels.

From our previous research (Boonmahitthisud & Chuayjuljit, 2012), NR/carboxylated styrene butadiene rubber (NR/XSBR) at the ratio of 80NR/20XSBR nanocomposites containing different loadings of carbon black (CB) at 3-9 phr and carbon nanotube (CNT) at 0.1-0.4 phr were prepared by latex compounding. The results revealed that the incorporation of either CB or CNT can improve mechanical properties and thermal stability. However, The CNT-filled 80NR/20XSBR nanocomposites provided a comparable tensile strength with that of CB-filled 80NR/20XSBR nanocomposites but at 25-fold lower loading. This is due to the fact that CNT has much higher aspect ratio and specific surface area, resulting in the stronger interfacial adhesion between CNT and rubber blend matrix as compared with CB.

In the rubber industries, utilization of inorganic nanoparticles also applies to NR in order to enhance the appropriated properties for various applications and value addition of NR such as montmorillonite (MMT) (Rezende et al., 2010), carbon nanotube (CNT) (Bokobza, 2007), titanium dioxide nanoparticles ($nTiO_2$) (Seentrakoon, Junhasavasdikul, & Chavasiri, 2013) and silica nanoparticles ($nSiO_2$) (Prasertsri & Rattanasom, 2011). $nSiO_2$ has been widely used as a reinforcing filler in the manufacture of rubber products due to its colorless, better reduction in heat build-up, high tear strength, and aging resistance, low rolling resistance, low cost of production and high available (Mokhothu, Luyt, Morselli, Bondioli, & Messori, 2015). However, the surfaces of $nSiO_2$ contain abundant hydrophilic silanol groups inducing the strong filler-filler interaction and high tendency to form self-aggregation of $nSiO_2$ inside the rubber matrix (Weng, Wei, Tang, Lin, & Guo, 2015) (Theppradit, Prasassarakich, & Poompradub, 2014). The aggregated $nSiO_2$ are not easily disintegrated by usual shear force during mixing resulting in making the rubber compounds more difficult to process, detrimental effects on the cure processes, none of the improvement in mechanical properties and reducing optical, mechanical and thermal properties of rubber products (Oberdisse, 2006; Ansarifar, Lim, & Nijhawan, 2004). To overcome the low dispersibility of $nSiO_2$ in rubber matrix, it is necessary to modify the surface of $nSiO_2$. Nowadays, the polymer-encapsulated inorganic nanohybrid particles have been obtained by encapsulating inorganic nanoparticles by various polymers such as polystyrene (PS) (Zeng, Yu, & Guo, 2004), poly(methyl methacrylate) (PMMA) (Chuayjuljit et al., 2014) and polyisoprene (Kongsinlark, Rempel, & Prasassarakich, 2012). These nanohybrid particles are expected to have reduce advantages in particle toxicity, mask taste and odor, improve stability, dispersibility and compatibility in the matrix (Zhang, Fu, Huang, & Zhou; (Akbarzadeh, Samiei, & Davaran, 2012). In addition, the combination of inorganic nanoparticles and polymers display synergistically enhanced properties of both components to one materials including high strength, chemical and thermal stability, optical, catalytic, electronic and magnetic properties possessed by inorganic nanoparticle components, whereas easy fabrication, lightweight, economic processing, durability and relatively low cost possessed by polymer components (Grothe, Kaskel, & Leuteritz, 2012). Therefore, the encapsulation of inorganic nanoparticles by polymers has been widely used in various applications such as military equipment, safety, garments, automotive, electronics, optical devices, agriculture, biomedical, cosmetics and additives (Table 1). Various polymerization methods are presently appropriate for the preparation of the nanohybrid particles including solution polymerization, dispersion polymerization, aerosol polymerization, conventional emulsion polymerization, inverse emulsion polymerization, miniemulsion polymerization, microemulsion polymerization and differential microemulsion polymerization (Table 2). The differential microemulsion polymerization will be presented as the alternative method for synthesizing polymer-encapsulated inorganic nanohybrid particles with average diameter below 50 nm by using a much lower amount of surfactant and providing a higher solid content compared with microemulsion polymerization.

PRE-TREATMENT OF $nSiO_2$ SURFACE BY A COUPLING AGENT AND ITS CHARACTERIZATION

In order to effectively encapsulate polymer onto $nSiO_2$ surface, the pre-treatment of $nSiO_2$ surface is necessary to turn $nSiO_2$ surface characteristic from hydrophilicity to organophilicity. Silane coupling agents have been considerable represented as a chemical bridge between $nSiO_2$ and polymer due to the remarkable molecule structures containing two types of reactive functional groups characterized by different kinds of reactivity as follows:

$OR\text{-}Si\text{-}(CH_2)_n\text{-}X$

OR refers to a hydrolyzable group which form chemical bonds with a hydroxyl group (-OH group) presenting in the $nSiO_2$ surface such as alkoxy, acryloxy, halogen or amine, whereas X refers a organo-fuctional group which can form covalent bonds with polymer such as epoxy, amino, methacryloxy and vinyl groups (Jerzy J. Chruściel & Leśniak, 2012). The reactive mechanism of silane coupling agents with $nSiO_2$ surface consists of 4 steps as shown in Figure 5 (Shin-Etsu, 2011). First, silane coupling agents are hydrolyzed to form silanol-containing species and the oligomers are formed through partial condensation. The oligomers are then bonded with –OH groups of $nSiO_2$ surfaces. Finally during drying or curing process, a covalent linkage is formed through a dehydration condensation reaction. Among the various silane coupling agents, vinyl-silane is an interesting coupling agent for polymer-encapsulated inorganic nanohybrid particles because it consists of double bond (C=C) reactive in radical polymerization at the end of its molecule such as 3-(trimethoxysilyl) propyl methacrylate (γ-MPS) (Fink, 2013). γ-MPS is widely used as a coupling agent which is directly bonded to naked $nSiO_2$ surfaces via interfacial reaction between methoxy groups of γ-MPS and –OH groups of $nSiO_2$ and also introduce double bonds onto the $nSiO_2$ surfaces for the further polymerization (Figure 6). From our previous research (Chuayjuljit & Boonmahitthisud, 2010) (Boonmahitthisud & Chuayjuljit, 2011) had described the pre-treatment of $nSiO_2$ with the γ-MPS molecules carried out according to Dow Corning Procedure. 4 g of γ-MPS was added to an acidified water which prepared by adjusting the pH of water (800 ml) to be 3-4 with acetic

Table 1. The various applications of the encapsulation of inorganic nanoparticles by polymers

Application	Properties	Inorganic Nanoparticle (core)	Polymer (Shell)	Literature Reviews
Military equipment	Electromagnetic interference shielding	Multi-walled carbon nanotubes	Poly(vinyl alcohol)/ Poly(acrylic acid)	(Yun et al., 2010)
	Electromagnetic and microwave absorption	Fe_3O_4	Poly(3, 4-ethylenedioxythiophene)	(Zhou et al., 2011)
Safety	Flame retardancy	Melamine phosphate	Polyamide 66	(Xiong, Chen, Zhao, Wang, & Wang, 2012)
	Sun protection	Titanium dioxide	Chitosan	(Chung, Jung, Lee, & Lee, 2010)
Garments	Antibacteria	Silver	Poly(2-aminoethyl methacrylate)	(Liu et al., 2014)
	Antistatic	Zinc oxide	Polyaniline	(Anand, Rani, Saxena, Bhandari, & Dhawan, 2015)
Automotive	Anti-scratch	Zirconia	Poly(methyl methacrylate)	(Hu, Zhou, & Wu, 2009)
Electronics	Supercapacitor	Carbon nanofiber	Polyaniline	(Jang, Bae, Choi, & Yoon, 2005)
	Conductivity	Gold	Polyaniline	(Xing et al., 2009)
Optical devices	Magnetic resonance imaging	Fe_3O_4	Poly[2-(methacryloyloxy)ethyl phosphorylcholine]	(Yuan et al., 2006)
Agriculture	Slow releasing	Attapulgite	Alginate-g-poly(acrylic acid-co-acrylamide)/humic acid	(Ni, Liu, Lü, Xie, & Wang, 2010)
Biomedical	Antimicrobial activity	Silver nanoparticles	polystyrene	(Betancourt-Galindo et al., 2012)
	Drug delivery	Methotrexate-immobilized iron oxide nanoparticle	Poly(ethylene glycol)	(Kohler et al., 2006)
Cosmetics	UV screening and antioxidant	Titanium dioxide	Poly(ethylene oxide)-block-poly(2-(dimethylamino)ethyl methacrylate)-block-polystyrene	(Du & Sun, 2014)
Additives	Strength	Silica	Polystyrene	(Chuayjuljit & Boonmahitthisud, 2010)

acid (CH_3COOH) and then stirred the γ-MPS solution to hydrolyze and form a clear homogeneous solution for a minimum of 30 min. Afterwards, the dry silica (13.4 g) was then added to the as-prepared γ-MPS solution with stirring for further 30 min. Then, the as-prepared γ-MPS solution was dried in vacuum oven at 120^{0}C for 12 h to form a covalent linkage by a dehydration condensation reaction. It is necessary to characterize the effective pre-treatment of $nSiO_2$ surfaces by γ-MPS molecules via Fourier transform infrared (FT-IR) spectroscopy over a wavenumber of 450-4000 cm^{-1}. Figure 7 shows the FT-IR spectral differences between naked $nSiO_2$ and γ-MPS-treated $nSiO_2$. The characteristic peaks of naked $nSiO_2$ (Figure 7a) appeared at 1100, 805 and 470 cm^{-1} are assigned to the vibration absorption of Si-O-Si groups. In addition, the broad band in the region 3700-3200 cm^{-1} and the peak at 1640 cm^{-1} that indicate the O-H stretching vibration of –OH groups on the naked $nSiO_2$ surfaces and O-H bending of absorbed water. As the γ-MPS molecules were bonded to $nSiO_2$ surfaces, new characteristic peaks at 1490, 1710 and 2940 cm^{-1} attributed to C=C, C=O and C-H, respectively appeared on the spectrum

Table 2. The polymerization techniques for the preparation of the nanohybrid particles

Polymerization Technique	Inorganic Nanoparticle (core)	Polymer (Shell)	Literature Reviews
Solution	Multi-walled carbon nanotubes	Poly(4-vinylpyridine)	(Hong, Kim, Hong, Jung, & Shim, 2008)
	Silica	Poly(methyl methacrylate)	(Yang & Nelson, 2004)
Dispersion	Silica	Poly(methyl methacrylate)	(E. J. Park, Kim, Hwang, Park, & Lim, 2007)
	Titanium dioixe	Poly(styrene-co-divinylbenzene)	(Kim, Kim, Ahn, Kang, & Suh, 2004)
Aerosol	Silica and gold	Poly(methyl methacrylate)	(Sigmund et al., 2014)
Conventional emulsion	Titanium dioxide	Poly(methyl methacrylate)	(de Oliveira, da Silva, Alves, de Oliveira, & dos Santos, 2005)
	Silica	Poly (glycidyl methacrylate)	(Zhong, Jian, & Zhaoxia, 2006)
Inverse emulsion	Iron oxide	Polyacrylamide	(Deng, Wang, Yang, Fu, & Elaïssari, 2003)
	Nickel	Polyacrylamide	(Romio et al., 2013)
Miniemulsion	Silica	Poly(methyl methacrylate)	(Qi, Bao, Weng, & Huang, 2006)
	Magnetic nanoparticles and zinc phthalocyanine	Poly(methyl methacrylate)	(Feuser et al., 2015)
Microemulsion	Multi-walled carbon nanotubes	Polyaniline	(Reddy, Sin, Yoo, Sohn, & Lee, 2009)
	Graphene	Polystyrene	(Patole et al., 2010)
Differential microemulsion	Silica	Poly(styrene butadiene)	(Tancharernrat, Rempel, & Prasassarakich, 2014)
	Silica	Polybutadiene	(Tancharernrat, Rempel, & Prasassarakich, 2015)

of γ-MPS-treated $nSiO_2$ (Figure 7b). This result confirmed that the $nSiO_2$ surfaces were successfully pre-treated by γ-MPS.

SYNTHESIS OF POLYMER-ENCAPSULATED INORGANIC NANOHYBRID PARTICLES

Among various methods to prepare polymer particles including emulsion, dispersion, seed emulsion, precipitation and suspension polymerization, emulsion polymerization provides simple and effective way to synthesize polymers and adapt for polymer-encapsulated inorganic nanohybrid particles with small particles in nanoscale (below 100 nm) as shown in Figure 8 (Daesun Hong & Chang, 2007). Emulsion polymerization is the heterogeneous free radical polymerization method which can be divided into 3 types based on the sizes of the droplets:

1. Conventional emulsion polymerization;
2. Mini-emulsion polymerization; and
3. Microemulsion polymerization.

Figure 5. The reactive mechanism of silane coupling agents with inorganic particles
Source: Shin-Etsu, 2011

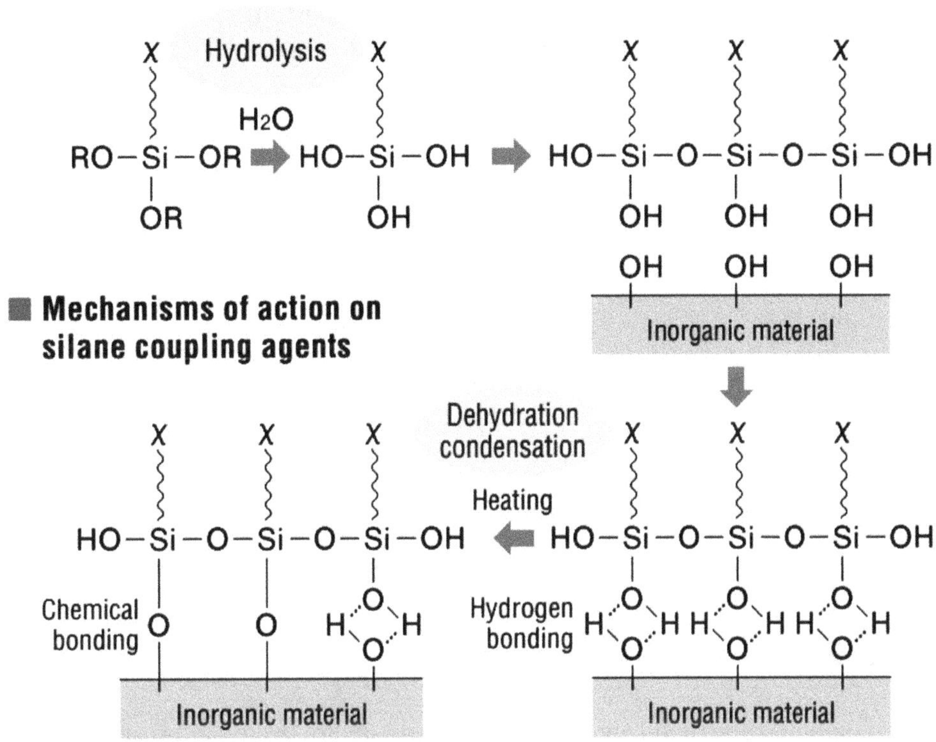

Figure 6. The surface of nSiO$_2$ pre-treated by γ-MPS
Source: Chuayjuljit & Boonmahitthisud, 2010

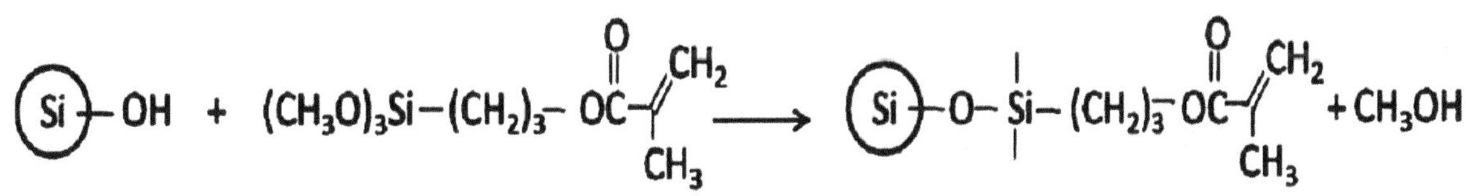

Figure 7. FT-IR technique shows the spectra of (a) naked nSiO$_2$ and (b) γ-MPS-treated nSiO$_2$
Source: Chuayjuljit & Boonmahitthisud, 2010

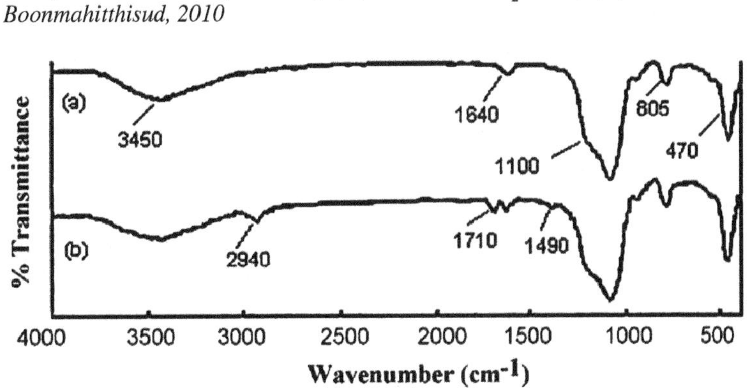

Figure 8. The average diameter size of particles from various polymerization
Source: Daesun Hong & Chang, 2007

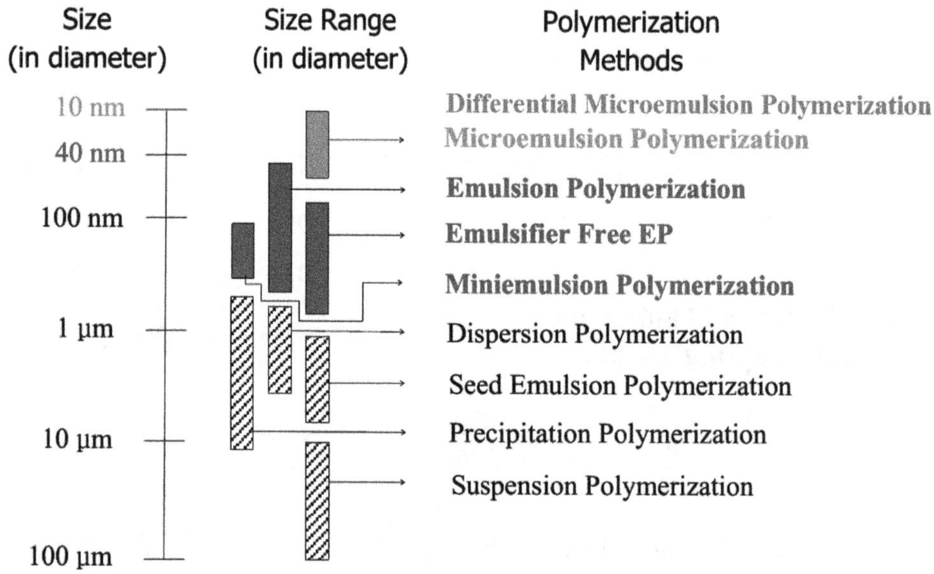

Conventional Emulsion Polymerization

A typical emulsion polymerization consists of monomer, initiator, water and surfactant. The polymerization is initiated by the addition of monomer, initiator and surfactant in the system. Most of monomer molecules are in the monomer droplet which 1±10 μm in diameter dispersed in water phase with the aid of surfactant and monomer swollen in micelle which 5-10 nm in diameter. Therefore, the particle nucleation can take place in 3 zones:

1. In monomer swollen in micelle or micellar nucleation
2. In aqueous or homogeneous nucleation and
3. In monomer droplet or droplet nucleation (Yamak, 2013).

However, the amount of polymerization occurring in the monomer droplets is regarded as being a very minor proportion of the whole because of their small surface area for diffusing of the free-radicals as shown in Figure 9. In the initial step, the free-radicals which dissociate from initiators react with a small amount of dissolved monomer to form oligo-radical. After that the single-radical or oligo-radical diffuse into monomer-swollen in micelle and start polymerizing. During polymerization, the growing polymer particles are then supplied by the monomer molecules from the monomer droplets by diffusion through the water phase. Finally, the polymerization is complete and the system now comprises a dispersion of polymer particles stabilized with the molecules of the original surfactant. However, the conventional emulsion polymerization provides polymer with average diameter more than 100 nm at the beginning of the system. Ding et al. (Ding, Zhao, Liu, Zhang, & Wang, 2004) had pre-treated $nSiO_2$ with oleic acid and then encapsulated $nSiO_2$ with PS via *'in situ'* emulsion polymerization. The obtained PS-encapsulated $nSiO_2$ (PS-$nSiO_2$) nanohybrids had an average particle from 180 to 250 nm.

Figure 9. The emulsion system before polymerization

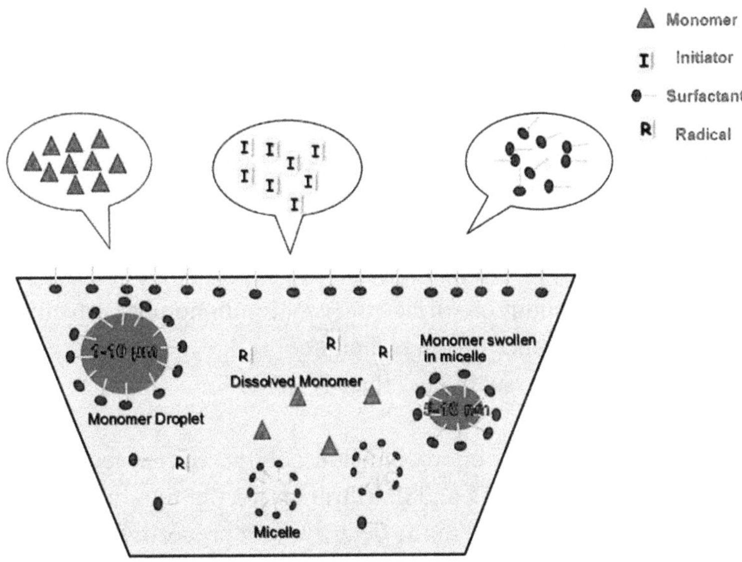

Mini-Emulsion Polymerization

Mini-emulsion polymerization consists of distributed monomer droplets in water phase by applying vigorous shear by means of ultrasonication or homogenizer using the combination of a classical surfactant and a co-surfactant such as a long chain fatty alcohol or alkane (cetyl alcohol or hexadecane). Polymerization starts within small monomer droplets with diameter in the range of 50 to 500 nm (Bradley & Grieser, 2002) . In the initial step, free-radicals diffuse into the small monomer droplets and start polymerizing as droplet nucleation. Then, the growing polymer particles are then supplied by the monomer molecules from within the polymerizing particles. The mini-emulsion polymerization provides polymer with average diameter same as the initial monomer droplets. Betancourt-Galindo et al. (Betancourt-Galindo et al., 2012) had presented the encapsulation of silver nanoparticles (nAg) with PS by means of *'in situ'* mini-emulsion polymerization. The nAg were pre-treated with 3-aminopropyl-trimethoxysilane (APTMS). The obtained PS-encapsulated nAg exhibited the spherical microsphere with average particle from 50 to 100 nm.

Microemulsion Polymerization

Unlike the opaque and milky conventional emulsion and mini-emulsion polymerization, microemulsion polymerization is thermodynamically stable, isotropic and optically transparent or translucent. The system can be formed spontaneously by mixing two immiscible phases, oil and water, due to the presence of high amounts of surfactant and co-surfactant such as a low molar mass alcohol (pentanol or hexanol) without applying any vigorous shear. The initial system consists of small droplets (monomer droplets and empty micelles) with average diameter below 1-10 nm. These small droplets exhibit large interfacial area (~10^5 m^2 dm^{-3}), hence, these droplets are effective in capturing free radicals (Al-Sabagh, Kabel, Noor El-Din, & Elsharaky, 2012). Therefore, the propagation occurs in these droplets. Until the termination, the monomer droplets are changed to be polymer micelles. Moreover, the unreacted monomer from polymer micelles or unreacted monomer droplets can diffuse to other monomer droplets or empty micelles. Finally, the reaction comprises polymer with small average diameter below 50 nm and high molecular weight in the range of 10^6-10^7 g mol^{-1}. Palkovits et al. (Palkovits et al., 2005) had synthesized PMMA-encapsulated $nSiO_2$ nanohybrids via reverse water-in-oil microemulsion polymerization with 26 nm in diameter. Therefore, the microemulsion polymerization is the most common method used to synthesize polymer particles with average diameter usually below 50 nm. However, the method has two major drawbacks as follows: (He, Pan, & Rempel, 2003)

1. Requirement much higher amount of surfactant (usually monomer/surfactant < 1), compared to the conventional emulsion polymerization technique.
2. Providing low polymer content, usually less than 10 wt%.

These drawbacks have obstructed the microemulsion polymerization technique from up-scaling to industrial level the high cost of surfactants and diafiltration for the removal of surfactants. Moreover, the excess surfactant in the system can lead to the deterioration properties of polymer.

INNOVATIVE POLYMERIZATION METHOD TO SYNTHESIZE POLYMER-ENCAPSULATED $nSiO_2$ NANOHYBRID PARTICLES AND ITS CHARACTERIZATION

As a result of the major drawbacks of microemulsion polymerization, a differential microemulsion polymerization method presents an attempt to significantly reduce the amount of surfactant and increase the polymer content in the microemulsion polymerization method for the preparation of further smaller polymer particles with average diameter below 50 nm. The key factor for overcoming these drawbacks is the prolongation of monomer addition resulting in the high weight ratio of monomer/surfactant throughout the reaction and also obtaining small nanoparticles. The polymerization is initiated by heating the mixture of initiator and surfactant in the reactor. Then, monomer is added to the reactor by gentle dropping throughout the reaction. After that the reaction reaches an aging period for complete polymerization of unreacted monomer. From our previous research (Chuayjuljit & Boonmahitthisud, 2010) (Boonmahitthisud A. & Chuayjuljit, 2011), the nanohybrid particles of PS-$nSiO_2$ were successfully prepared via *'in situ'* differential microemulsion polymerization. The procedure consists of two steps;

1. $nSiO_2$ surfaces were pre-treated with γ-MPS coupling agent as describes above;
2. Both of the polymerization and the encapsulation of styrene monomer onto the γ-MPS-treated $nSiO_2$ surfaces were simultaneously operated (Figure 10) by using sodium dodecylsulfate (SDS) and 2,2'-azobisisobutyronitrile (AIBN) as surfactant and oil-soluble initiator, respectively.

Table 3 provides the *'in situ'* differential microemulsion polymerization recipe for the preparation of PS-$nSiO_2$ nanolatex. The γ-MPS-treated $nSiO_2$, SDS, AIBN and water were mixed in a four-necked flask equipped with a reflux condenser, nitrogen gas inlet and dropping funnel for monomer feeding. After, the reaction temperature was raised to 70°C with stirring 250 rpm using a magnetic stirrer under the nitrogen atmosphere, the styrene monomer was gently dropped into the reactor over 1.5 h. Then, the reaction was maintained at 70⁰C with constant agitation for an additional hour. The resultant nanohybrids were calculated solid content (*%S*) by the gravimetric method from following equation:

$$\%\mathrm{S} = \frac{W1}{W2} \times 100 \quad (1)$$

where $W1$ and $W2$ are the weights of dried nanohybrids and nanolatex, respectively. The average diameter of the resultant nanohybrid particles was determined using dynamic light scattering (DLS). Moreover, it was necessary to confirm the encapsulation of $nSiO_2$ with PS by determining the functional groups of naked $nSiO_2$, γ-MPS-treated $nSiO_2$ and PS-$nSiO_2$ nanohybrids and the morphology of and PS-$nSiO_2$ nanohybrid particles using FT-IR and transmission electron microscopy (TEM), respectively.

Table 3. Formulation of the 'in situ' differential microemulsion polymerization for the preparation of PS-$nSiO_2$ nanolatex used in this research

Materials	Amount (g)
γ-MPS-treated $nSiO_2$	0.4
SDS	8
AIBN	0.12
Distilled water	60

Source: Chuayjuljit & Boonmahitthisud, 2010

Figure 10. The encapsulation of $nSiO_2$ by PS through γ-MPS as a silane coupling agent

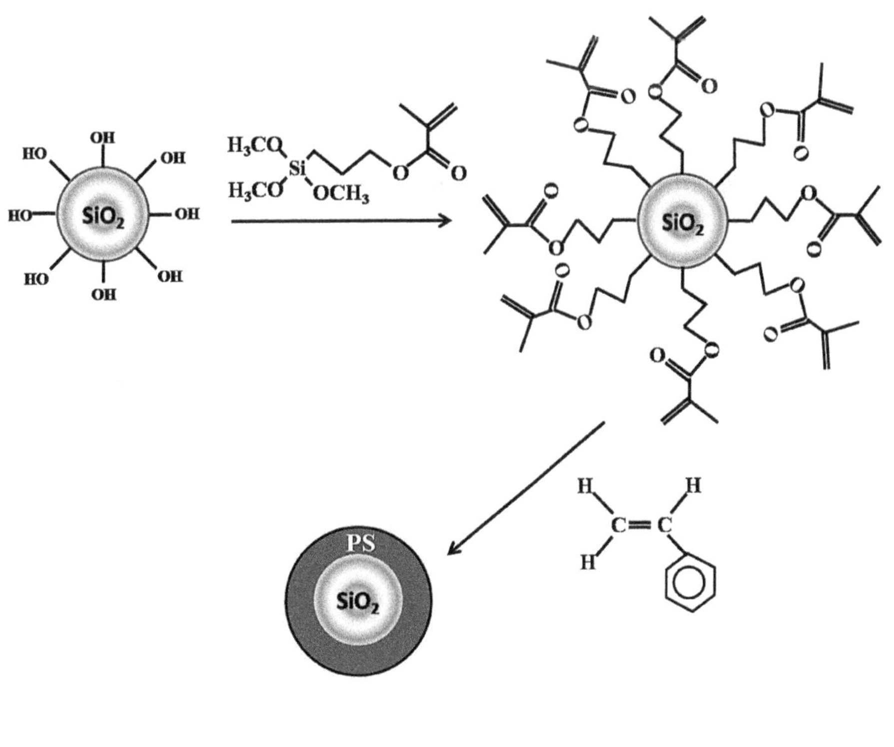

As reported from our earlier results, the resultant PS-nSiO_2 nanohybrids had a calculated solid content of about 22%. The average diameter of the nanohybrid particles was determined to be about 40 nm whereas the naked nSiO_2 was 10-20 nm in diameter. The TEM micrograph of PS-nSiO_2 nanohybrid particles in Figure 11 illustrates the spherical core-shell structure with nSiO_2 (in dark phase) as a core and PS (in grey phase) as a dense shell. Moreover, in order to demonstrate the successful pre-treatment and encapsulation of PS onto nSiO_2 surfaces, Figure 12 shows the FT-IR spectra of naked nSiO_2, γ-MPS-treated nSiO_2 and PS-nSiO_2 nanohybrids. As describe above, the nSiO_2 surfaces were pre-treated by γ-MPS to introduce double bonds on their surface. The FT-IR spectrum of PS-nSiO_2 nanohybrids has absorption peaks at 3025, 1610, 2920 and 2849, 1492 and 1452, 906 and 697 cm^{-1} assigned to C-H (aromatic), C=C (aromatic), -CH_2-CH_2, -C_6H_5 and –CH= (aromatic), respectively which are the characteristic peaks of PS. Moreover, the presence of absorption bands characteristic of nSiO_2 (1100, 805 and 470 cm^{-1}) are also observed in the spectrum of the PS-nSiO_2 nanohybrids. It is supported that PS have been achieve to encapsulate onto a silica nanoparticles via *'in situ'* differential microemulsion polymerization with the aid of γ-MPS coupling agent chemically anchored to the nSiO_2.

PROPERTIES OF NR AND RUBBER BLEND NANOCOMPOSITES REINFORCED WITH PS-nSiO_2 NANOHYBRID PARTICLES

NR is well-known as an interesting bio-based polymer which possesses the numerous outstanding properties. However, reinforcing nanofillers are necessarily added into NR in most cases in order to develop the appropriate properties for a wide range of applications. Polymer-encapsulated inorganic nanohybrid particles can provide not only the synergistic properties from polymer and inorganic nanoparticles but also the well-dispersion of nanoparticles in the rubber matrix. Therefore, the polymer-encapsulated inorganic nanohybrid particles have been chosen as the alternative reinforcing nanofillers for NR and rubber blend.

Figure 11. TEM image of the spherical core-shell structure of PS-nSiO_2 nanohybrid particles prepared via 'in situ' differential microemulsion polymerization
Source: Chuayjuljit & Boonmahitthisud, 2010

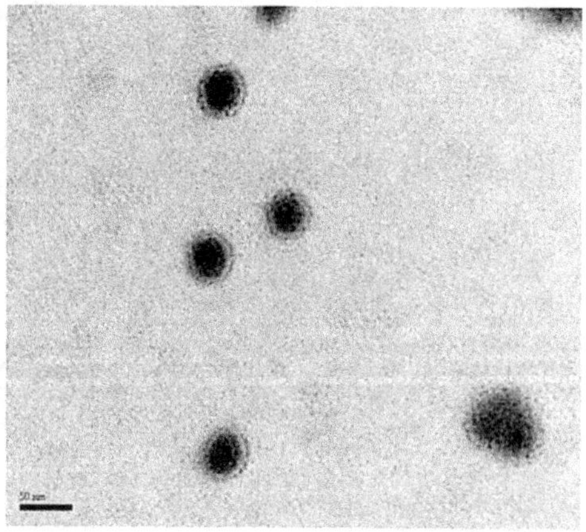

Figure 12. FT-IR spectra of (a) naked $nSiO_2$, (b) γ-MPS-treated $nSiO_2$ and (c) PS-$nSiO_2$ nanohybrids
Source: Chuayjuljit & Boonmahitthisud, 2010

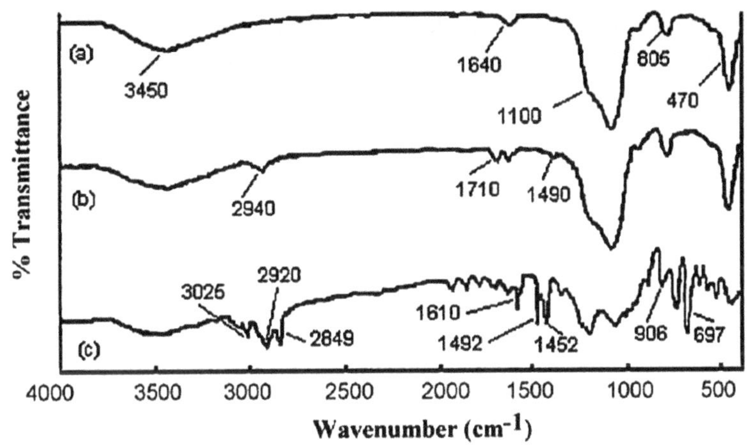

NR Reinforced with PS-Encapsulated $nSiO_2$ Nanohybrid Particles (Chuayjuljit & Boonmahitthisud, 2010)

The resultant PS-$nSiO_2$ nanohybrid particles via *'in situ'* differential microemulsion polymerization were used as the reinforcing nanofiller in NR latex. According to the *'in situ'* differential microemulsion polymerization method, NR latex can be mixed directly with the nanohybrid latex by latex compounding method. NR latex was first prepared as NR latex compound with curing ingredients including potassium oleate solution (20 wt%) 0.2 phr, potassium hydroxide solution (10 wt%) 0.5 phr, aqueous dispersions of sulfur (50 wt%) 1.5 phr, zinc oxide (50 wt%) 1.0 phr, zinc diethyl dithiocarbamate (50 wt%) 1.0 phr, CPL antioxidant (50 wt%) 1.0 phr and calcium carbonate (70 wt%) 30 phr by stirring at 150 rpm with a mechanical stirrer for 30 min. The as-prepared NR latex compound was then mixed with the nanohybrid latices at 3, 5, 7 and 9 parts per hundred of dried rubber (phr) whilst stirring for 3 h to ensure homogenization of the ingredients. The obtained NR nanocomposite latices were then cast into the thin sheet on a glass mold (20cm × 20cm × 0.15cm), air dried for 24 h and subsequently cured in an oven at 110°C for 3 h. The specimens were cut from the cured NR nanocomposite sheets for further characterization. The mechanical properties were evaluated by means of the tensile properties using a universal testing machine and dynamic mechanical properties using dynamic mechanical analyzer (DMA) under shear mode at a frequency of 1 Hz with a constant heating rate of 3°C/min over a temperature range of -100 to 150°C. The fracture surfaces of the neat NR and NR nanocomposites were also observed using a scanning electron microscopy (SEM) at a magnification of 500×. The specimens were fractured after soaking in liquid nitrogen. Moreover, the flammability of the neat NR and NR nanocomposites was estimated by the value of limited oxygen index (LOI) and the flame spread rate using a Limiting Oxygen Indexer and an Atlas 45° automatic flammability tester, respectively.

- Mechanical properties of the neat NR and NR nanocomposites including PS-$nSiO_2$ nanohybrid particles

The tensile properties of the neat NR and NR nanocomposites were determined in terms of tensile strength, modulus at 300% strain and elongation at break as shown in Figure 13. It can be seen in Figure 13a that there is an improvement in the tensile strength (a 20% increase as compared to the neat NR) with the incorporation of the PS-nSiO_2 nanohybrids at 3 phr. It is suggested that PS-nSiO_2 nanohybrids well-dispersed in the NR matrix and also can transfer the stress at the interfacial of the nanohybrids and NR. However, further increase the amounts of PS-nSiO_2 nanohybrids (i.e. at 5, 7 and 9 phr) led to the decrease in tensile strength, owing to the self-aggregation and poor-dispersion of the nanohybrids in the NR matrix. As expected, the inclusion of PS-nSiO_2 nanohybrids increased the modulus at 300% strain (Figure 13b) of the NR nanocomposites in a dose-dependent manner. It may be caused by the restriction of the NR chain movement and deformability during tensile loading by the stiffness PS-nSiO_2 nanohybrid particles. Moreover, the elongation at break (Figure 13c) was gradually reduced by the inclusion of the PS-nSiO_2 nanohybrids, which was a consequence of the restriction of NR chain mobility by the stiffness nanohybrids, as mentioned above. Figure 14 shows the dynamic mechanical properties for the neat NR and NR nanocomposites in term of the storage modulus (E') and loss tangent (Tan δ) versus temperature. In general, the addition of the stiffness particles will increase the E' of the composites (Figure 14a) (Boonmahitthisud & Chuayjuljit, 2012). In this case, the E' of the NR nanocomposites markedly decreased with increasing the PS-nSiO_2 nanohybrid contents, which indicates that the NR nanocomposites have some elastic response near their glass transition temperature (T_g) owing to the formation of semi-interpenetrating networks (semi-IPNs) nanostructure in the nanocomposites. In this semi-IPNs nanostructure, the uncrosslinked PS phase of the PS-nSiO_2 nanohybrids was entrapped in the crosslinked NR networks resulting in an increase in the free volume and also in the mobility of NR molecules. Moreover, the peaks for Tan δ (Figure 14b) within the temperature ranges of -70 to -10°C and 100 to 130°C have been attributed to the T_g of NR and PS at about -50°C and 110°C, respectively. The T_g of the NR nanocomposites shifted to lower temperature and also the peak height of Tan δ for the nanocomposites decreased with increase in the PS-nSiO_2 nanohybrids contents, indicated the elastic behavior of the nanocomposites (J.-G. Park, Ye, Topp, Misra, & Spencer, 2009). This result was in accordance with that of the E' observation.

- Morphology of the neat NR and NR nanocomposites including PS-nSiO_2 nanohybrid particles

The SEM micrographs of the neat NR and NR nanocomposites are illustrated in Figure 15. The SEM micrographs of the NR nanocomposites with various PS-nSiO_2 nanohybrid contents exhibited the quite rougher fracture surfaces than that of the neat NR. The surface roughness may respond to the elastic behavior induced by the semi-IPNs nanostructure, coinciding with the results obtained from the DMA.

- Flammability of the neat NR and NR nanocomposites including PS-nSiO_2 nanohybrid particles

The LOI values and flame spread rate of the neat NR and NR nanocomposites are shown in Figure 16 and 17, respectively. The higher LOI value and the lower flame spread rate of a given material, the lower is its flammability. The addition of the PS-nSiO_2 nanohybrids improve the flammability of the nanocomposites due to their higher LOI values and lower flame spread rate, as compared to the neat NR. However, all the specimens had LOI values below 20.8 which not enough to achieve flame retardancy (Ramesan, 2004). It is indicated that the PS-nSiO_2 nanohybrids can be used as a flame retardant in the NR products.

Figure 13. The tensile properties of the neat NR and NR nanocomposites filled with different PS-nSiO$_2$ loadings in terms of (a) tensile strength, (b) modulus at 300% strain and (c) elongation at break
Source: Chuayjuljit & Boonmahitthisud, 2010

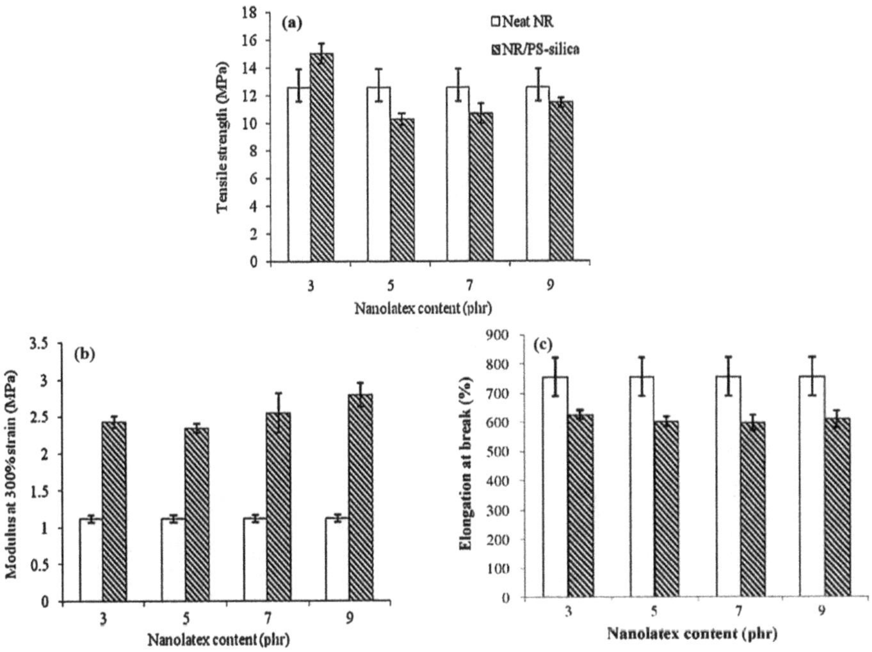

In summarized, the addition of an appropriate content of the PS-nSiO$_2$ nanohybrids into the NR matrix apparently improved the tensile strength, modulus at 300% strain and flammability of the obtained nanocomposites based on NR with PS-nSiO$_2$ nanohybrid particles. Moreover, the result from DMA exhibited some elastic responses of NR molecules near T_g increased with the incorporation of the PS-nSiO$_2$ nanohybrid particles due to the semi-IPNs nanostructure.

Rubber Blend Reinforced with PS-Encapsulated nSiO$_2$ Nanohybrid Particles (Boonmahitthisud A. & Chuayjuljit, 2011)

The obtained PS-nSiO$_2$ nanohybrid particles via *'in situ'* differential microemulsion polymerization were used as the reinforcing nanofiller for the rubber blend of NR and styrene butadiene rubber (SBR) at the ratio of 80NR/20SBR. The effects of the PS-nSiO$_2$ nanohybrid particles compared with those of nSiO$_2$ on mechanical properties and thermal stability were investigated. NR was first blended with SBR at dry weight ratio of 80/20 in the latex state by stirring at 150 rpm with a mechanical stirrer for 30 min. The 80NR/20SBR latex was then mixed with the curing agent including potassium oleate solution (10 wt%) 0.2 phr, potassium hydroxide solution (10 wt%) 0.5 phr, aqueous dispersions of sulfur (50 wt%) 1.5 phr, zinc oxide (50 wt%) 1.0 phr, zinc diethyl dithiocarbamate (50 wt%) 1.0 phr and Wingstay L antioxidant (50 wt%) 1.0 phr and also mixed with either nSiO$_2$ or the PS-nSiO$_2$ nanohybrids at the content of 0.1, 0.2, 0.3 and 0.4 phr, whilst stirring for an additional half hour. The obtained rubber blend nanocomposite latices were then cast into the thin sheet on a glass mold (20cm × 20cm × 0.15cm), air dried for 24 h and subsequently cured in an oven at 80^{0}C for 3 h. The specimens were cut from the cured rubber

Figure 14. The dynamic mechanical properties of the neat NR and NR nanocomposites filled with different PS-nSiO$_2$ loadings in terms of (a) storage modulus (E') and (b) loss tangent (Tan δ) versus temperature
Source: Chuayjuljit & Boonmahitthisud, 2010

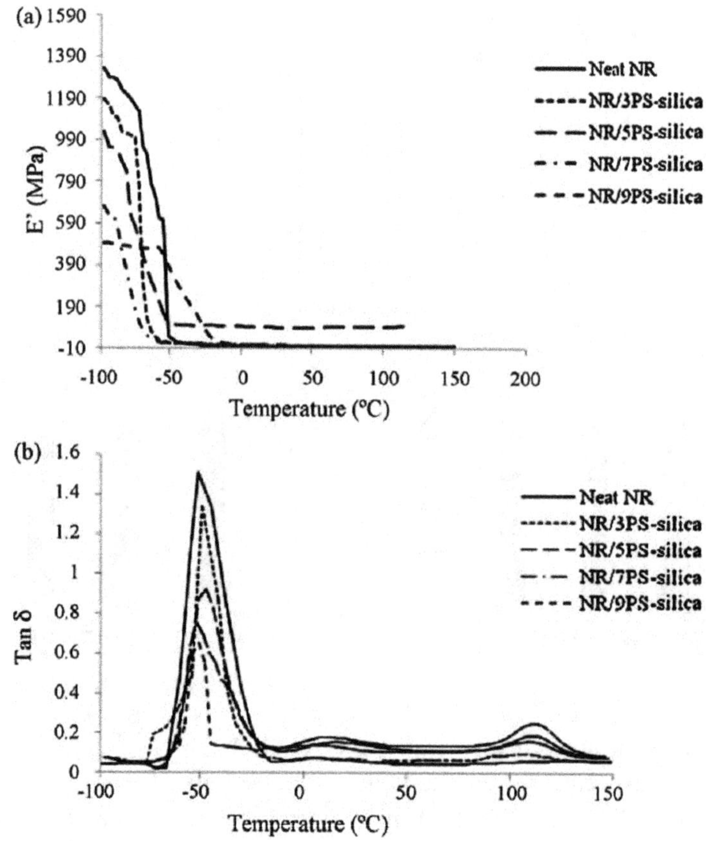

blend nanocomposite sheets for further characterization. The mechanical properties were evaluated by means of the tensile properties using a universal testing machine. Moreover, the thermal stability was determined using a thermogravimetric analyzer (TGA) at the temperature range of 50 to 1000°C with the heating rate of 20°C/min under a nitrogen atmosphere.

- Mechanical properties of the neat 80NR/20SBR and its nanocomposites including either nSiO$_2$ or PS-nSiO$_2$ nanohybrid particles

The tensile properties of the neat 80NR/20SBR and its nanocomposites with different loadings of either nSiO$_2$ or PS-nSiO$_2$ nanohybrids are shown in Figure 18. The tensile strength (Figure 18a) of the neat 80NR/20SBR was about 19 MPa. The addition of either nSiO$_2$ or PS-nSiO$_2$ nanohybrids exhibited higher tensile strength as compared to the neat 80NR/20SBR. With the addition of either nSiO$_2$ or PS-nSiO$_2$ nanohybrids at 0.2 phr of the 80NR/20SBR nanocomposites, an optimal tensile strength of about 21 and 24 MPa were observed, respectively. However, the tensile strength of the 80NR/20SBR nanocomposites with 0.3 and 0.4 either nSiO$_2$ or PS-nSiO$_2$ nanohybrid loadings were found to decrease as a consequence of the self-aggregation and poor-dispersion of the nanoparticles in the rubber blend matrix

Figure 15. SEM micrographs of the neat NR and NR nanocomposites filled with different PS-nSiO_2 loadings
Source: Chuayjuljit & Boonmahitthisud, 2010

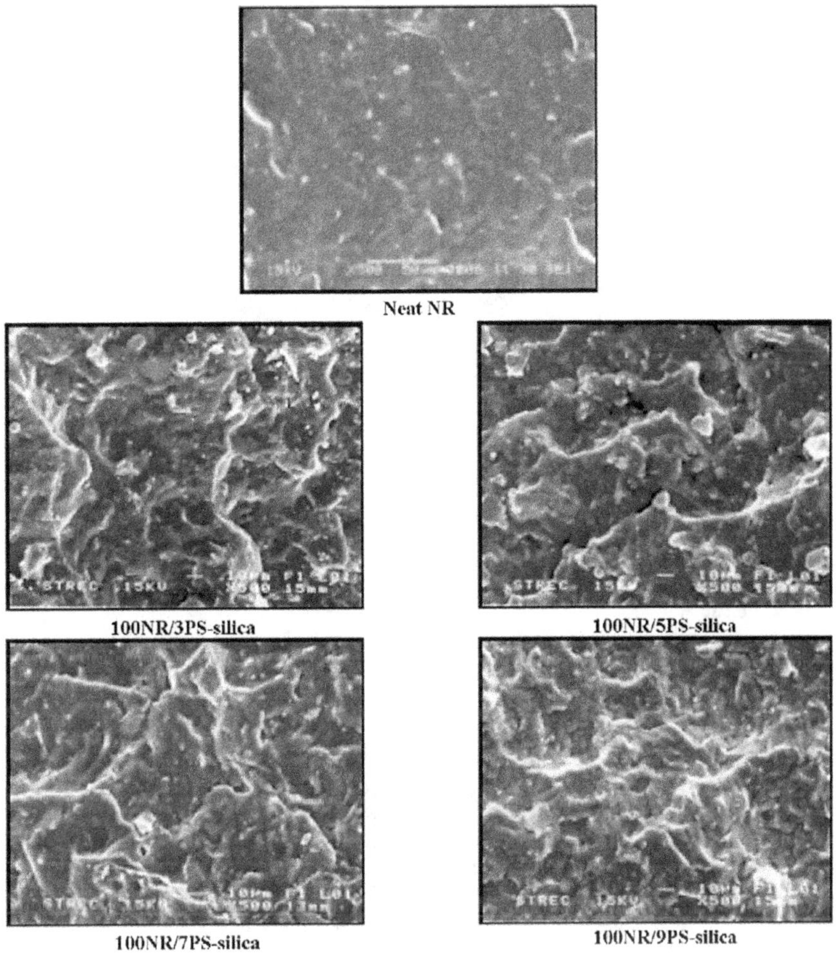

that led to a reduced stress transfer across the nanoparticles and rubber blend matrix interphase. From the figure, it can be observed that the PS-nSiO_2 nanohybrids exhibited reinforcing efficiency superior to nSiO_2 with the same nanofiller loadings. This may be due to the stronger interfacial adhesion between rubber blend matrix and the PS-nSiO_2 nanohybrids induced by PS molecules covered on the nSiO_2 surfaces that could increase the compatibility and avoid phase separation between nSiO_2 and rubber blend matrix. Figure 18b shows that all rubber blend nanocomposites exhibited improved modulus at 300% strain as compared to the neat 80NR/20SBR in a dose-dependent manner, owing to the restriction of the rubber chain movement by the stiffness nanoparticles. However, the rubber blend nanocomposites filled with the PS-nSiO_2 nanohybrids have lower modulus at 300% strain than those filled with the nSiO_2. This may be caused by the flexibility of PS molecules covered on the nSiO_2 surfaces that decrease the stiffening efficiency of the nSiO_2. As a result, the restriction of the rubber chain movement by addition of the nanoparticles and thus the flexibility of the rubber blend nanocomposites were reduced. As expected, all rubber blend nanocomposites exhibit lower elongation at break than that of the neat 80NR/20SBR.

Figure 16. LOI values of the neat NR and NR nanocomposites filled with different PS-nSiO$_2$ loadings
Source: Chuayjuljit & Boonmahitthisud, 2010

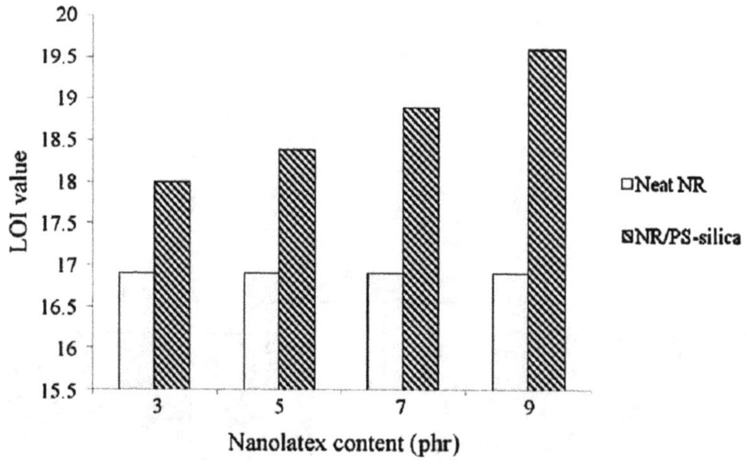

Figure 17. Flame spread rate of the neat NR and NR nanocomposites filled with different PS-nSiO$_2$ loadings
Source: Chuayjuljit & Boonmahitthisud, 2010

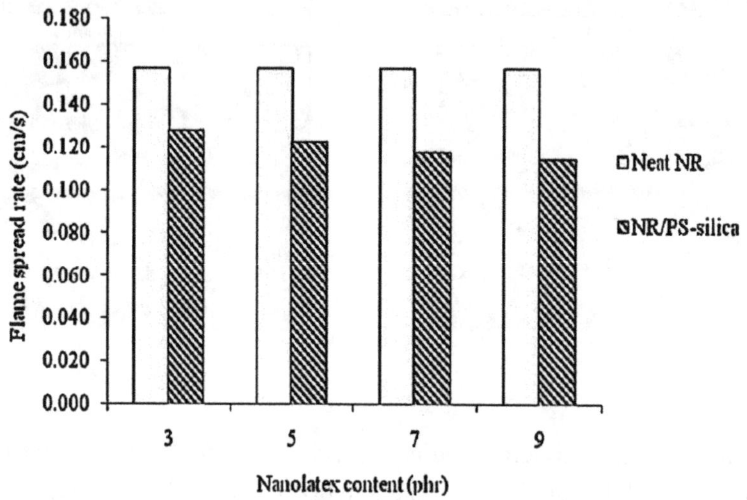

However, the elongation at break of the PS-nSiO$_2$-filled rubber blend nanocomposites is largely retained, giving a reduction of the elongation range of 1.2-16.4%, relative to the neat 80NR/20SBR.

- Thermal stability of the neat 80NR/20SBR and its nanocomposites including either nSiO$_2$ or PS-nSiO$_2$ nanohybrid particles

TGA was performed in order to evaluate the thermal stability of the neat 80NR/20SBR and its nanocomposites with either nSiO$_2$ or PS-nSiO$_2$ nanohybrids. Table 4 summarizes the TGA results in terms of onset temperature (T_{onset}) and end set temperature ($T_{end\ set}$). It can be observed that the inclusion of either nSiO$_2$ or PS-nSiO$_2$ nanohybrids into the 80NR/20SBR improved the thermal stability of all the

Figure 18. The tensile properties of the neat 80NR/20SBR and its nanocomposites filled with different either $nSiO_2$ or PS-$nSiO_2$ loadings in terms of (a) tensile strength, (b) modulus at 300% strain and (c) elongation at break
Source: Boonmahitthisud & Chuayjuljit, 2011

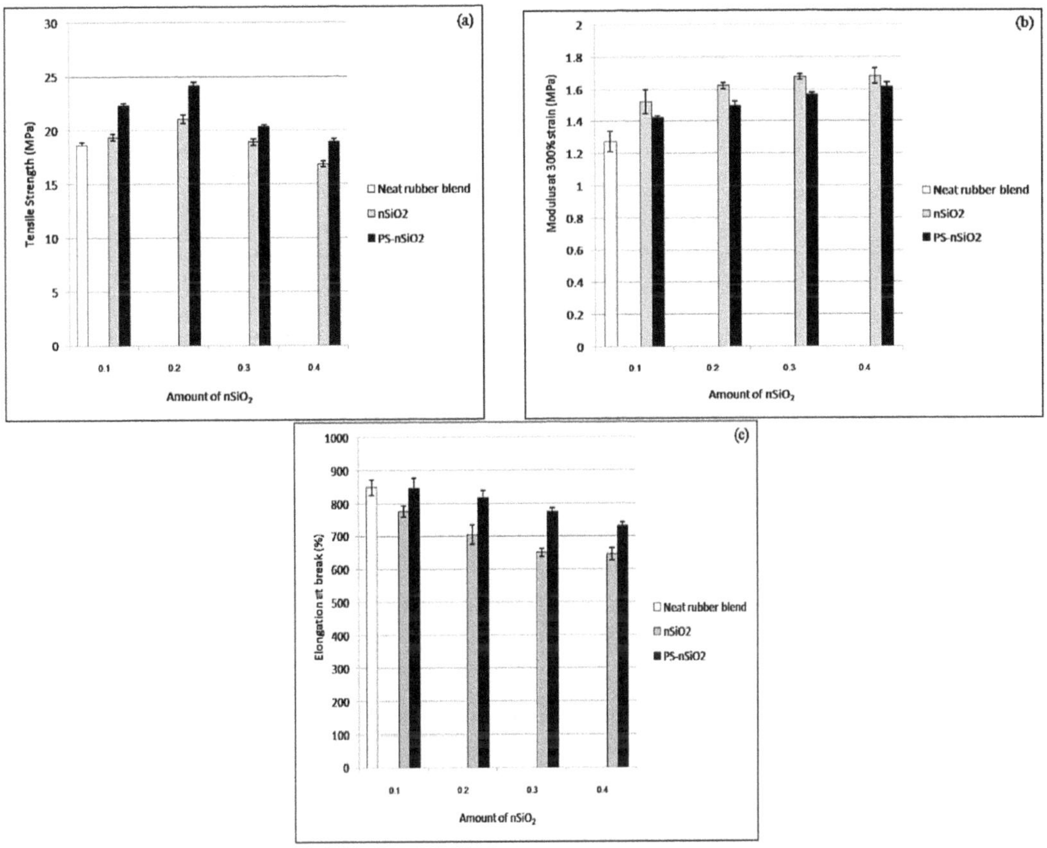

resultant nanocomposites in a dose-dependent manner compared to the neat 80NR/20SBR, as indicated by shifting towards higher T_{onset} and $T_{end\ set}$ in the range of 1-10^{0}C

In conclude, the inclusion of either $nSiO_2$ or PS-$nSiO_2$ nanohybrids enhanced the tensile strength, modulus at 300% strain and thermal stability of the 80NR/20SBR. A similar optimal tensile strength of the rubber blend nanocomposites was achieved with 0.2 phr both of $nSiO_2$ and PS-$nSiO_2$ nanohybrids, and so the PS-$nSiO_2$ nanohybrids exhibited reinforcing efficiency superior to $nSiO_2$ with the same nanofiller contents.

CONCLUSION

In this chapter, the use of PS-$nSiO_2$ nanohybrid particles as the alternative reinforcing nanofillers for NR and rubber blend nanocomposites has been summarized. We specially focus on the synthesis and characterization of PS-$nSiO_2$ nanolatex by means of *'in situ'* differential microemulsion polymerization, which is the innovation polymerization method to synthesize polymer particles with small average

Table 4. TGA results of the neat 80NR/20SBR and its nanocomposites filled with different either $nSiO_2$ or PS-$nSiO_2$ loadings

Sample	T_{onset} (°C)	$T_{end\ set}$ (°C)
Neat 80NR/20SBR	360	419
80NR/20SBR/0.1$nSiO_2$	361	422
80NR/20SBR/0.2$nSiO_2$	363	425
80NR/20SBR/0.3$nSiO_2$	364	426
80NR/20SBR/0.4$nSiO_2$	365	427
80NR/20SBR/0.1PS-$nSiO_2$	363	426
80NR/20SBR/0.2PS-$nSiO_2$	364	428
80NR/20SBR/0.3PS-$nSiO_2$	365	430
80NR/20SBR/0.4PS-$nSiO_2$	366	429

Source: Boonmahitthisud & Chuayjuljit, 2011

diameter below 50 nm using a much smaller surfactant content than that used in the microemulsion polymerization method. For improving the encapsulation of PS onto $nSiO_2$, the pre-treatment of $nSiO_2$ surfaces is necessary by using γ-MPS silane coupling agent. The encapsulation of $nSiO_2$ with PS has been proved by using FT-IR and TEM techniques. From the results, it was confirmed that the *'in situ'* differential microemulsion polymerization is the effective method to prepare the core-shell nanoparticles of PS-$nSiO_2$ with an average diameter of 40 nm. Furthermore, the incorporation of resultant PS-$nSiO_2$ nanohybrid particles into both NR and rubber blend (80NR/20SBR) matrices provides an enhancement in mechanical properties, thermal stability and flammability of NR and 80NR/20SBR nanocomposites. Moreover, the rubber nanocomposites filled with PS-$nSiO_2$ nanohybrid particles also show the elasticity of rubber molecules at low temperature (near T_g) as a consequence of the semi-IPNs nanostructure. Therefore, the addition PS-$nSiO_2$ nanohybrid particles as a reinforcing nanofiller to rubber nanocomposites great promise to not only improve the dispersibility of $nSiO_2$ in rubber matrix, which leads to an enhancement of the existing properties of the rubber nanocomposites but also provide specific properties that would expand applications and increase value addition of NR.

ACKNOWLEDGMENT

I would like to thank Professor Dr. Kobayashi Takaomi for giving me this opportunity to write a chapter and Associated Professor Saowaroj Chuayjuljit who has contributed to my knowledge. I also thank the Department of Materials Science, Faculty of Science, Chulalongkorn University for financial, material and instrument support.

REFERENCES

Akbarzadeh, A., Samiei, M., & Davaran, S. (2012). Magnetic nanoparticles: Preparation, physical properties, and applications in biomedicine. *Nanoscale Research Letters*, *7*(1), 1–13. doi:10.1186/1556-276X-7-144 PMID:22348683

Al-Sabagh, A. M., Kabel, K. I., Noor El-Din, M. R., & Elsharaky, E. A. (2012). Synthesis of polyalkyl-acrylate nanolatexes by microemulsion polymerization method. *Egyptian Journal of Petroleum*, *21*(2), 81–87. doi:10.1016/j.ejpe.2012.11.001

Anand, A., Rani, N., Saxena, P., Bhandari, H., & Dhawan, S. K. (2015). Development of polyaniline/zinc oxide nanocomposite impregnated fabric as an electrostatic charge dissipative material. *Polymer International*, *64*(9), 1096–1103. doi:10.1002/pi.4870

Ansarifar, A., Lim, H. P., & Nijhawan, R. (2004). Assessment of the effect of a bifunctional organosilane on the bound rubber and properties of some natural rubber compounds. *International Journal of Adhesion and Adhesives*, *24*(1), 9–22. doi:10.1016/S0143-7496(03)00095-2

Betancourt-Galindo, R., Cabrera Miranda, C., & Puente Urbina, B. A. (2012). Casta, #xf1, eda-Facio, A., . . . ndez, O. S. (2012). Encapsulation of Silver Nanoparticles in a Polystyrene Matrix by Miniemulsion Polymerization and Its Antimicrobial Activity. *ISRN Nanotechnology*, *5*. doi:10.5402/2012/186851

Bokobza, L. (2007). Multiwall carbon nanotube elastomeric composites: A review. *Polymer*, *48*(17), 4907–4920. doi:10.1016/j.polymer.2007.06.046

Boonmahitthisud A., & Chuayjuljit, S. (2011). Rubber blend of 80/20 NR/SBR reinforced with nanosilica and PS-encapsulated nanosilica. *Materials Science Forum*, *695*, 332-335.

Boonmahitthisud, A., & Chuayjuljit, S. (2012). NR/XSBR Nanocomposites with Carbon Black and Carbon Nanotube Prepared by Latex Compounding. *Journal of Metals, Materials and Minerals*, *22*, 77–85.

Bradley, M., & Grieser, F. (2002). Emulsion Polymerization Synthesis of Cationic Polymer Latex in an Ultrasonic Field. *Journal of Colloid and Interface Science*, *251*(1), 78–84. doi:10.1006/jcis.2002.8383 PMID:16290704

Camargo, P. H. C., Satyanarayana, K. G., & Wypych, F. (2009). Nanocomposites: Synthesis, structure, properties and new application opportunities. *Materials Research*, *12*(1), 1–39. doi:10.1590/S1516-14392009000100002

Chellaram, C., Murugaboopathi, G., John, A. A., Sivakumar, R., Ganesan, S., Krithika, S., & Priya, G. (2014). Significance of Nanotechnology in Food Industry. *APCBEE Procedia*, *8*, 109–113. doi:10.1016/j.apcbee.2014.03.010

Choi, S.-S. (2001). Improvement of properties of silica-filled styrene–butadiene rubber compounds using acrylonitrile–butadiene rubber. *Journal of Applied Polymer Science*, *79*(6), 1127–1133. doi:10.1002/1097-4628(20010207)79:6<1127::AID-APP170>3.0.CO;2-8

Chruściel, J. J., & Leśniak, E. (2012). Thermoplastic Elastomers. *InTech Journal*.

Chuayjuljit, S., & Boonmahitthisud, A. (2010). Natural rubber nanocomposites using polystyrene-encapsulated nanosilica prepared by differential microemulsion polymerization. *Applied Surface Science*, *256*(23), 7211–7216. doi:10.1016/j.apsusc.2010.05.053

Chuayjuljit, S., & Luecha, W. (2011). XSBR/NR Rubber Blends Filled with Polystyrene-Encapsulated Nanosilica Prepared by in situ Differential Microemulsion Polymerization. *Journal of Elastomers and Plastics*. doi: 10.1177/0095244311405001

Chuayjuljit, S., Sukasem, N., & Boonmahitthisud, A. (2014). Effects of Silica, Poly(methyl methacrylate) and Poly(methyl methacrylate)-Grafted-Silica Nanoparticles on the Physical Properties of Plasticized-Poly(vinyl chloride). *Polymer-Plastics Technology and Engineering*, *53*(2), 116–122. doi:10.1080/03602559.2013.843695

Chumeka, W., Pasetto, P., Pilard, J.-F., & Tanrattanakul, V. (2014). Bio-based triblock copolymers from natural rubber and poly(lactic acid): Synthesis and application in polymer blending. *Polymer*, *55*(17), 4478–4487. doi:10.1016/j.polymer.2014.06.091

Chung, K.-H., Jung, H.-Y., Lee, Y.-W., & Lee, K.-Y. (2010). Preparation of TiO2-loaded nanocapsules and their sun protection behaviors. *Journal of Industrial and Engineering Chemistry*, *16*(2), 261–266. doi:10.1016/j.jiec.2010.01.039

Comini, E., Baratto, C., Concina, I., Faglia, G., Falasconi, M., Ferroni, M., & Sberveglieri, G. et al. (2013). Metal oxide nanoscience and nanotechnology for chemical sensors. *Sensors and Actuators. B, Chemical*, *179*, 3–20. doi:10.1016/j.snb.2012.10.027

de Oliveira, M. A., da Silva, P. M. L. C., Alves, M. G., de Oliveira, C. P., & dos Santos, M. A. (2005). Encapsulation of TiO2 by emulsion polymerization with methyl metacrylate (MMA). *Polymer Bulletin*, *55*(6), 477–484. doi:10.1007/s00289-005-0457-3

Deng, Y., Wang, L., Yang, W., Fu, S., & Elaassari, A. (2003). Preparation of magnetic polymeric particles via inverse microemulsion polymerization process. *Journal of Magnetism and Magnetic Materials*, *257*(1), 69–78. doi:10.1016/S0304-8853(02)00987-3

Ding, X., Zhao, J., Liu, Y., Zhang, H., & Wang, Z. (2004). Silica nanoparticles encapsulated by polystyrene via surface grafting and in situ emulsion polymerization. *Materials Letters*, *58*(25), 3126–3130. doi:10.1016/j.matlet.2004.06.003

Du, J., & Sun, H. (2014). Polymer/TiO2 Hybrid Vesicles for Excellent UV Screening and Effective Encapsulation of Antioxidant Agents. *ACS Applied Materials & Interfaces*, *6*(16), 13535–13541. doi:10.1021/am502663j PMID:25059274

Duncan, T. V. (2011). Applications of nanotechnology in food packaging and food safety: Barrier materials, antimicrobials and sensors. *Journal of Colloid and Interface Science*, *363*(1), 1–24. doi:10.1016/j.jcis.2011.07.017 PMID:21824625

Feuser, P. E., Fernandes, A. C., Nele, M., Viegas, A. C., Ricci-Junior, E., Tedesco, A. C., & de Araújo, P. H. H. et al. (2015). Simultaneous encapsulation of magnetic nanoparticles and zinc phthalocyanine in poly(methyl methacrylate) nanoparticles by miniemulsion polymerization and in vitro studies. *Colloids and Surfaces. B, Biointerfaces*, *135*, 357–364. doi:10.1016/j.colsurfb.2015.07.067 PMID:26263221

Fink, J. K. (2013). *Silicones Reactive Polymers Fundamentals and Applications* (2nd ed.). Oxford, UK: William Andrew Publishing. doi:10.1016/B978-1-4557-3149-7.00008-5

Gallucci, D. (2015). *Southern Thailand hit by weak rubber prices.* Retrieved 4 February, 2016, from http://www.ft.com/intl/cms/s/3/ded417fe-2ec1-11e5-8873-775ba7c2ea3d.html#axzz3zrcxpTeq

González, N., Custal, M. À., Lalaouna, S., Riba, J.-R., & Armelin, E. (2016). Improvement of dielectric properties of natural rubber by adding perovskite nanoparticles. *European Polymer Journal*, *75*, 210–222. doi:10.1016/j.eurpolymj.2015.12.023

Grothe, J., Kaskel, S., & Leuteritz, A. (2012). *8.08 - Nanocomposites and Hybrid Materials A2 - Möller, Krzysztof MatyjaszewskiMartin. In Polymer Science: A Comprehensive Reference* (pp. 177–209). Amsterdam: Elsevier.

He, G., Pan, Q., & Rempel, G. L. (2003). Synthesis of Poly(methyl methacrylate) Nanosize Particles by Differential Microemulsion Polymerization. *Macromolecular Rapid Communications*, *24*(9), 585–588. doi:10.1002/marc.200390089

Hewitt, N. (2007). *Compounding Precipitated Silica in Elastomers.* Plastics Design Library: William Andrew.

Hong, S., Kim, M., Hong, C. K., Jung, D., & Shim, S. E. (2008). Encapsulation of multi-walled carbon nanotubes by poly(4-vinylpyridine) and its dispersion stability in various solvent media. *Synthetic Metals*, *158*(21–24), 900–907. doi:10.1016/j.synthmet.2008.06.023

Hong, D., & Chang, W. V. (2007). *Preparation of nano polystyrene beads by microemulsion polymerization.* Retrieved 12 February, 2016, from http://chems.usc.edu/assets/004/56944.pdf

Hu, Y., Zhou, S., & Wu, L. (2009). Surface mechanical properties of transparent poly(methyl methacrylate)/zirconia nanocomposites prepared by in situ bulk polymerization. *Polymer*, *50*(15), 3609–3616. doi:10.1016/j.polymer.2009.03.028

Huneau, B. (2011). Strain-induced crystallization of natural rubber: a review of x-ray diffraction investigations. *Rubber Chemistry and Technology*, *84*(3), 425–452. doi:10.5254/1.3601131

Jang, J., Bae, J., Choi, M., & Yoon, S.-H. (2005). Fabrication and characterization of polyaniline coated carbon nanofiber for supercapacitor. *Carbon*, *43*(13), 2730–2736. doi:10.1016/j.carbon.2005.05.039

Kaiyoorawong, S., & Yangdee, B. (2006). *Rights of rubber farmers in Thailand under free trade.* Retrieved 02 February, 2016, from http://wrm.org.uy/oldsite/countries/Thailand/Rights_of_rubber_farmers_in_Thailand.pdf

Killmann, W. (2001). *Non-forest tree plantations* (D. J. Mead, Ed.). Forestry Department.

Kim, M. K., Kim, C. A., Ahn, S. D., Kang, S.-R., & Suh, K. S. (2004). Density compatibility of encapsulation of white inorganic TiO2 particles using dispersion polymerization technique for electrophoretic display. *Synthetic Metals*, *146*(2), 197–199. doi:10.1016/j.synthmet.2004.07.002

Kohler, N., Sun, C., Fichtenholtz, A., Gunn, J., Fang, C., & Zhang, M. (2006). Methotrexate-Immobilized Poly(ethylene glycol) Magnetic Nanoparticles for MR Imaging and Drug Delivery. *Small*, *2*(6), 785–792. doi:10.1002/smll.200600009 PMID:17193123

Kongsinlark, A., Rempel, G. L., & Prasassarakich, P. (2012). Synthesis of monodispersed polyisoprene–silica nanoparticles via differential microemulsion polymerization and mechanical properties of polyisoprene nanocomposite. *Chemical Engineering Journal*, *193–194*, 215–226. doi:10.1016/j.cej.2012.04.008

Lee, J. W., & Kjeang, E. (2013). Nanofluidic fuel cell. *Journal of Power Sources*, *242*, 472–477. doi:10.1016/j.jpowsour.2013.05.129

Lindgren, G. (2003). *Impending world oil shortage*. Retrieved 02 February, 2016, from http://www.imc-cim.org/mmap/pdf/prod-lindgren-oil-e.pdf

Liu, H., Lv, M., Deng, B., Li, J., Yu, M., Huang, Q., & Fan, C. (2014). Laundering durable antibacterial cotton fabrics grafted with pomegranate-shaped polymer wrapped in silver nanoparticle aggregations. *Scientific Reports*, *4*, 5920. doi:10.1038/srep05920 PMID:25082297

Ma, J., Zhang, L.-Q., & Geng, L. (2010). *Manufacturing Techniques of Rubber Nanocomposites. In Rubber Nanocomposites* (pp. 21–62). John Wiley & Sons, Ltd.

Masoud S.N., & G., D. (2011). Polymeric Nanocomposite Materials. *InTech Journal*.

Matsui, S. (2003). Nanostructure fabrication using electron and ion beams A2 - Iwamoto, M. In K. Kaneto & S. Mashiko (Eds.), *Nanotechnology and Nano-Interface Controlled Electronic Devices* (pp. 3–20). Amsterdam: Elsevier. doi:10.1016/B978-044451091-4/50002-9

Mokhothu, T. H., Luyt, A. S., Morselli, D., Bondioli, F., & Messori, M. (2015). Influence of in situ-generated silica nanoparticles on EPDM morphology, thermal, thermomechanical, and mechanical properties. *Polymer Composites*, *36*(5), 825–833. doi:10.1002/pc.23002

MREPC. (2016). *Important Facts about Latex and Latex Gloves*. Retrieved 04 February, 2016, from http://www.latexgloves.info/manufacturing.php

Nair, A. B., & Joseph, R. (2014). 9 - Eco-friendly bio-composites using natural rubber (NR) matrices and natural fiber reinforcements A2 - Kohjiya, Shinzo. In Y. Ikeda (Ed.), *Chemistry, Manufacture and Applications of Natural Rubber* (pp. 249–283). Woodhead Publishing. doi:10.1533/9780857096913.2.249

Ni, B., Liu, M., Lü, S., Xie, L., & Wang, Y. (2010). Multifunctional Slow-Release Organic−Inorganic Compound Fertilizer. *Journal of Agricultural and Food Chemistry*, *58*(23), 12373–12378. doi:10.1021/jf1029306 PMID:21058723

Noreen, A., Zia, K. M., Zuber, M., Ali, M., & Mujahid, M. A critical review of algal biomass: A versatile platform of bio-based polyesters from renewable resources. *International Journal of Biological Macromolecules*. doi:10.1016/j.ijbiomac.2016.01.067 PMID:26808018

Oberdisse, J. (2006). Aggregation of colloidal nanoparticles in polymer matrices. *Soft Matter*, *2*(1), 29–36. doi:10.1039/B511959F PMID:16733636

Ono, Y. (2014). *Seven countries allying to halt rubber price slide.* Retrieved 03 February, 2016, from http://asia.nikkei.com/Politics-Economy/International-Relations/Seven-countries-allying-to-halt-rubber-price-slide

Palkovits, R., Althues, H., Rumplecker, A., Tesche, B., Dreier, A., Holle, U., & Kaskel, S. et al. (2005). Polymerization of w/o Microemulsions for the Preparation of Transparent SiO2/PMMA Nanocomposites. *Langmuir, 21*(13), 6048–6053. doi:10.1021/la050630k PMID:15952859

Park, E. J., Kim, W. S., Hwang, H. S., Park, C., & Lim, K. T. (2007). PMMA Encapsulated Silica Particles by Dispersion Polymerization in Supercritical Carbon Dioxide. *Macromolecular Symposia, 249-250*(1), 196–201. doi:10.1002/masy.200750332

Park, J.-G., Ye, Q., Topp, E. M., Misra, A., & Spencer, P. (2009). Water sorption and dynamic mechanical properties of dentin adhesives with a urethane-based multifunctional methacrylate monomer. *Dental Materials, 25*(12), 1569–1575. doi:10.1016/j.dental.2009.07.010 PMID:19709724

Patole, A. S., Patole, S. P., Kang, H., Yoo, J.-B., Kim, T.-H., & Ahn, J.-H. (2010). A facile approach to the fabrication of graphene/polystyrene nanocomposite by in situ microemulsion polymerization. *Journal of Colloid and Interface Science, 350*(2), 530–537. doi:10.1016/j.jcis.2010.01.035 PMID:20673907

Prasertsri, S., & Rattanasom, N. (2011). Mechanical and damping properties of silica/natural rubber composites prepared from latex system. *Polymer Testing, 30*(5), 515–526. doi:10.1016/j.polymertesting.2011.04.001

Qi, D., Bao, Y., Weng, Z., & Huang, Z. (2006). Preparation of acrylate polymer/silica nanocomposite particles with high silica encapsulation efficiency via miniemulsion polymerization. *Polymer, 47*(13), 4622–4629. doi:10.1016/j.polymcr.2006.04.024

Raj, S., Jose, S., Sumod, U. S., & Sabitha, M. (2012). Nanotechnology in cosmetics: Opportunities and challenges. *Journal of Pharmacy & Bioallied Sciences, 4*(3), 186–193. doi:10.4103/0975-7406.99016 PMID:22923959

Ramesan, M. T. (2004). Thermogravimetric analysis, flammability and oil resistance properties in natural rubber and dichlorocarbene modified styrene butadiene rubber blends. *Reactive & Functional Polymers, 59*(3), 267–274. doi:10.1016/j.reactfunctpolym.2004.02.005

Reddy, K. R., Sin, B. C., Yoo, C. H., Sohn, D., & Lee, Y. (2009). Coating of multiwalled carbon nanotubes with polymer nanospheres through microemulsion polymerization. *Journal of Colloid and Interface Science, 340*(2), 160–165. doi:10.1016/j.jcis.2009.08.044 PMID:19800632

Rezende, C. A., Bragança, F. C., Doi, T. R., Lee, L.-T., Galembeck, F., & Boué, F. (2010). Natural rubber-clay nanocomposites: Mechanical and structural properties. *Polymer, 51*(16), 3644–3652. doi:10.1016/j.polymer.2010.06.026

Romio, A. P., Rodrigues, H. H., Peres, A., Da Cas Viegas, A., Kobitskaya, E., Ziener, U., & Araújo, P. H. H. et al. (2013). Encapsulation of magnetic nickel nanoparticles via inverse miniemulsion polymerization. *Journal of Applied Polymer Science, 129*(3), 1426–1433. doi:10.1002/app.38840

Seentrakoon, B., Junhasavasdikul, B., & Chavasiri, W. (2013). Enhanced UV-protection and antibacterial properties of natural rubber/rutile-TiO2 nanocomposites. *Polymer Degradation & Stability*, *98*(2), 566–578. doi:10.1016/j.polymdegradstab.2012.11.018

Shahzadi, K., Wu, L., Ge, X., Zhao, F., Li, H., Pang, S., & Mu, X. et al. (2016). Preparation and characterization of bio-based hybrid film containing chitosan and silver nanowires. *Carbohydrate Polymers*, *137*, 732–738. doi:10.1016/j.carbpol.2015.11.012 PMID:26686186

Shin-Etsu, C. (2011). *Silane coupling agent*. Retrieved 10 February 2016, from http://www.shinetsusilicone-global.com/catalog/pdf/SilaneCouplingAgents_e.pdf

Sigmund, S., Akgün, E., Meyer, J., Hubbuch, J., Wörner, M., & Kasper, G. (2014). Defined polymer shells on nanoparticles via a continuous aerosol-based process. *Journal of Nanoparticle Research*, *16*(8), 1–9. doi:10.1007/s11051-014-2533-9

Tancharernrat, T., Rempel, G. L., & Prasassarakich, P. (2014). Preparation of styrene butadiene copolymer–silica nanocomposites via differential microemulsion polymerization and NR/SBR–SiO2 membranes for pervaporation of water–ethanol mixtures. *Chemical Engineering Journal*, *258*, 290–300. doi:10.1016/j.cej.2014.05.151

Tancharernrat, T., Rempel, G. L., & Prasassarakich, P. (2015). Synthesis of polybutadiene-silica nanoparticles via differential microemulsion polymerization and their hydrogenated nanoparticles by diimide reduction. *Polymer Degradation & Stability*, *118*, 69–81. doi:10.1016/j.polymdegradstab.2015.04.008

Theppradit, T., Prasassarakich, P., & Poompradub, S. (2014). Surface modification of silica particles and its effects on cure and mechanical properties of the natural rubber composites. *Materials Chemistry and Physics*, *148*(3), 940–948. doi:10.1016/j.matchemphys.2014.09.003

TRA. (2015a). *NR situation*. Retrieved 4 February, 2016, from http://www.thainr.com/en/index.php?detail=situation&selectmonth=01

TRA. (2015b). *Thai rubber statistics*. Retrieved 4 February, 2016, from http://www.thainr.com/uploadfile/20151103111354.pdf

Weng, P., Wei, Q., Tang, Z., Lin, T., & Guo, B. (2015). The influence of molybdenum disulfide nanoplatelets on the dispersion of nano silica in natural rubber composites. *Applied Surface Science*, *359*, 782–789. doi:10.1016/j.apsusc.2015.10.172

Wong Y. W. H., Leung, M. Y. S., Ku, S. K. A., & Lam, H. L. I. (2006). Selected applications of nanotechnology in textiles. *AUTEX Research Journal, 6*, 1-8.

Xing, S., Tan, L. H., Yang, M., Pan, M., Lv, Y., Tang, Q., & Chen, H. et al. (2009). Highly controlled core/shell structures: Tunable conductive polymer shells on gold nanoparticles and nanochains. *Journal of Materials Chemistry*, *19*(20), 3286–3291. doi:10.1039/b900993k

Xiong, W., Chen, L., Zhao, B., Wang, D.-y., & Wang, Y. (2012). Polyamide 6 with a flame retardant encapsulated by polyamide 66: Flame retardation, thermo-decomposition and the potential mechanism. *Chinese Journal of Polymer Science*, *30*(2), 297–307. doi:10.1007/s10118-012-1126-2

Yamak, H. B. (2013). *Emulsion Polymerization: Effects of Polymerization Variables on the Properties of Vinyl Acetate Based Emulsion Polymers. InTech Journal.*

Yang, F., & Nelson, G. L. (2004). PMMA/silica nanocomposite studies: Synthesis and properties. *Journal of Applied Polymer Science, 91*(6), 3844–3850. doi:10.1002/app.13573

Yuan, J. J., Armes, S. P., Takabayashi, Y., Prassides, K., Leite, C. A. P., Galembeck, F., & Lewis, A. L. (2006). Synthesis of Biocompatible Poly[2-(methacryloyloxy)ethyl phosphorylcholine]-Coated Magnetite Nanoparticles. *Langmuir, 22*(26), 10989–10993. doi:10.1021/la061834j PMID:17154575

Yun, J., Im, J. S., Lee, Y.-S., Bae, T.-S., Lim, Y.-M., & Kim, H.-I. (2010). pH and electro-responsive release behavior of MWCNT/PVA/PAAc composite microcapsules. *Colloids and Surfaces. A, Physicochemical and Engineering Aspects, 368*(1–3), 23–30. doi:10.1016/j.colsurfa.2010.07.010

Zeng, Z., Yu, J., & Guo, Z.-X. (2004). Preparation of Carboxyl-Functionalized Polystyrene/Silica Composite Nanoparticles. *Macromolecular Chemistry and Physics, 205*(16), 2197–2204. doi:10.1002/macp.200400201

Zhang, K., Fu, Q., Huang, Y., & Zhou, D. Encapsulation of inorganic particles by dispersion polymerization through ultrasonic irradiation. *Science in China Series B: Chemistry, 48*(6), 545-552. doi: 10.1360/042004-100

Zhong, Z., Jian, Y., & Zhaoxia, G. (2006). Synthesis of composite nanoparticles bearing epoxy functional groups: Encapsulation of silica by emulsion polymerization of GMA. *Journal of Wuhan University of Technology-Mater. Science Editor, 21*(2), 136–138. doi:10.1007/bf02840860

Zhou, W., Hu, X., Bai, X., Zhou, S., Sun, C., Yan, J., & Chen, P. (2011). Synthesis and Electromagnetic, Microwave Absorbing Properties of Core–Shell Fe3O4–Poly (3,4-ethylenedioxythiophene) Microspheres. *ACS Applied Materials & Interfaces, 3*(10), 3839–3845. doi:10.1021/am2004812 PMID:21913665

KEY TERMS AND DEFINITIONS

Encapsulation: A technology for the coating of inorganic particles with various polymers.

***'In situ'* Differential Microemulsion Polymerization:** The polymerization technique which has been developed for producing nanosized particles of less than 50 nm by using a much smaller amount of surfactant than that used in the conventional microemulsion polymerization.

Nanocomposites: Materials that result from the intimate mixture of two or more nano-scale phase materials.

Nanohybrid Materials: A material obtained through interaction of chemically different components, usually organic and inorganic.

Natural Rubber: An elastic material obtained from the latex sap of trees (*Hevea brasiliensis*) that can be vulcanized and finished into a variety of products.

Chapter 5
Poly (Lactic Acid) Generated for Advanced Materials

Pranut Potiyaraj
Chulalongkorn University, Thailand

ABSTRACT

The consumption of plastic products from petrochemical feedstock has increased sharply resulting in plastic waste problems while raw materials from fossil fuels tend to decrease rapidly. Researchers and the plastic industry have since proposed a sustainable solution through the development of bioplastics. Ideally, bioplastics which are synthesized from renewable bioresources normally render biodegradability in appropriate conditions. Polyester, one of the most diversely used synthetic polymers today, is an ideal choice for biodegradable polymers due to the relative ease of breaking ester linkages. Poly(lactic acid) or polylactide or PLA which is a thermoplastic polyester with many advantageous properties, for instance, environmentally friendly, biocompatibility, processability, and high chemical resistance is now available in the plastic market as a promising bioplastics. However, the cost of PLA is still much higher than that of general commodity plastics. In order to make PLA commercially competitive, advanced and innovative applications should thus be explored. In this chapter, technological background of PLA production as well as its economic situation is firstly reviewed. Then, the enhancement of PLA properties to suit advanced applications is illustrated. Some polymers used for blending with PLA along with some fillers utilized for the production of PLA composites are described. The chapter concludes with the degradation mechanism of PLA and the standard test methods.

INTRODUCTION

For the last century, plastic materials have been emerged as a key contributor to material industry development. Plastic products have been used as a common material in everyday life due to its competitive cost and acceptable performances comparing with ceramics or metals. Several new plastics have been researched and industrialized as materials for advanced applications, for example, biomedical devices, electronic parts, or construction materials. As a consequence, the plastic wastes, with the composition in municipal solid waste around 15 percent (SPI Bioplastics Council, 2012), lead to far more serious

DOI: 10.4018/978-1-5225-1971-3.ch005

environmental problems although several strategies have been implemented as solutions. Researchers thus try to develop a new class of plastics which can be degraded spontaneously in the natural condition by biological process. The use of these biodegradable plastics increases steadily, especially in packaging applications as a result of changing modern lifestyle.

Most of current commodity and engineering plastics nowadays were produced from petrochemical feedstock. Meanwhile, the depletion of fossil fuel has been recognized as a future challenge. The production process of petroleum-based plastics also generates greenhouse gas which creates critical environmental problems. The new environmental rules and regulations as well as the fluctuation in raw materials cost force the plastic industry to find a new starting block for the plastic production. Bioplastics are now a material of choices for the plastic industry as well as environmentally concerned consumers. As defined by IUPAC, bioplastics are bio-based polymers derived from the biomass or polymerized from monomer derived from the biomass. In 2012, bioplastics represent only around 0.4% of total plastics consumption as the price is 4-6 folds than that of petroleum based plastics. However, due to rapidly improved technologies, its cost is lower thus its demand growth is expected to exceed a compound annual growth rate of 25% for the period of 2012 to 2017. (Messenger, 2013) Global production of bioplastics is forecast to quadruple by 2017.

Second only to starch-based thermoplastics, poly(lactic acid) or polylactide (PLA) currently shares most of the bioplastic market. Carothers and colleagues at DuPont invented PLA in 1932, however, the process was too expensive and not until 1952 that DuPont patented the process in 1952. Due to its high production cost, it was initially produced only for biomedical applications, with the approval of the US Food and Drug Administration (FDA), such as sutures, implants, and controlled drug release. The advanced technology in lactic acid production has lowered the cost of PLA production and significantly increased interest in PLA for other applications as its claim to be a bioplastic with low carbon footprint. Similar to other bioplastics, PLA has a narrower processing window than that of other common plastics. Although several new grades of PLA are now commercially available but the processability of PLA is still a challenge. Furthermore, its brittleness is the major setback of PLA, thus several researchers have currently focused on enhancing its mechanical performance through polymer blending and polymer composite techniques.

BACKGROUND

A New Promising Bioplastics

PLA is a biodegradable thermoplastic aliphatic polyester synthesis from lactic acid which is derived from renewable biomass resources, such as corn starch, tapioca roots, or sugarcane. Its repeating unit is shown in Figure 1. Lactic acid has two optical isomers, i.e. L-lactic acid and D-lactic acid. A mixture of the two isomers is called DL-lactic acid. The polymerization of optically pure L- and D-lactic acids gives isotactic homopolymers of PLLA and PDLA, respectively. PLLA and PDLA are crystalline polymers with a melting temperature (T_m) around 180°C. The polymerization of DL-lactic acid leads to the formation of poly(DL-lactic acid) of which the degree of crystallinity is controlled by the ratio of D to L enantiomers used.

Figure 1. Repeating unit of PLA

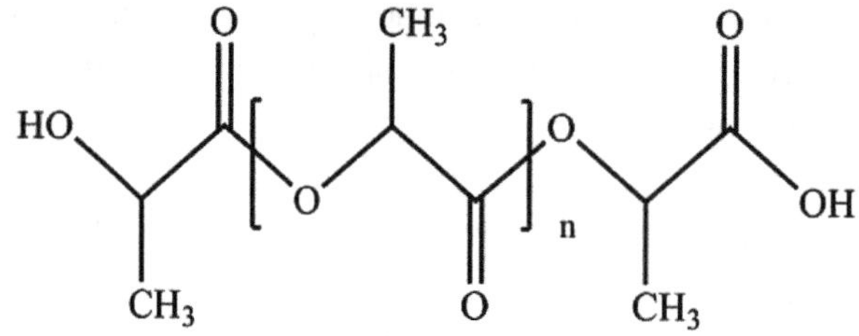

Production Processes

The direct polymerization of lactic acid (α-hydroxypropionic acid) through condensation reactions forming an ester bond between the carboxyl group and the methyl group is shown in Scheme I. The reaction typically carries out under 200°C equipped with the distillation system to removed water by product in the presence of appropriate catalyst. However, high molecular weight PLA are difficult to produce with direct polymerization due to unfavorable equilibrium in the usual condensation reaction. In this technique, the average molecular weight achieved is up to 50,000 g/mol thus giving PLA with poor properties. Some techniques were proposed by researcher to increase molecular weight of PLA synthesized by direct polymerization. (Kim & Kim 1999) The two-step polymerization consists of prepolymerization and polymerization steps, as shown in Scheme II, yields a higher molecular weight PLA but still limited by the equilibrium reaction of polycondensation due to hydrolysis of ester bonds. (Lopes *et al.* 2014)

To be used for industrial applications, e.g. fibers, films and bottles, the number of repeating units (*n*) should be between 100 and 200 in the case of poly(ethylene terephthalate) or PET. Nevertheless, *n* should be between 700 and 1400 in the case of PLA which is much higher. Therefore, the requirements on raw material purity and production technology are critical. The initially commercialized process for the production of PLA was through catalytic ring opening polymerization (ROP) of lactide as shown in Scheme III. (Hu *et al.*, 2016) This process involves the conversion of lactic acid into a low molecular weight oligomer which is depolymerized by the back-biting reaction to obtain lactide, a cyclic dimer of lactic acid. The ring opening polymerization of lactide in metal-catalyzed reaction to produce poly(lactic acid) is then followed. This is still the most largely used process in PLA production.

Figure 2. Scheme I

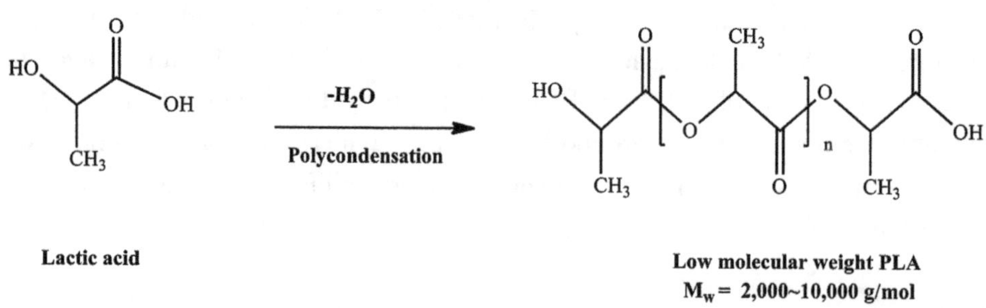

Figure 3. Scheme II

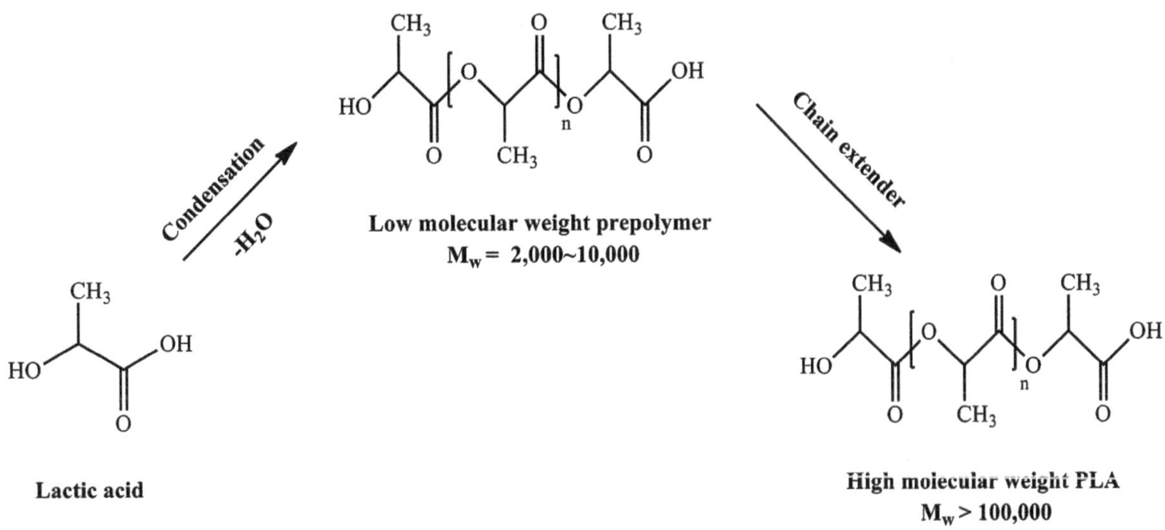

Figure 4. Scheme III

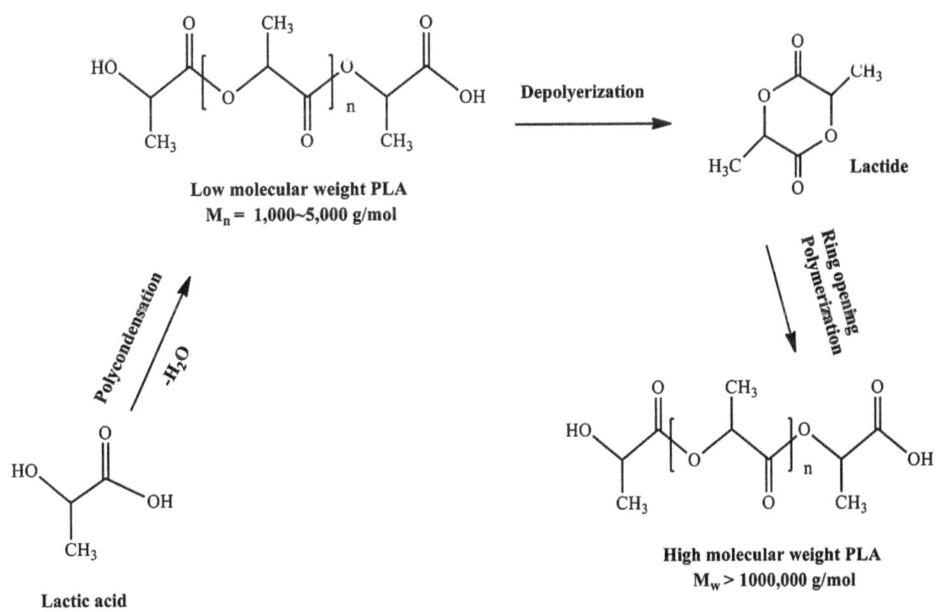

Economic Situation

PLA contributes to above 35% of biodegradable bioplastics market. Around 60% is accounted for packaging application of which 75% for film and plastic bags. It is forecasted that textiles are the fastest growing segment, in terms of revenue from 2014 to 2020. These increasing demands are due to national environmental protection initiatives of several countries including Japan, Germany, China, Korea and Brasil. PLA is considered as a novel material for broader range of value-added applications, e.g. 3D printing, nonwoven textile, medical implants as well as electronic parts.

The global PLA market is projected to reach USD 5.2 Billion by 2020. Government regulations and growing Environmental awareness is the key driving force that keeps Europe and North America as the largest consumer market for PLA. Asia Pacific is the fastest growing region for PLA market, at the fastest rate of 19.5% in terms of revenue from 2014 to 2020, although price sensitive consumers are the limitation of its growth. In addition, PLA requires specific industrial composting system to get completely degraded.

Commercial Products

Cargill Dow LLC, a joint venture between Cargill and Dow Chemicals, was founded in 1997 to commercialize PLA. In 2005, Cargill acquired the company and changed its name to NatureWorks LLC. Seven year later, Thailand's largest chemical producer, PTT Chemical Public Company Limited (PTT Chemical) invested US$150 million in NatureWorks LLC and became the joint venture. The company produces the NatureWorks® proprietary PLA at a Cargill facility in Nebraska, which is then marketed globally under the Ingeo® brand. Currently, it is the top producer in market share at about 31% of global PLA production.

In Europe, Corbion Purac offers lactides and PLA resins for general purpose and high performance bioplastics under the tradename of Purapol®. The company produces lactic acid and lactide in Gorinchem (Netherlands) and Rayong (Thailand). Sugar cane from Thailand is the primarily raw material while the use of other non-food raw materials is planned to lower the cost.

Other commercial products include Ecovio® which is the tradename of bioplastic products from BASF, Loopla® from Belgian-based Galactic SA, Biofront® from Teijin, Ecoloju™ and PLABio™ from Mitsubishi. Recently, there are also several smaller manufacturers of PLA operated in China.

DEGRADATION OF PLA

Biodegradation Mechanism

Apart from being a bio-based polymer, PLA is also a compostable plastic. While biodegradable plastics are broadly defined as plastics that decompose by the action of naturally occurring microorganisms, usually bacteria, fungi, and algae, compostable plastics must be able to biodegrade completely under specific composting conditions. The American Society for Testing and Materials (ASTM) publishes a standard ASTM D6400 entitled "Standard Specification for Compostable Plastics" to define compostable plastic as "a plastic that undergoes degradation by biological processes during composting to yield CO_2, water, inorganic compounds, and biomass at a rate consistent with other known compostable materials and leaves no visible distinguishable or toxic residue". (ASTM, 2009) Generally, PLA undergoes a 2-step degradation process. First, the moisture and heat in the compost break the polymeric chains apart, through hydrolysis reaction, creating oligomers as well as lactic acid. Microorganisms in compost and soil consume the oligomers and monomer as nutrients. Since lactic acid is widely found in nature, a large number of organisms metabolize lactic acid. The end result of composting is carbon dioxide, water and humus, a soil nutrient. This degradation process is temperature and humidity dependent. The PLA products will take up to 180 days to completely degrade in a commercial composting facility. In home composting facility, it may take longer time.

Standard Test Method

This section briefly described standard test method to verify that a plastic is compostable according to ASTM D6400. This ASTM standard is very similar to the European standard EN 13432. The standard also refers to the Guideline 208 "Terrestrials Plant, Growth Test" by Organization for Economic Development (OECD). For a plastic to receive the new logo of the US Composting Council it must comply all of the requirements of the ASTM D6400 as follows:

1. **Disintegration During Composting:** A compostable plastic must disintegrate during composting such that any remaining residuals are not readily distinguishable. Quantitatively, no more than 10% of its original dry weight should be remain after sieving on a 2.0 mm sieve.
2. **Inherent Biodegradation:** A compostable plastic must demonstrate a satisfactory rate of degradation under the test period not longer than 180 days. For example, for a homopolymer, 60% of the organic carbon must be converted to carbon dioxide by the end of test period. This requirement is similar to the requirements of ISO 14855 and DIN V49000, but differs from the stricter EN 13432 standard, which requires 90% biodegradation in that time. ASTM D5338 "Standard Test Method for Determining Aerobic Biodegradation of Plastic Materials Under Controlled Composting Conditions, Incorporating Thermophilic Temperatures" can be followed to investigate the biodegradation rate.
3. **No Adverse Impact on Ability of Compost to Support Plant Growth:** Additionally, the compostable plastic must not introduce unacceptable levels of heavy metals or other toxic substances in to the environment during decomposition. To satisfy this requirement, it must have concentration of regulated heavy metals less than 50% of those prescribed for sludges or composts in the country where the compostable plastic is sold. Furthremore, the germination rate and the plant biomass of the sample composts must be no less than 90% of that the corresponding blank composts for two different plant species following OECD Guideline 208.

Other Disposal Methods

Beside composting in natural environment, other disposal methods used for commodity plastics can also be applied to PLA. Those are landfilling, combustion and recycling. Landfilling is the least suitable method as a large amount of plastics from municipal solid waste is currently piled up in the limited land uses. Combustion also yields back energy but there are still more research and development to carry out to lower the cost thus make this method commercially feasible. Recycling seems to be more prospect candidate. The environmental impact of recycling PLA is over 50 times better than composting and 16 times better than combusting PLA. However, due to narrow processing windows and low heat resistance, the recycling process must be investigated technologically.

INNOVATIVE PLA BLENDS AND COMPOSITES

Toughening of PLA

New biotechnology to produce lactic acid and polymerization technology to synthesize PLA have progressed rapidly thus the cost of PLA resin reduces significantly. This make PLA economically competitive with traditional petroleum-based polymers. Although several new applications of PLA have been expanded recently, its usages in some fields are still limited primarily due to the its brittleness. Many techniques have been proposed and developed by researchers.

Theoretically, the variation in stereochemistry, molecular weight, and crystallinity of pristine PLA can improve its ductility and impact resistance. However, such improvements are not practically sufficient to satisfy the requirements of most applications. (Liu & Zhang, 2011) Copolymerization as well as the addition of impact modifiers are then of interests to the industry.

Copolymerization

Copolymerization technique is extensively used to adjust the properties of manufactured plastics to meet specific requirements those are unattainable by a homopolymer. Mechanical, physical as well as other properties of a copolymer can be customized by designing of copolymer molecular chains. For PLA, copolymerization can be adapted through polycondensation of lactic acid with other monomers or ring-opening copolymerization of lactic acid with other cyclic monomer. The copolymerization can be conducted through ionic, coordination, or free radical reaction.

Addition of Impact Modifier: Plasticization

Plasticizers in various forms are generally used in the plastic industry to improve the processability of plastics and to improve flexibility and ductility. For instance, poly(vinyl chloride) (PVC) is normally formulated with the presence of plasticizers to obtain flexible products. A preferred plasticizer should be able to space chains of polymers apart thus increasing the free volume. As a consequence, the glass transition temperature (T_g) of plasticized plastics are lowered significantly. In general, liquid low molecular weight plasticizers, especially phthalate esters are used to improve the ductility and impact resistance of brittle plastics. However, these low molecular weight plasticizers tend to migrate towards the surface of products due to their high mobility to cause the deterioration in its long-term thermal and mechanical stability. To overcome this migration problem, higher molecular weight plasticizers are now being studied as a replacement. These high molecular weight plasticizers are however more difficult to disperse in polymer comparing with the lower molecular weight plasticizers.

For PLA, biodegradability and toxicity of plasticizers may be another important concerns. The plasticizing efficiency of a plasticizer depends on its miscibility with the host polymer which in turn can be considered from the solubility parameters and the molecular weight. The miscibility of a plastic with the same plasticizer decreases with increase in the molecular weight of the plasticizer as a result of high entropy of mixing. Researchers have successfully employ some monomers and oligomers as a potential plasticizer, i.e. polyethylene glycol (PEG) and citrate esters. (*Li* et al., 2015)

Polymer Blending

Brittleness of plastics can be conveniently and economically improved through polymer blending technique. For example, PS is toughened with some elastomers in order to obtain high impact polystyrene (HIPS). PLA has also blended with either biodegradable and nonbiodegradable flexible polymers. The earlier works dealing with PLA blending involve the solution blending technique. However, the melt blending techniques are commercially preferred while the immiscibility between the two polymers is its major upheaval.

High density polyethylene/poly(lactic acid) blends with/without compatibilizer were prepared by melt blending technique. It was found that, with the presence of maleic anhydride grafted high density polyethylene as a compatibilizer, at the ratio 80/20, the blend possessed better mechanical properties than those without a compatibilizer. (Madhu *et al.*, 2016)

The effect of the addition of poly(styrene-co-glycidyl methacrylate) copolymer on the properties of meltblended PLA/PMMA was studied. The preparation of these PLA-based blends shows an impact resistance improvement of about 60% compared with the neat PLA and maintained similar tensile strength and modulus values as well as glass transition temperature to neat PLA. The addition of only 3 phr of copolymer to PLA/PMMA blend improved the impact resistance almost 100%. As the amount of copolymer in the blend increased then higher was average molecular weight and polydispersity index. With the copolymer, the impact resistance, elongation at break and thermal stability were improved while tensile strength and elastic modulus remained almost unaltered. (Anakabe *et al.*, 2016)

PLA/polystyrene (PS) blends were prepared by melt mixing and the blends were determined for ther mechanical properties. The results indicated that the blends containing 50wt% and 75wt% PLA exhibited improved tensile strength and elongation at break as compared to neat PS. The stress transfer calculations of PLA/PS blends revealed that this blend exhibited a high degree of compatibility. (Hamad *et al.*, 2016)

Blends with varied ratio of PLA and thermoplastic polyurethane (TPU) were prepared by melt blending. TPU started to degrade immediately due to degradation of the low molecular weight plasticizer in the polymer while PLA exhibited a short period after which degradation progressed rapidly and was almost complete after 70 days. Blends with a co-continuous phase morphology degraded at a higher rate than the rest of the samples due to the higher exposure of the TPU phase in these blends. (Jašo *et al.,* 2015) PLA and TPU were melt-blended to form shape memory bio-based blends. The shape memory behaviors of PLA/TPU blends predeformed under different temperatures were investigated. The overall annealing effect on the shape recovery ratios were quite effective for both PLA/TPU blends of 80/20 and 60/40 without sacrificing the shape fixing ratios at the optimum predeformation temperature of 80 °C, attributing to the increased crystallinity of PLA and homogenized phase domains of TPU. (Lai *et al.*, 2016)

PLA/polycaprolactone (PCL) blend films were prepared by solvent casting with and without compatibilizer (PLA grafted with maleic anhydride). The blends showed lower thermal diffusivity compared to pure polymers with the highest value when the compatibilizer was used. The cold crystallization with the highest value of enthalpy of fusion was also observed for the compatibilized sample. (Correa-Pacheco *et al.*, 2015) PLA/PCL were also prepared by melt mixing in a twin-screw co-rotating extruder in order to increase the low intrinsic elongation at break of PLA for packaging applications. PCL enhanced the ductile behavior of composites as well as provided higher degradation rates of blends. (Ferri et al., 2016a)

Blown films of PLA/poly (butylene adipate-co-terephthalate) (PBAT) blends were prepared with the presence of a multifunctional epoxy chain extender, Joncryl ADR 4368. The mechanical properties of the blends were generally improved to the dramatically morphology changes caused by the chain extender. (Arruda *et al.*, 2015) The miscibility in PLA/PBAT blends was studied and the results show a limited one-way partial miscibility of PBAT in the PLA-rich phase with the glass transition temperature of the PLA-rich phase in PBAT showing a drop as compared to the neat PLA. The morphology of the dispersed phase shows that the dispersed phase exists in a fibril form indicating partial miscibility. (Dil *at al.*, 2015) A/PBAT/OMMT composites were prepared using an internal batch mixer. With 1 wt% of OMMT, the nanoparticles were located at the PLA-PBAT interface which was favorable for morphology stabilization. It was also found that, with suitable mixing technique, the average PBAT droplet-diameter was reduced thus improved properties of the composites (Nofar *et al.*, 2016)

PLA was blended with poly(butylene succinate) (PBS), a bio-degradable polymers at various ratios. The ductility of the blends was better than that of neat PLA with over 250% elongation when 10 wt% of PBS was added. The results also indicated a co-continuous phase morphology at some ratios. (Deng & Thomas, 2015) Properties of PLA/PBS blends were improved by the addition of polyethylene glycol (PEG). It was found that PEG acted as a plasticizer for the PLA phase in the phase-separated blends, increasing the elongation at break in all blends and decreasing the glass transition of the PLA phase. The tensile strength and Young's modulus both decreased when the amount of PEG increased while the elongation at break increased. Scanning electron microscope images showed that the fracture mode of PLA changed from brittle to ductile with the addition of PEG. (Pivsa-Art *et al.*, 2015)

Apart from conventional melt blending technique, researchers also investigate the reactive blending technique to improve polymer miscibility. (Lim *et al.*, 2016; Zeng *et al.*, 2015) Biodegradable blown PLA/PBAT blend films were prepared using epoxy functionalized-poly(lactide) (EF-PLA) reactive modifiers to improve the compatibilization. The epoxy groups on the EF-PLA modifiers react with PBAT forming an *in situ* copolymer that localizes at the blend interphase resulting in compatibilization effect. The blends have improved processability and mechanical properties (Schneider *et al.* 2015)

The addition of a polyolefin elastomer e.g. poly (ethylene-co-octene) or POE in PLA matrix can effectively improve its brittleness. POE was modified with glycidyl methacrylate in order to improve the compatibility between the two polymers. The results pointed out that the impact strength of PLA markedly increased while tensile and flexural properties of the blends were slightly lower than that of neat PLA. It was also observed that the tensile and flexural properties were slightly higher when the modified POE was used rather than those with unmodified POE which indicated the improved interfacial compatibility between two polymers. (Wongsampanwech & Potiyaraj, 2014)

Polymer Composites

The introduction of fillers into polymer matrix theoretically improves the rigidity of plastics. However, there are some reports that the addition of carbon fiber, calcium carbonates as well as clay (montmorillonite) with some modification can also increase the impact strength of PLA.

PLA Composites

There are quite a few bioplastics which are made commercially available comparing with petroleum-based plastics. As a consequence, there are needs to extend the range of properties of bioplastics to suit the end-use requirements. One of the most convenience technique is to prepare polymer composites with various kinds of fillers. But even with traditional polymers, the preparation of polymer composites still has a great challenge in obtaining a good distribution of the reinforcing filler as well as good filler-matrix interfacial adhesion. In general, a coupling agent, existing in several forms, is employed to increase interfacial adhesion. In the case of PLA composites, the same problems have to be tackled.

There are some negative consequences arising from the addition of reinforcing fillers. High levels of fillers can significantly increase melt viscosity during processing causing shear heating and degradation and affecting the ability to fill thin parts. Appearance problems are also a potential with flow lines, poor colorability, and opacity being among the issues. When considering applications that require biodegradability, the effect of fillers on the rate or extent of degradability should be evaluated.

PLA Filled with Inorganic Fillers

When PLA is filled with inorganic fillers such as talc, mica, glass, etc., the system is formulated with multiple other components or additives to optimize a balance of physical properties, processing characteristics, and appearance for specific end use applications. It is common to utilize the concentrated compounds or masterbatch approach for more efficient dispersion, where the filler or antiblocking agent is first dispersed at a higher concentration, often in another polymer and then dilute to the required concentration in the final products. Both talc and mica can be used to increase the modulus of PLA with 10-30% added. Mica is effective at lower levels, but talc also acts as a nucleating agent, which is advantageous in fast cycle injection molding applications.

Flame retarded PLA composites have been prepared by combining PLA with a variety of fillers, flame retardants, and other additives. For example, high levels (50-150 parts) of surface treated silanes/titinates) metal hydroxides were combined with additional fillers (talc, kaolin, mica, glass, etc.) and toughening agents to produce a composition for domestic appliances. Most fillers increase the stiffness of PLA with little benefit to toughness.

Composites based on green high-density polyethylene, PLA and nanosized calcium carbonate ($CaCO_3$) were prepared. PLA and $CaCO_3$ levels influenced the crystallization of HDPE while HDPE matrix induced PLA dispersed phase crystallization. The solvent extraction method was efficient at revealing the morphologies of the composites, indicating a well distributed PLA dispersed phase in the green HDPE matrix. Additionally, $CaCO_3$ content contributed to reducing the particle domains of PLA by decreasing the interfacial tension between HDPE and PLA. With respect to mechanical properties, the results showed that modulus of elasticity and tensile strength at yield are affected by PLA content. (Oliveira *et al.*, 2016)

The effects of silica concentration and the preferential adsorption between silica and the component on the thermo-mechanical and rheological properties were studied by preparing nanosilica filled PLA/PMMA blends with varying silica concentrations via melt blending technique. The results of indicate that PLA molecules were selectively adsorbed on the surface of the nanosilica thus their mechanical properties improved. (Hao *et al.*, 2016)

PLA/LLDPE blends were reinforced with rice husk silica by melt-mixing in a twin- screw extruder and processed into film by a chill roll cast extruder. The result show that the addition 0.5 and 1 wt% rice husk silica improved the elongation at break of PLA/LLDPE composites increases significantly. (Sareeladdanon & Potiyaraj, 2014)

The use of commercial silica (SiO_2) and $CaCO_3$ as nanofillers as well as compatibilizers in PLA/LDPE blends was investigated. The calculations of the minimal interfacial energy and optimal wetting abilities indicated that SiO_2 filler was a better potential compatibilizer than $CaCO_3$ for the studied PLA/LDPE blends. The significantly improved morphology of the ternary PLA/LDPE/SiO_2 blend was found to give an improved strength, toughness, and crystallinity. (Vrsaljko *et al.*, 2014)

PLA sheets reinforced with organically modified montmorillonite were manufactured through reactive extrusion-calendering using a masterbatch approach using a reactive agent. While calendered sheets exhibited a homogenous and preferential distribution of clay particles in MD, the coexistence of mixed structures, involving tactoids of various sizes as well as intercalated clay layers was observed. However, a higher and finer dispersion of clay particles was achieved through clay-polymer tethering via chain extender molecules. Under tensile loading, the composites exhibited the presence of several cavities at the clay/PLA interface resulting from a poor interfacial interaction between clays and the polymer. However, with a reactive agent, the composites exhibited multiple cavitation processes which slightly enhanced shear flow of the polymeric matrix. (Cailloux et al., 2016)

PLA/PCL/organically modified montmorillonite (OMMT) nanocomposites were melt-processed in a twin-screw extruder under high shear conditions. As a result of the processing conditions employed, the OMMT layers located in the less compatible PCL phase in all the ternary nanocomposites. The morphology of the PLA/PCL blend evolved from "sea-island" to co-continuous upon the addition of OMMT. The addition of OMMT to the PLA/PCL blend fully compensated the loss in stiffness and oxygen barrier performance produced by PCL in PLA; the nanocomposite with 3% OMMT showed the same modulus and permeability values as those of pure PLA. Moreover, the ductile behavior of the PLA/PCL blend remained constant even in the nanocomposite containing 5% OMMT. (Urquijo *et al.*, 2016) PL

Carbon-Based Composites

Carbon in several forms are used as reinforcing fillers for PLA. PLA/carbon nanofiber composites were successfully prepared and their properties were investigated. The tensile modulus shows a rise of almost 50% at 15 wt% of carbon nanofiber content. It was also revealed that the glass transition is not significantly affected the presence of carbon nanofiber. The presence of CNF favored crystallization, especially during the cooling cycle. (Anwer & Naguib, 2016) The continuous carbon fiber reinforced PLA composite was manufactured by the 3D printing technique in order to analyze the improvement of this new processing method. The experimental results demonstrated that the tensile strength and flexural strengths of modified carbon fiber reinforced composites were13.8% and 164% higher than original carbon fiber reinforced samples. (Li *et al.*, 2016)

The carbon nanotubes (CNTs) as shown in Figure 3 have been investigated as a reinforcing filler because of their high aspect ratio, excellent electrical conductivity, chemical stability, and superior mechanical properties. These properties observed at the nanoscale have motivated researchers to utilize carbon nanotubes as advanced filler materials in nanocomposites. CNTs are categorized as single-walled nanotubes (SWNTs) and multiwalled nanotubes (MWNTs). MWNTs are like a series of single-walled tubes nested within one another. In MWNTs only the outer wall generally contributes significantly to the electrical and mechanical properties when used in composites. In comparison to MWNTs, single-walled nanotubes are highly electrically conducting or semiconducting and have outstanding strength and very large surface area per unit mass, and also they have unique optical properties. However, their applications as additives in polymer matrixes are limited as CNTs are usually agglomerate due to van derWaals force thus difficult to disperse in polymer matrix, leading to defects in the composites and limiting the efficiency of CNTs on polymer matrix. Although many methods have been proposed, significant challenge in developing high-performance PLA/CNT nanocomposites is still how to enhance dispersion and alignment of CNTs in the PLA matrix. (Akbari et al. 2015) MWNTs in PLA/PS blends were prepared through masterbatch in a two-step process resulting in a double percolated morphology where PS percolates the PLA phase, and MWCNTs percolate the PS phase. The bicontinuous morphology and the high selectivity of MWCNT localization in the system were observed. (Nasti *et al.*, 2016)

Another class of carbon-based fillers are graphite and graphene. Graphite is one of the only three naturally occurring allotropes of carbon, besides diamond and amorphous carbon. The differences between the three naturally occurring allotropes are diamond having a diamond lattice crystalline structure, graphite having a honeycomb lattice structure while amorphous carbon is without a crystalline structure. Graphite is a very brittle material thus cannot be used as a structural material although it shows the superior ability to conduct electricity and heat, also highly resistant to chemical attack and self-lubricating. Graphite is made of hundreds of thousands of layers of graphene. Graphene is a single and tightly packed layer of carbon atoms that are bonded together in a hexagonal honeycomb. It is an allotrope of carbon in the structure of a plane of sp^2-bonded atoms. Graphene sheets stack to form graphite with an interplanar spacing of 0.335 nm and the carbon–carbon length in graphene is about 0.142 nm. A team of researchers from the University of Manchester successfully extracted single-atom-thick crystallites from bulk graphite using Scotch tape or mechanical cleavage technique. They were awarded Nobel Prizes in Physics in 2010. Graphene becomes one of the strongest materials, known with a breaking strength over 100 times greater than hypothetical steel with similar thickness. Its thermal conductivity is significantly higher than graphene oxide, MWCNTs as well as SWCNTs. Furthermore, graphene also shows excellent electric charge transport. Currently, graphene can be produced by chemical vapor deposition (CVD), mechanical cleavage and liquid-phase exfoliation, cutting open carbon nanotubes (CNT), carbon dioxide reduction and graphite conversion.

Graphite-based composites consisting of PLA and poly(ε-caprolactone) (PCL) immiscible blends are developed by means of preliminary dispersion of a high surface area graphite in the molten PCL by applying a sonication treatment. As a consequent of this processing, graphite turns out to be dispersed in the polymer matrix at a sub-micrometer level and acts as a nucleating agent for the PCL crystallization. The graphite/PCL is subsequently introduced in PLA through melt blending. SEM characterization demonstrates that the presence of graphite modifies the morphology of the blend. The filler is observed to ameliorate significantly the compatibility of PLA/PCL blends by increasing the interface adhesion between the two polymer phases thus enhancing the mechanical properties of the blend, improving the elongation at break and the Young's modulus. (Forouharshad *et al.* 2016)

Behavior of graphene nanoplate (GNP) and graphene oxide (GO) in solution mixed PLA/PMMA blends were investigated. Depending on the mixing sequence, the GNPs were localized in the both phases and interface through migration to reach thermodynamics equilibrium. However, GO localization was significantly affected by the mixing sequence due to better interaction with the polymer phases. (Paydayesh *et al.* 2016)

The same drawback exists with the preparation of graphene-based polymer nanocomposites as the inclination of incomplete exfoliation and restacking of graphene back to graphite can easily occur. The mechanism for interaction in polymer/graphene nanocomposites relies on the interfacial adhesion between fillers and matrix. The incorporation of nanostructures into polymer can usually be done by three main strategies; *in situ* intercalative polymerization, solution intercalation, and melt intercalation.

For the *in situ* intercalative polymerization, graphene particles are first dispersed in liquid monomer or solution, followed by the polymerization. The process assists in raising the interlayer spacing and exfoliates the layered structure of graphite so a high level of dispersion of graphene-based filler is achieved without a prior exfoliation step. Functionalized graphene can also improve the initial dispersion in the monomer. This method thus needs to be achieve at the upstream step.

The solution intercalation is based on solvent system, in which, the polymer or pre-polymer is dissolved and graphene or modified graphene layers are dispersed. The common suitable solvents for graphene or modified graphene are water, acetone, chloroform, tetrahydrofuran, dimethyl formamide, and toluene, which bring the weak forces to stack the layers together. When the solvent is evaporated, the polymer and graphene are left to form nanocomposites. Basically, this process involves the mixing of colloidal suspensions of graphene-based material with polymers in the same solvent with some exfoliation steps such as ultrasonication. This method is a preferred fabrication technique due to the ease of processing, but the solvent evaporation is not so economically effective. The co-precipitation technique is proposed to assist the solvent removal issue.

Graphene-based polymer nanocomposites can be fabricated through melt intercalation, in which, graphene is mixed when polymer is heated to molten state. The process avoids the use of toxic solvent and it is applicable using a twin screw extruder, a two-roll mill or an internal mixer, which offers large-scale processing. This method is suitable for almost all thermoplastic polymers. However, since graphene has a tendency to restack and agglomerate in polymer matrix, the aggregated graphene shows similar properties to graphite rather than graphene itself. The graphene functionalized through modification is consequently preferred for melt intercalation. (Norazlina & Kamal, 2015)

Nanocomposites based on PLA containing 3 phr of modified montmorillonites and different amounts of graphene were prepared using melt blending process. The addition of montmorillonites into PLA increased melt viscosity and storage modulus of graphene/PLA composites. The synergetic effect between montmorillonite and graphene attributed to the better dispersion of fillers in the PLA matrix. PLA nanocomposites show effective improvement of mechanical properties when compared with the PLA. (Bouakaz *et al.*, 2015)

Green Composites

Although PLA is a biodegradable plastic, the ability of the products to be reused or recycled is still crucial. To be as environmentally friendly as possible, the use of either non-renewable or renewable natural resources should be optimized as materials demand is growing steadily. Furthermore, the reuse and recycling must be carried out in order to overcome the higher cost of PLA. Nevertheless, the use of conventional reinforcing fillers to produce PLA composites leads to the complication of reuse, recycling or biodegradation of the composites. Scientists thus try to develop reinforcing fillers which will be able to degrade or recycling, most of them are natural fibers.

Fiber-reinforced plastics (FRP) have been used in a wide range of applications, especially where high strength and stiffness are required, unfortunately, traditional FRP composites often pose considerable problems with respect to their reuse or recycling at the end of their usable lifetime, mainly because of the complication of the separation method of the nonbiodegradable fibers and matrixes. This led to the development of green composites of natural fibers and/or biodegradable resin. At present, a variety of partially biodegradable and green composites have been developed with fairly good mechanical properties using different natural fibers (e.g. flax, ramie, hemp, etc.) and biodegradable polymers.

The main components of green composites are matrix and reinforcing filler. The matrix is the continuous phase which isolates the fillers from one another and holds them in place. A good matrix should be able to transfer the load onto the fillers and evenly distribute stress concentration. The reinforcing fillers are the second main component which normally enhance the mechanical properties of the matrix. Due to presence of different components, the interfacial adhesion between phases should be improved to facilitate stress transfer. Various surface treatments of fillers, e.g. silane, acetone and alkali treatment, as well as coupling agents are frequently employed to improve the wettability of the fillers.

In general, mechanical properties of composites depend on a number of parameters such as percentage of filler content, interfacial characteristics between filler and matrix, filler aspect ratio, surface modification of fillers and addition of coupling agents to enhance the compatibility between filler and matrix.

The tensile strength of green composites can be enhanced by reducing the stress concentration. The filler characteristics like wettability of fillers into matrix and filler loading are responsible for improving tensile stiffness of green composites. There are reports that PLA-based green composites possess better stiffness than neat PLA and PLA is not degraded by the compounding process. An important mechanical property of green composites is impact strength which is generally lower when compared with glass fiber composites. To obtain enhanced toughness, the level of adhesion, interfacial bonding, fiber pullout and energy absorption are crucial.

The research and development work carried out by researchers and technologist has shown that green fiber composites are comparable or even superior to synthetic fiber composites. (*Bajpai* et al., 2014) Natural fibers like flax, nettle, jute, sisal, kenaf, etc. are important raw materials not only for the textile industry but also for modern eco-friendly composites used in different areas of application, ranging from everyday life to hi-tech application. Still more research and development is required for the extraction, characterization and property modification using various surface modification techniques of natural fillers. Also, research for improving the adhesion characteristics of matrix and natural fillers through physico-chemical treatment of fillers is ongoing. The batch-to-batch variation in natural filler composition and quality can lead to consistency problems in the final products.

Micro fibers from cereal husk was employed as a reinforcing filler for PLA composites with a filler load of up to 40 wt.%. The removal of unwanted materials on the fiber surface and some of the amorphous portions from the fiber surface were achieved without damaging the fiber surface when using the enzyme modification. It was found that the modification facilitated the improved adhesion and as well as the increment of modulus, failure strains and strength properties of the resulting PLA composites. (Mamun & Bledzki, 2013)

Highly exfoliated, biodegradable thermoplastic starch (TPS)/ PLA/montmorillonite nanocomposites were prepared by an eco-friendly approach, involving *in situ* gelatinization of potato starch in presence of dispersed nanoclay followed by melt mixing with PLA. The morphological analysis revealed that montmorillonite was selectively dispersed into the thermoplastic starch in a highly delaminated manner. An increase in mechanical as well as thermomechanical properties was evident in the presence of PLA and more influenced in the presence of clay. (Ayana *et al.*, 2014) PLA/TPS blends were prepared by melt extrusion process by the addition of an aliphatic/aromatic biodegradable polyester that provides good resistance to aging and moisture. This elongation at break of this blend slightly improved but phase separation was observed due to the lack of strong interactions between the two polymers. This was improved by the incorporation of small amounts of maleinized linseed oil (MLO). The decrease of the glass transition temperature indicated that MLO provides plasticizing-compatibilizing effect as illustrated by the SEM micrographs suggesting an increase in compatibility. (Ferri *et al.*, 2016b) These improved mechanical properties along with its biodegradability might lead to a green composite in the area of packaging.

The preparation of nanocrystalline cellulose (NCC) filled PLA green composites via a continuous route. It combines several separated steps, including extraction of NCC, surface acetylation of NCC, and final composite preparation, without traditional freeze drying. The obtained PLA composites were then studied and the results reveal that surface acetylation of NCC can improve its affinity to PLA evidently. The phase interface adhesion also plays an important role in the mechanical behavior of PLA/NCC composites. (Xu *et al.*, 2016)

PBS-*g*-cellulose nanocrystal (CNC) was prepared through *in situ* polymerization and was utilized as a filler in PBS/PLA blends in the presence of dicumyl peroxide (DCP). The addition of DCP and PBS-*g*-CNC decreased the PBS dispersed phase PLA matrix and improve the dispersion of PBS-*g*-CNC in both PBS and PLA phases. The crystallinity and impact strength of the composites increased markedly. The addition of DCP also increased the thermal stability of composites due to of the crosslink effect of DCP on PBS/PLA matrix. (Zhang & Zhang, 2016)

Properties of PLA and gelatinized corn and tapioca starches (GS) blend films at various amounts in the presence of methylenediphenyl diisocyanate and polyethylene glycol as a compatibilizer was investigated. The results indicated that the presence of small amount of MDI significantly increased the tensile properties of the blends compared with the uncompatibilized blends. This is attributed to an improvement of the interfacial interaction between PLA and GS phases, as evidenced by the morphological results. (Phetwarotai *et al.*, 2012; Phetwarotai *et al.*, 2013)

APPLICATIONS OF PLA

As a matter of fact, bioplastics have found its industrial usages primarily in food packaging application. PLA can also be used as a decomposable packaging material, either by extrusion or injection techniques. Films, bags, disposable cups, and other disposable tableware have now been made from PLA. The higher price of raw material limits its consumption volume comparing with traditional commodity plastics. PLA can also be processed into fibers, yarns and fabrics. These textile products have many potential uses, for example as upholstery, disposable garments, awnings, and diapers.

PLA products have been approved by the US Food and Drug Administration (FDA) for human clinical applications. Its biocompatibility in the human body allows PLA to be used for biomedical applications. PLA is fabricated into tissue engineering scaffolds using electrospinning and 3D printing techniques. Nevertheless, most of the research is *in vitro* rather than *in vivo*. Its drawbacks include hydrophobic nature and slow hydrolytic degradation. PLA is also used as a delivery tools to achieve the targeted delivery of bioactive compounds to areas in the body to maximize therapeutic potential and minimize side-effects. With its excellent biocompatibility, biodegradability, mechanical strength, PLA can be used to produce dosage forms such as pellets, microcapsules, microparticles, or nanoparticles. Membrane applications (e.g. wound covers), implants and medical devices, and dermatological treatments are among the biomedical applications which PLA can be utilized. (Xiao *et al.*, 2012)

A compatible PLA/PBAT blend films were prepared using a solvent-casting method. The water vapor permeability and oxygen transfer rate of the blend films were varied depending on their blending ratio. The blend films also exhibited antifogging effect with reduced quality changes in the food packaging application. (Wang *et al.*, 2016)

There are several plastics which are being explored for 3D Printing, however the two dominant plastics are ABS and PLA. PLA responds to moisture by generating bubbles or spurting at the nozzle as well as discoloration as well as lower mechanical properties. Comparing with ABS, PLA undergoes more of a phase-change when heated and becomes much more liquid. If actively cooled, much sharper details can be seen on printed corners without the risk of cracking or warp. The increased flow can also lead to stronger binding between layers, improving the strength of the printed part. These characteristics as well as its low warping on parts make it a popular material for 3D printing.

Lignin/PLA based carbon fibers having high tensile strength and modulus were prepared by melt spinning of lignin/PLA blends followed by the thermal stabilization and carbonization of the as-spun fibers. Due to the hydrogen-bonding interaction between lignin and PLA phases, the tensile modulus of the blend-based carbon fibers relative to that from lignin increased. However, the tensile strength of Lignin/PLA based carbon fibers decreased because of the presence of voids that were caused by the volatilization of PLA during the carbonization processes. (Wang *et al.*, 2015)

The increasing amount of waste from electronic gadget which has relatively lower life-cycle becomes a concern nowadays. PLA delivers performance in consumer electronics comparable to PP and ABS with the ability of biodegradaion. Furthermore, static dissipative biopolymer composites have been prepared by adding conducting components such as polypyrrole or graphene particles into PLA. It was found that the resistivity of the composites is in optimal electrostatic discharge (ESD) protection range, offering potential in electronic packaging applications where ESD is a major concern. PLA/graphene nanocomposites with 0–15 wt% (0–9.1 vol%) of nanofiller were prepared. Electromagnetic properties and electromagnetic interference shielding effectiveness of the nanocomposites were determined. It was observed that graphene did not alter its magnetic permeability but enhanced its electrical permittivity and conductivity markedly leading to higher shielding effectiveness. (Kashi *et al.*, 2016)

Composite fibers from PLA reinforced with graphene nanoparticles were prepared. Graphene was prepared into masterbatch with poly(butylene succinate) and poly(lactic acid) via the co-precipitation technique. The obtained masterbatch was then used to prepared composite compounds at various ratios. The compounds were processed into fiber by the melt spinning method. The results indicate that, if graphene at the optimal content was used, the tensile strength increased. PLA fiber with 1.2 phr gave the increasing tensile strength around 42.7% comparing with the unfilled fiber. However, when the graphene content increased beyond the optimal point, the tensile strength decreased due to agglomeration of graphene particles. The electrical resistivity gradually decreased when the amount of graphene in the composite fibers increased. Although the composite fibers did not show notable electrical conductivity behavior, they can be used for the applications which anti-static property is needed. (Poompiew & Potiyaraj, 2016)

FUTURE RESEARCH DIRECTIONS

Although PLA has been discovered for almost a century ago, its usage properties are still limited and research works are still needed. Researchers are focused on how to take more control on PLA structure during the polymerization process so that the desired processability and properties can be reached while the cost of production can be competed to that of other conventional plastics. For the time being, the brittleness of PLA is still the main objective of its research and development while its new value-added applications should be explored in order to match the high price for example as a biomaterial or in an electronic application. The shape memory effect of PLA has been investigated. Polymer blend nanocomposites based on thermoplastic polyurethane (PU) elastomer, PLA and surface modified carbon nanotubes were prepared via simple melt mixing process. Shape memory studies of the PU/PLA/modified CNT composites exhibit a remarkable recoverability of its shape at lower applied dc voltages, when compared to pure or pristine CNT loaded system. (Raja *et al.*, 2013) The shape memory effect of PLA can be used to design prototype systems for a wide range of biomedical applications, including self-folding devices, microactuators for treating stroke, cardiovascular stents, self-tightening sutures, and minimally invasive/

self-fitting tissue engineering scaffolds. (Xu & Song, 2015) While the use of 2D materials, especially graphene has been the current focused of materials technologist.

Reinforcement of PLA with other materials creates huge opportunities for future. As natural fillers are low cost and degradable, it's reinforcement with PLA leads to a material which is economical as well as ecological. New technologies are used to produce these green composites so as to give it more strength and stiffness as well as many other features suitable for specific applications. A major problem to the commercialization of green composites is the high cost associated with bioplastics. Future attempts in developing cheaper production techniques of these biodegradable matrixes through faster and more efficient process, would certainly be beneficial. Another area of research interest is green nanocomposites which is an emerging attention. (NatureWorks, 2007)

CONCLUSION

After being discovered for almost a century ago, poly(lactic acid) or PLA is now globally emerged as one of the most consumed bioplastics due to current environmental concerns and fluctuated fossil-fuel price. New synthesis techniques have been explored in order to obtain PLA with desired molecular weight and structure to suit demanding requirements. At the present, PLA can be processed using common large-scale polymer processing instruments. However, the mechanical strength, especially its toughness, is still its downside. Several researchers are now seeking resolution to this shortcoming. The copolymerization technique can improve the property of PLA intrinsically but may be difficult to utilize economically. The addition of toughening agent in the form of plasticizers, polymer blends, or polymer composites can also be accomplished effortlessly. However, due to its higher cost than conventional plastics, although it is currently much lower than in previous years, one should try to extend its applications, apart from common usages, to more advance purposes, for instance, in biomedical or in electrical uses.

REFERENCES

Akbari, A., Majumder, M., & Tehrani, A. (2015). Polylactic acid (PLA) carbon nanotube Nanocomposites. In Handbook of Polymer Nanocomposites. Processing, Performance and Application – Volume B: Carbon Nanotube Based Polymer Composites (pp. 283-298). Springer-Verlag.

Anakabe, J., Huici, A. M. Z., Eceiza, A., & Arbelaiz, A. (2016). The effect of the addition of poly(styrene-co-glycidyl methacrylate) copolymer on the properties of polylactide/ poly(methyl methacrylate) blend. *Journal of Applied Polymer Science*, *133*(37), 43935–43945. doi:10.1002/app.43935

Anwer, M. A. S., & Naguib, H. E. (2016). Study on the morphological, dynamic mechanical and thermal properties of PLA carbon nanofibre composites. *Composites. Part B, Engineering*, *91*, 631–639. doi:10.1016/j.compositesb.2016.01.039

Arruda, L. C., Magaton, M., Bretas, R. E. S., & Ueki, M. M. (2015). Influence of chain extender on mechanical, thermal and morphological properties of blown films of PLA/PBAT blends. *Polymer Testing*, *43*, 27–37. doi:10.1016/j.polymertesting.2015.02.005

ASTM. (2009). *ASTM D6400-04 Standard Specification of Compostable Plastics*. West Conshohocken, PA: ASTM International.

Ayana, B., Suin, S., & Khatua, B. B. (2014). Highly exfoliated eco-friendly thermoplastic starch (TPS)/poly (lactic acid)(PLA)/clay nanocomposites using unmodified nanoclay. *Carbohydrate Polymers*, *110*, 430–439. doi:10.1016/j.carbpol.2014.04.024 PMID:24906776

Bajpai, P. K., Singh, I., & Madaan, J. (2014). Development and characterization of PLA-based green composites: A review. *Journal of Thermoplastic Composite Materials*, *227*(1), 52–81. doi:10.1177/0892705712439571

Bijarimi, M., Ahmad, S., Rasid, R., Khushairi, M. A., & Zakir, M. (2016). Poly(lactic acid) / Poly(ethylene glycol) blends: Mechanical, thermal and morphological properties. AIP Conference Proceedings, 1727, 020002-1-02002-9.

Bouakaz, B. S., Pillin, I., Habi, A., & Grohens, Y. (2015). Synergy between fillers in organomontmorillonite/graphene-PLA nanocomposites. *Applied Clay Science*, *116-117*, 69–77. doi:10.1016/j.clay.2015.08.017

Cailloux, J., Hakim, R. N., Santana, O. O., Bou, J., Abt, T., Sánchez-Soto, M., & Maspoch, M. L. et al. (2016, September). Reactive extrusion: A useful process to manufacture structurally modified PLA/o-MMT composites. *Composites. Part A, Applied Science and Manufacturing*, *88*, 106–115. doi:10.1016/j.compositesa.2016.05.024

Correa-Pacheco, Z. N., Jiménez-Pérez, J. L., Sabino, M. A., Cruz-Orea, A., & Loaiza, M. (2015). Photothermal and morphological characterization of PLA/PCL polymer blends. *Applied Physics. A, Materials Science & Processing*, *120*(4), 1323–1329. doi:10.1007/s00339-015-9309-7

de Oliveira, A. G., da Silva, A. L. N., de Sousa, A. M. F., Leite, M. C. A. M., Jandorno, J. C., & Escócio, V. A. (2016). Composites based on green high-density polyethylene, polylactide and nanosized calcium carbonate: Effect of the processing parameter and blend composition. *Materials Chemistry and Physics*, *181*, 344–351. doi:10.1016/j.matchemphys.2016.06.068

Deng, Y., & Thomas, N. L. (2015). Blending poly(butylene succinate) with poly(lactic acid): Ductility and phase inversion effects. *European Polymer Journal*, *71*, 534–546. doi:10.1016/j.eurpolymj.2015.08.029

Dil, E. J., Carreau, P. J., & Favis, B. D. (2015). Morphology, miscibility and continuity development in poly(lactic acid)/poly(butylene adipate-co-terephthalate) blends. *Polymer*, *68*, 202–212. doi:10.1016/j.polymer.2015.05.012

Ferri, J. M., Fenollar, O., Jorda-Vilaplana, A., García-Sanoguera, D., & Balart, R. (2016a). Effect of miscibility on mechanical and thermal properties of poly(lactic acid)/polycaprolactone blends. *Polymer International*, *65*(4), 453–463. doi:10.1002/pi.5079

Ferri, J. M., Garcia, D. G., Sánchez-Nacher, L., Fenollar, O., & Balart, R. (2016b). The effect of maleinized linseed oil (MLO) on mechanical performance of poly(lactic acid)-thermoplastic starch (PLA-TPS) blends. *Carbohydrate Polymers*, *147*, 60–68. doi:10.1016/j.carbpol.2016.03.082 PMID:27178909

Forouharshad, M., Gardella, L., Furfaro, D., Galimberti, M., & Monticelli, O. (2015). A low-environmental-impact approach for novel biocomposites based on PLLA/PCL blends and high surface area graphite. *European Polymer Journal*, *70*, 28–36. doi:10.1016/j.eurpolymj.2015.06.016

Hamad, K., Kaseem, M., Deri, F., & Ko, Y. G. (2016). Mechanical properties and compatibility of poly-lactic acid/polystyrene polymer blend. *Materials Letters*, *164*, 409–412. doi:10.1016/j.matlet.2015.11.029

Hao, X., Kaschta, J., Pan, Y., Liu, X., & Schubert, D. W. (2016). Intermolecular cooperativity and entanglement network in a miscible PLA/PMMA blend in the presence of nanosilica. *Polymer*, *82*, 57–65. doi:10.1016/j.polymer.2015.11.029

Hu, Y., Daoud, W. A., Cheuk, K. K. L., & Lin, C. S. K. (2016). Newly developed techniques on polycondensation, ring-opening polymerization and polymer modification: Focus on poly(lactic acid). *Materials (Basel)*, *9*(3), 133–147. doi:10.3390/ma9030133

Jašo, V., Glenn, G., Klamczynski, A., & Petrović, Z. S. (2015). Biodegradability study of polylactic acid/thermoplastic polyurethane blends. *Polymer Testing*, *47*, 1–3. doi:10.1016/j.polymertesting.2015.07.011

Kashi, S., Gupta, R. K., Baum, T., Kao, N., & Bhattacharya, S. N. (2016). Morphology, electromagnetic properties and electromagnetic interference shielding performance of poly lactide/graphene nanoplatelet nanocomposites. *Materials & Design*, *95*, 119–126. doi:10.1016/j.matdes.2016.01.086

Kim, S. H., & Kim, Y. H. (1999). Direct condensation polymerization of lactic acid. *Macromolecular Symposia*, *144*(1), 277–287. doi:10.1002/masy.19991440125

Lai, S.-M., Wu, W.-L., & Wang, Y.-J. (2016). Annealing effect on the shape memory properties of poly-lactic acid (PLA)/thermoplastic polyurethane (TPU) bio-based blends. *Journal of Polymer Research*, *23*(5), 99–111. doi:10.1007/s10965-016-0993-6

Li, F.-J., Liang, J.-Z., Zhang, S.-D., & Zhu, B. (2015). Tensile properties of polylactide/poly(ethylene glycol) blends. *Journal of Polymers and the Environment*, *23*(3), 407–415. doi:10.1007/s10924-015-0718-7

Li, N., Li, Y., & Liu, S. (2016). Rapid prototyping of continuous carbon fiber reinforced polylacticacid composites by 3D printing. *Journal of Materials Processing Technology*, *238*, 218–225. doi:10.1016/j.jmatprotec.2016.07.025

Lim, S.-W., Choi, M.-C., Jeong, J.-H., Park, E.-Y., & Ha, C.-S. (2016). Toughening poly(lactic acid) (PLA) through *in situ* reactive blending with liquid polybutadiene rubber (LPB). *Composite Interfaces*, *23*(8), 807–818. doi:10.1080/09276440.2016.1175168

Liu, H., & Zhang, J. (2011). Research progress in toughening modification of poly(lactic acid). *Journal of Polymer Science. Part B, Polymer Physics*, *49*(15), 1051–1083. doi:10.1002/polb.22283

Lopes, M. S., Jardini, A. L., & Filho, R. M. (2014). Synthesis and characterizations of poly (lactic acid) by ring-opening polymerization for biomedical applications. *Chemical Engineering Transactions*, *38*, 331–336.

Madhu, G., Bhunia, H., Bajpai, P. K., & Nando, G. B. (2016). Physico-Mechanical Properties and Biodegradation of Oxo-degradable HDPE/PLA Blends. *Polymer Science, Series A*, *58*(1), 57–75. doi:10.1134/S0965545X16010077

Mamun, A., & Bledzki, A. K. (2013). Micro fibre reinforced PLA and PP composites: Enzyme modification, mechanical and thermal properties. *Composites Science and Technology*, *78*, 10–17. doi:10.1016/j.compscitech.2013.01.013

Messenger, B. (Ed.). (2013). *Analysis: Bioplastics sector set for rapid growth up to 2017. In Waste Management World*. Vienna, Austria: Industriemagazin Verlag.

Nasti, G., Gentile, G., Cerruti, P., Carfagna, C., & Ambrogi, V. (2016). Double percolation of multi-walled carbon nanotubes in polystyrene/polylactic acid blends. *Polymer*, *99*, 193–203. doi:10.1016/j.polymer.2016.06.058

NatureWorks. (2007). *Technology Focus Report: Polylactic Acid Containing Fillers and Fibers*. Minnetonka, MN: NatureWorks.

Nofar, M., Heuzey, M. C., Carreau, P. J., & Kamal, M. R. (2016). Effects of nanoclay and its localization on the morphology stabilization of PLA/PBAT blends under shear flow. *Polymer*, *98*, 353–364. doi:10.1016/j.polymer.2016.06.044

Norazlina, H., & Kamal, Y. (2015). Graphene modifications in polylactic acid nanocomposites: A review. *Polymer Bulletin*, *72*(4), 931–961. doi:10.1007/s00289-015-1308-5

Paydayesh, A., Arefazar, A., & Jalaliarani, A. (2016). A morphological study on the migration and selective localization of graphene in the PLA/PMMA blends. *Journal of Applied Polymer Science*, *133*(34), 43799–43810. doi:10.1002/app.43799

Phetwarotai, W., Potiyaraj, P., & Aht-Ong, D. (2012). Characteristics of biodegradable polylactide/gelatinized starch films: Effects of starch, plasticizer, and compatibilizer. *Journal of Applied Polymer Science*, *126*(S1), E162–E172. doi:10.1002/app.36736

Phetwarotai, W., Potiyaraj, P., & Aht-Ong, D. (2013). Biodegradation of polylactide and gelatinized starch blend films under controlled soil burial conditions. *Journal of Polymers and the Environment*, *21*(1), 95–107. doi:10.1007/s10924-012-0530-6

Pivsa-Art, W., Fujii, K., Nomura, K., Aso, Y., Ohara, H., & Yamane, Y. (2015). The effect of poly(ethylene glycol) as plasticizer in blends of poly(lactic acid) and poly(butylene succinate). *Journal of Applied Polymer Science*, *133*(8), 43044–43053.

Poompiew, N., & Potiyaraj, P. (2016, January) *Preparation and properties of poly (lactic acid) reinforced with graphene nanoparticle*. Paper presented at International Conference on Applied Sciences, Fukuoka, Japan.

Raja, M., Ryu, S. H., & Shanmugharaj, A. M. (2013). Thermal, mechanical and electroactive shape memory properties of polyurethane (PU)/poly (lactic acid) (PLA)/CNT nanocomposites. *European Polymer Journal*, *49*(11), 3492–3500. doi:10.1016/j.eurpolymj.2013.08.009

Sareeladdanon, S., & Potiyaraj, P. (2014). Mechanical properties of PLA/LLDPE films reinforced with silica from rice husk. *Advanced Materials Research*, *1025-1026*, 221–226. doi:10.4028/www.scientific.net/AMR.1025-1026.221

Schneider, J., Manjure, S., & Narayan, R. (2016). Reactive modification and compatibilization of poly(lactide) and poly(butylene adipate-co-terephthalate) blends with epoxy functionalized-poly(lactide) for blown film applications. *Journal of Applied Polymer Science, 133*(16), 43310–43318. doi:10.1002/app.43310

SPI Bioplastics Council. (2012). *Bioplastics Industry Overview Guide*. Washington, DC: The Society of the Plastics Industry.

Urquijo, J., Dagréou, S., Guerrica-Echevarría, G., & Eguiazábal, J. I. (2016). Structure and properties of poly(lactic acid)/poly(ε-caprolactone) nanocomposites with kinetically induced nanoclay location. *Journal of Applied Polymer Science, 133*(33), 43815–43826. doi:10.1002/app.43815

Vrsaljko, D., Macut, D., & Kovačević, V. (2014). Potential role of nanofillers as compatibilizers in immiscible PLA/LDPE blends. *Journal of Applied Polymer Science, 132*(6), 41414–41428.

Wang, L. F., Rhim, J. W., & Hong, S. I. (2016). Preparation of poly(lactide)/poly(butylene adipate-co-terephthalate) blend films using a solvent casting method and their food packaging application. *LWT-. Food Science and Technology (Campinas.), 68*, 454–461.

Wang, S., Li, Y., Xiang, H., Zhou, Z., Chang, T., & Zhu, M. (2015). Low cost carbon fibers from bio-renewable Lignin/poly(lactic acid) (PLA) blends. *Composites Science and Technology, 119*, 20–25. doi:10.1016/j.compscitech.2015.09.021

Wongsampanwech, S., & Potiyaraj, P. (2014). Toughening of poly(lactic acid) by blending with poly (ethylene-co-octene). *Advanced Materials Research, 1025-1026*, 461–465. doi:10.4028/www.scientific.net/AMR.1025-1026.461

Xiao, L., Wang, B., Yang, G., & Gauthier, M. (2012). Poly(lactic acid)-based biomaterials: Synthesis, modification and applications. In D. N. Ghista (Ed.), *Biomedical Science, Engineering and Technology* (pp. 247–282). Rijeka, Croatia: Intech Europe. doi:10.5772/23927

Xu, C., Chen, J., Wu, D., Chen, Y., Lv, Q., & Wang, M. (2016). Polylactide/acetylated nanocrystalline cellulose composites preparedby a continuous route: A phase interface-property relation study. *Carbohydrate Polymers, 146*, 58–66. doi:10.1016/j.carbpol.2016.03.058 PMID:27112851

Xu, J., & Song, J. (2015). Polylactic acid (PLA)-based shape-memory materials for biomedical applications. In L. Yahia (Ed.), *Shape Memory Polymers for Biomedical Applications* (pp. 197–217). London, UK: Woodhead Publishing. doi:10.1016/B978-0-85709-698-2.00010-6

Zeng, J.-B., Lia, K.-A., & Du, A.-K. (2015). Compatibilization strategies in poly(lactic acid)-based blends. *RSC Advances, 5*(41), 32546–32565. doi:10.1039/C5RA01655J

Zhang, X., & Zhang, Y. (2016). Reinforcement effect of poly(butylene succinate) (PBS)-graftedcellulose nanocrystal on toughened PBS/polylactic acid blends. *Carbohydrate Polymers, 140*, 374–382. doi:10.1016/j.carbpol.2015.12.073 PMID:26876864

KEY TERMS AND DEFINITIONS

Biodegradable Plastics: Plastics that decomposes naturally by the action of living microorganisms. This is achieved when microorganisms break down the structure of the polymers resulting in less environmentally toxic substances for example carbon dioxide, water, inorganic compounds and biomass. Biodegradable plastics can be either petroleum-based plastics or bioplastics.

Bioplastics: Plastics synthesized from renewable biomass resources, such as agricultural products and microorganisms in order to reduce greenhouse effect of the common petroleum-based plastics. Most of bioplastics are biodegradable providing a suitable condition.

Graphene: A single layer of carbon bonded together in a hexagonal honeycomb lattice. It is an allotrope of carbon in the structure of a plane of sp^2 bonded atoms with a molecule bond length of 0.142 nanometers.

Green Composites: Composite materials that compose wholly, or in part, of components derived from renewable resources.

Poly (Lactic Acid) or Poly Lactide (PLA): A thermoplastic aliphatic polyester synthesized from renewable resources, such as corn, tapioca, or sugarcane. PLA, which is discovered by Carothers and his colleagues at DuPont in 1932, is also a biodegradable plastic.

Polymer Blends: Macroscopically homogeneous mixtures in which two or more different polymers are blended to achieve various property combinations of the resulting materials. It is normally considered as a more cost-effective way to obtain materials with required properties than by synthesis of new polymers.

Polymer Composites: Multicomponent materials comprising different phases, in which at least one component is a polymer and at least one phase is a continuous phase.

Chapter 6
Inorganic-Organic Composite Materials from Liquid Natural Rubber and Epoxidised Natural Rubber Derivatives:
Prospects and Applications

Wan Ahmad Kamil Mahmood
Universiti Sains Malaysia, Malaysia

Mohammad Hossein Azarian
Universiti Sains Malaysia, Malaysia

ABSTRACT

Organic-Inorganic composite materials (OICs) are used to describe the group of materials synthesized from polymers and inorganic metal alkkoxides. The interests in these materials arised from the need to 'combine' the physical properties of inorganic glass materials and polymers such that the resultant OICs have the strength of the inorganic glass and flexibiliy of polymeric materials. Sol-gel technique have been the technique of choice due to much of its advantages, in particular the low temperature reaction. This is very important when natural rubber and its derivatives are used as the polymer component of the OICs. Work in our laboratory has demonstrated that OICs form liquid natural rubber (LNR) and 50% epoxidised natural rubber (ENR-50) can be prepared from various metal alkoxides, such silicon, zirconium and titanium. The OICs can be prepared as flexible transparent films, nanofibers and nano-beads. This Chapter will describe the preparation techniques and the properties of these OICs from various compositions of one and more metal alkoxides in both LNR and ENR-50. The applications of these materials in PANI will be briefly described.

DOI: 10.4018/978-1-5225-1971-3.ch006

INTRODUCTION

Inorganic-Organic- composite materials (IOCs) is used to describe the group of materials synthesized from polymers and inorganic metal alkkoxides. The interests in these materials arised from the need to 'combine' the physical properties of inorganic glass materials and polymers such that the resultant IOCs have the strength of the inorganic glass and flexibility of polymeric materials. Sol-gel have been the technique of choice, due to much of its advantages, in particular the low temperature reaction. This is very important when natural rubber and its derivatives are used as the polymer component of the IOCs. It has been demonstrated that IOCs form liquid natural rubber (LNR) and 50% epoxidised natural rubber (ENR-50) can be prepared from various metal alkoxides, such silicon, zirconium and titanium. The IOCs can be prepared as flexible transparent films, nanofibers and nanobeads. This Chapter describes the preparation techniques and the properties of these IOCs from various compositions of one and more metal alkoxides in both LNR and ENR-50. The applications of these materials in polyaniline(PANI) also are briefly described.

Inorganic-Organic Composite Materials and Sol-Gel Techniques

The interest to combine the properties of inorganic and organic components in order to create new composite materials has been a long time challenge to chemist, particularly material and inorganic chemist. IOC materials are prepared or synthesized to seek new materials with improved properties. Organic polymers possess some limitations such as low thermal stability and poor mechanical, electrical and optical properties due to their intrinsic nature. While inorganic materials are usually known for their excellent thermal, electrical, optical and magnetic properties, addition within polymer components can further improve their properties and enhance their applications. IOCs can generally be prepared or synthesized by combining organic polymers and inorganic components in a chemical reaction. For example, inorganic metal alkoxides, which are usually used as the inorganic component, form the inorganic part of the network in the composites. These composite materials have become an important part of material science research to prepare new materials with excellent mechanical, thermal, electronic and optical properties It is noted that commercial IOCs have been part of the manufacturing industry since 1950s (Sanchez, Julián, Belleville, & Popall, 2005).

There are several routes to prepare IOCs material, but the most prominent is by the sol-gel process which offers unique advantages as followed: homogeneous multi-component system can be easily obtained, temperature required for material processing is considerably low and the rheological properties of the sol allows the formation of thin film, fiber and composite by different techniques such as dip coating, spinning and solvent casting.

The sol-gel process can be define as the formation of metal oxide networks through polycondensation reactions of a precursor in a polymer solution liquid phase (Livage, Henry, & Sanchez, 1988; Livage, 1997). Sol is a suspension of particles of one phase in another major phase (polymer solution). When particles in a sol form continuous network, the sol loses its mobility and becomes a soft solid therefore the sol has formed gel. The sol-gel process begins with a solution of metal alkoxide precursors [$M(OR)_4$)] and water, where M is a transition metal network-forming element, and R is an alkyl group. Hydrolysis and condensation of the alkoxide are the two fundamental steps to produce a network in presence of polymer matrix. For the most transition metal oxide precursors the hydrolysis and condensation reactions are too fast which resulting in loss of morphological and structural control in the final oxide material.

To overcome this problem the addition of water can be eliminated and reaction may carry out under reduced pressure.

The first sol-gel process began in 1845 by Ebelman (Ebelman, 1845), when he reported the formation of transparent material as a result of hydrolysis of an ester of silicic acid. Later in 1960's, Shroeder (Shroeder, 1962) deposited transparent coating on glass surface in order to improve refractive index of glass by using thin layer of titanium butoxide as a precursor. Sol-gel science has been investigated extensively since the mid of 1970 till now when variety of IOCs formed from metal alkoxide solution (West, 1990). Hashim et al. (Hashim, Kawabata, & Kohjiya, 1995) have employed the sol-gel technique to prepare silica-reinforced vulcanizates using tetraethoxysilane (TEOS) and epoxidized natural rubber (ENR). The sol-gel vulacanizates obtained were more rigid and stronger than typical sulfur-cured ENR vulcanizated that contained comparable amount of silica. The stress-strain and dynamic mechanical property analysis suggested that the chemicals bonds are formed between silica network particles and ENR polymer matrix. Zareba-Grodz et al. (Zarba-Grodź, Miśta, Strek, Bukowska, Hermanowicz, & Maruszewski, 2004) synthesized an inorganic–organic hybrid by sol–gel process and found that the organic and inorganic phases were formed from two interpenetrating polymeric networks. The organic phase was a copolymer of acrylamide and 2-hydroxyethylmethacrylate while the inorganic phase was a silica polymer.

Natural Rubber and Its Modified Derivatives

Natural rubber (NR) or poly (*cis*-1, 4-isoprene) is an elastic biopolymer obtained from more than 2000 different plant species belongs to the families of *Euphorbiaceae* or *Compositaceae* spurge. However *Hevea brasiliensis* is the only significant commercial source of NR which named the rubber tree. NR is a main constituent of the coagulated milky latex obtained during healing of wounds caused by mechanical injury of rubber trees. In fact *Hevea brasiliensis* trees used NR to poison the insects or to rapidly heal wounds. The world production of natural rubber in 1998 was about 6.6 million tons which more than 70% of this rubber was produced in Thailand, Indonesia and Malaysia (Shah, Hasan, Shah, Kanwal & Zeb, 2013). Latex obtained from rubber trees is consists of 50-70% (wt/wt) water, 25-35% (wt/wt) polyisoprene, 1-1.8% (wt/wt) protein, 1-2% (wt/wt) carbohydrates, 0.4-1.1% (wt/wt) neutral lipids, 0.5-0.6% (wt/wt) polar lipids, 0.4-0.6% (wt/wt) inorganic components, 0.4% (wt/wt) amino acids, amides and etc (Rose & Steinbüchel, 2005). NR is an important biopolymer in the rubber industry especially for the manufacturing of tyres due to its outstanding properties (high elasticity and high tensile strength). However, it has some disadvantages such as low abrasion and flame resistance which derived from its unsaturated sites and resultant non-polar properties. To overcome these drawbacks and improving of NR properties, its chemical modification emerges as an important method. The chemical modification of NR is not only to improve its undesirable properties but to transforming it into new polymeric materials for specific applications. For instance epoxidation of NR which can be carried out by using formic acid and hydrogen peroxide, resulted in epoxidized natural rubber (ENR) with better performance in terms of oil resistance, gas permeability and very good compatibility with polyaniline and zirconia. Natural Rubber (NR) compounding or final processing is quite difficult due to its very high Mw (up to 10^6 Da) and its resultant high viscosity. The NR viscosity may can be reduced by depolymerisation of rubber to a shorter polymeric chains to decrease the MW. A lower molecular weight NR can be obtained by chemically modification of NR without introduction of new molecules. The reaction can be carried out in the presence of chemical reagent using heat or light sources. The process makes the high molecular weight NR into polymer with low molecular weight (less than 100,000 Da) known as liquid natural rubber (LNR). Simply due to the high reactivity of NR polymer chains, its other chemical modification such as cyclisation, halogenation and graft copolymerization with various vinyl monomers is also possible which is out of the scope of this chapter.

Epoxidized Natural Rubber

Epoxidized natural rubber (ENR) is a chemically modified NR which the epoxide groups resulting from epoxidation, are randomly introduced to the polymer backbone. Reacting natural rubber with peroxy formic acid can produce ENR which also called epoxidized cis-1,4-polyisoprene. Figure 1 shows the chemical structure of ENR.

Three types of ENR are commercially available; ENR 10, ENR 25 and ENR 50 where 10, 25 and 50 mole % of epoxide group incorporated into the natural rubber chain respectively. The properties of ENR change gradually with increasing epoxidation level. For instance, the glass transition temperature T_g goes up linearly as the level of epoxidation increases. Owing to the polar nature of the oxirane groups, ENR demonstrates various interesting properties, e.g. low gas permeability or oil resistance, as well as enhanced compatibility with other polymers. Epoxidation of natural rubber can effect the polarity and glass transition temperature (T_g) of polymer. The polarity of ENR is proportional to mole % of epoxidation and increases with rise in epoxide contents. The glass transition temperature increases by approximately 1 °C for every 1 mole percent of epoxidation and from -60 for ENR-10 reaches to -25 for ENR-50. ENR can be tailored for various usages such as: tires, seals, coating and paints, cosmetic and adhesives. Tanrattanakul et al. (Tanrattanakul, Sungthong, & Raksa, 2008) reported the use of ENR as a toughening agent for nylon 6. NR was used to compare with ENR and polymer blends were prepared by using a twin-screw extruder. It was found that NR slightly decreased impact strength of nylon 6, whereas this property increased up to six fold by blending with ENR. More recently Xu et al. (Xu, Jia, Luo, Jia, & Peng, 2015) have reported the use of ENR as an interfacial modifier to improve the mechanical properties of NR/silica composites. The ENR/Silica composite was prepared by using an open mill and the interfacial interaction of ENR with silica was investigated by FTIR, XPS, TEM, XRD and stress–strain testing. The results clearly indicated that the ring-opening reaction occurs between the epoxy groups of ENR chains and Si-OH groups of silica surfaces and the covalent bonds are formed. The bond formation can improve the dispersion of silica in the rubber matrix and enhance the interfacial combination between components. Besides the commercial usage, ENR further has the potential to be applied as the polymer matrix in advanced materials. For instance, Jo et al. (O. Jo, Saha, Kim, Ho, & Kim, 2015) have prepared a nanocomposite consisting of ENR and functionalized multiwalled carbon nanotubes by eco-friendly melt blending method. Nanocomposites were found to exhibit greater thermal conductivity and

Figure 1. Molecular structure of epoxidized natural rubber

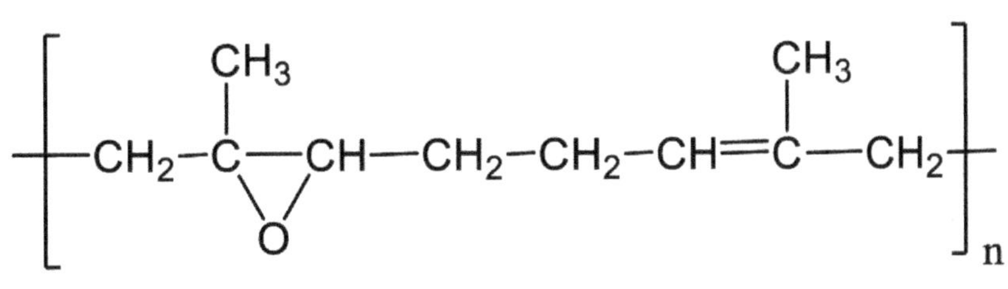

excellent gas barrier properties. Results indicated that the presence of epoxy moieties of ENR provided a stronger network formation between aminosilane treated MWCNTs surface and rubber matrices.

SYNTHESIS OF ZIRCONIA NETWORKS WITHIN ENR-50 FILM BY SOL-GEL ROUTE

Sol-gel technique was employed to synthesize epoxidized natural rubber/zirconia hybrid films by varying the addition of zirconium(IV)-n-propoxide in the range from 10 to 50 wt.% during the synthesis process. Zirconia has its unique physical and chemical properties such as wide optical transparency range, high refractive index, high thermal stability and excellent chemical resistance. Thus it may be of importance to confirm the interaction between zirconia and ENR-50, as this understanding is essential for theoretical and application point of view. Furthermore, it would be interesting to improve the morphology, optical transparency, thermal stability and glass transition temperature of ENR-50/zirconia hybrid films by the addition of zirconium(IV)-n-propoxide. It is also motivating to adjust the refractive index of ENR-50/ zirconia hybrid materials. Epoxidized natural rubber with 50 mol% epoxidation (ENR-50, Malaysian Rubber Board, Kuala Lumpur, Malaysia) was used as the polymer matrix. Zirconium (IV)-n-propoxide solution with 70 wt.% in 1-propanol (Fluka) was used as a filer. The hybrids were prepared by dissolving ENR-50 in THF by maintaining a constant rubber to solvent ratio 1:100 (W/V). Zirconium(IV)-n-propoxide with three different wt.% of 10, 30, and 50 with respect to the rubber was added to the solvated rubber solution under vigorous stirring at room temperature. The mixture was stirred 4 hours at room temperature for homogenization and then sonicated for 2 hours. The resultant solution was casting on petri dish for gelation under controlled condition to avoid the formation of pore samples by rapid evaporation of solvent.

Interaction of Zirconia Network With ENR-50

The interaction nature of zirconia networks within ENR-50 was done by means FTIR, UV-Visible spectroscopy and DSC analysis. Figure 2 shows the FTIR spectra of the ENR-50, three prepared samples of ENR-50/zirconia films and zirconium(IV)- n-propoxide. The FTIR spectra of ENR-50 and zirconium(IV)-n-propoxide are provided in the same figure for comparison with the reaction products. The spectra of ENR-50 (Figure 2(a)) exhibited the most characteristic band at 875 cm^{-1} corresponding to the vibration of epoxy groups. The other characteristic peaks were described in (Mahmood, Khan & Azarian, 2013). It has been observed that with the increase of zirconium(IV)-n-propoxide (10, 30 and 50 wt% — samples b, c and d), the intensity of broad ——OH group bands at 3400–3430 cm^{-1} increases which is accompanied by the decreasing of epoxide peak of ENR-50, and can be attributed to the partial ring opening of epoxide group. The peak identified at 1638–1648 cm^{-1} may correspond to the C——O——C or C=O groups, resulting from the opening of epoxide ring. The broad peak observed at the region of 400 to 800 cm^{-1} can be attributed to the Zr——O——Zr stretching. It is important to note that the intensity of these peak due to the groups of C——O——C or C=O, and Zr——O——Zr increased with the increasing of wt% of zirconium (IV)-n-propoxide addition. Thus, these results indicate the significant interaction between ENR-50 and zirconium(IV)-n-propoxide which is occurred due to the hydrogen bonding formation.

UV–Vis absorption spectra of the samples obtained from ENR-50 and ENR-50/zirconia hybrid samples prepared from 10, 30 and 50 wt% of zirconium(IV)-n-propoxide addition are shown in Figure 3. It is clearly observed that the absorption at 292 nm due to n–π* transition of the nonbonding electron of the epoxide groups in ENR-50 is shifted to 296, 297 and 302 nm after 10, 30 and 50 percent addition of zirconium(IV)-n-propoxide, indicating the significant interaction between ENR-50 and zirconia. Thus, these results further support the observation that the interaction between ENR-50 and zirconia increases with the increasing of zirconium(IV)-n-propoxide addition and maximum interaction occurs for 50 wt% addition. Hence, the nature of interaction between ENR-50 and zirconium(IV)-n-propoxide is probably due to the formation of hydrogen bond between oxygen of epoxide group from ENR-50 and hydrogen of hydroxyl group from zirconia networks, as illustrated in Figure 4.

The glass transition temperature (T_g) of ENR-50 and ENR-50/ zirconia materials obtained using DSC measurements, as shown in Figure 5. Second heating scan was performed for all samples to get accurate T_g values. For ENR-50, the T_g was found −21.71 °C, while for hybrid samples, b, c and d, the T_g values were −20.66 °C, −20.00 °C and −19.74 °C, respectively. These results revealed that the T_g of the ENR-50/zirconia samples increased with the increase of zirconia addition and reaches to more positive values compared to ENR-50 sample. Such increment of the T_g values indicates that the mobilization of rubber macromolecules has been restricted due to the addition of zirconia and thus suggests the hydrogen bonding interaction between ENR-50 and zirconia network which becomes stronger with the higher addition of zirconia. These results are consistent with the report of natural rubber composite preparation with the addition of carbon nanotubes by Sui et al (Sui, Zhong, & Yang, 2008).

Figure 2. FTIR spectra of ENR-50, ENR-50/zirconia and zirconium(IV)-n-propoxide. Spectrum (a) and (e) correspond to ENR-50 and zirconium(IV)-n-propoxide, respectively. The spectra of ENR-50/zirconia hybrid samples are presented by b, c and d

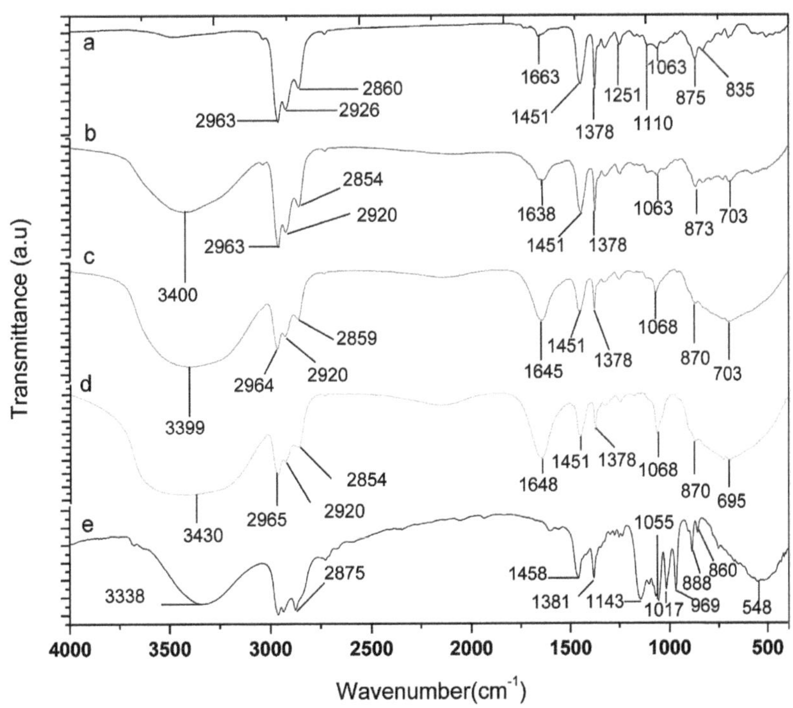

Figure 3. UV–Vis spectra of ENR-50 and ENR-50/zirconia hybrid samples

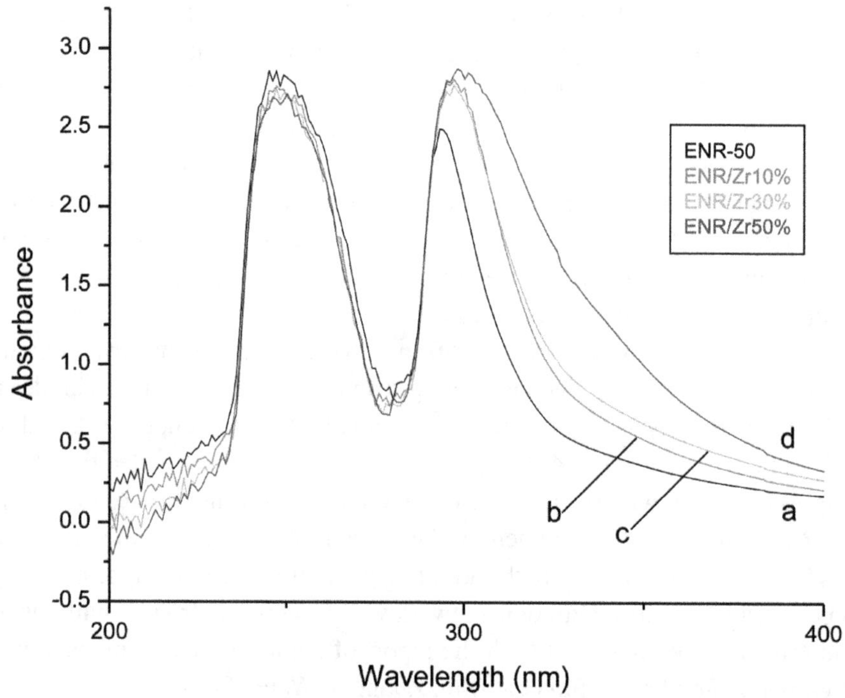

Figure 4. Possible mechanism of bonding between ENR-50 and zirconia

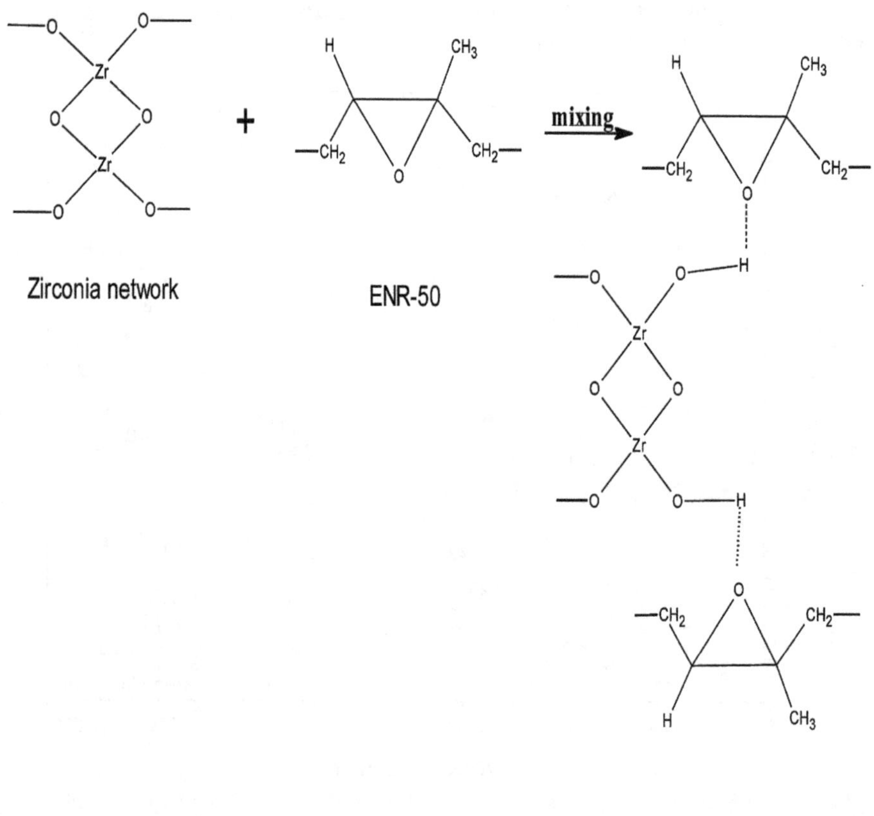

Figure 5. DSC heating scan curves of ENR-50 (a) and ENR-50/zirconia hybrid film samples (b, c and d)

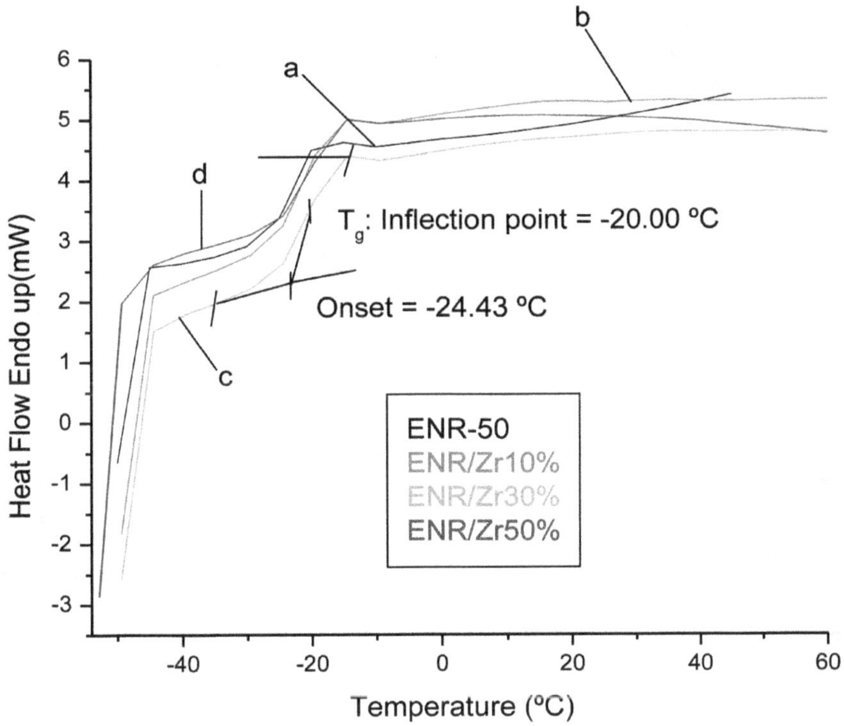

Morphology, Thermal and Optical Properties Study of ENR-50/Zirconia

Figure 6 shows the typical SEM images of the ENR-50/zirconia hybrid films prepared with 30 wt% addition of zirconium(IV)-n-propoxide. In this image, dark background phase represents the ENR-50 matrix and the bright phase corresponds to the zirconia particles that are embedded in ENR-50 matrix. For 10 wt% zirconium(IV)-n-propoxide addition, most of the zirconia nanoparticles individually distributed amongst the ENR-50 matrix. When the zirconium(IV)-n-propoxide addition increased to 30 wt%, the density of the zirconia nanoparticles in the ENR-50 matrix becomes higher and forms spherical shape structure as presented in Figure 6. In addition, the zirconia nanoparticles are uniformly and homogeneously distributed throughout the polymer matrix. When the zirconium(IV)-n-propoxide addition further increased to 50 wt%, the zirconia nanoparticles are being severely aggregated in most areas of the ENR-50 matrix and aggregation dominates compared to the lower zirconium(IV)-n-propoxide addition.

Thus to obtain ENR-50/zirconia hybrid film with uniformly and homogeneously distributed zirconia, the optimum zirconium (IV)- n-propoxide addition could be the 30 wt%. The thermal stability of films was tested (PE TGA7 1991 thermogravimetric analyzer) under a flow of nitrogen atmosphere. DTG curves of ENR-50 and ENR-50/zirconia films are presented in Figure 7. It was found that the ENR/zirconia films have higher thermal stability than neat ENR-50. Clear enhancement of the maximum decomposition temperature in going from ENR-50 (410 °C) to ENR-50/zirconia films(10%: 446 °C, 30%: 454 °C and 50%:475 °C) was observed. Moreover, the decomposition temperature of ENR-50/zirconia hybrid materials increases with the increase of zirconium(IV)-n-propoxide addition. The reason behind of these results is due to the protective effect of zirconia network against organic ENR-50 matrix (Laachachi, Cochez, Leroy, Ferriol, & M. Lopez-Cuesta, 2007; Laachachi, Cochez, Leroy, Ferriol, Lopez-Cuesta, 2007).

Figure 6. SEM image of ENR-50 composited with 30 wt% zirconium(IV)- n-propoxide

From UV–Visible spectroscopy, the optical transparency of the prepared ENR-50/zirconia hybrid film samples was observed. The absorbance of the three samples started to became zero after the wavelength 400 nm, which demonstrates the very good optical transparency in the visible light range. The optical photograph of ENR-50/zirconia hybrid film samples also were captured in order to observe their optical transparency, as shown in Figure 8 the films have thickness of 150 μm and were placed on the characters printed paper. It can be observed that the prepared hybrid films exhibited excellent optical transparency regardless of the wt% of the zirconium(IV)-n-propoxide addition. Furthermore, the transparency of these films indicates the homogeneity nature in the hybrid films.

Refractive indices of the hybrid materials were measured (Atago 3261 Automatic Digital RX-5000α Microphotonics Ellipsometer with LED light source) at the wavelength of 589 nm. Figure 9 shows the refractive indices of the ENR-50 and ENR-50/ zirconia hybrid films. The refractive index of the ENR-50 and its composition with 10, 30 and 50% zirconia were found 1.472, 1.482, 1.495 and 1.513, respectively. Thus, it was found that the refractive indices of the hybrid materials linearly increased with the increase of zirconium(IV)-n-propoxide addition. These results are consistent with the report of epoxy resin/ zirconia hybrid materials synthesis using zirconium alkoxide (Ochi, Nii, & Harada, 2011). The reason lies in that the interaction and the dispersion of zirconia networks to ENR-50 matrix become stronger with the increase of zirconium(IV)-n-propoxide addition. Furthermore, the zirconia network has high refractive index of about 2.2 which plays an important role for the increasing of refractive index of hybrid materials (Liu & Ueda, 2009; Lü & Yang, 2009).

Figure 7. DTG curves of ENR-50 (a) and ENR-50/zirconia hybrid film samples (b, c and d)

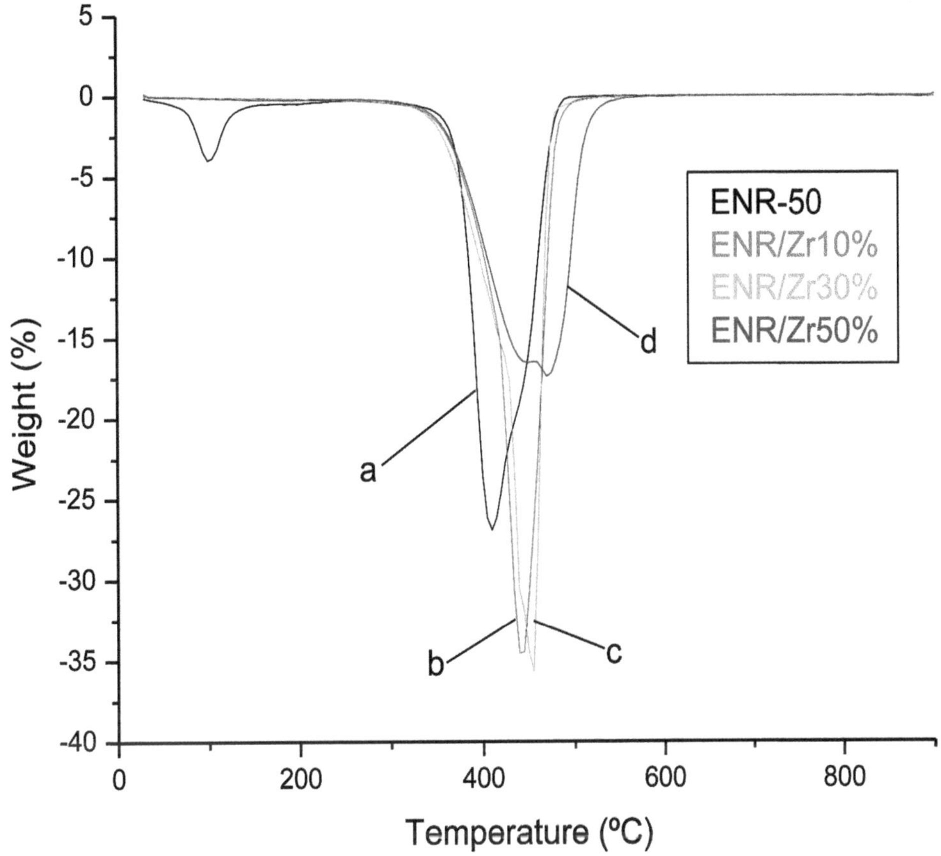

Figure 8. Optical images of ENR-50/zirconia hybrid film samples prepared with 10 (a), 30 (b) and 50 wt % (c) of zirconium(IV)-n-propoxide

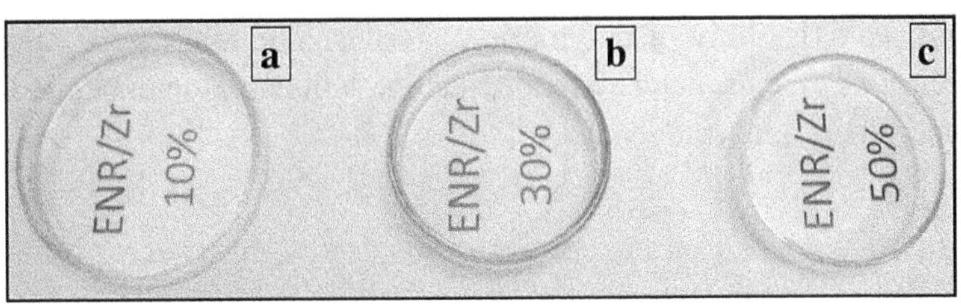

Figure 9. Refractive indices of ENR-50 and ENR-50/zirconia hybrid films

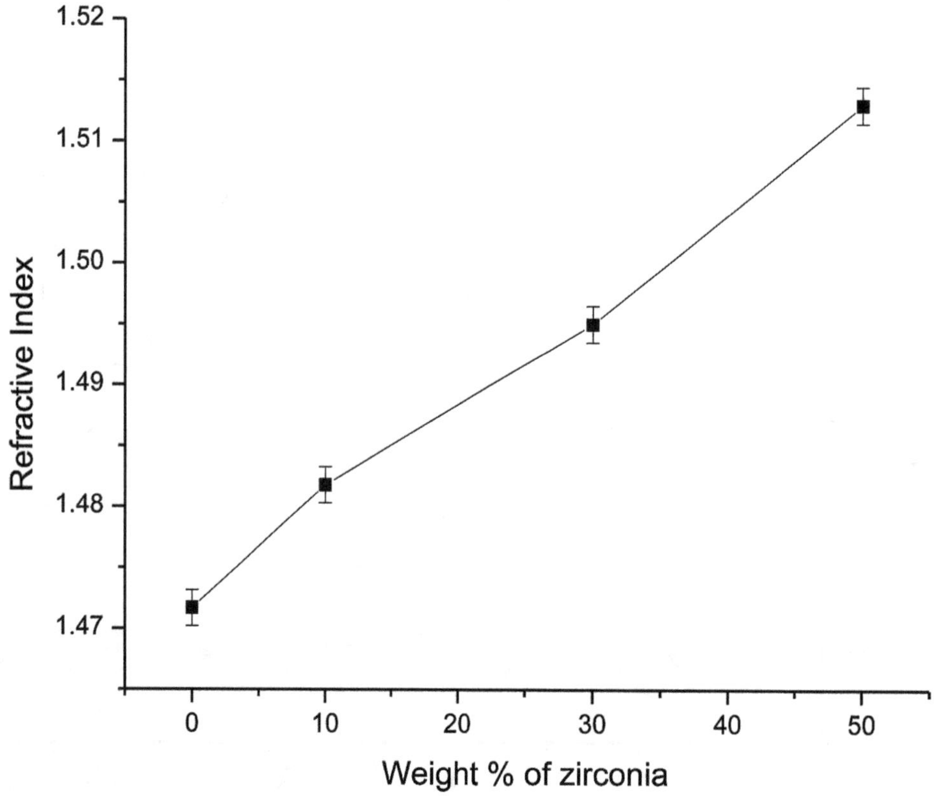

CONCLUSION

This chapter reviewed organic-Inorganic composite materials (OICs) synthesized from polymers and inorganic metal alkkoxides with sol-gel technique. The OICs have characteristics having both the strength of the inorganic glass and flexibiliy of polymeric materials. The OICs were formed from LNR and ENR-50 in the presence of various metal alkoxides such as silicon, zirconium and titanium for flexible transparent films, nanofibers and nanobeads.

REFERENCES

Ali Shah, A., Hasan, F., Shah, Z., Kanwal, N., & Zeb, S. (2013). Biodegradation of natural and synthetic rubbers: A review. *International Biodeterioration & Biodegradation*, *83*, 145–157. doi:10.1016/j.ibiod.2013.05.004

Ebelman, M. (1845). Untersuchungen uber die Verbindung der Borsaure und Kieselsaure mit Aether. *Ann. Chem*, *57*(3), 319–355. doi:10.1002/jlac.18460570303

Hashim, A. S., Kawabata, N., & Kohjiya, S. (1995). Silica reinforcement of epoxidized natural rubber by the sol-gel method. *Journal of Sol-Gel Science and Technology*, *5*(3), 211–218. doi:10.1007/BF00487018

Jo, J. O., Saha, P., Kim, N. G., Chang Ho, C., & Kim, J. K. (2015). Development of nanocomposite with epoxidized natural rubber and functionalized multiwalled carbon nanotubes for enhanced thermal conductivity and gas barrier property. *Materials & Design*, *83*, 777–785. doi:10.1016/j.matdes.2015.06.045

Laachachi, A., Cochez, M., Leroy, E., Ferriol, M., & Lopez-Cuesta, J. M. (2007, January). Fire retardant systems in poly(methyl methacrylate): Interactions between metal oxide nanoparticles and phosphinates. *Polymer Degradation & Stability*, *92*(1), 61–69. doi:10.1016/j.polymdegradstab.2006.09.011

Laachachi, A., Leroy, E., Cochez, M., Ferriol, M., & Lopez Cuesta, J. M. (2005, August). Use of oxide nanoparticles and organoclays to improve thermal stability and fire retardancy of poly(methyl methacrylate). *Polymer Degradation & Stability*, *89*(2), 344–352. doi:10.1016/j.polymdegradstab.2005.01.019

Liu, J., & Ueda, M. (2009). High refractive index polymers: Fundamental research and practical applications. *Journal of Materials Chemistry*, *19*(47), 8907. doi:10.1039/b909690f

Livage, J. (1997). Sol-gel processes. *Current Opinion in Solid State and Materials Science*, *2*(2), 132–138. doi:10.1016/S1359-0286(97)80057-5

Livage, J., Henry, M., & Sanchez, C. (1988). Sol-gel chemistry of transition metal oxides. *Progress in Solid State Chemistry*, *18*(4), 259–341. doi:10.1016/0079-6786(88)90005-2

Lü, C., & Yang, B. (2009). High refractive index organic–inorganic nanocomposites: Design, synthesis and application. *Journal of Materials Chemistry*, *19*(19), 2884. doi:10.1039/b816254a

Mahmood, W. A. K., Rahman Khan, M. M., & Azarian, M. H. (2013). Sol-gel synthesis and morphology, thermal and optical properties of epoxidized natural rubber/zirconia hybrid films. *Journal of Non-Crystalline Solids*, *378*, 152–157. doi:10.1016/j.jnoncrysol.2013.07.002

Maruszewski, K. (2004). Synthesis and properties of an inorganic-organic hybrid prepared by the sol-gel method. *Opt. Mater. (Amst)*, *26*(2), 207–211. doi:10.1016/j.optmat.2003.11.022

Ochi, M., Nii, D., & Harada, M. (2011, September). Preparation of epoxy/zirconia hybrid materials via in situ polymerization using zirconium alkoxide coordinated with acid anhydride. *Materials Chemistry and Physics*, *129*(1–2), 424–432. doi:10.1016/j.matchemphys.2011.04.034

Rose, K., & Steinbüchel, A. (2005). Biodegradation of natural rubber and related compounds: Recent insights into a hardly understood catabolic capability of microorganisms. *Applied and Environmental Microbiology*, *71*(6), 2803–2812. doi:10.1128/AEM.71.6.2803-2812.2005 PMID:15932971

Sanchez, C., Julián, B., Belleville, P., & Popall, M. (2005). Applications of hybrid organic-inorganic nanocomposites. *Journal of Materials Chemistry*, *15*(35–36), 3559–3592. doi:10.1039/b509097k

Shroeder, H. (1962). Properties and Applications of Oxide Layers Deposited on Glass from Organic Solutions, Opt. *Acta Int. J. Opt.*, *9*(3), 249–254. doi:10.1080/713826428

Sui, G., Zhong, W., & Yang, X. (2008). Preparation and properties of natural rubber composites reinforced with pretreated carbon nanotubes. *Polymers for Advanced Technologies*, *19*(August), 1543–1549.

Tanrattanakul, V., Sungthong, N., & Raksa, P. (2008). Rubber toughening of nylon 6 with epoxidized natural rubber. *Polymer Testing*, *27*(7), 794–800. doi:10.1016/j.polymertesting.2008.05.013

West, N. K. (1990). Properties and Applications of Oxide Layers Deposited on Glass from Organic Solutions. *Chemical Reviews*, *90*, 33–72.

Xu, T., Jia, Z., Luo, Y., Jia, D., & Peng, Z. (2015). Interfacial interaction between the epoxidized natural rubber and silica in natural rubber/silica composites. *Applied Surface Science*, *328*, 306–313. doi:10.1016/j.apsusc.2014.12.029

Zarba-Grodź, I., Miśta, W., Strek, W., Bukowska, E., Hermanowicz, K., & Maruszewski, K. (2004). Synthesis and properties of an inorganic-organic hybrid prepared by the sol-gel method. *Opt. Mater. (Amst)*, *26*(2), 207–211. doi:10.1016/j.optmat.2003.11.022

Chapter 7
Clay Minerals Converted to Porous Materials and Their Application:
Challenge and Perspective

Pham Thanh Huyen
Hanoi University of Science and Technology, Vietnam

ABSTRACT

Clay minerals can be used as raw materials for the production of various industrial products. However, most bentonite and kaolinite deposits contain a significant quantity of non-clay mineral impurities. These impurities often affect the quality of clay minerals for adsorption and catalytic application. Therefore, in order to be used as adsorbents and catalysts, those clay minerals need some beneficiation, activation processing to improve their properties or conversion to a porous materials. In this chapter, an overview of the current state, the properties, the beneficiation, activation as well as the conversion of bentonite, kaolinite... to porous materials such as pillared clays, zeolites and their intended applications were presented. In addition, in this review, the challenges and difficulty in the conversion of bentonite and kaolinite to porous materials were also discussed.

INTRODUCTION

Clays have attracted great attention for sustainable materials or environmental materials. Clay minerals such as bentonite and kaolinite have characteristics of high surface area, layered structure, chemical and mechanical stability, cation exchange ability, abundant in nature and low cost and can be used for many industrial applications. The structure and composition of these clays usually influence their physical and chemical properties and determine the applications of the clay minerals *(Muray, 1991)*. Industrial uses of bentonite depend on quality and quantity of their smectites (especially montmorillonite content), as well as the type, valence and the amount of exchangeable cations *(Önal & Sarikaya, 2007)*. Most bentonite contains a significant quantity of non-clay mineral impurities or other clay minerals, therefore,

DOI: 10.4018/978-1-5225-1971-3.ch007

bentonite needs some beneficiation to the activation processing to improve their properties or conversion to a porous materials.

In order to increase the surface area and acidity of the clay minerals, as well as eliminate several mineral impurities, acid treatment process is needed. Treatments of clay minerals with inorganic acids of rather high concentration and usually at high temperature are known as acid activation. The acid treatment changes surface area and porous structure of the clays. Key parameters which influence the acid treatment process are: the present of other clay minerals and non-clay minerals, the chemical composition, the type of cations between the layers, the type of the acid, the activating temperature and time, acid to bentonite ratio and so on *(Nguetnkam, Kamga, Villiéras, Ekodeck, Razafitianamaharavo, & Yvon, 2005;Bhattacharyya & Gupta, 2008;Amari, Chlendi, Gannouni, & Bellagi, 2010)*. However at high acid concentration, aluminum component in the framework can also be dissolved and leads to the collapse of the clay structure, resulting in the decrease of the surface area. So the activation reaction condition should vary depending on the treatment condition. The cations between the layers of montmorillonite could be exchanged with large polyoxocations which is then converted to oxide clusters upon calcination to form pillared clays. Zeolites are crystalline aluminosilicates that have uniform microporous system. Therefore this kind of clays can be used as adsorbents or catalyst supports. Most research papers described zeolites synthesized from pure and expensive chemicals containing organic aluminum and silicium (such as aluminum isopropoxide, tetraethyl orthosilicate TEOS) and organic templates (tetramethylammonium hydroxide TMAOH, tetramethylammonium bromide TMABr and so on) in alkaline media on certain conditions *(Don, Huyen, Hong, 2013)*. Recently, kaolinite-based zeolites are reported using cheap natural materials compared to conventional system which are therefore more cost efficient *(Johnson & Arshad, 2014)*.

Background

Bentonite is composed mostly of montmorillonite which are made up of two silica tetrahedral sheets with a central alumina octahedral sheet. The tetrahedral and octahedral sheets are combined in such way that the tips of the tetrahedra of and one of the hydroxyl layers of the octahedral sheet form a common layer (Figure 1). The atoms in this layer, which are shared to both sheets, become oxygen instead of hydroxyl.

In order to create porous structure and increase the surface area and acidity of the clay minerals, as well as eliminate the mineral impurities, inorganic acids such as H_2SO_4, HCl *(Pawar, Lalhmunsiama, Bajaj & Lee, 2016)*, HNO_3*(Ishaq, Sultan, Ahmad, Ullah, Yaseen, & Amir,, 2015)* can be used. However, up to now, few work using HNO_3 has been reported.

An alternative way to use clays as a porous materials is pillared clays. Pillared clays (PILCs) are layered materials, prepared by substituting the small mono and divalent interlayered cations in the smectite layers, with large polyoxocations or charged colloidal particles. Upon heating, the cationic pillars form oxide clusters that insert between the clay layers, creating a well defined two-dimensional pore structure with high specific surface area *(Lenarda, Ganzerla, Storaro, Enzo, & Zanoni, 1994;Gandia, Vicente, & Gil, 2000;Bineesh, Kim, Kim, & Park, 2011)*. The cross-linking of oligocations to the clay layers creates a stable texture, which supply the solid with a developed network of micro- and mesopores. PILCs have high acidity, which is due to the parent materials and to the exchangeable cations, an oxyhydroxide that provides Brönsted and Lewis acidity, which is easily tunable by changing the synthesis conditions *(Kou, Mendioroz, Salerno, & Muñoz, 2003)*. Therefore PILCs have been used as supports and bifuntional catalysts *(Gil, Korili, Trujillano, & Vicente, 2011)*.

Figure 1. Structure of montmorillonite
Source: Bhattacharyya & Gupta, 2008

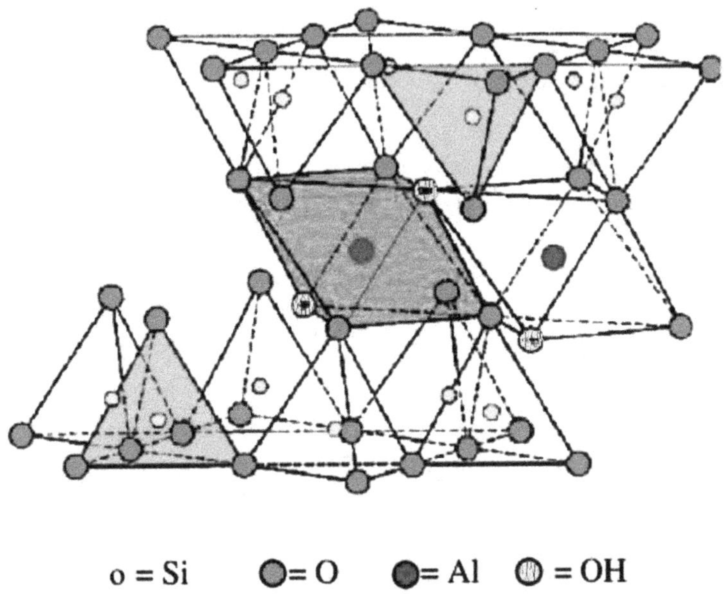

Kaolinite has a 1:1 layer structure with the basic unit consisting of a tetrahedral sheet of SiO_4 and an octahedral sheet with Al^{+3} as the octahedral cation. Both the sheets combine to form a common layer having the tips of the silica tetrahedra points towards the octahedral layer. The tetrahedral layer is inverted over the octahedral layer with the apical 'O' atoms being shared by the two layers (Figure 2).

Kaolinite structrure is quite stable. Therefore, it must be converted to other unstable form, metakaolin which is dehydroxylated form of kaolinite and can be used as precursor for zeolite synthesis. The

Figure 2. Structure of kaolinite
Source: Bhattacharyya & Gupta, 2008

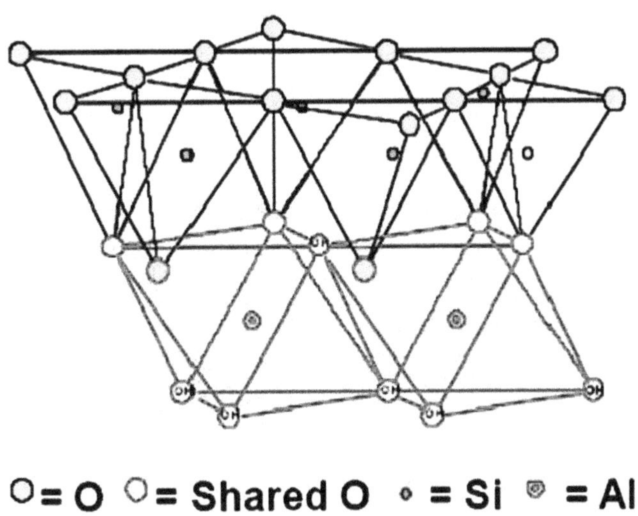

hydrothermal synthesis technique is the main technique used to synthesize all of the zeolites *(Johnson & Arshad, 2014).*

MAIN FOCUS OF THE CHAPTER

In this chapter, the procedure of the conversion of bentonite to porous materials, pillared clays and kaolinite to zeolite are overviewed in addition with their applications of the porous clays.

Acid Activation of Bentonite in Procedure and Applications

Before acid activation, the raw bentonite should be purified by aqueous dispersion and then decantation to remove heavy minerals and contaminants. In these processes, hydrocyclones can be used to remove effectively the impurities from clay slurry *(Clarey, Edwards, Tsipursky, Beall, & Eisenhour, 2000).* The activation can be carried out by refluxing the clay at desired temperature (usually 70-100°C) for 2 – 10 hours in the presence of acid (Table 1). After the acid activation, the solids are separated and washed with water until excess Cl^- and SO_4^{2-} ions are not detectable. Recently, in order to decrease the reaction time, the activation was assisted with microwave irradiation *(Foletto, Paz, & Gündel, 2013;Korichi, Elias, Mefti, & Bensmaili, 2012).* By choosing suitable irradiation power, the surface area of the clay could increase. Microwave heating was more powerful and can achieve the same results with a shorter time compared to the conventional heating method. But the severe condition of irradiation power could cause the complete destruction of clay structure.

During acid treatment the octahedral central atoms as well as tetrahedral Al could be removed from the clay structure. The composition of the clay layers affected their stability during acid activation: trioctahedral layers (where most of the sites occupied by divalent cations, e.g. Mg^{2+}) dissolved much faster than dioctahedral (where two-thirds of the sites are occupied mainly by trivalent cations, e.g. Al^{3+} or Fe^{3+}). The exchangeable cations were initially replaced by protons, followed by partial dissolution of tetrahedral and octahedral sheets, creating new acid sites in the structure, thus making the clay more porous, increasing the cation exchange capacity (CEC), and changing the chemical and mineralogical characteristics of bentonite. *(Nguetnkam, Kamga, Villiéras, Ekodeck, Razafitianamaharavo, & Yvon, 2005;Babaki, Salem & Jafarizad, 2008;Korichi, Elias, Mefti, & Bensmaili., 2012).*

$$2Fe^{3+} + 3H_2SO_4 \rightarrow Fe_2(SO_4)_3 + 6H^+$$
$$Ca^{2+} + H_2SO_4 \rightarrow CaSO_4 + 2H^+$$

In the next step, clay layers were dissolved and amorphous materials were formed and the consequence of this was the increase of specific surface area. The reaction conditions for acid activation of bentonite are listed in Table 1.

The elemental analysis of samples can be performed by X-ray fluorescence (XRF). As mentioned earlier, there're many impurities existed naturally in clays which should be eliminated by acid activation. It was found that the relative contents of the undissolved elements in the bentonite increased in the order of Ca^{2+}, Na^+, Fe^{3+}, Mg^{2+}, Ti^{4+}, Al^{3+}, S^{3+}, P^{5+}, Si^{4+}, and K^+ by the increase in HCl molarities *(Bendou & Amrani, 2014).* The K^+ cations between 2:1 layers of illite and the Si^{4+} cations at tetrahedral sites of smectite and illite are not dissolved by acid activation. The increase in relative contents of these cations

Table 1 Reaction condition for acid activation of bentonite

Acid species	T (°C)	Time (h)	Liquid:solid (v/m)	Acid concentration	BET area (m2/g) Raw→Activated bentonite	References
H_2SO_4	80	2	5/1	8N	111 → 212	*Nguetnkam, Kamga, Villiéras, Ekodeck, Razafitianamaharavo & Yvon, 2005*
H_2SO_4	97	6	Dry method	40 – 45%	25 → 285	*Noyan, Onal & Sarikaya, 2007 , Noyan, 2007 ,*
H_2SO_4	97	6	20/1	40%	NA	*Önal & Sarikaya, 2007*
H_2SO_4	90	2	10/1	5M	20 → 108*	*Babaki, Salem & Jafarizad., 2008*
H_2SO_4	90	16	NA	98% (acid/ clay=0,3)	75 → 273	*Vijayakumar,* Nagendrappa & Prakash, *2009*
H_2SO_4	96,2	6,93	5,98/1	32,94%	69 → 195	*Amari, Chlendi, Gannouni & Bellagi., 2010*
H_2SO_4	microwave 800W, 15 min		5/1	6N	16 → 325	*Foletto, Paz & Gündel, 2013*
H_2SO_4	65	10	16/1	6N	92 → 294	*Pawar, Lalhmunsiama, Bajaj & Lee, 2016*
HCl	95	6	2/1	1N	15 → 38	*Kirali & Lac., 2006*
HCl	98-105	6	2,8/1	2M	173	*Korichi, Elias, Mefti & Bensmaili, 2012*
	Microwave 500W, 20 min		2.8/1	2M	165	
HCl	RT	24	20/1	0.4M	508 →837	*Bendu & Amrani, 2014*
HCl	100	10 mins	1/10	0.1M	25 → 84	*Toor,* Jin, Dai & Vimonses, *2015*
HCl	80	2	5/1	1M	94 → 86	*Ullaha, Hussain., Gul, Khan & Bangasha, 2015*
HNO_3	70	3	6/1	0,1N	NA	*Ishaq, Sultan, Ahmad, Ullah, Yaseen & Amir,, 2015*

*The surface area was measured by methylene blue

depends on the dissolution of other cations by the activation. The relative $S^{3+}$and P^{5+} contents were decreased slowly by the acid activation because the S cations might be located either on octahedral sites or tetrahedral sites. When the relative S content increased, these cations were located most probably at the tetrahedral sites in the 2:1 layers of smectite similar to Si^{4+} cations *(Bendou & Amrani, 2014)*. Also, the formation of amorphous materials can be easily determined by XRD with a significant increase in the background noise *(Nguetnkam, Kamga, Villiéras, Ekodeck, Razafitianamaharavo, & Yvon, 2005)*. The specific surface area is usually measured by gas adsorption technique (using BET method). Since other minerals such as kaolinite, quartz and illite are insoluble and cannot be removed these ions during acid activation. Therefore, the specific surface area is an average value that does not allow determining the contributions of different minerals present in the mixture *(Nguetnkam, Kamga, Villiéras, Ekodeck, Razafitianamaharavo, & Yvon, 2005)*. One possible way to take advantage of the differences in solid/gas interactions between a molecular probe and the solid was proposed that case of a kaolinite/smectite mixture could adsorb water vapor. It was found out that by increasing smectite content, the adsorbed water content of the smectite-kaolinite mixtures at given relative humidity increases *(Likos & Lu, 2002)*.

It was also evaluated by FT-IR spectroscopy that the partial dissolution of the octahedral sheets can be confirmed with the decreased intensity of –OH bending vibration band at 914 cm^{-1} (deformation δOH of AlAl–OH) and the bands at 530 cm^{-1} (Si–O–Al), 1030 cm^{-1} (Si–O), and 3430 cm^{-1} (adsorbed H_2O) (Figure 3) *(Amari, Chlendi, Gannouni, & Bellagi, 2010)*. The formation of amorphous silica can also be proved by FTIR with the progressive widening of the 1098 cm^{-1} band and the 1020 cm^{-1} shoulder, as appeared with 3743 cm^{-1} band (elongation vibration νOH of free Si–OH groups), and the increase of the 799 cm^{-1} band *(Nguetnkam, Kamga, Villiéras, Ekodeck, Razafitianamaharavo, & Yvon, 2005)*.

It is also important to evaluate the surface acidity of the acid-activated clay which was measured using the so-called 'boil-out' volumetric titration method, and was expressed in milligrams of KOH per gram of the sample *(Pawar, Lalhmunsiama, Bajaj, & Lee, 2016)*. For example on more specifically, 1 g of the sample (previously dried at 110°C for 6 h) was boiled and stirred for 15 min with 20 mL of distilled water. Then, the clay water suspension was titrated with a 0.01 N potassium hydroxide (KOH) solution using phenolphthalein as an indicator. The acidity was then determined in mg of KOH used per gram of clay by using following equation:

$$Acidity = \left[\left(V\ x\ N\ x\ 56.10\right) / \ Wt.\right]\ x\ 100$$

where V is the volume of KOH used in the titration (mL), N is the normality of the potassium hydroxide solution, and Wt. is the weight of clay (g).

Figure 3. Infrared spectra of raw bentonite (E_o) and activated bentonite (E_{op})
Source: Amari, Chlendi, Gannouni, & Bellagi, 2010

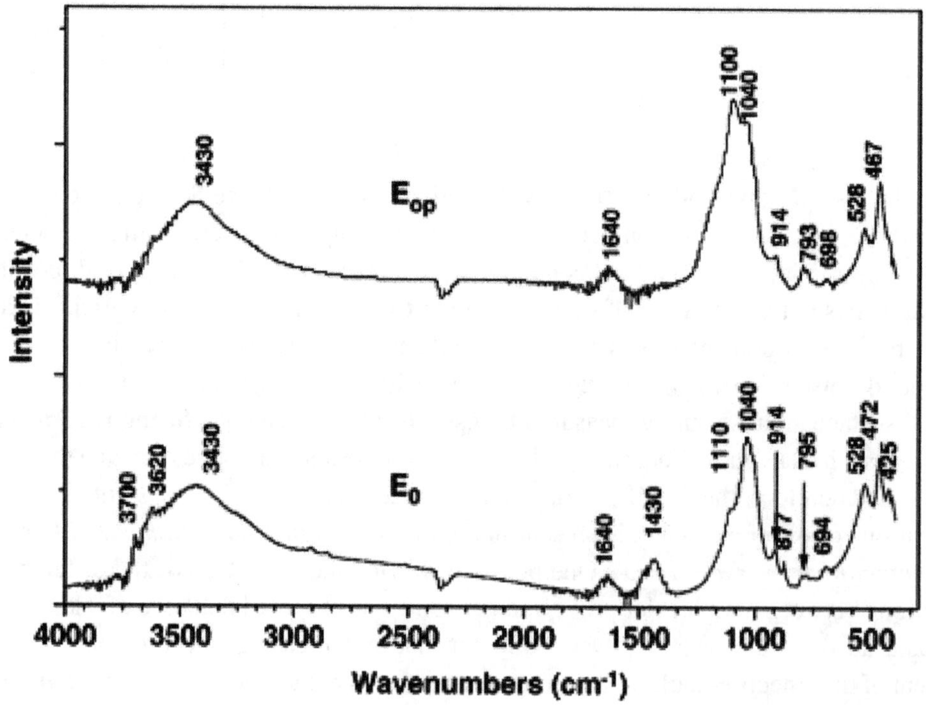

In the acidic activated clay, the Brønsted and Lewis acid sites can be determined by pyridine-FTIR *(Pawar, Lalhmunsiama, Bajaj, & Lee, 2016)*. This confirmed that FT-IR spectra showed the bands at 1600, 1545, 1492, and 1446 cm^{-1} (Figure 4) resulted from bentonite adsorbed pyridine. The bands at 1545–1548 cm^{-1} were assigned to the Brønsted pyridine sites and those at 1445–1447 cm^{-1} and 1600 cm^{-1} were assigned to the hydrogen bonding between the pyridine atom and the –OH groups of the clay mineral. A strong band at 1492 cm^{-1} indicated the combined presence of all of the acid sites, i.e., Lewis pyridine + Bronsted pyridine + Hydrogen. Unmodified bentonite after pyridine adsorption shows week acid sites and broad band due to –OH at 1638 cm^{-1}, which was attributable to the presence of interlayer water.

As porous materials, bentonites are known and applied for several fields. Bentonites are the most important among solids that contain natural mesopore and also contain a small amount of micropores. The acid activation affects the micro- and meso-pores of bentonites and therefore influences their physico-chemical properties of bentonites such as absorption and catalytic activity. Acid activated bentonites are usually used in production of selective adsorbents, bleaching earth, catalyst supports and so on.

Noyan et al. and Foletto et al used the acid activated bentonites for the bleaching of soybean oil. They confirmed that the bleaching ability was increased with increasing surface area, porosity and surface acidity of bentonite, depending more on the mesopore size *(Noyan, Onal & Sarikaya, 2007;Foletto, Paz, & Gündel, 2013)*. The bleaching capacity of the bentonite for cotton oil was determined with different conditions. The highest bleaching capacity obtained was 74.2% for bentonite activated at 0.5 solid to liquid ratio, contact time for 6 h, 1 N HCI concentration and 4% moisture *(Kirali & Lac, 2006)*.

Ishaq et al. investigated the adsorptive desulfurization of dibenzothiophene on the acid treated bentonite impregnated with magnetite *(Ishaq, Sultan, Ahmad, Ullah, Yaseen, & Amir, 2015)*. The high sulfur adsorption capacity of bentonite loaded with magnetite nanoparticles could be attributed to the catalytic activity of magnetite for the co-conversion or destruction of the sulfur species.

Figure 4. Pyridine adsorption FT-IR spectra of acid treated bentonite
Source: Pawar, Lalhmunsiama, Bajaj, & Lee, 2016

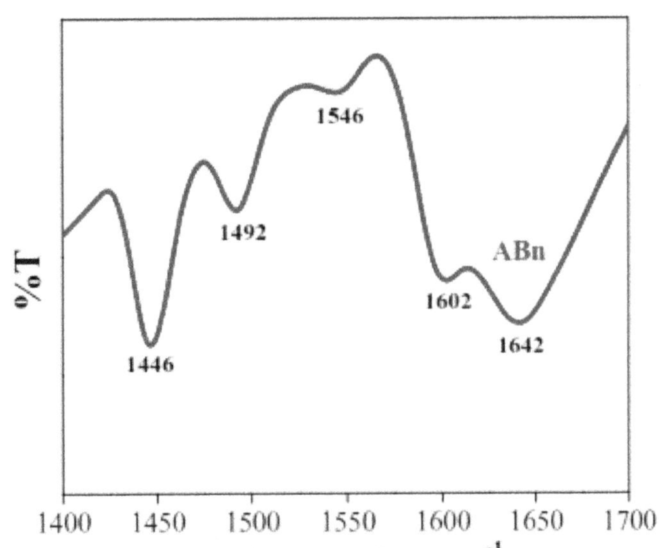

Experimental results revealed that bentonite modified by acid (0.1 M HCl) for 20 min was followed by thermal activation at 100°C could significantly be enhanced in the surface and adsorption properties for removal of waste dye water. This activation method could be cost-effective in the approach for the clay modification to enhance their adsorption capacity and applicability and useful for promising to be adsorbent to wastewater treatment. For example, bentonite activated with sulphuric acid is suitable for adsorbing toluene and the acid-activation enhanced toluene adsorption distinctly *(Amari, Chlendi, Gannouni, & Bellagi, 2010)*. The specific surface area increased from 69 m^2/g to 195 m^2/g by acid activation. Adsorption isotherms of toluene on acid-activated bentonite were correlated with the Langmuir, Freundlich, Langmuir–Freundlich, and Toth models. The Langmuir–Freundlich equation gave a good correlation of toluene adsorption at all temperatures.

Removal of Cu(II) and Pb(II) from aqueous solution by activated bentonite was studied as a low-cost adsorbent *[Pawar., Lalhmunsiama, Bajaj & Lee., 2016]*. The adsorption processes were showed a better fit to the pseudo-second order kinetic model. It was observed that intra-particle diffusion was important in the removal of these two metal ions.

Conversion Bentonite to Pillar Clay in The Procedure, Characterization and Application

Pillar clays are obtained from smectite clay minerals composed mostly of by following three-steps in synthesis procedure:

1. Polymerization of a multivalent cation (such as Al^{3+}, Ga^{3+}, Fe^{3+}, and Cr^{3+}), leading to polycations;
2. Intercalation of these polycations into the interlayer space of smectite clays, involving the substitution of natural exchangeable charge-compensating cations; and
3. Calcination at moderate temperatures (Figure 5). The latter step transforms the polycations into stable oxi-hydroxide phases named pillars, the solids obtained thus being called pillared clays *(Bachir, Lan,* Mereacre, Powell, Koch, & Weidler, *2009*; *Gil, Korili, Trujillano, & Vicente, 2011)*.

Another way was to obtain thermal stable porous materials having the range of 14–22Å from the cationic layered clays, when ionic surfactants was interacted with the inorganic ions in mostly aqueous silicate species through following three steps:

1. The cationic surfactant and neutral amine co-surfactants were intercalated in the interlayer space of the host clay forming micelle structures;
2. Silica pillars were created by in-situ polymerization of a silica source around the micelle structures, and then;
3. The organic surfactants were removed from the material by calcination. This process obtained porous pillar clays with a high specific surface area and combined micro and mesoporous structures *(Galarneau, Barodawalla, & Pinnavaia, 1994)*.

In order to increase the surface acidity of pillar clay, the obtained materials were activated with sulfuric acid (Figure 6) *(Jeenpadiphat & Tungasmita, 2013)*. In the H_2SO_4-activated pillar bentonite sample (H-PilB-HT-RC) the treatment formed a smaller BET surface area, total pore volume and pore diameter than the corresponding non-H_2SO_4-activated bentonite. These might be resulted in the acid corrosion of some parts of the pillar bentonite structure leading to the shrinkage of the interlayer. However, the H_2SO_4 treatment was very important in catalytic heterogeneous acid. For example the H_2SO_4-activation increased almost 20-fold compared to the non- acidified pillar samples.

PILCs are usually characterized by their basal spacing as measured by XRD, accessible specific surface area (SSA) and porosity measured by N_2 adsorption and acidity as measured by TPD-NH_3 and/ or FTIR pyridine method. The basal spacing of smectite intercalated with Al Keggin cations is about 1.8 nm, corresponding to Δd of 0.9 nm (Figure 7). If a smaller Δd value is observed, pillaring is not considered as achieved *(Vicente,* Gil, & Bergaya, *2013)*

At the oxide/water interface adsorption mechanism was described as the surface complex formation to heavy metal ions. The uptake of heavy metals by clay minerals is more complicated via ion exchange reactions and by the formation of inner-sphere complexes through ≡Si-O^- and ≡Al-O^-groups at the clay particle edges *(McBride, 1994;Barbier, Duc, & Petit-Ramel, 2000;Yan, Shan, Wen, & Owens, 2008).* The chemical nature of metal - clay interaction was changed with increasing pH. At low pH values cation exchange was dominating process, whereas at high pH values, the uptake of heavy metal ions was accompanied by release of hydrogen ions created by displacement of protons from surface hydroxyl groups.

Bouchenafa-Saïb et al. studied the pillaring of Algerian montmorillonite with polyhydroxymetal such as aluminum (Al-PILC) and chromium (Cr-PILC) and its application in adsorption of cadmium pollutant in the water. It has been shown that a high temperature and a long time of metallic polymerization (before the intercalation) are in favor to a good pillaring. Indeed, the chromium polymer resulted by the

Figure 5. General preparation procedure of pillared clay

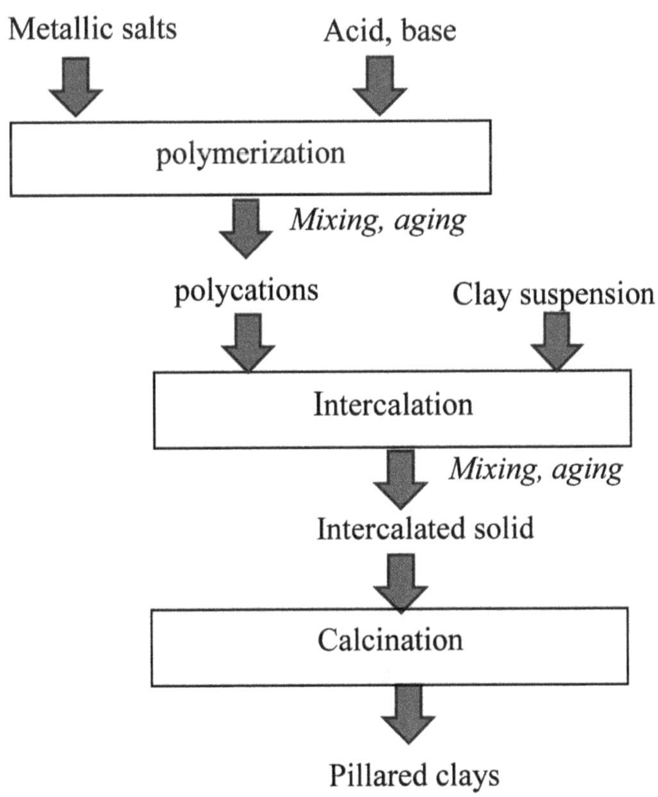

Figure 6. The mechanism of formation of the silica pillar bentonites with the CTAB surfactant alone (PilB-T and PilB-T-C), or also with the co-surfactant HDA (PilB-HT, PilB-HT-C and PilB-HT-RC), and the H2SO4-activated pillar bentonite (H-PilB-HT-RC)
Source: Jeenpadiphat & Tungasmita, 2013

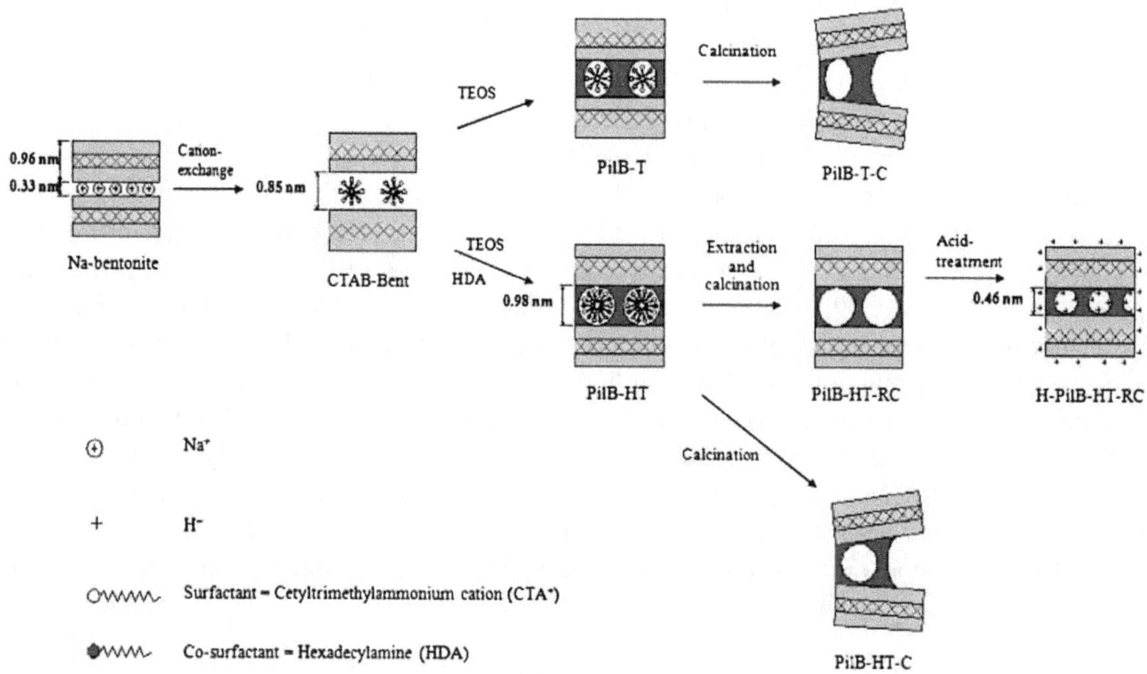

Figure 7. Structure of an Al-PILC
Source: Vicente, Gil & Bergaya, 2013

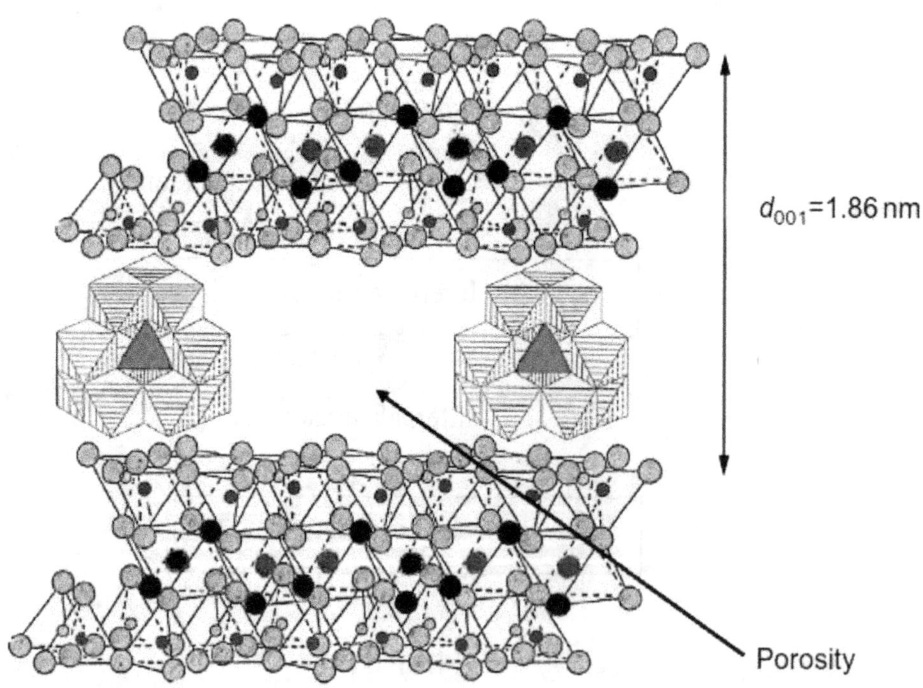

hydrolysis under 95°C for 36 h leads (after intercalation and burn-off at 150°C) to a pillared clays with an interesting and stable textural properties (*d*001 = 19.13Å, surface area of 196 m2/g when calcined at 300°C). The effect of ratio Cr/Al shows that its increase is not favorable to the expansion of clay layers. The highest Cd adsorption yield is about 91%. In the Na-montmorrilonite and Al-montmorrilonite, the adsorption was carried out in two possible ways:

1. Non-specific and
2. Specific adsorption (Figure 8) *(Bouchenafa-Saïb, Khouli, & Mohammedi, 2007).*

At pH <5, only the $Al(OH_2)_2^+$ species were present, but, between pH 5 and 7 the $AlOH(OH_2)$ groups were predominant. In the non-specific case the adsorption taken place without the proton H^+ elimination, when the non-specific was accompanied by H^+ separation. Because of the low attraction between the metal and the Cr-PILC, the specific adsorption is probably carried out only on the Al pillared clays. Nevertheless, the non-selective adsorption on this adsorbent was noticed. In the case of the Na-Mont support, the ionic groups of the layers of the clay constitute the sites of adsorption. For this reason, when Cr-PILC was used for removing Cd^{2+} in water, the sites such as Mg(OH)s, SiOH and Al(OH) adsorbed Cd^{2+} ion and the silanol groups of the side of the clay layers were transformed to $SiO\text{-}Na^+$ form.

Al_{13}-pillared acid-activated Na^+-montmorillonite (Al_{13}-PAAMts) was prepared under optimized conditions and used for Cd^{2+} adsorption *(Yan, Shan, Wen, & Owens 2008).* In the Al_{13}-PAAMts, basal spacing, specific surface area and total pore volume were increased after pillaring. Adsorption of Cd^{2+} on these pillared montmorillonites increased with increasing pH. The Al_{13}-PAAMt had a higher affinity

Figure 8. Solid–liquid adsorption
Source: Bouchenafa-Saïb, Khouli, & Mohammedi, 2007

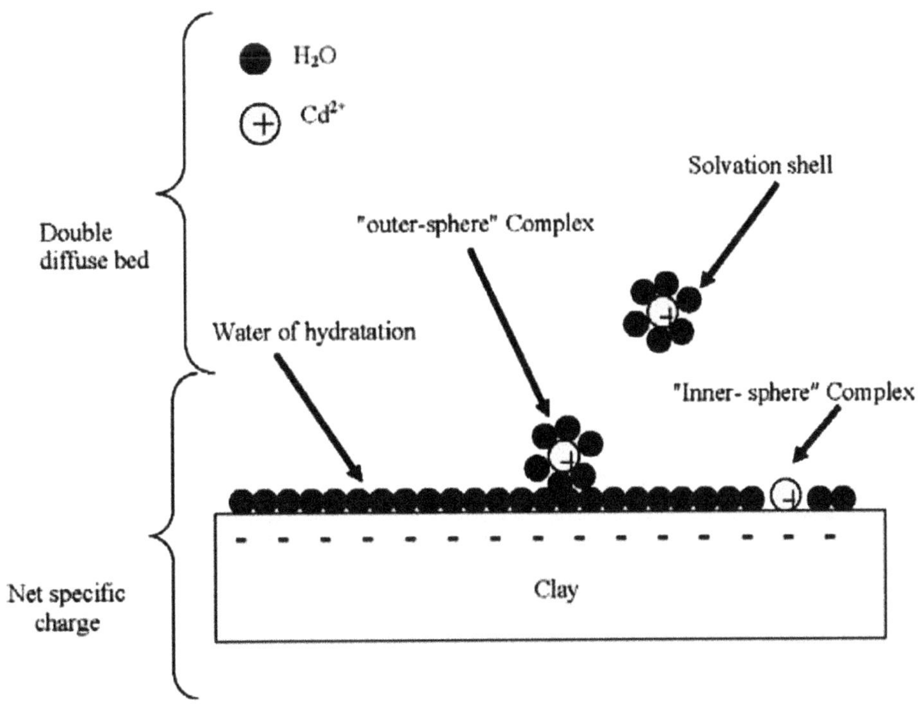

for Cd^{2+} than either Al_{13}-pillared Na^+-montmorillonite (Al_{13}-PMt) or Na^+-montmorillonite (Na^+-Mt). The adsorption kinetics of Al_{13}-PAAMt and Al_{13}-PMt were faster than that of Na^+-Mt by following a pseudo-second-order model in the experimental data, but hysteretic desorption was observed.

Arfaoui et al. used aluminum pillaring, acid activation and pillared clays followed by acid activation and reported adsorbent behavior of removal of chromium ions from aqueous solution. They found out that the maximum of adsorption differed depending on the treatment. The retention of chromium was correlated with the variation of the values of the cation exchange capacities and not with the values of specific surfaces *(Arfaoui, Frini-Srasra & Srasra 2008)*.

Mao et al. synthesized titanium-functionalized silica-pillared clays (TiO_2-SPCs) through post synthetic route for the removal of Cr(VI) ions from aqueous solutions under different temperatures and initial concentrations (Figure 9) *(Mao, Zhu, Li, Yao, & Kong 2014)*. The TiO_2-SPCs showed very high efficiency and sensitivity for removing Cr(VI), depending on the titanium content influences the adsorption capacity and efficiency. While relative lower titanium caused the smaller adsorption capacity, if the titanium of the adsorbent was too high, the adsorption efficiency would be decreased. This was because that the TiO_2 particles formed the gallery pores and some pore channels where the adsorption active sites were blocked.

Introduction of organic precursors or organometallic molecules into the interlayer region of pillared clays, PICLs, became high hydrophobicity that could effectively remove dyes from aqueous solutions *(Zermane, Boura, Baudu, & Basly 2010;Cheknane, Zermane, Baudu, Bouras, & Basly, 2012)*. The pore structure as well as the surface chemistry of the adsorbents was important in the adsorption capacity of

Figure 9. Proposed mechanism for titanium modification on silica-pillared clay process and Cr(VI) adsorption on TiO_2-SPCs
Source: Mao, Zhu K., Li B., Yao C., & Kong, 2014

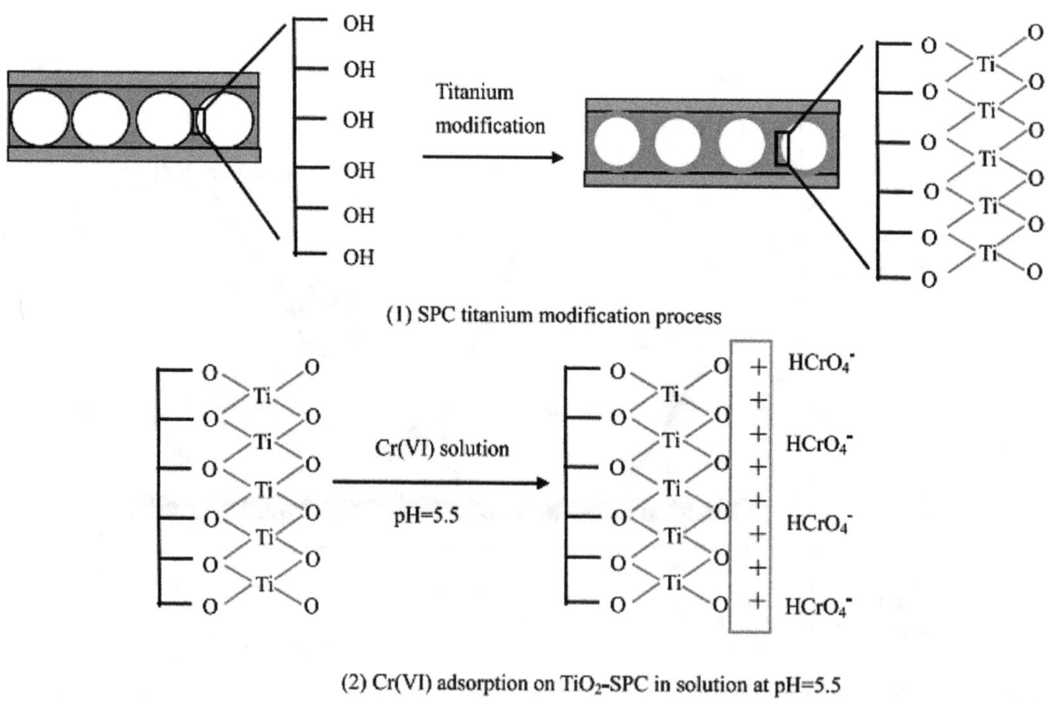

reactive dyes. The adsorbent had a net negative charge on their surface and the electrostatic attraction between the negatively charged surface and the positively charged dye molecules could be attractive to the adsorption. Therefore, the adsorption capacities should increase when the interactions were repulsive. But, that was decreased, when they are attractive *(Gil, Korili, Trujillano, & Vicente, 2011)*. The point of zero charge for the adsorbents was determined from mass titrations *(Ip, Barford, & McKay, 2009;Gil, Korili, Trujillano, & Vicente, 2011)*.

Zermane et al. prepared iron surfactant-modified pillared clay (Fe-SMPM) for good adsorbent to basic yellow 28 dye (BY28) and 4-nitrophenol (4-NP) *(Zermane, Bouras, Baudu, & Basly, 2010)*. *Gil et al.* synthesized for Al-PILC and Zr-PILC and evaluated their adsorption capacity for the removal of Orange II and Methylene Blue from aqueous solutions *(Gil, Korili, Trujillano, & Vicente, 2011)*. Both clays were found to have the same adsorption capacity when Orange II was used as adsorbent, whereas the adsorption capacity of Zr-PILC was higher than that of Al-PILC for Methylene Blue. These were discussed with variations in the diffusion into micropores of the PICLs.

Hou et al. prepared the iron-pillared bentonite (Fe-Ben) by ion-exchange using the natural bentonite (GZ-Ben) *(Hou, Ma, Zhang, Tang, Fan, & Wan, 2011)*. The results showed that the d(001) value and surface area of the bentonite material increased after iron pillaring and the Fe-Ben adsorbed much more Rhodamine B (RhB) than GZ-Ben.

In a surfactant-modified aluminium-pillared montmorillonite (CTAB-Al-Mont-PILC) the granule removed Basic Yellow 28 (BY 28) and Basic Green 4 (BG 4) *(Cheknane, Zermane, Baudu, Bouras, & Basly, 2012)*. The adsorption kinetics of the two dyes, studied at pH 6 and 150 mg L^{-1}, was followed with the pseudo-first order kinetic model having rate constants of 2.5–4.2 $x10^{-2}$ min^{-1}.

As applied by hydroxy-aluminum pillared bentonite to adsorb Basic Fuchsin (BF), Basic Green (BG) and Acid Turquoise Blue A (ATBA). The adsorption capacity of BF and BG reached 6.5 mg/g and 9.9 mg/g. Also the adsorption capacity changed insignificantly with the increase of pH from 2 to 8. The adsorption rate of BF and BG was faster than that of ATBA and a pseudo-second-order model in the experimental data well *(Hao, Yan, Yu, Yang, Yu, Shan & Du, 2014)*.

Conversion of Kaolin to Zeolite

It is known that raw kaolin is produced from natural deposits and is present in complex mixtures of different minerals. Besides kaolinite, it always contains other impurities, especially quartz and iron minerals that affect the color of kaolin. To minimize the amount of quartz and other impurities, raw kaolin was pretreated for synthesis of zeolites *(Ayele, Pariente, Chebude, & Díaz, 2015;Mohiuddin, Isa, Mdleleni, Sincadu, Key, & Tshabalala, 2016)*. In the purification, separation and sedimentation could remove quartz, and ultrasonic suspension and magnetic separation could remove iron (Figure 10). However it was difficult to remove all quartz content, and therefore quartz usually contaminated the final zeolite products.

It was known that kaolinite was used as the source of alumina and silica in the zeolite synthesis by calcination between 700 °C and 1000 °C. This process produced metakaolin by dehydroxylation process as the following reaction *(Johnson & Arshad, 2014)*:

$$\underset{\text{Kaolinite}}{\mathrm{Si_2Al_2O_5}(OH)_4} \rightarrow \underset{\text{Metakaolinite}}{Al_2O_3.2SiO_2} + 2H_2O \tag{1}$$

Figure 10. Flow diagram for the procedure of the mechanical purification of the raw kaolin

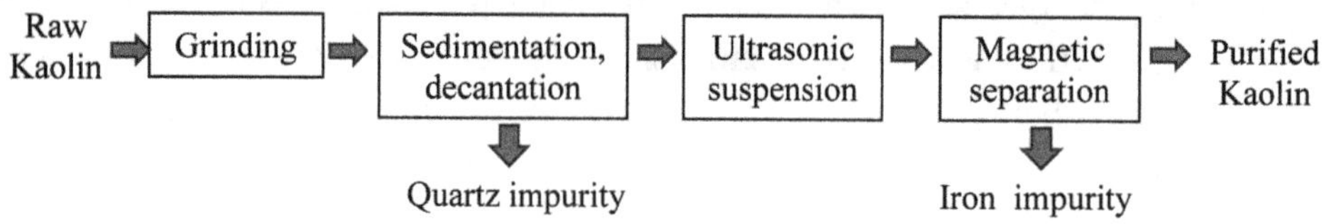

In the synthesis reaction four hydroxyl groups were transformed to two water molecules, which then left two oxygen anions in the material, as shown in reaction (2):

$4(OH^-) \rightarrow 2H_2O + 2O^{2-}$. (2)

Dehydration by thermal treatment converts kaolin to semicrystalline metakaolin, which was much more reactive than the starting *material (Akolekar, Chaffee, & Howe, 1997; Ayele, Pariente, Chebude, & Díaz, 2015).* The metakaolin then was modified by hydrothermal treatment in NaOH to form an amorphous aluminosilicate gel which was useful precursor for zeolite *(Cristóbal, Luengo, & Vizcayno, 2010).* The general reaction of reactants such as polymeric silica and alumina was followed with below equations to a microporous zeolite crystalline framework.

$T–OH + {}^-O–T \leftrightarrow T–O–T + OH^-$ (3)
$T–OH + HO–T \leftrightarrow T–O–T + H_2O$ (4)

In a reversible mechanism of the reaction (3) the breaking and making of T–O–T were occurred, where T was Al or Si. Whereas, the reaction (4) was the condensation reaction that formed the basic framework structure of silicate and aluminosilicate units by means of electrostatic and van der Waals. This reaction is catalyzed by hydroxyl ions *(Johnson & Arshad, 2014).* The synthesis procedure for zeolite from kaolin was summarized as followed (Figure 11).

It was noted that zeolites were metastable phases, and slight changes in the synthesis conditions brought about a pollution of the desired product by the co-crystallization of other phases with a similar composition but with completely different properties (for instance, zeolite A and zeolite X) *(Bebon Colson, Marro, Klein, & Renzo, 2002;Loiola, Andrade, Sasaki, & da Silva, 2012).* In several parameters such as NaOH concentration, solid/liquid ratio, Na_2O/SiO_2 ratio, kaolinite content, the presence of impurities in the starting material, crystallization time, the gel formation condition and gel ageing time, the formation of zeolites were influenced in their specific type, crystallinity and particle size. The difference in chemical reactivity of kaolins from around the world of various geological areas also affected the synthesis conditions of various zeolites and must be optimized for particular kaolin *(Maia, Neves, Angélica, & Pöllmann, 2015;Mohiuddin, Isa, Mdleleni, Sincadu, Key, & Tshabalala, 2016).* Low silica zeolites such as zeolite A and X were easily synthesized from kaolinite due to the similarity in Si/Al ratios of zeolite and starting kaolinite precusor. However, synthesis of high silica zeolites such as Y and ZSM-5 required addition of an external silica source or de-aluminum of kaolinite to obtain a favorable Si/Al ratio (*Feng, Li, & Shan, 2009;Pan, Lu, Wang, Chen, Wang, & Yan, 2014).*

The 13X zeolite was synthesized from low-grade natural kaolin via alkali fusion followed by hydrothermal treatment (50°C/2h for homogenization, 90°C/8h for crystallization). Further prolonging in the

Figure 11. The synthesis procedure for zeolite from kaolin

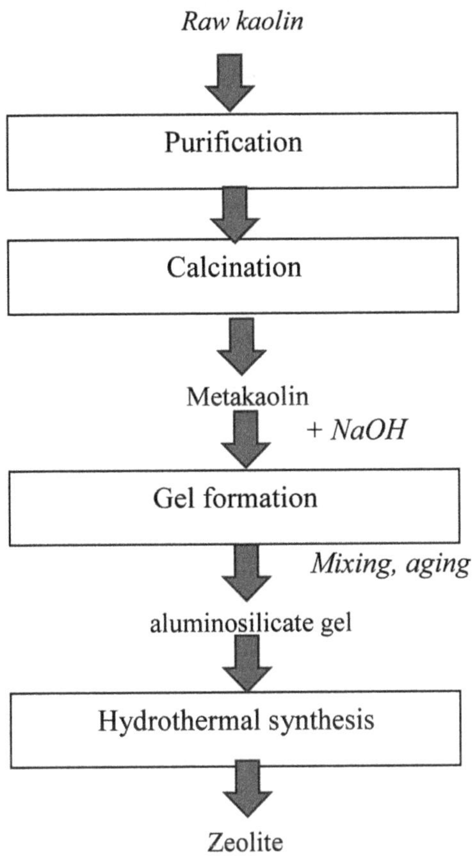

crystallization time might transform 13X zeolite into more stable zeolite hydroxysodalite as confirmed by XRD (Figure 12). The crystallinity of the 13X zeolite sample increased while crystallization temperature was raised from 70 to 90°C and then decreased when the crystallization temperature up to 100°C *(Ma, Yan, Alshameri, Qiu, Zhou, & Li, 2014).*

Zeolites with high degrees of purity and structural order can be synthesized from kaolin waste from the Capim (KC) and Jari (KJ) regions *(Maia, Neves, Angélica, & Pöllmann, 2015).* The optimal conditions for the synthesis of zeolite A with high purity and degree of structural order were different for each residue. The optimal conditions were at 110 °C with Na/Al ratio of 1.64 for both residues and synthesis time of 18 h for KC and 20 h for KJ.

An increase in kaolinite content led to pure ZSM-5 phase formation at the chosen crystallization conditions. Higher kaolinite content shortened the induction time, increased the nucleation and crystallization rate of ZSM-5. Because of a higher content of active silicate and aluminate anions originating from the metakaolinite. The easy solubility into the gel medium and the formation of the primary building units (PBU's) were necessary for nucleus formation and crystal growth in a shorter time period. The high quartz and low alumina contents seemed to hinder the crystallization of ZSM-5 under certain crystallization conditions *(Mohiuddin, Isa, Mdleleni, Sincadu, Key, & Tshabalala, 2016).* Raw kaolin obtained from Grahams town, South Africa was beneficiated by using wet sieving and filtration prior

Figure 12. XRD patterns of the as-synthesized products obtained at different crystallization times (a) 0 h, (b) 1 h, (c) 4 h, (d) 6 h, (e) 8 h, (f) 12 h, and (g) 24 h
Source: Ma, Yan, Alshameri, Qiu, Zhou, & Li, 2014

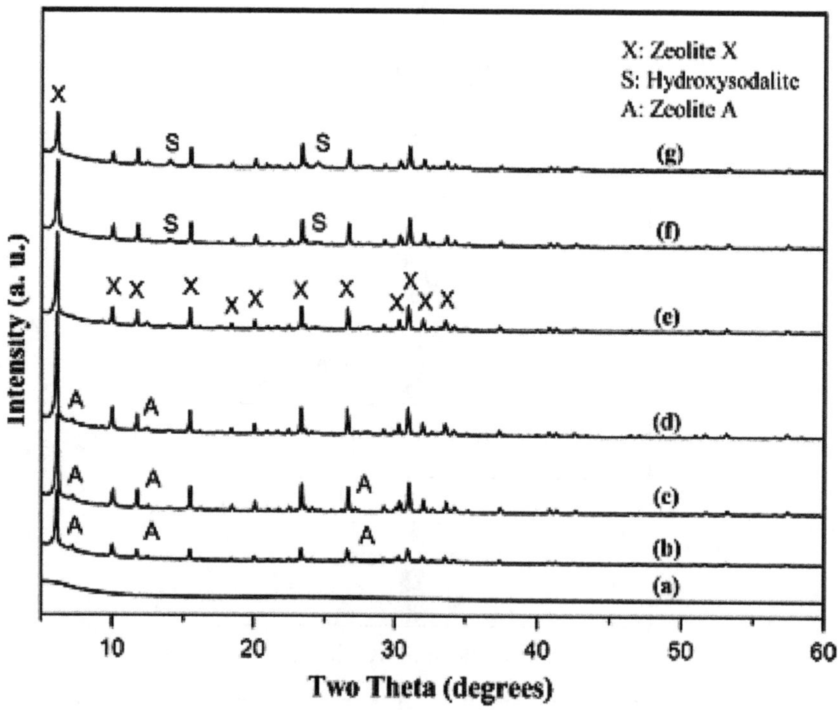

the hydrothermal synthesis of ZSM-5 zeolite at different temperatures (120–190 °C) and times (24–96 h) under autogenous pressure. The optimum condition for synthesizing ZSM-5 from this clay was 150 °C for 48h. Increasing crystallization time led to the change of morphology, porosity and surface area of ZSM-5. For example, as synthesized at 150 °C for 24 h, high external surface area were obtained in the zeolite. Microporosity further developed when time increased and the ZSM-5 became more crystalline.

Ayele et al. synthesized Zeolite A from natural kaolin from Ethiopia *(Ayele, Pariente, Chebude, & Díaz, 2015)*. Cubic crystal with rounded edge of zeolite A (Figure 13) with molar ratio SiO_2/Al_2O_3 of 2.04 and Na_2O/SiO_2 of 0.44 was obtained with 90% crystallinity in NaOH 3M when the solid/liquid ratio was 1.25 g/25 ml for the crystallization time of 3h and the gel formation condition at 50°C in 1 h and gel ageing time of 3h.

It was reported that microwave (MW) could be used to shorten the processing time. In the MW system, an initial short exposure and ageing were important for the crystallization of zeolite A. Dissolution of metakaolin in alkali was enhanced before heating for the nucleation and crystallization of zeolites. Another major advantage was the inhibition of the impurity phases. The temperature gradient is even throughout the reaction media. Therefore formation of the strange phases in the product is inhibited compared to the conventional heating *(Chandrasekhar & Pramada, 2008)*. In these works, the obtained zeolites had a uniform pore size, high specific surface area, thermal stability, tunable pores and acidities, and are therefore useful in various petrochemical processes.

Figure 13. SEM images for synthesized zeolite A and commercial zeolite A (SZA)
Source: Ayele, Pariente, Chebude, & Díaz, 2015

However, in some cases the average pore size was low and limited their use in processes, such as organic waste treatment, and heavy crude oil and bio-oil upgrading. To overcome this disadvantage, nanocrystalline materials with a bimodal pore system (mesopores and micropores) were synthesized by the connection of microporous and mesoporous channels in a highly ordered manner *(Alaba, Sani, & Daud, 2015)*. Don et al. successfully synthesized nanozeolite NaX from Vietnamese kaolin *(Don, Huyen, & Hong, 2013)*. Several factors that influenced the particle size, crystallinity of nanozeolite NaX which was obtained at low water content, low crystallizing temperature. At optimized synthesis conditions, nano NaX has surface area of 573 m^2g^{-1}, external surface area of 92 m^2g^{-1}, pore distribution at 0.81 and 10.8 nm; crystallinity of 92%, crystal size of 25 nm and thermal stability up to 789°C.

The synthesized conditions for conversion of kaolinite to zeolites are summarized in Table 2.

Zeolites synthesized from kaolin were good candidate for metal removal in waste water. Zeolites A and X prepared from Egyptian kaolin were highly efficient in heavy metal removal. The optimum contact time for removal of Cd, Cu, Pb and Zn as the concentration of 20 mg/L at constant temperature of 25±0.1 °C and pH of 7.5 ±0.2, was 30 min, and for Ni removal was 60 min for both zeolite types *(Jamil, Ibrahim, El-Maksoud, & El-Wakeel, 2010)*. Zeolite 4A (LTA) was successfully synthesized from Brazil kaolin by a hydrothermal method, and used as water softener for reduction of 95% Ca^{2+} in the water hardness *(Loiola, Andrade, Sasaki, & da Silva, 2012)*. In their work, Loiola et al. measured the BET surface area of the zeolite 4A by N_2 adsorption, so it was only 9m^2/g (*Table 2*). This is a big mistake because N_2 diameter is larger than the pore size of 4A, therefore the BET result in this work is not the true value for this sample.

Melo et al. prepared 4A zeolites from kaolin in order to obtain 5A zeolites through ionic exchange. And then the zeolite 5A was used as an adsorbent for arsenic ions. It was also observed that 94% of arsenic removed *(Melo, Riella, Kuhnen, Angioletto, Melo, Bernardin, Rocha, & Silva, 2012)*. As used in low cost Na-Y zeolites were prepared from Egyptian kaolins collected from different sources to remove Pb^{2+}, Fe^{2+}, Zn^{2+}. It was found out that the zeolite contents in the prepared samples were proportional to the kaolinite contents in the corresponding raw kaolin. For all zeolites, in Pb^{2+} loading experiments, the relation between adsorbed and desorbed equivalents of positive charges was non-stoichiometric loading. The loading process involves competition between three processes:

Table 2 Synthesized condition for conversion of kaolinite to zeolites

Type of Zeolite	Kaolin Activation	Zeolite Preparation	Zeolite Characterizations	References
Y	Thermal activation: 850°C, 2h Activation with H2SO4: 90°, 3h Calcination at 550oC, 2h	- Gel formation with NaOH, RT, aging 24-72h. - Hydrothermal crystallization: 100°C, 16–24h	Crystallite sizes ~ 68-91 nm, Si/Al ratio~ 1,39-1,61	*Alaba, Sani & Daud, 2015*
13X	Alkali fusion with NaOH 200oC, 4h	- Gel formation of fused mixture with H2O at 50oC, 2h - crystallization: 90oC, 8h	BET 326 m2/g	*Ma, Yan, Alshameri, Qiu, Zhou & Li, 2014*
4A	Calcination at 600oC, 3h	- Gel formation with NaOH: 50oC, 1h, aging 3h - Hydrothermal crystallization: 100°C, 3h.	Si/Al ~ 1, crystallinity ~ 90%	*Ayele, Pariente, Chebude & Díaz, 2015*
4A	Calcination at 700oC, 2h	- Gel formation with NaOH. MW irradiation (85oC) for 2 min, ageing 20 h, RT. - Hydrothermal crystallization: 85oC, 2h (microwave)	Si/Al ~ 1 Single phase crystallinity ~ 100%	*Chandrasekhar & Pramada, 2008*
4A	Calcination at 900oC, 2h	- Gel formation with NaOH - Hydrothermal crystallization: 70oC, 2h and then 30oC, 18h	Single phase BET 9m^2/g (N_2 adsorption)	*Loiola, Andrade, Sasaki & Silva, 2012*
4A	Calcination at 600-700oC, 2h	- Gel formation with NaOH - Hydrothermal crystallization: 110oC, 18-20h	High purity	*Maia, Neves, Angélica & Pöllmann, 2015*
5A	Calcination at 925oC, 2h	- Gel formation with NaOH - Hydrothermal crystallization: 80oC, 2-3.5h - ion exchange with CaCl2 to get 5A	Cation exchange Capacity (CEC): 183.40 meq/100 g	*Melo, Riella, Kuhnen, Angioletto, Melo, Bernardin, Rocha & Silva, 2012*

1. Precipitation of some Pb^{2+} cations as hydroxides,
2. Ion exchange and
3. Chemisorption of Pb cations on zeolite surfaces.

During the Pb^{2+} loading processes, the amount of loaded Pb^{2+} was always higher than the released Na^+ ion, suggesting the predominance of the chemisorption mechanism. While adsorption was the predominant mechanism during the loading of Fe^{2+} into zeolite Y, ion exchange was predominant in case of Zn^{2+}. During the loading process, zeolitic framework was damaged due to low pH and/or due to long contact with the metal ion solutions *(EL-Mekkawi & Selim, 2014;EL-Mekkawi, Ibrahim, & Selim, 2015)*.

FUTURE RESEARCH DIRECTIONS

In acid activated bentonite, pillarer clays and kaolinite-based zeolites in reality, more attention should be paid to scale up the preparation process, especially in the treatment of the acid/basis liquor, which remained after removal the solid products. Since all the minerals are not pure, it's difficult to purify the starting clays. Therefore, the influence of impurities and other minerals should be controlled.

CONCLUSION

Acid activation and the conversion of bentonite to pillared clays and kaolin to zeolites are complicated processes. Many parameters have to be well controlled to obtain the good final products.

- **For Porous Bentonite:** The particular clay mineral, the present of other clay minerals and non-clay minerals, the chemical composition, the type of cations between the layers, the type of the acid, the activating temperature and time, acid to bentonite ratio… influence the surface area and porous structure of the clays in acid activation process.
- **For Pillar Clays:** Type, concentration of polyoxocations, pH, aging time and temperature, calcination temperature… determine the basal spacing, specific surface area, porosity and acidity of pillared clays.
- **For Zeolite:** The formation of specific type, crystallinity, particle size, surface area, acidity… of obtained zeolites depend on the NaOH concentration, solid/liquid ratio, Na_2O/SiO_2 ratio, kaolinite content, the presence of impurities in the starting material, crystallization time, the gel formation condition and gel ageing time …

Acid activated bentonite, pillarer clays and kaolinite-based zeolites are potential candidate for environment applications, especially for dye and heavy metal removal.

REFERENCES

Achma, R. B., Ghorbel, A., Dafinov, A., & Medina, F. (2008). Copper-supported pillared clay catalysts for the wet hydrogen peroxide catalytic oxidation of model pollutant tyrosol. *Applied Catalysis A, General*, *349*(1-2), 20–28. doi:10.1016/j.apcata.2008.07.021

Akolekar, D., Chaffee, A., & Howe, R. F. (1997). The transformation of kaolin to low-silica X zeolite. *Zeolites*, *19*(5-6), 359–365. doi:10.1016/S0144-2449(97)00132-2

Alaba, P. A., Sani, Y. M., & Daud, W. M. A. W. (2015). Synthesis and characterization of hierarchical nanoporous HY zeolites from acid-activated kaolin. *Chinese Journal of Catalysis*, *36*(11), 1846–1851. doi:10.1016/S1872-2067(15)60962-7

Amari, A., Chlendi, M., Gannouni, A., & Bellagi, A. (2010). Optimised activation of bentonite for toluene adsorption. *Applied Clay Science*, *47*(3-4), 457–461. doi:10.1016/j.clay.2009.11.035

Arfaoui, S., Frini-Srasra, N., & Srasra, E. (2008). Modelling of the adsorption of the chromium ion by modified clays. *Desalination*, *222*(1-3), 474–481. doi:10.1016/j.desal.2007.03.014

Ayele, L., & Pariente, J. P. (2015). Synthesis of zeolite A from Ethiopian kaolin. *Microporous and Mesoporous Materials*, *215*, 29–36. doi:10.1016/j.micromeso.2015.05.022

Babaki, H., Salem, A., & Jafarizad, A. (2008). Kinetic model for the isothermal activation of bentonite by sulfuric acid. *Materials Chemistry and Physics*, *108*(2-3), 263–268. doi:10.1016/j.matchemphys.2007.09.034

Bachir, C., Lan, Y., Mereacre, V., Powell, A. K., Koch, C. B., & Weidler, P. G. (2009). Magnetic titanium-pillared clays (Ti-M-PILC): Magnetic studies and Mossbauer spectroscopy. *Clays and Clay Minerals*, *57*(4), 433–443. doi:10.1346/CCMN.2009.0570404

Barbier, F., Duc, G., & Petit-Ramel, M. (2000). Adsorption of lead and cadmium ions from aqueous solution to the montmorillonite: Water interface. *Colloids and Surfaces. A, Physicochemical and Engineering Aspects*, *166*(1-3), 153–159. doi:10.1016/S0927-7757(99)00501-4

Bebon, C., Colson, D., Marrot, B., Klein, J. P., & Renzo, F. D. (2002). Synthesis of zeolites: Study and application of a new process of homogeneous shaking out of the medium to minimize the shear rate during the crystallization. *Microporous and Mesoporous Materials*, *53*(1-3), 13–20. doi:10.1016/S1387-1811(02)00318-9

Bendou, S., & Amrani, M. (2014). Effect of Hydrochloric Acid on the Structural of Sodic-Bentonite Clay. *Journal of Minerals and Materials Characterization and Engineering*, *2*(05), 404–413. doi:10.4236/jmmce.2014.25045

Bhattacharyya, K. G., & Gupta, S. S. (2008). Adsorption of a few heavy metals on natural and modified kaolinite and montmorillonite: A review. *Advances in Colloid and Interface Science*, *140*(2), 114–131. doi:10.1016/j.cis.2007.12.008 PMID:18319190

Bineesh, K. V., Kim, D.-K., Kim, M.-I., & Park, D.-W. (2011). Selective catalytic oxidation of H_2S over V_2O_5 supported on TiO_2-pillared clay catalysts in the presence of water and ammonia. *Applied Clay Science*, *53*(2), 204–211. doi:10.1016/j.clay.2010.12.022

Bouchenafa-Saïb, N., Khouli, K., & Mohammedi, O. (2007). Preparation and characterization of pillared montmorillonite: Application in adsorption of cadmium. *Desalination*, *217*(1-3), 282–290. doi:10.1016/j.desal.2007.03.007

Chandrasekhar, S., & Pramada, P. N. (2008). Microwave assisted synthesis of zeolite A from metakaolin. *Microporous and Mesoporous Materials*, *108*(1-3), 152–161. doi:10.1016/j.micromeso.2007.04.003

Cheknane, B., Zermane, F., Baudu, M., Bouras, O., & Basly, J. P. (2012). Sorption of basic dyes onto granulated pillared clays: Thermodynamic and kinetic studies. *Journal of Colloid and Interface Science*, *381*(1), 158–163. doi:10.1016/j.jcis.2012.05.045 PMID:22721789

Clarey, M., Edwards, J., Tsipursky, S. J., Beall, G. W., & Eisenhour, D. D. (2000), Method of manufacturing polymer-grade clay for use in nanocomposites, US patent 6,050,509.

Cristóbal, A. G. S., Castelló, R., Luengo, M. A. M., & Vizcayno, C. (2010). Zeolites prepared from calcined and mechanically modified kaolins A comparative study. *Applied Clay Science*, *49*(3), 239–246. doi:10.1016/j.clay.2010.05.012

Don T. N., Huyen P.T, & Hong N.K.D. (2013). Synthesis, characterization and application of nanozeolite NaX from Vietnamese kaolin. *Adv. Nat. Sci.: Nanosci. Nanotechnol., 4.*

Dong, H., Chen, Y., Sheng, G., Li, J., Cao, J., Li, Z., & Li, Y. (2016). The roles of a pillared bentonite on enhancing Se(VI) removal by ZVI and the influence of co-existing solutes in groundwater. *Journal of Hazardous Materials*, *304*, 306–312. doi:10.1016/j.jhazmat.2015.10.072 PMID:26561754

EL-Mekkawi, D. M., Ibrahim, F. A., & Selim, M. M.. (2015). Stability and thermal transformation studies of Zn^{2+}- and Fe^{2+} - loaded zeolite Y prepared from Egyptian kaolin. *Solid State Sciences*, *48*, 294–299. doi:10.1016/j.solidstatesciences.2015.09.002

EL-Mekkawi, D. M., & Selim, M. M.. (2014). Removal of Pb^{2+} from water by using Na-Y zeolites prepared from Egyptian kaolins collected from different sources. *Journal of Environmental Chemical Engineering*, *2*(1), 723–730. doi:10.1016/j.jece.2013.11.014

Feng, H., Li, C., & Shan, H. (2009). Effect of calcination temperature of kaolin microspheres on the in situ synthesis of ZSM-5. *Catalysis Letters*, *129*(1-2), 71–78. doi:10.1007/s10562-008-9794-9

Foletto, E. L., Paz, D. S., & Gündel, A. (2013). Acid-activation assisted by microwave of a Brazilian bentonite and its activity in the bleaching of soybean oil. *Applied Clay Science*, *83–84*, 63–67. doi:10.1016/j.clay.2013.08.017

Franco, F., Benítez-Guerrero, M., Gonzalez-Triviño, I., Pérez-Recuerda, R., Assiego, C., Cifuentes-Melchor, J., & Pascual-Cosp, J. (2016). Low-cost aluminum and iron oxides supported on dioctahedral and trioctahedral smectites: A comparative study of the effectiveness on the heavy metal adsorption from water. *Applied Clay Science*, *119*, 321–332. doi:10.1016/j.clay.2015.10.035

Galarneau, A., Barodawalla, A., & Pinnavaia, T. J. (1994). Porous clay heterostructures formed by gallery-templated synthesis. *Nature*, *374*(6522), 529–531. doi:10.1038/374529a0

Gandia, L. M., Vicente, M. A., & Gil, A. (2000). Preparation and characterization of manganese oxide catalysts supported on alumina and zirconia-pillared clays. *Applied Catalysis A, General*, *196*(2), 281–292. doi:10.1016/S0926-860X(99)00479-2

Gil, A., Korili, S. A., Trujillano, R., & Vicente, M. A. (Eds.). (2010). *Pillared Clays and Related Catalysts*. Springer. doi:10.1007/978-1-4419-6670-4

Gil, A., Korili, S. A., Trujillano, R., & Vicente, M. A. (2011). A review on characterization of pillared clays by specific techniques. *Applied Clay Science*, *53*(2), 97–105. doi:10.1016/j.clay.2010.09.018

Gyftopoulou, M. E., Millan, M., Bridgwater, A. V., Dugwell, D., Kandiyoti, R., & Hriljac, J. A. (2005). Pillared clays as catalysts for hydrocracking of heavy liquid fuels. *Applied Catalysis A, General*, *282*(1-2), 205–214. doi:10.1016/j.apcata.2004.12.012

Hao, Y., Yan, L., Yu, H., Yang, K., Yu, S., Shan, R., & Du, B. (2014). Comparative study on adsorption of basic and acid dyes by hydroxy-aluminum pillared bentonite. *Journal of Molecular Liquids*, *199*, 202–207. doi:10.1016/j.molliq.2014.09.005

Hou, M.-F., Ma, C.-X., Zhang, W.-D., Tang, X.-Y., Fan, Y.-N., & Wan, H.-F. (2011). Removal of rhodamine B using iron-pillared bentonite. *Journal of Hazardous Materials*, *186*(2-3), 1118–1123. doi:10.1016/j.jhazmat.2010.11.110 PMID:21168960

Ip, A. W. M., Barford, J. P., & McKay, G. (2009). Reactive Black dye adsorption/desorption onto different adsorbents: Effect of salt, surface chemistry, pore size and surface area. *Journal of Colloid and Interface Science*, 33732–33738. PMID:19523648

Ishaq, M., Sultan, S., Ahmad, I., Ullah, H., Yaseen, M., & Amir, A. (2015). (in press). Adsorptive desulfurization of model oil using untreated, acid activated and magnetite nanoparticle loaded bentonite as adsorbent. *Journal of Saudi Chemical Society*. doi:10.1016/j.jscs.2015.02.003

Jamil, T. S., Ibrahim, H. S., El-Maksoud, I. H. A., & El-Wakeel, S. T. (2010). Application of zeolite prepared from Egyptian kaolin for removal of heavy metals: I. Optimum conditions. *Desalination*, *258*(1-3), 34–40. doi:10.1016/j.desal.2010.03.052

Jeenpadiphat, S., & Tungasmita, D. N. (2013). Acid-activated pillar bentonite as a novel catalyst for the esterification of high FFA oil. *Powder Technology*, *237*, 634–640. doi:10.1016/j.powtec.2013.02.001

Johnson, E. B. G., & Arshad, S. E. (2014). Hydrothermally synthesized zeolites based on kaolinite: A review. *Applied Clay Science*, *97–98*, 215–221. doi:10.1016/j.clay.2014.06.005

Kirali, E. G., & Lac, O. (2006). Statistical modelling of acid activation on cotton oil bleaching by Turkish bentonite. *Journal of Food Engineering*, *75*(1), 137–141. doi:10.1016/j.jfoodeng.2005.06.010

Korichi, S., Elias, A., Mefti, A., & Bensmaili, A. (2012). The effect of microwave irradiation and conventional acid activation on the textural properties of smectite: Comparative study. *Applied Clay Science*, *59-60*, 76–83. doi:10.1016/j.clay.2012.01.020

Kou, M. R. S., Mendioroz, S., Salerno, P., & Muñoz, V. (2003). Catalytic activity of pillared clays in methanol conversion. *Applied Catalysis A, General*, *240*(1-2), 273–285. doi:10.1016/S0926-860X(02)00466-0

Lenarda, M., Ganzerla, R., Storaro, L., Enzo, S., & Zanoni, R. (1994). Bifunctional catalysts from pillared clays: Vapour phase conversion of propene to acetone catalyzed by iron and ruthenium containing alurninum pillared bentonites. *Journal of Molecular Catalysis*, *92*(2), 201–215. doi:10.1016/0304-5102(94)00064-6

Likos, W. J., & Lu, N. (2002). Water vapor sorption behavior of smectite-kaolinite mixtures. *Clays and Clay Minerals*, *50*(5), 553–561. doi:10.1346/000986002320679297

Loiola, A. R., Andrade, J. C. R. A., Sasaki, J. M., & da Silva, L. R. D. (2012). Structural analysis of zeolite NaA synthesized by a cost-effective hydrothermal method using kaolin and its use as water softener. *Journal of Colloid and Interface Science*, *367*(1), 34–39. doi:10.1016/j.jcis.2010.11.026 PMID:22074689

Ma, Y., Yan, C., Alshameri, A., Qiu, X., Zhou, C., & Li, D. (2014). Synthesis and characterization of 13X zeolite from low-grade natural kaolin. *Advanced Powder Technology*, *25*(2), 495–499. doi:10.1016/j.apt.2013.08.002

Maia, A. A. B., Neves, R. F., Angélica, R. S., & Pöllmann, H. (2015). Synthesis, optimisation and characterisation of the zeolite NaA using kaolin waste from the Amazon Region. Production of Zeolites KA, MgA and CaA. *Applied Clay Science*, *108*, 55–60. doi:10.1016/j.clay.2015.02.017

Mao, H., Zhu, K., Li, B., Yao, C., & Kong, Y. (2014). Synthesis of titania modified silica-pillared clay (SPC) with highly ordered interlayered mesoporous structure for removing toxic metal ion Cr(VI) from aqueous state. *Applied Surface Science*, *292*, 1009–1019. doi:10.1016/j.apsusc.2013.12.113

McBride, M. B. (1994). *Environmental Chemistry of Soils*. New York: Oxford University Press.

Melo, C. R., Riella, H. G., Kuhnen, N. C., Angioletto, E., Melo, A. R., Bernardin, A. M., & Silva, L. et al. (2012). Review: Synthesis of 4A zeolites from kaolin for obtaining 5A zeolites through ionic exchange for adsorption of arsenic. *Materials Science and Engineering B*, *177*(4), 345–349. doi:10.1016/j.mseb.2012.01.015

Mohiuddin, E., Isa, Y. M., Mdleleni, M. M., Sincadu, N., Key, D., & Tshabalala, T. (2016). Synthesis of ZSM-5 from impure and beneficiated Grahamstown kaolin: Effect of kaolinite content, crystallisation temperatures and time. *Applied Clay Science*, *119*, 213–221. doi:10.1016/j.clay.2015.10.008

Murray, H. H. (1991). Overview - clay mineral applications. *Applied Clay Science*, *5*(5-6), 379–395. doi:10.1016/0169-1317(91)90014-Z

Nguetnkam, J. P., Kamga, R., Villiéras, F., Ekodeck, G. E., Razafitianamaharavo, A., & Yvon, J. (2005). Assessment of the surface areas of silica and clay in acid-leached clay materials using concepts of adsorption on heterogeneous surfaces. *Journal of Colloid and Interface Science*, *289*(1), 104–115. doi:10.1016/j.jcis.2005.03.053 PMID:16009222

Noyan, H., Onal, M., & Sarikaya, Y. (2007). The effect of sulphuric acid activation on the crystallinity, surface area, porosity, surface acidity, and bleaching power of a bentonite. *Food Chemistry*, *105*(1), 156–163. doi:10.1016/j.foodchem.2007.03.060

Önal, M. (2007). Swelling and cation exchange capacity relationship for the samples obtained from a bentonite by acid activations and heat treatments. *Applied Clay Science*, *37*(1-2), 74–80. doi:10.1016/j.clay.2006.12.004

Önal, M., & Sarikaya, Y. (2007). Preparation and characterization of acid-activated bentonite powders. *Powder Technology*, *172*(1), 14–18. doi:10.1016/j.powtec.2006.10.034

Pan, F., Lu, X., Wang, Y., Chen, S., Wang, T., & Yan, Y. (2014). Synthesis and crystallization kinetics of ZSM-5 without organic template from coal-series kaolinite. *Microporous and Mesoporous Materials*, *184*, 134–140. doi:10.1016/j.micromeso.2013.10.013

Pawar, R. R., Lalhmunsiama, , Bajaj, H. C., & Lee, S.-M. (2016). Activated bentonite as a low-cost adsorbent for the removal of Cu(II) and Pb(II) from aqueous solutions: Batch and column studies. *Journal of Industrial and Engineering Chemistry*, *34*, 213–223. doi:10.1016/j.jiec.2015.11.014

Perelomov, L., Sarkar, B., Rahman, M. M., Goryacheva, A., & Naidu, R. (2016). Uptake of lead by Na-exchanged and Al-pillared bentonite in the presence of organic acids with different functional groups. *Applied Clay Science*, *119*, 417–423. doi:10.1016/j.clay.2015.11.004

Toor, M., Jin, B., Dai, S., & Vimonses, V. (2015). Activating natural bentonite as a cost-effective adsorbent for removal of Congo-red in wastewater. *Journal of Industrial and Engineering Chemistry*, *21*, 653–661. doi:10.1016/j.jiec.2014.03.033

Ullaha, Z., Hussain, S., Gul, S., Khan, S., & Bangasha, F. K. (2015). Use of HCl-modified bentonite clay for the adsorption of Acid Blue 129 from aqueous solutions. *Desalination and Water Treatment*, 1–10.

Vicente, M. A., Gil, A., & Bergaya, F. (2013). Pillared Clays and Clay Minerals. In E. Bergaya & L. Lagaly (Eds.), *Handbook of Clay Science* (pp. 525–557). Elsevier Ltd. doi:10.1016/B978-0-08-098258-8.00017-1

Vijayakumar, B., Nagendrappa, G., & Prakash, B. S. J. (2009). Acid Activated Indian Bentonite, an Efficient Catalyst for Esterification of Carboxylic Acids. *Catalysis Letters*, *128*(1-2), 183–189. doi:10.1007/s10562-008-9729-5

Wang, J.-Q., Huang, Y.-X., Pan, Y., & Mi, J.-X. (2014). Hydrothermal synthesis of high purity zeolite A from natural kaolin without calcination. *Microporous and Mesoporous Materials*, *199*, 50–56. doi:10.1016/j.micromeso.2014.08.002

Yan, L., Shan, X., Wen, B., & Owens, G. (2008). Adsorption of cadmium onto Al_{13}-pillared acid-activated montmorillonite. *Journal of Hazardous Materials*, *156*(1-3), 499–508. doi:10.1016/j.jhazmat.2007.12.045 PMID:18243531

Zermane, F., Bouras, O., Baudu, M., & Basly, J.-P. (2010). Cooperative coadsorption of 4-nitrophenol and basic yellow 28 dye onto an iron organo–inorgano pillared montmorillonite clay. *Journal of Colloid and Interface Science*, *350*(1), 315–319. doi:10.1016/j.jcis.2010.06.040 PMID:20638666

Zuo, S., & Zhou, R. (2006). Al-pillared clays supported rare earths and palladium catalysts for deep oxidation of low concentration of benzene. *Applied Surface Science*, *253*(5), 2508–2514. doi:10.1016/j.apsusc.2006.05.119

Chapter 8
Geopolymer Sourced with Fly Ash and Industrial Aluminum Waste for Sustainable Materials

Sujitra Onutai
Nagaoka University of Technology, Japan

Sirithan Jiemsirilers
Chulalongkorn University, Thailand

Takaomi Kobayashi
Nagaoka University of Technology, Japan

ABSTRACT

Geopolymer is alumino-silicate material which can apply for many applications due to that geopolymers have several attractive properties of high strength, low permeability, high acid resistance, hazardous materials, and immobilization of toxic materials. This chapter presents synthesis of the dense and the porous structures of geopolymer by using fly ash and industrial waste as the raw materials. Aluminum hydroxide waste (Al-waste) and fly ash (FA) were used to synthesis the dense geopolymers for cement materials. The Al-waste based geopolymer influenced the geopolymer strength, when sodium hydroxide (NaOH) concentration was changed at different curing temperatures. As preliminary microwave oven was exposed in the lower NaOH paste of geopolymer, the successful synthesis of geopolymer cement was obtained. In addition, porous fly ash geopolymers was achieved using a household microwave oven. The geopolymer paste was cured within 1 min by using a microwave oven at different output power. Porous geopolymers were formed immediately as ued at 850 W power of the microwave oven.

INTRODUCTION

Geopolymers are known as sustainable materials with eco-friendly process, since these simply can use wastes for raw materials, and have excellent mechanical properties (Kumar & Kumar, 2011; Nazari & Sanjayan, 2015). Usually, the geopolymer materials are activated in a high alkali solution (Davidovits,

DOI: 10.4018/978-1-5225-1971-3.ch008

2008) because alumino silicate geopolymers consist of tetrahedral AlO_4 and SiO_4 units that occur easily in a polycondensed polymer framework in high-alkali conditions (Davidovits, 1991). In geopolymer materials prepared at low temperatures with high alkali solutions, generally alumino-silicate materials such as slag (Karakoç et al., 2014), metakaolin (Rovnaník, 2010a), and fly ash (Ryu, Lee, Koh, & Chung, 2013) were mixed with alkaline solution. For example, sodium or potassium hydroxide is necessary for dissolving silicon and aluminium atoms from raw materials of natural minerals or industrial waste. The silicate alkaline solution includes:

- A binder,
- Alkaline activator, and
- A dispersant or plasticizer (Abdullah, Jamaludin, et al., 2012a; Nazari, Bagheri, & Riahi, 2011; Somna, Jaturapitakkul, Kajitvichyanukul, & Chindaprasirt, 2011).

The basic forms of the chemical structure have silicon-aluminate 3D structures with cross-linked chain bonds together in the geopolymer matrix (Songpiriyakij, Kubprasit, Jaturapitakkul, & Chindaprasirt, 2010). The condensation reaction of geopolymers proceeded greatly at room temperature (Puertas, Martínez-Ramírez, Alonso, & Vázquez, 2000). At elevated temperatures of 40–95 °C, geopolymerization is improved in the strength of specimens (Bakharev, 2005; Chindaprasirt, Chareerat, & Sirivivatnanon, 2007; Panias, Giannopoulou, & Perraki, 2007; Swanepoel & Strydom, 2002). Dense geopolymer materials having apparent density higher than 1 g/cm^3 were applied for construction materials (Andini et al., 2008). Several reports described the use of metakaolin and fly ash as starting materials in geopolymerization (Duxson, Fernández-Jiménez, Lukey, et al., 2007; Komnitsas & Zaharaki, 2007). Moreover, industrial wastes of various types were used to synthesize geopolymers materials such as:

- Aluminum slag,
- Grey cast iron slag, and
- Blast furnace slag.

Among them, the use of etching surface waste, aluminum hydroxide, is interesting in the raw material for geopolymers. The Al-waste has high aluminum contents, making it suitable for the synthesis of better Al-source of geopolymers. Therefore, it is extremely attractive to investigate the geopolymerization of fly ash and Al-waste as a sustainable source for cement substitute materials.

In contrast, foam geopolymers were synthesized in high- or low-temperature processing conditions for widely diverse applications such as in carriers of catalyzer, adsorbents, sensors, hot gas filters, solid–liquid separation processes, and thermal insulators (López, Sugita, & Kobayashi, 2014a; Strozi Cilla, Raymundo Morelli, & Colombo, 2014; Studart, Gonzenbach, Tervoort, & Gauckler, 2006; H. Wang, Li, Wang, & Yan, 2015). In high-temperature processing, the raw materials and foaming agents (aluminium powder or hydrogen peroxide) were mixed with silica fume in situ foaming techniques. After geopolymerization, the geopolymer sample was sintered at 1000-1200 °C to produce in the high strength of porous geopolymer (Abdullah, Jamaludin, et al., 2012b). Moreover, a low temperature process achieved geopolymerization in an alkaline environment with a foaming agent for the decomposed into the water and oxygen. In this case, aluminium and silicon in fumed silica also initiated the release of hydrogen gas, leading to a porous structure. Porous geopolymer was cured in a conventional oven at 40-70 °C for 24–48 h in a long curing period (Böke, Birch, Nyale, & Petrik, 2015; Masi, Rickard, Vickers, Bignozzi,

& van Riessen, 2014; Rickard, Vickers, & van Riessen, 2013; Somaratna, Ravikumar, & Neithalath, 2010; Vaou & Panias, 2010). Heat curing was applied to improve the reaction and the strength of geopolymer materials. Recently, microwave energy was used for innovative tool in the chemical reactions, thereby creating an inverse temperature profile (Gubb et al., 2011). The benefits of microwave heating also include instantaneous, very rapid heat-up time, and easy to control (Chindaprasirt, Rattanasak, & Taebuanhuad, 2013). The application of microwave oven cure for dense geopolymers prepared from FA and Al-waste was introduced (Onutai, Jiemsirilers, Thavorniti, & Kobayashi, 2015).

Background

As a construction materials, the portland cement and concrete are mostly essential and widely used in the world. The demand of using cement and concrete has continuously increased and tend to grow up in the future (Chindaprasirt et al., 2007). In the estimation, the usage of cement was approximately 2540 million tons (Mt) in 2006. The cement production in 2050 will be estimated around 3680 Mt to 4830 Mt (Hasanbeigi, Price, & Lin, 2012). In this circumstance, many problems are occurred in the cement industry of environmental fields. The greenhouse gases have been liberated into the atmosphere for production process such emitting and becomes the main problem of the global warming, by carbon dioxide emission during the production processes (Flower & Sanjayan, 2007). The amount of carbon dioxide is released into the atmosphere with approximately 0.8 tons/ 1tons in the cement industry (Alaa M. Rashad, 2011; Heath, Paine, & McManus, 2014; Pangdaeng, Phoo-ngernkham, Sata, & Chindaprasirt, 2014). In order to reduce the emission in the cement processes, many efforts have been paid to find the other cementitious materials to replace cement (Davidovits, 1994; Rovnaník, 2010a). Therefore, development of sustainable material is needed for cement and concrete.

In 1950s, Victor Glukhovsky et al. developed alkali activated systems containing calcium silicate hydrated and alumino-silicate phases, which were used to build a tall building. These alkaline alumino-silicate cementatious systems are called "soil silicate" (Komnitsas & Zaharaki, 2007). After 1972, Joseph Davidovits (Davidovits, 1982, 1984; Davidovits, 2008) reported another form of cementitious materials using silicon and aluminum activated in a high alkali solution. The new materials are called "Geopolymer". The gepolymers are inorganic polymers having three dimensional alumino-silicates in the cementitious materials (Chindaprasirt et al., 2007; Cioffi, Maffucci, & Santoro, 2003; Davidovits, 1991). These materials have frame work of sialate and having silica-alumina networking in the molecular framework. Moreover, geopolymer materials are easily applied for coating and adhesive layers as binder for fiber composites, waste encapsulation (Davidovits, 2008). In addition, cement materials for concrete were proposal for geopolymers produced by condensation reaction. The polymeric process of alumino-silicate and alkali activators is one of other options for alternative green cement. This is due to reduction of approximately 80% of CO_2 emission to the atmosphere as compared with process for the cement. Moreover, geopolymer has an advantage in the production at low temperature and low pressure by using alumino-silicate sources. Also, it can be prepared from natural source in nature and industry such as metakaolin, fly ash, blast furnace slag etc. with a strong alkaline solution (Abdulkareem, Mustafa Al Bakri, Kamarudin, Khairul Nizar, & Saif, 2014b; Rovnaník, 2010b). Geopolymers have excellent properties such as high mechanical properties, low shrinkage, fire resistance and durability (Duxson, Fernández-Jiménez, Provis et al., 2007). Summarized geopolymer sourced from waste materials for many applications:

- As concrete and cement (Pavithra et al., 2016),
- Composites (Roviello, Ricciotti, Ferone, Colangelo, & Tarallo, 2015),
- Immobilisation of toxic and radioactive waste (Xu & Van Deventer, 2002),
- Adsorbent for heavy metal ion (Li, Wang, & Zhu, 2006) and
- Fire resistant and protection matrix (Sakkas, Panias, Nomikos, & Sofianos, 2014).

In general chemical structure of geopolymer, sialate (silicon-oxo-aluminate) segments are main chemical framework (Figure 1). This consists of SiO_4 and AlO_4 tetrahedra linked alternately by sharing all oxygen atoms. Alkali cations or positive ions such as Na^+, K^+ and Ca^{++} are charged-balanced of the negative charge of Al^{3+} in IV-fold coordination (Davidovits, 1994). A polymeric structure of (poly) sialate formed constitutes the main building blocks of geopolymeric structure and the empirical formula of polysialates is displayed asfollow (Hanjitsuwan et al., 2014a): M_n $[(\text{-Si-O}_2)_z\text{-Al-O}_2\text{-}]_n$, wH_2O, where M is monovalent cation such as sodium or potassium, n is degree of polycondensation, z is amount of $Si\text{-}O_2$ (1, 2, 3 or higher) and w is amount of water.

It is common that geopolymerization can start from room temperature containing silica and alumina unit to be oligomers (dimer, trimer) which provide the actual unit structures of the three dimensional macromolecular edifice. The geopolymerization stages can be supposed into three stages:

1. Deconstruction,
2. Polymerization and
3. Stabilization (Blackstock, Neill, & Mcintosh, 2012; Muñiz-Villarreal et al., 2011; Yao, Zhang, Zhu, & Chen, 2009).

In geopolymer, the polymerized sialates are linkaged together by the type of sialate (Si-O-Al) and siloxo (Si-O-Si). The chemical structure of polysialate geopolymers includes the classified of polysialates, which are separated into four elementary units according to terms of the Si/Al atomic ratio as followed (Davidovits, 1982, 1984; Davidovits, 1994):

1. For the units, first is sialate and poly(sialate) at Si:Al = 1. These have -Si-O-Al-O- included in the chain and ring polymers from the polycondensation reaction of monomer, ortho-sialate $(OH)_3\text{-Si-O-Al-}(OH)_3$.
2. In the second unit, sialate- siloxo and poly(sialate- siloxo) at Si:Al = 2, (-Si-O-Al-O-Si-O-) are the chain and ring polymer from the result of condensation between ortho-silicic acid $Si(OH)_4$. These chains have three isomorphs as a linear, mono-siloxo-sialate and 3 cycles.
3. In the third unit, sialate- disiloxo and poly(sialate- disiloxo) at Si:Al = 3, (-Si-O-Al-O-Si-O-Si-O-) are described from the result of condensation reaction between two ortho-silicic acid $Si(OH)_4$. The sialate units are corresponded to form six isomorphs: 2 linear, 2 branched and 2 cycles.
4. Finally for fourth unit, sialate link and poly(sialate- multisiloxo) unit at Si:Al > 3 which are chains from bridge Si-O-Al between two poly(siloxonate), poly(silanol) or poly(sialate) chains.

The geopolymerization process involves:

1. Deconstruction,
2. Polymerization and

Figure 1. Chemical structure of polysialates

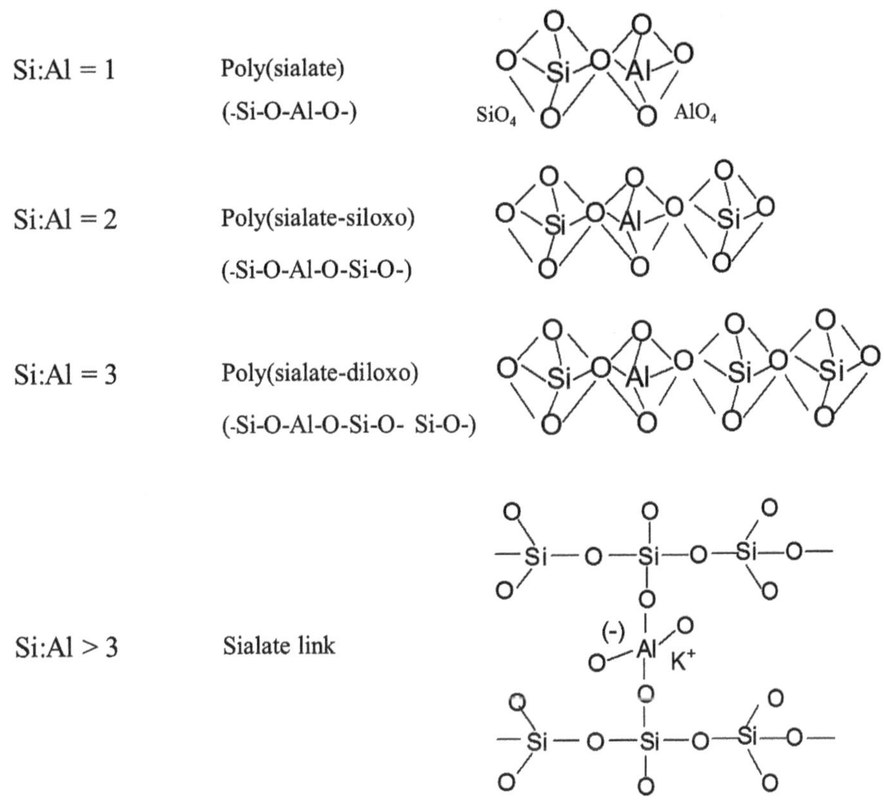

3. Stabilization (Yao et al., 2009; Zhang, Zhu, Zhou, & Wang, 2016).

In the deconstruction process, leaching of the solid alumino-silicate materials by alkaline hydrolysis produces aluminate and silicate species (Comrie & Kriven, 2006). The dissolution of the solid particles of the source for geopolymerization occurs at the surface. In high pH alkali solutions, the Si-O-Si, Al-O-Al and Al-O-Si of aluminosilicate source are dissolved (Somna et al., 2011). The alkali metal cations from the activated solution play important role in the geopolymerization and balance out the charge of the aluminium in form of four-fold coordination with oxygen (Lecomte et al., 2006). Then, the polymerization process occurs in the solution. The neighbouring reactive groups formed as Si-O-Na and O-H-Al and small silicate molecules from the alkali silicate thereafter undergo condensation reaction with the liberation of alkali hydroxide to form the preliminary unit for the geopolymerization. In the stabilization, the application of mild heat typically between room temperature up to 90 °C causes formation of the preliminary units in the terms of polycondensation. This can form rigid chains or nets of oxygen bonded aluminate and silicate tetrahedral which are similar to cement chemical structure. The higher "curing" temperature produces stronger geopolymer cements as each of the hydroxyl groups in these macromolecules are capable of condensing with one from a neighbouring molecule. It is theoretically possible for any one silicon to be bonded via an oxygen bond and to four neighbouring silicon or aluminium sites. As a result, very rigid polymer network is formed.

Several reports describe the use of metakaolin and fly ash as starting materials in geopolymerization. Moreover, industrial wastes of various types are used to synthesize geopolymer materials. For example, in cement and concrete, Pavithra et al.(Pavithra et al., 2016) developed the geopolymer concrete with fly ash. Moreover, Yeonho et al. (Park, Abolmaali, Kim, & Ghahremannejad, 2016) studied the feasibility of geopolymer concrete to which crumb rubber from recycled tires. Geopolymer cement and concrete showed low shrinkage after drying, less cracks and high strength. Therefore, the production cost of the geopolymer cement and concrete could be in 10-30% cheaper than Portland cement production.

In the adsorption of heavy metal application, geopolymer can be used as the materials for absorbing heavy metal from waste water. In recent years, fly ash based geopolymer was used for Cu and Pb removal from an aqueous solution. The absorption capacity was reached to 92 mg/g and 80 mg/g of Cu and Pb, respectively (Al-Zboon, Al-Harahsheh, & Hani, 2011; S. Wang, Li, & Zhu, 2007). Moreover, the geopolymer not only absorbed heavy metal but also could absorbed dye from waste water. The absorption capacity of fly ash based geopolymer demonstrated approximately 0.12 mmol/g (Li et al., 2006). Geopolymer adsorbents were prepared from silica and metakaolin in different Al and Si components (López et al., 2014a; López, Sugita, Takaya, & Kobayashi, 2014b) and applied for removal of metal ions, Cs^+ and Pb^{2+}, from other heavy metal ions mixture (López, Sugita, Takaya, & Kobayashi, 2014c). In the application for immobilization heavy metal and radioactive waste, geopolymer was used to immobilize, stabilize and adsorb low amount of radioactive waste in pure and mixed form (Van Jaarsveld, Van Deventer, & Lorenzen, 1997). The results indicated that the fly ash based geopolymer had high property to immobilize lead ion (Lee et al., 2016). Geopolymers acted as the binder to convert semi solid waste to adhesive solid and also immobilized hazardous waste by locking in the three dimensional framework. Moreover, the geopolymer materials could be apply for composites materials, hybrid composite materials for geopolymer and epoxy melamine resins were designed to produce a novel material showing high compatibility between the organic and inorganic phases. In the applications, requiring high temperature resistance is for kitchen tools and plates, coatings and flooring (Roviello et al., 2015). In addition, the geopolymer materials can be used for fire protection possessing excellent physicochemical, thermal and mechanical properties. So, fire resistant geopolymers were developed by the potassium based geopolymers (K-geopolymer) prepared with mixing ferronickel (FeNi) slag, doped with pure alumina (Sakkas et al., 2014).

UTILIZATION OF ALUMINIUM WASTE BASED GEOPOLYMER

Since geopolymers consist of aluminum-silica framework in the chemical structure, the source of Al becomes important especially in cases of waste reuse. Every year, aluminum consumption is increasing for many applications such as anodizing processes for producing decorative and protective films on aluminum and alloys. After etching step, Al-waste in gibbsite phase ($Al(OH)_3$) is wasted: typically about 360 tons/year from various industries is discarded in landfills (Álvarez-Ayuso, 2009; Colonetti, Hobold Kammer, & De Noni Junior, 2014; Ribeiro & Labrincha, 2008). Consumption of fossil fuels in power plants for energy production is increasing, leaving fly ash waste. About 3 million tons/year of such industrial waste are produced by large and small power plants. These wastes should be regenerated to produce novel materials. If people can recycle this, then such material becomes sustainable products that provide environmental, social, and economic benefits. Especially, as renewable resources of both aluminum hydroxide waste and fly ash, these seem to be applicable for reuse in concrete materials.

Therefore, approximately 13,500 million tons of Portland cement are manufactured annually using processes that incorporate high heating energy of 1400–1600 °C for calcining raw materials (Detphan & Chindaprasirt, 2009).

Aluminum hydroxide waste (Al-waste) from the aluminum frame industry and fly ash (FA) from power plant were used as the raw materials of geopolymers. Their chemical compositions are presented in Table 1 and Figure 2 shows SEM micrograph of the raw materials for a) Al-waste and b) FA. The Al-waste particles were irregularly shaped, with diameter of about 100 μm. The particle distribution of the Al-waste was 40–110 μm. In contrast, the FA contained spherical particles with about 40 μm diameter in the fine powder. The laser-scattered particle distribution of fly ash showed that the powder size was 0.2–130 μm. Geopolymerization was conducted for Al-waste and FA, when the wastes were mixed with aqueous NaOH solutions of 5, 10, and 15 M concentrations. Al-waste were used for 0, 10, 20, 40, and 60 wt% for the Al-waste at different NaOH concentrations. After geopolymerization precess, the geopolymer bulk were cured at room temperature, 60 °C and 80 °C for 24 h.

Table 1. Chemical composition of fly ash and aluminum hydroxide waste

Chemical Compound	Fly Ash (%)	Aluminum Waste (%)
SiO_2	31.94	0.18
Al_2O_3	18.37	63.28
Fe_2O_3	17.39	-
CaO	19.84	0.01
MgO	1.86	0.31
SO_3	5.13	< 0.01
Na_2O	1.20	-
K_2O	3.05	0.05
MnO	0.07	2.05
P_2O_5	0.16	0.01
TiO_2	0.40	0.01
L.O.I	1.47	34.02

Figure 2. SEM micrograph of a) Aluminum hydroxide waste and b) Fly ash with particle size distribution

a) b)

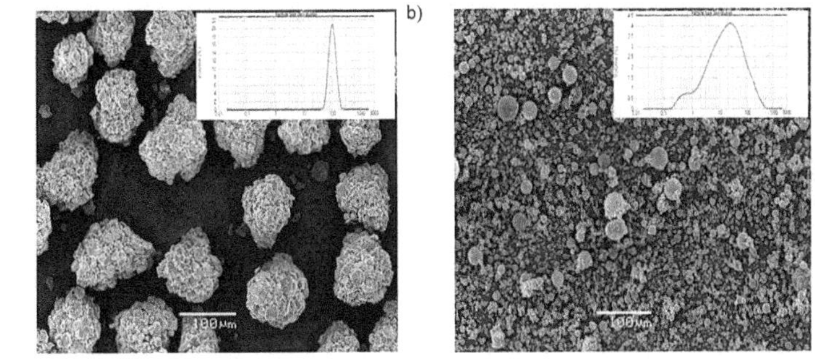

As mentioned, geopolymerization in alkaline solution is known to increase the mechanical property of the resultant geopolymers. The properties depended strongly upon silicate and aluminum ratios and on the curing temperature (Görhan & Kürklü, 2014; Hanjitsuwan et al., 2014b; Nazari et al., 2011; Somna et al., 2011). It is important for the present study to examine the dependence of the curing temperature and the NaOH concentration. The flexural strength and compressive strength for each specimen was observed in geopolymers prepared using different aluminium hydroxide contents and NaOH concentrations. Actually, a tendency was seen by which the geopolymer strength increased as the NaOH concentration increased. For the flexural strength (Figure 4a), the high strength in 40 wt% of the Al-waste contents was obtained in 5, 10 and 15 M of NaOH concentration. This might be because of the fact that the higher concentrations of NaOH are able to dissolve both silica and alumina in a homogeneous slurry paste. Especially, high strength appeared at 80 °C, meaning that the curing temperature affected the geopolymer strength. After curing at higher temperature, increased bulk strength for the 40 wt% of Al-waste was observed with the maximum flexural strength of 6.05 MPa. A similar temperature tendency was also apparent for compressive strength (Figure 3b). The high compressive strength of 40 MPa was obtained for 40 wt% of Al-waste used in 10 M of NaOH concentration at 80 °C. As a result, slurry containing high alkaline solution was preferred for the condensation reaction to form Si–O and Al–O in the geopolymers. As reported (Görhan & Kürklü, 2014), this preference was mainly attributable to the excellent dissolution of silica and alumina components in the slurry. Therefore, high temperatures enhanced the geopolymerization in the Si–O–Al formation by the condensation. Regarding to development of compressive strength in Portland cement pastes and concrete specimens, the strength increased with increased curing temperature (Chindaprasirt et al., 2007; Somna et al., 2011). However, when the NaOH concentration was 15 M, the compressive strength started to drop. Excess NaOH was the cause of the compressive strength decrease attributable to the alumino-silicate gel precipitation at the early stage (Somna et al., 2011). On the other hand, the temperature was higher than 80 °C, water in the geopolymer slurry was evaporated to produce a porous structure (López et al., 2014a). In addition, the mass loss was capable of forming cracks in the geopolymer matrix because of the thermal shrinkage (Abdulkareem, Mustafa Al Bakri, Kamarudin, Khairul Nizar, & Saif, 2014a). The apparent bulk density of the geopolymer specimens increased when increasing the amount of the Al-waste to 40 wt%. However, as the Al-waste was added with 60 wt%, the apparent bulk density decreased. As seen in Figure 4, the micro CT-x ray technique showed the interior structure of geopolymer matrix in different amounts of the Al-waste. These geopolymers were prepared in the 40 wt% of the Al-waste at 80 °C with 10 M NaOH concentration. The inside geopolymer exhibited numerous cracks and pores for Figure 4a in the no addition of the Al-waste. In cases Figure 4b and 4d for 20 and 60 wt% of the Al-waste contents, the inside structures had fewer cracks, but round pores existed in these geopolymers morphological structure. In contrast, a denser structure was observed in the 40 wt% of the Al-waste (Figure 4c). Especially, in Figure 6d for 60 wt% the Al-waste, many pores were visible in the geopolymer. This result suggests that excess Al-waste caused the decrease of the geopolymer strength.

FLY ASH BASED GEOPOLYMERS

Fly ash is one of source material for geopolymer (Table 2) and is generally by-product obtained from industrial processes. Fly ash is commonly composed of a fine powder consisting mostly of spherical glassy particles (Figure 2) and is produced in coal fired power stations. Fly ash has pozzalonic proper-

Figure 3. Comparision of a) flexural strength and b) compressive strength of geopolymer paste at age of 7 days

a)

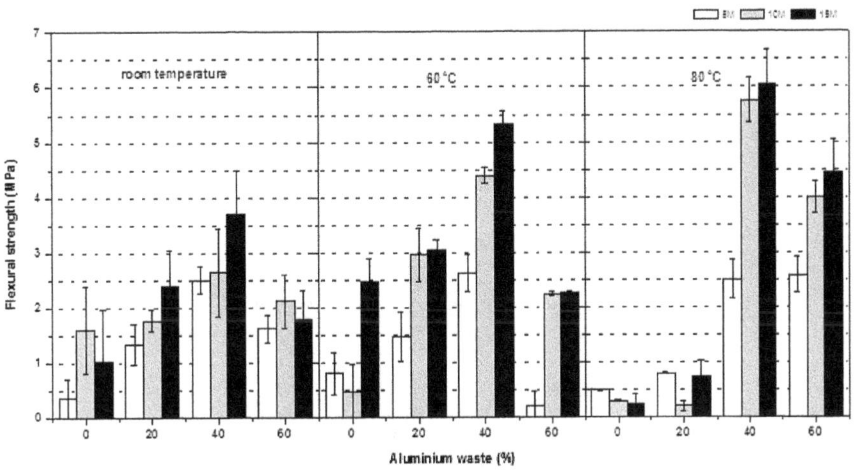

b)

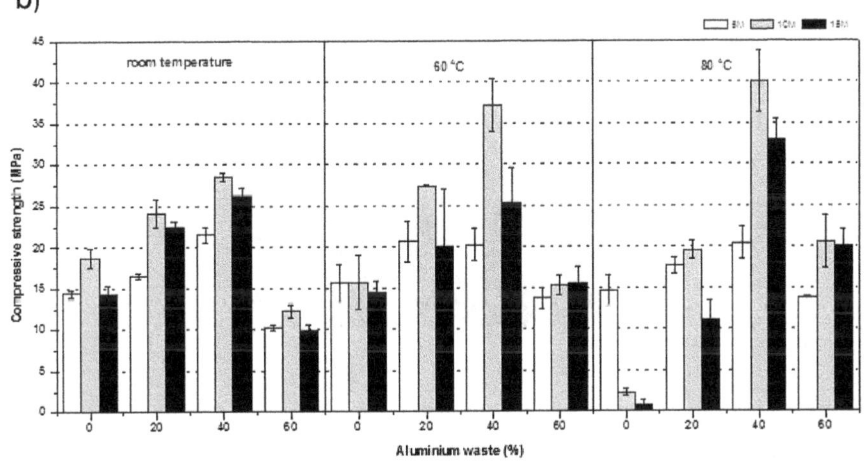

ties and for example reacts with lime to form cementitious compounds. It is known as a supplementary cementitious materials (Pangdaeng et al., 2014). So the mechanical property of fly ash based geopolymers were improved. The use of dense fly ash based geopolymer concrete led to many improvements in overall concrete performance. Some key benefits are included: reduced water content, improved workability, reduced segregation, low heat of hydration, improved finishing, lower drying shrinkage, increased durability, improved long term strength, good carbonation protection, improved resistance to sulfate attack and improved resistance to alkali silica reactivity as listed below.

Moreover, the improvement porous structure of fly ash based geopolymers also was reported for application as shown in Table 3. Moreover, for example, the porous structure of geopolymer materials was observed when microwave method was used as heating source (Onutai, Jiemsirilers, Thavorniti, & Kobayashi, 2016). Fly ash was obtained from Mae Moh Power Plant in Lamphang, Thailand for used as

Figure 4. X-ray CT images of geopolymer sepecimens with different amounts of Al-waste on structure of geopolymer specimens: a) 0 wt %, b) 20 wt %, c) 40 wt % and d) 60 wt % Al-waste with 10 M NaOH and cured at 80 °C for 24h

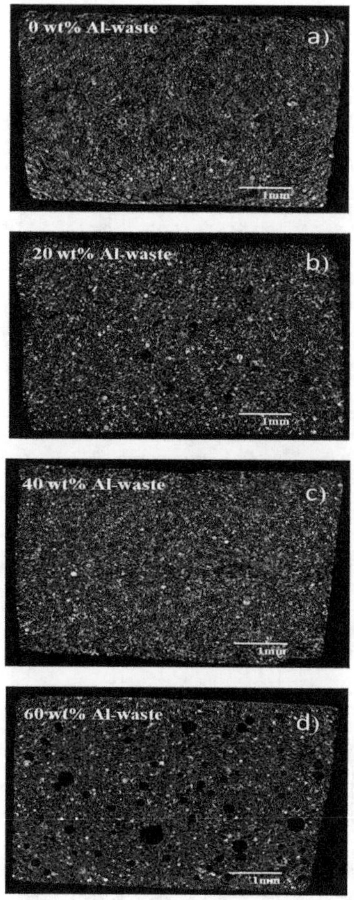

a raw materials. For geopolymerization, sodium hydroxide solutions and sodium silicate solutions were used as activators. The solid fraction in geopolymer mixture was 65 wt%. The ratio of Na_2SiO_3/NaOH solution and SiO_2/Al_2O_3 were fixed at 2.5 and 2.69 at 2-15 M NaOH, respectively. The geopolymer paste was cured by heating in a microwave oven at the different power setting (200-850 W) for 1 min. During microwave heating, the geopolymers temperature were detected by using an infrared thermal scanner as presented in Figure 5. The temperature development was the function of microwave power and sodium hydroxide concentrations. The specimen temperature increased with an increasing incident power and sodium hydroxide concentration. The temperature reached to 65-130 °C within 1 min. The aqueous paste of fly ash and sodium silicate contained different NaOH concentrations of 2 M-15 M was cured in the household microwave oven and then the geopolymer paste adsorbed microwave energy. The microwave heating occurred depending upon the NaOH concentration due to the dipolar polarization and ionic conduction mechanisms. Polar molecules such as water produce heat because the water absorbs microwaves as a result of the polarization induced by the external oscillating electric field (Vollmer, 2004). Therefore, the conduction mechanism from charged particles such as Na^+ and OH^- can oscil-

Table 2. Summarized synthesis dense fly ash based geopolymer

Raw Material	Conditions	Compressive Strength	References
Fly ash (paste)	NaOH solution = 10 M Na_2SiO_3/NaOH ratio = 1.5 Curing temp. at 65 °C for 48 h	35.0 MPa at 2 days	(Chindaprasirt, Jaturapitakkul, Chalee, & Rattanasak, 2009)
Fly ash (paste)	NaOH solution = 14 M Curing temp. at 25 °C	15.0 MPa at 7 days 25.5 MPa at 60 days	(Somna et al., 2011)
Fly ash (paste)	NaOH solution = 18 M Na_2SiO_3/NaOH ratio = 0.67 Curing temp. at 40 °C for 24 h	56.0 MPa at 7 days	(Hanjitsuwan et al., 2014a)
Fly ash (paste)	NaOH solution = 9 M Na_2SiO_3/NaOH ratio = 1 Curing temp. at 60 °C for 24 h	53.0 MPa at 91 days	(Jang & Lee, 2016)
Fly ash (mortar)	NaOH solution = 10 M Na_2SiO_3/NaOH ratio = 0.67 Curing temp. at 60 °C for 72 h	52.0 MPa at 7 days	(Chindaprasirt et al., 2007)
Fly ash (mortar)	NaOH solution = 6 M Na_2SiO_3/NaOH ratio = 2 Curing temp. at 85 °C for 24 h	22.0 MPa at 7 days	(Görhan & Kürklü, 2014)
Fly ash (paste) (mortar) (lightweight concrete)	NaOH solution = 12 M Na_2SiO_3/NaOH ratio = 1 Curing temp. at 40 °C for 24 h NaOH solution = 12 M Na_2SiO_3/NaOH ratio = 1 Curing temp. at 40 °C for 24 h NaOH solution = 12 M Na_2SiO_3/NaOH ratio = 1 Curing temp. at 40 °C for 24 h	44.0 MPa at 28 days 47.0 MPa at 28 days 18.0 MPa at 28 days	(Abdulkareem et al., 2014b)
Fly ash (paste) Fly ash-rice husk and bark ash (paste)	NaOH solution = 18 M Na_2SiO_3/NaOH ratio = 2.5 Curing temp. at 25 °C FA:RHBA = 40:60 NaOH solution = 18 M Na_2SiO_3/NaOH ratio = 2.5 Curing temp. at 60 °C for 24 h	17.0 MPa at 7 days 34.0 MPa at 28 days 49.2 MPa at 7 days 56.0 MPa at 28 days	(Songpiriyakij et al., 2010)
Fly ash-sawdust (paste)	FA:Sawdust = 80:20 NaOH solution = 10 M Na_2SiO_3/NaOH ratio = 8 Curing temp. at 40 °C for 24 h	55.0 MPa at 14 days	(Duan, Yan, Zhou, & Luo, 2016)
Fly ash (paste) Fly ash-metakaolin (paste)	NaOH solution = 10 M Na_2SiO_3/NaOH ratio = 1.6 Curing temp. at 20 °C FA:MK = 80:20 NaOH solution = 10 M Na_2SiO_3/NaOH ratio = 1.6 Curing temp. at 20 °C	57.0 MPa at 28 days 80.0 MPa at 28 days	(Duan, Yan, & Zhou, 2016)
Fly ash-silica fume (concrete)	NaOH solution = 14 M Na_2SiO_3/NaOH ratio = 1 Curing temp. at 100 °C for 72 h	73.0 MPa at 28 days	(Okoye, Durgaprasad, & Singh, 2016)

Table 3. Summarized synthesis porous fly ash based geopolymer

Raw Material	Method	Conditions	References
Fly ash	added aluminium powder = 1.5-5.0 wt%	NaOH solution = 8 M Na_2SiO_3/NaOH ratio = 2.5	(Sanjayan, Nazari, Chen, & Nguyen, 2015)
Fly ash	added aluminium powder = 0.06-0.12 wt%	NaOH solution = 2.8-3.2 g. Curing temp. at 80 °C for 12 h	(Hlaváček, Šmilauer, Škvára, Kopecký, & Šulc, 2015)
Fly ash	added sodium hypochlorite = 0.40-0.60 wt%	NaOH: FA= 0.20 wt% Curing temp. at 60 °C for 24 h and 80 °C for 48 h	(Böke et al., 2015)
Fly ash-metakaolin	gel casting	KOH solution = 15 M K_2SiO_3/KOH ratio = 2 Curing temp. at 80 °C for 5 h	(Strozi Cilla, Colombo, & Raymundo Morelli, 2014)
Fly ash	added superplasticizer = 2 volume%	NaOH solution = 12 M Na_2SiO_3/NaOH ratio = 2.5 Curing temp. at 60 °C for 24 h	(Abdullah, Hussin, et al., 2012)
Fly ash	added sheet glass powder = 8 wt% added surfactant solution = 10 wt %	NaOH solution = 10 wt% Curing temp. = 105 °C for 12 h Sintering = 1050 °C for 2 h	(Zhao et al., 2010)

late back and forth under electric force producing an electric current to heats up neighboring materials through molecular collisions of charged species (Anwar et al., 2015; Vollmer, 2004). Moreover, the heat of geopolymers are increased during geopolymerization as a fact that the geopolymerization process is exothermic chemical reaction (Davidovits, 1991; Jumrat, Chatveera, & Rattanadecho, 2011). When the geopolymer paste was processed to the solid geopolymer, microwave heating was able to achieve higher temperatures in a very short time depending on the NaOH concentration. During the process, the volume of geopolymer specimen are immediately increased. Consequently, the water was vaporized easily at temperatures higher than 100 °C. Vaporization of the water might influence the morphology of the geopolymers. Bubbles from evaporation of water were fixed inside the geopolymer sample along with the general heat curing process, producing a porous geopolymer structure (Hanjitsuwan et al., 2014a).

The porous geopolymer morphology in the microstructure was compared in each sample synthesized in a conventional oven at 80 °C, with household microwave at 200 W and 850 W power for 1 min (Figure 6). The morphological analysis elucidated that the microstructural changes occurred with heating and curing at 80 °C and microwave curing. All samples showed that the ash particles remained in the geopolymer matrix. During the dense morphology of the matrix was created, the rough surface of the geopolymers was observed for the paste having 10 M and 15 M NaOH concentrations. At 850 W of microwave curing, the tendency was extremely effective at higher NaOH concentrations, as depicted in Figures. 6i) and 6l). In the geopolymers cured with 15 M of NaOH concentration, the apparent density was increased markedly when the microwave was operated at 500-850 W. The values of apparent bulk density showed about 0.9 g/cm^3 for the 15 M NaOH concentration at 500-850 W, while these were 1.4-1.5 g/cm^3 for the 2 M and 5 M NaOH concentrations. When the NaOH concentration was increased, the strong alkali rapidly increased the interior temperature, especially in 10 M and 15 M materials at 700 and 850 W. Therefore, the bubbles of the water evaporation seemed to be increased in the bulk porous structure, thereby decreasing the apparent density.

Figure 5. Temperature profile of geopolymers in different power of microwave 1 min and pictures of the obtained geopolymer at 850 W

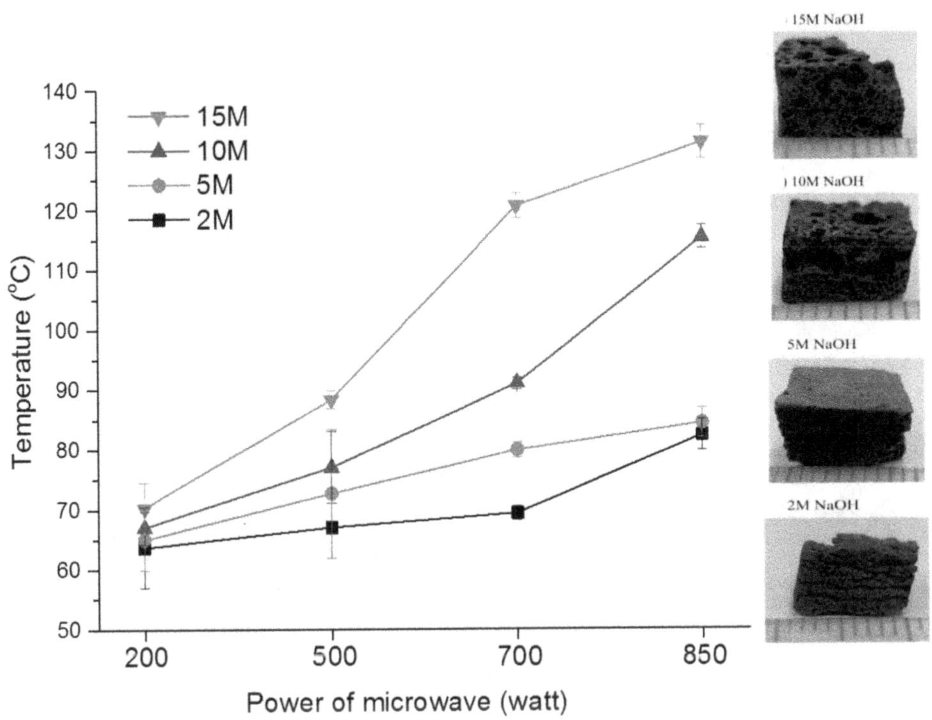

CONCLUSION

In this chapter authors conclude the reuse fly ash and industrial wastes, Al-waste, for producing geopolymers which are sustainable and green materials. In their waste materials, geopolymers were synthesized from fly ash and aluminium waste with different proportions of raw materials for alternative cement. Moreover, when different NaOH concentration and different curing temperature was changed, effect of the alkali was described. Fly ash based geopolymer and synthesis porous geopolymer using microwave radiation from household microwave were also reviewed. The results showed that porous geopolymer was produced at 10 or 15 M NaOH with 850 W of microwave radiation power, due to that the high power of microwave radiation increased the slurry temperature and affected water evaporation in the geopolymer paste. The bulk and porous geopolymers were successfully produced by using FA and Al-wastes.

Figure 6. Microstructure of geopolymer sample synthesized by curing at 80 °C 24 h and using microwave radiation at power 200 and 850 W for 1 min

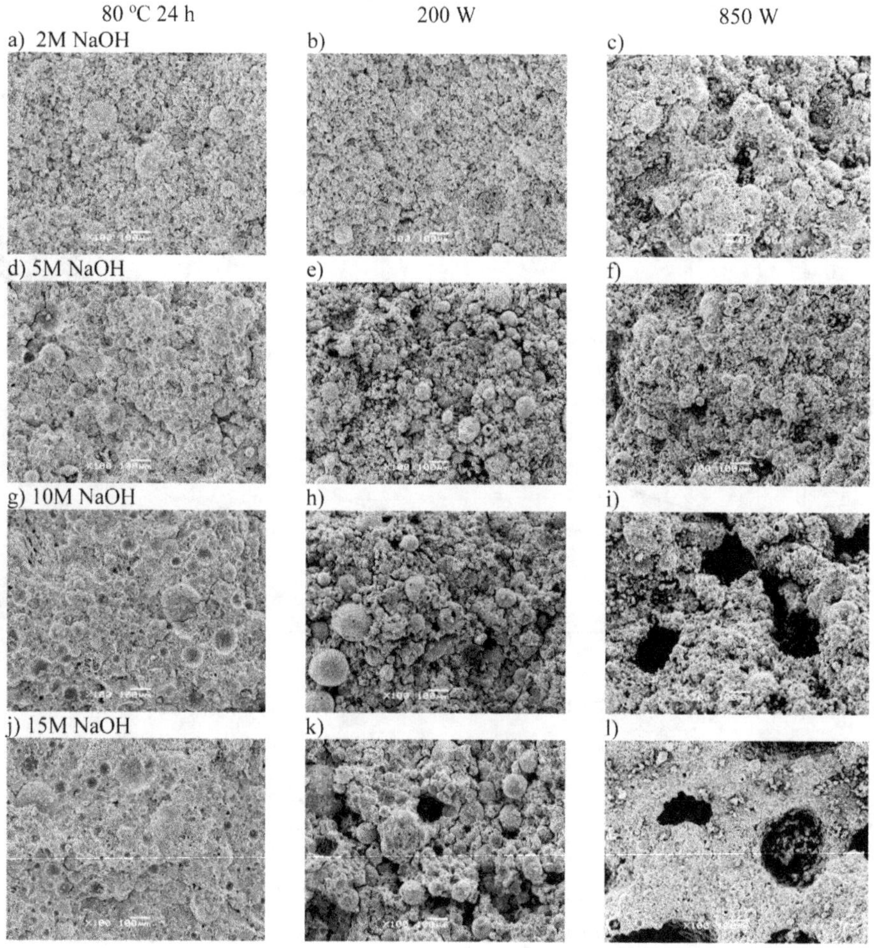

REFERENCES

Abdulkareem, O. A., Mustafa Al Bakri, A. M., Kamarudin, H., Khairul Nizar, I., & Saif, A. A. (2014). Effects of elevated temperatures on the thermal behavior and mechanical performance of fly ash geopolymer paste, mortar and lightweight concrete. *Construction & Building Materials*, *50*, 377–387. doi:10.1016/j.conbuildmat.2013.09.047

Abdullah, M. M. A. B., Hussin, K., Bnhussain, M., Ismail, K. N., Yahya, Z., & Abdul Razak, R. (2012). Fly Ash-based Geopolymer Lightweight Concrete Using Foaming Agent. *International Journal of Molecular Sciences*, *13*(6), 7186–7198. doi:10.3390/ijms13067186 PMID:22837687

Abdullah, M. M. A. B., Jamaludin, L., Hussin, K., Bnhussain, M., Ghazali, C. M. R., & Ahmad, M. I. (2012). Fly Ash Porous Material using Geopolymerization Process for High Temperature Exposure. *International Journal of Molecular Sciences*, *13*(4), 4388–4395. doi:10.3390/ijms13044388 PMID:22605984

Al-Zboon, K., Al-Harahsheh, M. S., & Hani, F. B. (2011). Fly ash-based geopolymer for Pb removal from aqueous solution. *Journal of Hazardous Materials*, *188*(1–3), 414–421. doi:10.1016/j.jhazmat.2011.01.133 PMID:21349635

Alaa, M., & Rashad, S. R. Z. (2011). The effect of activator concentration on the residual strength of alkali-activated fly ash pastes subjected to thermal load. *Construction & Building Materials*, *25*(7), 3098–3107. doi:10.1016/j.conbuildmat.2010.12.044

Álvarez-Ayuso, E. (2009). Approaches for the treatment of waste streams of the aluminium anodising industry. *Journal of Hazardous Materials*, *164*(2-3), 409–414. doi:10.1016/j.jhazmat.2008.08.054 PMID:18823703

Andini, S., Cioffi, R., Colangelo, F., Grieco, T., Montagnaro, F., & Santoro, L. (2008). Coal fly ash as raw material for the manufacture of geopolymer-based products. *Waste Management (New York, N.Y.)*, *28*(2), 416–423. doi:10.1016/j.wasman.2007.02.001 PMID:17382528

Anwar, J., Shafique, U., Waheed-uz-Zaman, , Rehman, R., Salman, M., Dar, A., & Ashraf, S. et al. (2015). Microwave chemistry: Effect of ions on dielectric heating in microwave ovens. *Arabian Journal of Chemistry*, *8*(1), 100–104. doi:10.1016/j.arabjc.2011.01.014

Bakharev, T. (2005). Geopolymeric materials prepared using Class F fly ash and elevated temperature curing. *Cement and Concrete Research*, *35*(6), 1224–1232. doi:10.1016/j.cemconres.2004.06.031

Blackstock, J. M., Neill, J., & Mcintosh, J. A. (2012). *Geopolymeric structural building units and methods of manufacture thereof.* Google Patents.

Böke, N., Birch, G. D., Nyale, S. M., & Petrik, L. F. (2015). New synthesis method for the production of coal fly ash-based foamed geopolymers. *Construction & Building Materials*, *75*, 189–199. doi:10.1016/j.conbuildmat.2014.07.041

Chindaprasirt, P., Chareerat, T., & Sirivivatnanon, V. (2007). Workability and strength of coarse high calcium fly ash geopolymer. *Cement and Concrete Composites*, *29*(3), 224–229. doi:10.1016/j.cemconcomp.2006.11.002

Chindaprasirt, P., Jaturapitakkul, C., Chalee, W., & Rattanasak, U. (2009). Comparative study on the characteristics of fly ash and bottom ash geopolymers. *Waste Management (New York, N.Y.)*, *29*(2), 539–543. doi:10.1016/j.wasman.2008.06.023 PMID:18715775

Chindaprasirt, P., Rattanasak, U., & Taebuanhuad, S. (2013). Resistance to acid and sulfate solutions of microwave-assisted high calcium fly ash geopolymer. *Materials and Structures*, *46*(3), 375–381. doi:10.1617/s11527-012-9907-1

Cioffi, R., Maffucci, L., & Santoro, L. (2003). Optimization of geopolymer synthesis by calcination and polycondensation of a kaolinitic residue. *Resources, Conservation and Recycling*, *40*(1), 27–38. doi:10.1016/S0921-3449(03)00023-5

Colonetti, E., Hobold Kammer, E., & De Noni Junior, A. (2014). Chemically-bonded phosphate ceramics obtained from aluminum anodizing waste for use as coatings. *Ceramics International*, *40*(99, Part A), 14431–14438. doi:10.1016/j.ceramint.2014.06.039

Comrie, D. C., & Kriven, W. M. (2006). *Composite Cold Ceramic Geopolymer in a Refractory Application. In Advances in Ceramic Matrix Composites IX* (pp. 211–225). John Wiley & Sons, Inc.

Davidovits, J. (1982). *Mineral polymers and methods of making them*. Google Patents.

Davidovits, J. (1984). *Synthetic mineral polymer compound of the silicoaluminates family and preparation process*. Google Patents.

Davidovits, J. (1991). Geopolymers - Inorganic polymeric new materials. *Journal of Thermal Analysis, 37*(8), 1633–1656. doi:10.1007/BF01912193

Davidovits, J. (1994). Properties of geopolymer cements. *First international conference on alkaline cements and concretes, 1*, 131-149.

Davidovits, J. (2008). *Geopolymer Chemistry & Applications*. Institut Géopolymère 16 rue Galilée F-02100 Saint-Quentin France.

Detphan, S., & Chindaprasirt, P. (2009). Preparation of fly ash and rice husk ash geopolymer. *International Journal of Minerals. Metals and Materials, 16*(6), 720–726. doi:10.1016/S1674-4799(10)60019-2

Duan, P., Yan, C., & Zhou, W. (2016). Influence of partial replacement of fly ash by metakaolin on mechanical properties and microstructure of fly ash geopolymer paste exposed to sulfate attack. *Ceramics International, 42*(2, Part B), 3504–3517. doi:10.1016/j.ceramint.2015.10.154

Duan, P., Yan, C., Zhou, W., & Luo, W. (2016). Fresh properties, mechanical strength and microstructure of fly ash geopolymer paste reinforced with sawdust. *Construction & Building Materials, 111*, 600–610. doi:10.1016/j.conbuildmat.2016.02.091

Duxson, P., Fernández-Jiménez, A., Lukey, G. C., Palomo, A., Provis, J. L., & Van Deventer, J. S. J. (2007). Geopolymer technology: The current state of the art. *Journal of Materials Science, 42*(9), 2917–2933. doi:10.1007/s10853-006-0637-z

Flower, D. J. M., & Sanjayan, J. G. (2007). Green house gas emissions due to concrete manufacture. *The International Journal of Life Cycle Assessment, 12*(5), 282–288. doi:10.1007/s11367-007-0327-3

Görhan, G., & Kürklü, G. (2014). The influence of the NaOH solution on the properties of the fly ash-based geopolymer mortar cured at different temperatures. *Composites. Part B, Engineering, 58*(0), 371–377. doi:10.1016/j.compositesb.2013.10.082

Gubb, T. A., Baranova, I., Allan, S. M., Fall, M. L., Shulman, H. S., & Kriven, W. M. (2011). *Microwave Enhanced Drying and Firing of Geopolymers. In Developments in Strategic Materials and Computational Design II* (pp. 35–43). John Wiley & Sons, Inc. doi:10.1002/9781118095393.ch4

Hanjitsuwan, S., Hunpratub, S., Thongbai, P., Maensiri, S., Sata, V., & Chindaprasirt, P. (2014). Effects of NaOH concentrations on physical and electrical properties of high calcium fly ash geopolymer paste. *Cement and Concrete Composites, 45*, 9–14. doi:10.1016/j.cemconcomp.2013.09.012

Hasanbeigi, A., Price, L., & Lin, E. (2012). Emerging energy-efficiency and CO2 emission-reduction technologies for cement and concrete production: A technical review. *Renewable & Sustainable Energy Reviews, 16*(8), 6220–6238. doi:10.1016/j.rser.2012.07.019

Heath, A., Paine, K., & McManus, M. (2014). Minimising the global warming potential of clay based geopolymers. *Journal of Cleaner Production*, *78*, 75–83. doi:10.1016/j.jclepro.2014.04.046

Hlaváček, P., Šmilauer, V., Škvára, F., Kopecký, L., & Šulc, R. (2015). Inorganic foams made from alkali-activated fly ash: Mechanical, chemical and physical properties. *Journal of the European Ceramic Society*, *35*(2), 703–709. doi:10.1016/j.jeurceramsoc.2014.08.024

Jang, J. G., & Lee, H. K. (2016). Effect of fly ash characteristics on delayed high-strength development of geopolymers. *Construction & Building Materials*, *102*(Part 1), 260–269. doi:10.1016/j.conbuildmat.2015.10.172

Jumrat, S., Chatveera, B., & Rattanadecho, P. (2011). Dielectric properties and temperature profile of fly ash-based geopolymer mortar. *International Communications in Heat and Mass Transfer*, *38*(2), 242–248. doi:10.1016/j.icheatmasstransfer.2010.11.020

Karakoç, M. B., Türkmen, İ., Maraş, M. M., Kantarci, F., Demirboğa, R., & Uğur Toprak, M. (2014). Mechanical properties and setting time of ferrochrome slag based geopolymer paste and mortar. *Construction & Building Materials*, *72*(0), 283–292. doi:10.1016/j.conbuildmat.2014.09.021

Komnitsas, K., & Zaharaki, D. (2007). Geopolymerisation: A review and prospects for the minerals industry. *Minerals Engineering*, *20*(14), 1261–1277. doi:10.1016/j.mineng.2007.07.011

Kumar, S., & Kumar, R. (2011). Mechanical activation of fly ash: Effect on reaction, structure and properties of resulting geopolymer. *Ceramics International*, *37*(2), 533–541. doi:10.1016/j.ceramint.2010.09.038

Lecomte, I., Henrist, C., Liégeois, M., Maseri, F., Rulmont, A., & Cloots, R. (2006). (Micro)-structural comparison between geopolymers, alkali-activated slag cement and Portland cement. *Journal of the European Ceramic Society*, *26*(16), 3789–3797. doi:10.1016/j.jeurceramsoc.2005.12.021

Lee, S., van Riessen, A., Chon, C.-M., Kang, N.-H., Jou, H.-T., & Kim, Y.-J. (2016). Impact of activator type on the immobilisation of lead in fly ash-based geopolymer. *Journal of Hazardous Materials*, *305*, 59–66. doi:10.1016/j.jhazmat.2015.11.023 PMID:26642447

Li, L., Wang, S., & Zhu, Z. (2006). Geopolymeric adsorbents from fly ash for dye removal from aqueous solution. *Journal of Colloid and Interface Science*, *300*(1), 52–59. doi:10.1016/j.jcis.2006.03.062 PMID:16626729

López, F. J., Sugita, S., & Kobayashi, T. (2014). Cesium-adsorbent Geopolymer Foams Based on Silica from Rice Husk and Metakaolin. *Chemistry Letters*, *43*(1), 128–130. doi:10.1246/cl.130851

López, F. J., Sugita, S., Takaya, M., & Kobayashi, T. (2014). Geopolymers Using Rice Husk Silica and Metakaolin Derivatives; Preparation and Their Characteristics. *Journal of Materials Science and Chemical Engineering*, *2*(05), 35–43. doi:10.4236/msce.2014.25006

López, F. J., Sugita, S., Takaya, M., & Kobayashi, T. (2014). Metakaolin-Based Geopolymers for Targeted Adsorbents to Heavy Metal Ion Separation. *Journal of Materials Science and Chemical Engineering*, *2*(07), 16–27. doi:10.4236/msce.2014.27002

Masi, G., Rickard, W. D. A., Vickers, L., Bignozzi, M. C., & van Riessen, A. (2014). A comparison between different foaming methods for the synthesis of light weight geopolymers. *Ceramics International, 40*(9, Part A), 13891–13902. doi:10.1016/j.ceramint.2014.05.108

Muñiz-Villarreal, M. S., Manzano-Ramírez, A., Sampieri-Bulbarela, S., Gasca-Tirado, J. R., Reyes-Araiza, J. L., Rubio-Ávalos, J. C., & Amigó-Borrás, V. et al. (2011). The effect of temperature on the geopolymerization process of a metakaolin-based geopolymer. *Materials Letters, 65*(6), 995–998. doi:10.1016/j.matlet.2010.12.049

Nazari, A., Bagheri, A., & Riahi, S. (2011). Properties of geopolymer with seeded fly ash and rice husk bark ash. *Materials Science and Engineering A, 528*(24), 7395–7401. doi:10.1016/j.msea.2011.06.027

Nazari, A., & Sanjayan, J. G. (2015). Synthesis of geopolymer from industrial wastes. *Journal of Cleaner Production, 99*, 297–304. doi:10.1016/j.jclepro.2015.03.003

Okoye, F. N., Durgaprasad, J., & Singh, N. B. (2016). Effect of silica fume on the mechanical properties of fly ash based-geopolymer concrete. *Ceramics International, 42*(2, Part B), 3000–3006. doi:10.1016/j.ceramint.2015.10.084

Onutai, S., Jiemsirilers, S., Thavorniti, P., & Kobayashi, T. (2015). Aluminium hydroxide waste based geopolymer composed of fly ash for sustainable cement materials. *Construction & Building Materials, 101*(Part 1), 298–308. doi:10.1016/j.conbuildmat.2015.10.097

Onutai, S., Jiemsirilers, S., Thavorniti, P., & Kobayashi, T. (2016). Fast microwave syntheses of fly ash based porous geopolymers in the presence of high alkali concentration. *Ceramics International, 42*(8), 9866–9874. doi:10.1016/j.ceramint.2016.03.086

Pangdaeng, S., Phoo-ngernkham, T., Sata, V., & Chindaprasirt, P. (2014). Influence of curing conditions on properties of high calcium fly ash geopolymer containing Portland cement as additive. *Materials & Design, 53*, 269–274. doi:10.1016/j.matdes.2013.07.018

Panias, D., Giannopoulou, I. P., & Perraki, T. (2007). Effect of synthesis parameters on the mechanical properties of fly ash-based geopolymers. *Colloids and Surfaces. A, Physicochemical and Engineering Aspects, 301*(1–3), 246–254. doi:10.1016/j.colsurfa.2006.12.064

Park, Y., Abolmaali, A., Kim, Y. H., & Ghahremannejad, M. (2016). Compressive strength of fly ash-based geopolymer concrete with crumb rubber partially replacing sand. *Construction & Building Materials, 118*, 43–51. doi:10.1016/j.conbuildmat.2016.05.001

Pavithra, P., Srinivasula Reddy, M., Dinakar, P., Hanumantha Rao, B., Satpathy, B. K., & Mohanty, A. N. (2016). A mix design procedure for geopolymer concrete with fly ash. *Journal of Cleaner Production, 133*, 117–125. doi:10.1016/j.jclepro.2016.05.041

Puertas, F., Martínez-Ramírez, S., Alonso, S., & Vázquez, T. (2000). Alkali-activated fly ash/slag cements: Strength behaviour and hydration products. *Cement and Concrete Research, 30*(10), 1625–1632. doi:10.1016/S0008-8846(00)00298-2

Ribeiro, M. J., & Labrincha, J. A. (2008). Properties of sintered mullite and cordierite pressed bodies manufactured using Al-rich anodising sludge. *Ceramics International*, *34*(3), 593–597. doi:10.1016/j.ceramint.2006.12.005

Rickard, W. D. A., Vickers, L., & van Riessen, A. (2013). Performance of fibre reinforced, low density metakaolin geopolymers under simulated fire conditions. *Applied Clay Science*, *73*, 71–77. doi:10.1016/j.clay.2012.10.006

Roviello, G., Ricciotti, L., Ferone, C., Colangelo, F., & Tarallo, O. (2015). Fire resistant melamine based organic-geopolymer hybrid composites. *Cement and Concrete Composites*, *59*, 89–99. doi:10.1016/j.cemconcomp.2015.03.007

Rovnaník, P. (2010). Effect of curing temperature on the development of hard structure of metakaolin-based geopolymer. *Construction & Building Materials*, *24*(7), 1176–1183. doi:10.1016/j.conbuildmat.2009.12.023

Ryu, G. S., Lee, Y. B., Koh, K. T., & Chung, Y. S. (2013). The mechanical properties of fly ash-based geopolymer concrete with alkaline activators. *Construction & Building Materials*, *47*(0), 409–418. doi:10.1016/j.conbuildmat.2013.05.069

Sakkas, K., Panias, D., Nomikos, P. P., & Sofianos, A. I. (2014). Potassium based geopolymer for passive fire protection of concrete tunnels linings. *Tunnelling and Underground Space Technology*, *43*, 148–156. doi:10.1016/j.tust.2014.05.003

Sanjayan, J. G., Nazari, A., Chen, L., & Nguyen, G. H. (2015). Physical and mechanical properties of lightweight aerated geopolymer. *Construction & Building Materials*, *79*, 236–244. doi:10.1016/j.conbuildmat.2015.01.043

Somaratna, J., Ravikumar, D., & Neithalath, N. (2010). Response of alkali activated fly ash mortars to microwave curing. *Cement and Concrete Research*, *40*(12), 1688–1696. doi:10.1016/j.cemconres.2010.08.010

Somna, K., Jaturapitakkul, C., Kajitvichyanukul, P., & Chindaprasirt, P. (2011). NaOH-activated ground fly ash geopolymer cured at ambient temperature. *Fuel*, *90*(6), 2118–2124. doi:10.1016/j.fuel.2011.01.018

Songpiriyakij, S., Kubprasit, T., Jaturapitakkul, C., & Chindaprasirt, P. (2010). Compressive strength and degree of reaction of biomass- and fly ash-based geopolymer. *Construction & Building Materials*, *24*(3), 236–240. doi:10.1016/j.conbuildmat.2009.09.002

Strozi Cilla, M., Colombo, P., & Raymundo Morelli, M. (2014). Geopolymer foams by gelcasting. *Ceramics International*, *40*(4), 5723–5730. doi:10.1016/j.ceramint.2013.11.011

Strozi Cilla, M., Raymundo Morelli, M., & Colombo, P. (2014). Effect of process parameters on the physical properties of porous geopolymers obtained by gelcasting. *Ceramics International*, *40*(88, Part B), 13585–13590. doi:10.1016/j.ceramint.2014.05.074

Studart, A. R., Gonzenbach, U. T., Tervoort, E., & Gauckler, L. J. (2006). Processing routes to macroporous ceramics: A review. *Journal of the American Ceramic Society*, *89*(6), 1771–1789. doi:10.1111/j.1551-2916.2006.01044.x

Swanepoel, J. C., & Strydom, C. A. (2002). Utilisation of fly ash in a geopolymeric material. *Applied Geochemistry*, *17*(8), 1143–1148. doi:10.1016/S0883-2927(02)00005-7

Van Jaarsveld, J. G. S., Van Deventer, J. S. J., & Lorenzen, L. (1997). The potential use of geopolymeric materials to immobilise toxic metals: Part I. Theory and applications. *Minerals Engineering*, *10*(7), 659–669. doi:10.1016/S0892-6875(97)00046-0

Vaou, V., & Panias, D. (2010). Thermal insulating foamy geopolymers from perlite. *Minerals Engineering*, *23*(14), 1146–1151. doi:10.1016/j.mineng.2010.07.015

Vollmer, M. (2004). Physics of the microwave oven. *Physics Education*, *39*(1), 74–81. doi:10.1088/0031-9120/39/1/006

Wang, H., Li, H., Wang, Y., & Yan, F. (2015). Preparation of macroporous ceramic from metakaolinite-based geopolymer by calcination. *Ceramics International*, *41*(99, Part A), 11177–11183. doi:10.1016/j.ceramint.2015.05.067

Wang, S., Li, L., & Zhu, Z. H. (2007). Solid-state conversion of fly ash to effective adsorbents for Cu removal from wastewater. *Journal of Hazardous Materials*, *139*(2), 254–259. doi:10.1016/j.jhazmat.2006.06.018 PMID:16839666

Xu, H., & Van Deventer, J. S. J. (2002). Geopolymerisation of multiple minerals. *Minerals Engineering*, *15*(12), 1131–1139. doi:10.1016/S0892-6875(02)00255-8

Yao, X., Zhang, Z., Zhu, H., & Chen, Y. (2009). Geopolymerization process of alkali–metakaolinite characterized by isothermal calorimetry. *Thermochimica Acta*, *493*(1–2), 49–54. doi:10.1016/j.tca.2009.04.002

Zhang, Z. H., Zhu, H. J., Zhou, C. H., & Wang, H. (2016). Geopolymer from kaolin in China: An overview. *Applied Clay Science*, *119*(Part 1), 31–41. doi:10.1016/j.clay.2015.04.023

Zhao, Y., Ye, J., Lu, X., Liu, M., Lin, Y., Gong, W., & Ning, G. (2010). Preparation of sintered foam materials by alkali-activated coal fly ash. *Journal of Hazardous Materials*, *174*(1–3), 108–112. doi:10.1016/j.jhazmat.2009.09.023 PMID:19775812

KEY TERMS AND DEFINITIONS

Apparent Density: The mass per unit volume (or the weight per unit volume) of a material, taking into consideration any voids.

Compressive Strength: The maximum compressive stress that, under a gradually applied load, a given solid material can sustain without fracture.

Hazardous Materials: Any item or agent (biological, chemical, radiological, and/or physical), which has the potential to cause harm to humans, animals, or the environment, either by itself or through interaction with other factors.

Industrial Waste: The waste produced by industrial activity which includes any material that is rendered useless during a manufacturing process such as that of factories, industries, mills, and mining operations.

Mechanical Property: A property that involves a relationship between stress and strain or a reaction to an applied force.

Sustainable Materials: Materials used throughout our consumer and industrial economy that can be produced in required volumes without depleting non-renewable resources and without disrupting the established steady-state equilibrium of the environment and key natural resource systems.

Chapter 9
Sustainable Cooling Research Using Activated Carbon Adsorbents and Their Environmental Impact

Ahmed M. Elsayed
University of Bath, UK & Alexandria University, Egypt

Saad Mahmoud
University of Birmingham, UK

Hassan J. Dakkama
University of Birmingham, UK & Middle Technical University, Iraq

Raya Al-Dadah
University of Birmingham, UK

Waseem Kaialy
University of Wolverhampton, UK

ABSTRACT

Conventional vapour compression system is one of the most utilized cycles in refrigeration and air conditioning systems, due to its compact size, the relatively low running cost, the high coefficient of performance and the wide range of the operating temperatures. Nevertheless, the system suffers from the high initial cost and the high-energy consumption. Unlike the vapour compression cycle, adsorption heat pumps have the advantage of utilizing waste heat reducing the energy consumption and the carbon emissions. Activated carbon is a porous adsorbent material that can be efficiently used in low temperature adsorption refrigeration systems. This chapter reviews the recent developments in the compact adsorption cooling systems using activated carbon regarding the enhancement of the material properties, the design of the sustainable adsorption systems and their environmental and cost perspectives.

DOI: 10.4018/978-1-5225-1971-3.ch009

INTRODUCTION

Air-conditioning systems in commercial buildings and households consume 15% of electricity produced. Such significant energy consumption leads to the depletion of fossil fuel resources and increasing the emissions of carbon dioxide and thus increasing the greenhouse effect. Adsorption refrigeration systems utilise waste heat from automobile exhaust gas, solar energy and process waste heat and can help in producing cooling effect without environmental harmful effects. So they are concerned to low global warming potential and zero ozone depletion potential. Although absorption cooling systems between liquid and vapour have become commercially available technology in the last few decades, they suffer from corrosion, toxicity and the potential crystallisation of the working fluids. In contrast, adsorption cooling between solid and vapour systems offers a number of advantages including the lower energy consumption without circulating pump, low temperature heat sources, and increased operation reliability due to reduced number of moving parts.

Activated carbon materials have been used with various refrigerants in adsorption systems. The activated carbon is mainly obtained by gasifying the char using an oxidizing agent and the raw material for wood and coconut shell. After gasifying, it exhibits high porosity and a high BET surface area of 500-3000 m^2/g and, therefore, is commonly used for low temperature (< 0 °C) cooling systems because of the properties of its high uptake, large surface area and pore volumes (Tamainot -Telto, & Critoph, 1996; Meunier, 1985; El-Sharkawy, Kuwahara, Saha, Koyama, & Ng, 2006). Compared to activated carbon granular or powder, its fiber (ACF) has a larger surface area and more uniform pore structure but suffers from the high thermal contact resistance between the fibres and the heat exchanger surface (Yeo, Tan, & Abdullah, 2012). Modified activated carbon adsorbents have also been produced with composites (Wang, Wu, Wang, & Wang, 2006) or acid-treatment (Yeo, Tan, & Abdullah, 2012), enhancing uptake capacity as compared to their original form (Yang, 2003). The composites are formed through impregnating salts on the porous material for an enhanced performance (Aristov, 2007).

This chapter provides a comprehensive review of different classes of activated carbon used for refrigerants and their material characterisation, sustainable adsorption cycles, environmental impact of adsorption cooling and the cost performance are presented as compared to those of other systems in addition with their fundamentals of adsorption cooling systems. Also, the classification of activated carbon materials and their recent researches are reviewed in the point of view of sustainable adsorption cooling for the configurations and environmental concerns.

1. The first section introduces the fundamentals of adsorption cooling systems, the thermodynamic cycle analysis as well as the parameters used to assess their performance as Coefficient of Performance (COP) and Specific Cooling Power (SCP).
2. The second section describes the different techniques used to measure the properties of activated carbons for various refrigerants. In terms of adsorption kinetics, adsorption isotherms, the techniques used for their measurements and the models developed for the adsorption kinetics are explained.
3. The third section is concerned with sustainable cycle configurations used with activated carbons for low temperature cooling for enhancing heat recovery, mass recovery and cascading. Lastly adsorption systems were evaluated from environmental and economic perspectives.

FUNDAMENTALS OF ADSORPTION SYSTEMS

The cooling cycle is composed of a two-heat source and two heat sink with thermodynamic cycle and is operated at three temperatures of low, medium and high levels as shown in Figure 1. Here, the lower temperature effects evaporation and the higher one is for regeneration. The cycle also operates between two pressures, (i.e, the evaporator pressure (P_e) and the condenser pressure (P_c)), and two refrigerant/adsorbent concentration levels.

Figure 2 shows sorption Clapeyron diagram including refrigeration and heat engine cycles. In the process direction 1-2 with the pre-heating process, the adsorption bed with the high refrigerant concentration is heated at high temperature to desorb the refrigerant, resulting in increasing the bed pressure from P_e to P_c for the desorption. During the process 2-3, the temperature increases using the same temperature level and connected to the condenser including direction 2-3`. This allows the released refrigerant from desorption bed to be condensed and returned back to evaporator through 3`-4`. In the process 3-4, meaning the pre-cooling process, the adsorption bed of the low refrigerant concentration at level in 3 is cooled down using intermediate temperature level and reducing the bed pressure from P_c to P_e. The adsorption bed of the low concentration level and low pressure is then cooled through 4-1 while being connected to the evaporator and adsorbing the refrigerant vapour to achieve the cooling effect at the evaporation temperature at the process 4`-1.

The cooling load or heat of evaporation (Q_e), the coefficient of performance (COP) and the specific cooling power (SCP) are important parameters used to assess the performance of adsorption system (Elsayed, Al-Dadah, Mahmoud, Elshaer, & Bowen, 2014). The three terms are defined below: The

Figure 1. A schematic representation of the sorption cycle

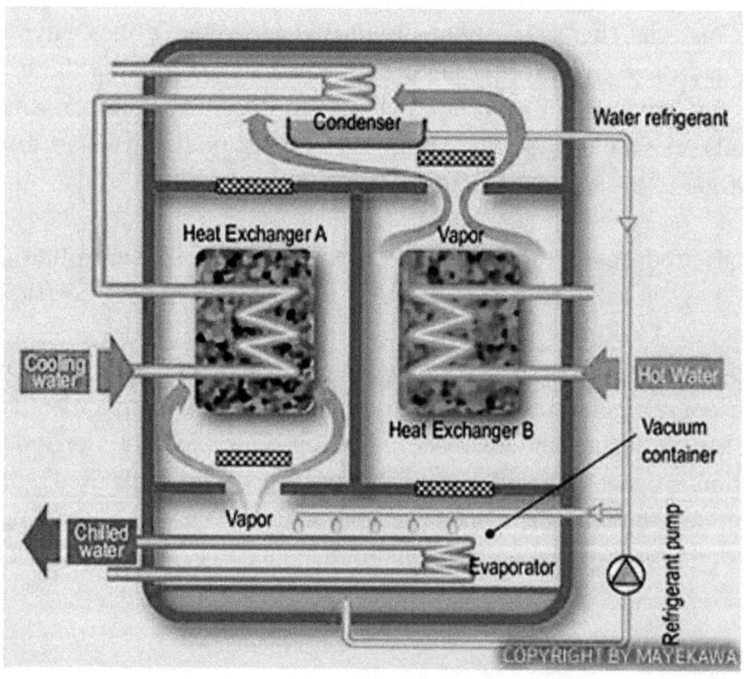

Figure 2. Sorption Clapeyron diagram including for refrigeration and heat engine cycles

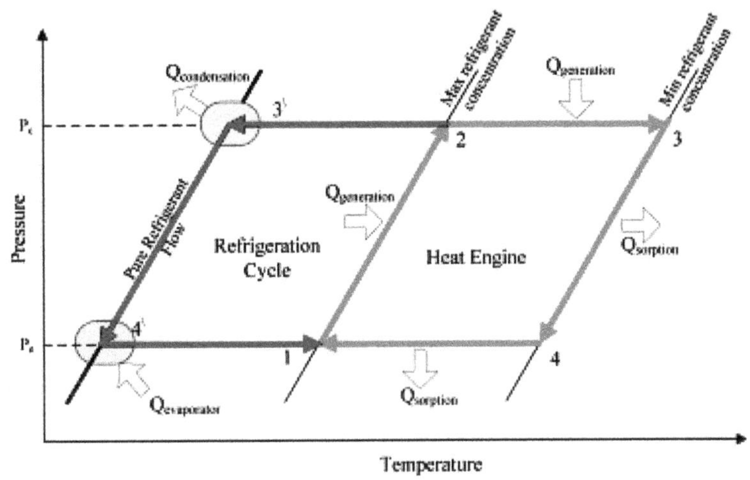

cooling load (CL) is the cooling capacity produced from the evaporator during the evaporation process in the direction 4-1:

$$Q_e = \left(1 / t_{cycle}\right) h_{fg} \int_0^{tcycle} \left(\frac{dm_{ref,evap}}{dt} \right) dt \text{ [kW]}$$

In which t_{cycle} is the cycle time, h_{fg} is the latent heat of evaporation and dm_{ref}/dt is evaporated refrigerant flow. The COP is the ratio between the cooling effect and the energy consumed in the regeneration process.

$$COP = Q_e / Q_{heating} \text{ [-]}$$

The consumed heat during pre-heating in the process 1-2 and the desorption 2-3 processes:

$$Qheating = (1 / t_{cycle}) m_{w,hot} Cp_{w,hot} \int_0^{tcycle} (T_{hot,w,in} - T_{hot,w,out}) dt \text{ [kW]}$$

where $T_{hot,w,in}$ and $T_{hot,w,out}$ are the hot water inlet and outlet temperature to desorber, respectively. Specific Cooling Power (SCP) is the cooling effect produced per kilograms of dry adsorbent material.

$$SCP = Q_e / M_{adsorbent} \text{ [kW/kg]}$$

Usually COP of adsorption chillers are less than 0.6 and high performing chiller are desired to have SCP as much as 1000 W/kg, cooling capacity of adsorption chillers are usually from few kilowatts up to 500 kW.

CLASSIFICATION OF ACTIVATED CARBON MATERIALS AND RECENT RESEARCH

Activated Carbons Used for Adsorption and Desorption Cycles

Activated carbon has been used as an amorphous solid material for the above mentioned cycle. The material has a black colour and looks like charcoal with granular or powder. The preparation includes two stages through the carbonization of carbonaceous raw material as treated in an inert atmosphere at temperatures below 800 °C and then the activation of the carbonized product (Bansal, & Goyal, 2005). In the first stage, the cross-bonds between carbon atoms are broken down in the absence of oxygen, leading to the production of adsorbent with poor adsorbing characteristics because the pores are blocked by the carbonization by-products (Abdullah, Tan, & Limb, 2011). The role of the activation process is enhancing the porosity through cleaning the pores in three stages, namely:

1. Carbonized surface cleaning and exposing to activating agent, then
2. Burning the elementary crystal, and
3. The oxidation process reducing the total micro-pore volume.

Activated carbon is classified with their forms in powder, pellet, and granular, also with original carbonaceous source as olive stones (Spahis, Addoun, & Mahmoudi, 2007), oil palm biomass (Abdullah, Tan, & Limb, 2011), guava seed, or coconut-shell, and finally with basis on carbon activation methods by gases or chemicals (Yang, 2003). Activated carbons are nonpolar or minor polar in the properties that enable adsorbing nonpolar or slightly polar organic molecules such as ammonia, ethanol and methanol more than other sorbents. Another advantage is that the heat of adsorption is generally lower in case of activated carbon rather than in other sorbents (Yang, 2003). The uptake of different carbon reflects the adsorption characteristics of activated carbon with ethanol, carbon dioxide, R134a and methanol including Maxsorb III-ethanol pair (El-Sharkawy, Saha, Koyama, He, Yap, & Ng, 2008), Granular activated carbon/R134a pair (Askalany, Salem, Ismail, Ali, & Morsy, 2012), Clapeyron diagram for 208C/CO2 (Zhong, Critoph, & Thorpe, 2006) and Maxsorb III/methanol (El-Sharkawy, Hassan, Saha, Koyama, & Nasr, 2009). Slow kinetics such as Maxsorb III–ethanol pairs is recommended for solar adsorption cooling and the carbon dioxide pairs need to operate at high pressures with no available commercial adsorption units in the market.

Generally, the adsorbent material can be modified through chemical treatment by acid impregnation for improving their properties by physical heating and biologically bacterial treatment (Yeo, Tan, & Abdullah, 2012). In addition, the material can be used with composite systems to improve the performance. The composites are formed by impregnating hygroscopic salt to the pores of adsorbent material acting as a host. With composite adsorbents (Ullah, Saidur, Ping, Akikur, & Shuvo, 2013), the heat transfer rate and mass transfer were reduced by salt swelling and agglomeration. The selection of composite adsorbent was operated with a multi-stage selection in the cycle uptake, the thermal conductivity, the low specific heat, density and compactness of the system. The operating pressure was taken into account by Wang et al. (Wang, Metcalf, Critoph, Thorpe, & Tamainot-Telto, 2012). The impregnation of sodium silicate and calcium chloride in activated carbon showed lower cycle efficiency, as the refrigerant uptake was less than 0.1 of composite adsorbent dry mass, and the stability of the impregnated salt was increased through using binder (Daou, K., Wang, R.Z. & Xia, 2006). Also, the impregnation time significantly affected the composite performance (Wu, Wan, & Zhu, 2007). In another factor affecting with the particle size, the smaller size used for preparing consolidated composite made more stable and anti-fracture (Eun, Song, Han, Lee, & Kim, 2000). Even though the impregnation process confined the adsorbent pores causing a reduction in the cumulative pore volume (Tso, Chao, & Fu, 2012), some cases found the increased micro-pores volume (Oike, Huang, Watanabe, Hayashi, Oshima, Hasatani, & Kobayashi, 2007). Through the confinement of activated carbon by calcium chloride/ silica gel, the surface area and pore volume were reduced (Tso & Chao, 2012) in comparison to those of activated carbon having the larger pore volume and surface area for better adsorption capacity (Chen, Wang, Oliveira, & Hu, 2009).

Characterization of Adsorbent Materials

Several studies in adsorption materials showed their physical properties for particle size, porosity, permeability, density, specific heat capacity, surface area, pore volume distribution and thermal conductivity. The adsorbent materials affect the uptake and the adsorption rate of any adsorbents (Yang, 2003; Królicki, Wilk, Zajączkowski, & Kaczmarczyk, 2006). Therefore, measuring such property importantly takes a paramount in the development and application as following below.

The material porosity was generally measured using mercury intrusion measurements (Jänchen, Schumann, Thrun, Brandt, Unger, & Hellwig, 2012). The mercury porosimetry was used depending on that mercury penetration into the pores by capillary action by the application of external pressure. The required equilibration to pressure is inversely proportional to the size of the pores. From the pressure versus intrusion data, the instrument generates volume and size distributions. The porosimeter analysis was recommended for macro-porous adsorbents, while gas adsorption was used for meso-pores and micro-pores measurements where nitrogen or carbon dioxide was injected into the pores instead of mercury (Krause, 2009).

Regarding the particle size, different techniques were utilized by sieving method where several mesh sizes were fitted with placing the large or finer meshes. Each mesh holds adsorbent particles above certain size and then the weight of each sieve size is measured to get the size distribution of the powder (Shi, 2015). A technique such as sedimentation is a method based on that the material particles are uniformly dispersed in a liquid. When it is allowed to settle, the particle size is taken as a function of time (Oyarzún, 1999). In optical and electrical sensing zone method (Oyarzún, 1999), the electric impedance generates voltage with amplitude proportional to particle volumes in an electrolyte, when they pass through an orifice. Another technique is to use scanning electron microscopy (SEM) (Elsayed, Al-Dadah, Mahmoud, Elshaer, & Bowen, 2014). The SEM signals are derived from electron-sample interactions reveal information including the particle size. While in laser diffraction method for the size distribution of 0.02-2000μm range, a laser beam passed through a dispersed particulate sample reflects the beam scatter depending the particle size distribution for measuring the angular variation in intensity of the light scattered (Elsayed, Elsayed, Al-Dadah, Mahmoud, Elshaer, & Kaialy, 2016). The angular scattering intensity is then used to calculate the size of the particles responsible for creating the scattering pattern (McGlinchey, 2005). Finally, photon correlation spectroscopy can be applicable for 1nm to 5μm, which measures with Brownian motion of fine particles as a function of time (Seville & Wu, 2016). Since a laser beam is scattered by particles in suspension, the diffusion of particles causes rapid fluctuations in scattering intensity around a mean value at a certain angle. From the scattered light intensity signal, the information of the motioned particles and subsequently their size is obtained (Seville & Wu, 2016). Table 1 summarizes the advantages and disadvantages of each technique.

Permeability is a measure to porous properties of the materials. Permeability allows fluids to pass through its porous media by ensuring the mass transfer of the refrigerant (Meunier, 1993). When the porous adsorbent had low permeability showing high pressure gradient to the refrigerant and meaning high mass transfer resistance (Sahoo, John, Newalkar, Choudhary, & Ayappa, 2011). The experimental pressure dropped and flowed the velocity, usually correlating with the permeability using Darcy model (Fujioka & Suzuki, 2013) or Ergun model (Wang, Tamainot-Telto, Thrope, Critoph, & Metcalf, 2011). There was report on an optimal particle diameter for the maximum cooling power output (Hongyu, Zhaohong, Haoran, Noriyuki K., Dandan, Mitsuhiro, & Huafang, 2014). This involved pressure differential sensor and velocity measurement device across the adsorbent material packed bed as shown in Figure 3.

Generally, density is defined as the mass divided by volume. But the determination of the density of porous materials including the pore spaces within the material particles correctly needs to perform in the volume measurement. As seen in Table 2 the bulk and true densities of activated carbon materials were summarized by Henninger et al. They well evaluated bulk density by methanol adsorption on activated carbon for thermally driven chillers (Henninger, Schicktanz, Hugenell, Sievers, & Henning, 2012). When calibrated graduated beakers was used and filled with adsorbent, it was weighted with accurate balance. On the other hand, the true density, which is defined as the density of the pure solid material excluding the pore spaces in the adsorbent samples, is determined using helium pycnometer where the helium is filled in the pores of the sample, thus removing its volume. There was tendency that bulk density is less than the true density for porous materials.

The specific heat is the amount of heat required to raise the temperature of unit mass of the adsorbent sample by one degree Celsius. Highly conductive materials exhibit low specific heat capacity which is measured through differential scanning calorimetry (DSC). The effective specific heats of the adsorbent / refrigerant pair were determined at different uptake values under isosetric conditions (Critoph, 1996; Tamainot-Telto, Metcalf, & Critoph, 2009). Here, the refrigerant specific heat was selected to liquid or gas refrigerant phase to fit measurements.

Through these characters of the porous materials, surface properties are associated with surface area, pore size and volume of any materials. These were determined eventually with gaseous uptake in different methods applied for their surface properties (Yang, 2003). The isothermal adsorption of nitrogen provides the surface area of porous materials usually can be measured either with the Langmuir's theory (Bansal & Goyal, 2005) or Brunauer Emmet and Teller theory (Chan, Chao, Sze-To, & Hui, 2012). In the Langmuir model based by the maximum amount of usually nitrogen adsorbed per gram of adsorbent for the formation of monolayer, the molecular weight of the adsorbate and knowing the approximate contact area of an adsorbate molecule, surface area of adsorbent are calculated. For more reliable and common method, the adsorption of nitrogen, CO_2, argon or krypton was carried out on the adsorbent samples using BET model (Elsayed, Al-Dadah, Mahmoud, Elshaer, & Bowen, 2014) by knowing the size and the number of the adsorbed gas molecules. The Langmuir surface area was always higher than the BET surface area due to the Langmuir equation derived around the monolayer being the limit of adsorption, while the BET equation is based on the concept of multilayer adsorption in terms of the monolayer. Other methods for determining the surface area were reported by free surface energy method and the uniform circular capillaries method (Bartell & Bower, 1952). The surface area depended on the size of the pores was measured as based on density functional theory (DFT) where the pore size can be determined by the observed peaks in the pore size distribution (Henninger, Schicktanz, Hugenell, Sievers, & Henning, 2012).

Table 1. Advantages and disadvantages of various techniques for particle size distribution measurement

Technique	Advantages	Disadvantages
Sieving	• Easy of operation • Wide size range • Inexpensive	• Reproducibility • Wear/damage in use or cleaning • Irregular/agglomerated particles • Rod-like particles: overestimate of under-size • Labour intensive-
Sedimentation	• Equipment required can be relatively simple and inexpensive. • Can measure a wide range of sizes with accuracy and reproducibility	• Sedimentation analyses must be carried out at concentrations which are sufficiently low for interactive effects between particles to be negligible so that their terminal falling velocities can be taken as equal to those of isolated particles. • Large particles create turbulence, are slowed and are recorded undersize. Particle re-aggregation during extended measurements. • Careful temperature control is necessary to suppress convection currents. The lower limit of particle size is set by the increasing importance of Brownian motion for progressively smaller particles. • Particles have to be completely insoluble in the suspending liquid
Electrical Sensing Zone	• True volume distribution • High resolution • Wide range of measurement: particle diameter from approx. 0.5-400μ	• Needs calibration • Medium must be an electrolyte • Low particle concentration • Errors with porous particles • Orifice blocking troublesome • Particles below minimum detectable size go unnoticed • Difficult with high density material
Laser Diffraction	• Non-intrusive: uses a low power laser beam • Fast: typically <3minutes to take a measurement and analyse. • Precise and wide range up to 64 size bands can be displayed covering a range of up to 100,000:1 in size. Simple to use. • Absolute measurement: No calibration is required, the instrument is based on fundamental physical properties.	• expenses • volume measurement all other outputs are numerical transformations of this basic output form, assuming spherical particles • must be a difference in refractive indices between particles and suspending medium
PCS	• Non-intrusive • Fast • Nanometre size rang	• Vibration, temperature fluctuations can interfere with analysis • Restricted to solid in liquid or liquid in liquid samples • Expenses • Need to know R.I. values and viscosity

Thermal conductivity is an indication value for degree to conducting heat of the materials and describes the transport of energy in the form of heat through a body, resulting in a temperature gradient. It is crucial to control the heat transfer to the adsorbent materials during the adsorption and desorption phases and the operating cycle time. The conductivity is expressed in terms of measured heat flux and adsorbent thickness in terms of Fourier law (Wang, Wu, Wang, & Wang, 2006b). It is determined either using steady-state or transient techniques. The transient techniques are considered for taking into account the short cyclic time in adsorption cooling application as measured with facilities for the conductivity of dry adsorbent (Çağlar, 2012) or wet adsorbent under vacuum (Dawoud, Sohel, Freni, Vasta, & Restuccia, 2011; Freni, Tokarev, Restuccia, Okunev, & Aristov, 2002; Gurgel, Oliveira, Marques Ailson, Oliveira, & Tavares, 2005). In the dry measurement approach, an effective thermal conductivity model

Figure 3. Permeability test facility
Source: Zhao, Wang, Wan, Ma, & Jiang, 2013

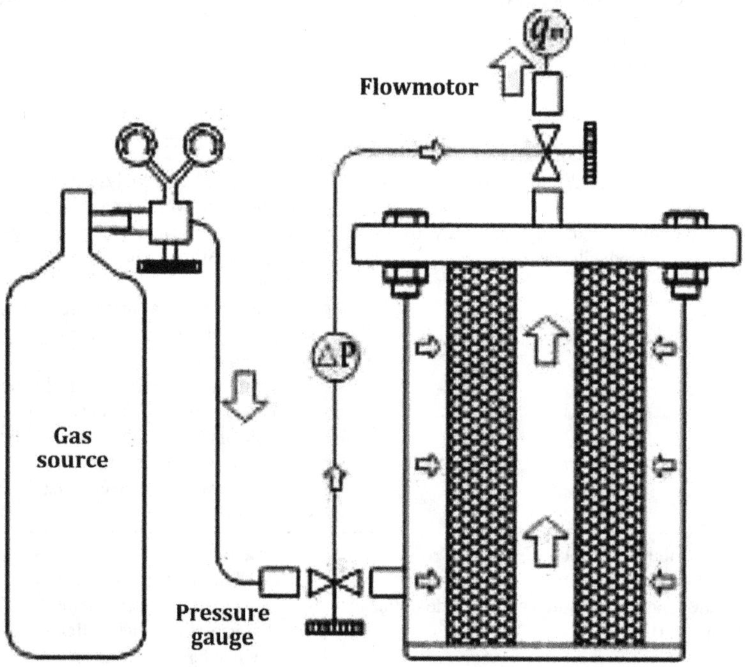

Table 2. Physical characteristics of activated carbon samples

Sample	Cp		Bulk Density g cm^{-3}	True Density g cm^{-3}	Calc. Porosity ε Form Hg Intrusion
	T/°C	J g^{-1} K^{-1}			
G32-H	40 80 120	0.86 0.95 1.05	0.37±0.04	2.174±0.002	0.77
RUTGERS CG1-3	40 80 120	0.84 0.95 1.05	0.37±0.04	2.139±0.002	0.87
Norit R 1 Extra	40 80 120	0.88 0.95 1.04	0.42±0.03	2.259±0.007	0.84
Norit RX 3 Extra	40 80 120	0.83 0.97 1.04	0.37±0.03	2.169±0.001	0.75
CarboTech C40/1	40 80 120	0.78 1.0 1.06	0.38±0.03	2.142±0.001	0.82
Carbotech A35/1	40 80 120	0.86 0.95 1.05	0.33±0.03	2.191±0.002	0.88

Source: Henninger et al., 2012

was combined for the adsorbent and refrigerant conductivity by using the facilities including infrared detector (Zhao, Wang, Wan, Ma, & Jiang, 2013; Jiang, Wang, & Wang, 2014; Zhao, Wang, Wang, & Yu, 2014) or the transient plane source (TPS) (Hu, Juan-Juan, & Ze-Shao, 2009). The TPS conductivity of wet adsorbent material using hot wire method with refrigerant under vacuum was utilized by different researcher (Wang, Wu, Wang, & Wang, 2006b; Dawoud, Sohel, Freni, Vasta, & Restuccia, 2011; Freni, Tokarev, Restuccia, Okunev, & Aristov, 2002). The thermal conductivity of the sample was controlled by the operating temperatures and pressures. Measuring the thermal conductivity through the wet approach is more desirable and accurate compared to the dry one as the latter approach requires assumptions of mixing rules such as Maxwell-Eucken between the refrigerant and the solid material. Those assumptions can cause significant uncertainty in the effective adsorption bed thermal conductivity (Carson, 2002).

ACTIVATED CARBON/ REFRIGERANT ADSORPTION CHARACTERIZATION

The major parameters that evaluate the suitability and the performance of any adsorbent material are its:

- Adsorption capacity,
- Selectivity,
- Ability to regenerate,
- Kinetics,
- Performance stability,
- Regeneration temperature, and
- Cost (Yang, 2003).

As seen in Table 3, many refrigerants were used for activated carbon in adsorption systems such as ammonia, methanol, ethanol and carbon dioxide, etc. One of the most common refrigerants was tested with activated carbon and its composites for ammonia, which exhibited a high latent heat of vaporization (Critoph, 1996; Anyanwu & Ogueke, 2005; Wang, Wang, & Oliveira, 2009a). As utilized at low temperatures, ice making application was suitable, however, flammable and toxic ammonia was limiting to capability. Other refrigerants reported were hydrocarbons such as butane (Critoph, 1996) and propane (Mugnier & Goetz, 2001). The main drawbacks of such refrigerants are the low latent heat and the modest uptake making their performance very low and unattractive for adsorption systems. Methanol was reasonable for the latent heat with 1224 kJ/kg and methanol/activated carbon working pair observed an uptake nearly two folds of water/silica gel, which had a similar cooling capacity (San & Lin, 2008). In ethanol moderate latent heat was nearly three times larger than that of R134a, carbon dioxide, butane and propane and nearly 80% equalled to that of methanol. Ethanol is relatively less toxic than other solvents, but flammable and can operate under a low desorption temperature (<120 °C), meaning that the ethanol adsorption systems can be operated at low temperatures. Carbon dioxide as alternative refrigerant is natural, non-toxic, non-flammable, but it has very high operating pressure which can reach 70 bar. Most of the research were carried out to only developing adsorption isotherms (Jribi, El-Sharkawy, Koyama, & Saha, 2010b; Jribi, Chakraborty, El-Sharkawy, Saha, & Koyama, 2009) and theoretical cycle analysis (Jribi, Koyama, & Saha, 2010a). Although CO_2 was used for a pressure swing adsorption cooling to the activated carbon (Anupam, Chatterjee, Halder, & Sarkar, 2011a), the achieving temperatures was below zero and a large number of cycles with compression combined with adsorption, bed cooling, desorption combined with expansion was required. The R134a was investigated through a number of researches (Loh, El-Sharkawy, Ng, & Saha, 2009; Banker, Prasad, Dutta, & Srinivasan, 2010; Loh, Shah, Chakraborty, NG, & Chun, 2010), but, high global warming potential and thus their future usage was likely to be restricted (Ramji, Leo, Tan, & Abdullah, 2014). Figure 4 shows the temperature distribution of chilled fluid in evaporator with time for different refrigerants based on activated carbon. Water /activated carbon showed higher SCP and COP compared to other working pairs.

Adsorption Isotherms

Adsorption isotherms give the adsorption uptake through plotting the quantity of adsorbed adsorbate against the relative pressure of the adsorbate at a constant temperature. Different experimental methods are utilized to develop the adsorbent materials isotherms including:

- Gas flow,
- Volumetric, and
- Gravimetric methods (Jiayou, 2003; Kumar, 2011).

In gas flow technique, helium or nitrogen is used as carrier gases and the partial pressure of the adsorbate is determined by a gas flow meter. The volume adsorbed by the adsorbent is determined from the peak area in the adsorption/desorption chart recorded by a potentiometer with time.

In volumetric technique, the adsorbent is placed in an evacuated chamber and the charging chamber contains a charge of refrigerant. Based on the volumes of the two chambers, the piping volume and the adsorbent porosity, the remaining charge in the charging chamber could be determined using ideal gas relations for kinetics and equilibrium adsorption isotherms (Figure 5) (Habib, Saha, Rahman, Chakraborty, Koyama, & Ng, 2010).

In gravimetric technique, the weight of sample measured by a highly sensitive microbalance was located inside the vacuum system and isolated from the surroundings (Elsayed, AL-Dadah, Mahmoud, & Kaialy, 2014). The sample was heated in a furnace through the gas purged atmosphere into the system allowing the adsorption process to take place. The balance measures the weight change in the adsorbent mass directly as illustrated in Figure 6. Thus the isotherms can be obtained directly at different pressures and temperatures.

Isotherms data of different pairs were measured to different operating temperatures and pressures for various activated carbons pairs as tested (El-Sharkawy, Saha, Koyama, He, Yap, & Ng, 2008) for ethanol/ Maxsorb-III with mean particle of 72 micron diameter using thermo-gravimetric analyser (TGA) reacting chamber as shown in Figure 7. The isotherm of composite adsorbents NB and NA was determined at an evaporating temperature of 27°C using high-vacuum gravimetric method (Cui, Tao, Chen, Guo, & Yao, 2005). Adsorption uptake of ethanol/ HCl was treated by zeolite through the vapour exposure to the refrigerant and continuously measured at different exposing times for the mass change (Ivanova, Kostova, & Koumanova, 2010). For activated carbon fibres (A-20) and (A-15), the isotherms were developed through a special designed apparatus where the amount of ethanol adsorbed at equilibrium condition, the refrigerant level inside the evaporator was measured by using a digital microscope with ±0.01 mm resolution prior and after each experiment. The mass of ethanol evaporated was calculated by using the liquid level difference (El-Sharkawy, Kuwahara, Saha, Koyama, & Ng, 2006).

Figure 4. Temperature distribution of chilled fluid in evaporator during cycle time for different refrigerants based on activated carbon

Source: Ramji, Leo, Tan, & Abdullah, 2014

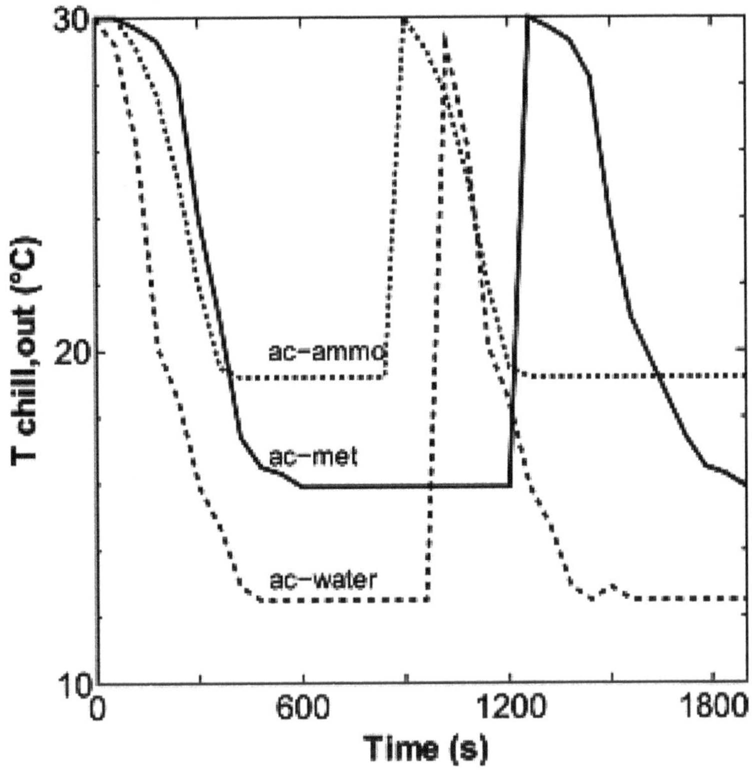

Table 3. Summary of the refrigerants, working temperatures and Specific Cooling Power (SCP) of various adsorption refrigeration systems based on the activated carbon as an adsorbent

Author	Pair	Tevap [˚C]	Tcond/ Tads[˚C]	Tdes [˚C]	SCP [W/kg]	Specifications
Ismail et al. **(2014)**	R32 R507 R290 R134a	-5 to 20	30	85		At high chilling temperatures and low ambient temperatures, R-32 produced highest specific cooling capacities
Attalla et al. **(2014)**	Granular GAC/ R134a pair (Characterization)		20-60	60 to 105		A circular tube heat exchanger with fins was adopted as an adsorber bed The mass of refrigerant was estimated from the difference of the test section mass at the beginning and end of each test *Large uptake of R134a (up to 1.92) was observed with GAC activated carbon after 1200 sec.
Miyazaki et al. **(2014)**	Chemically treated Maxsorb/Ethanol (characterization)	14	30	80	140	The time length of adsorption process should be longer than that of the desorption process for the maximum SCP SCP of 140 W/kg with COP of 0.48 was predicted by simulation with the temperature conditions of 80oC for hot water, 30oC for cooling water, and 14oC for chilled water.
Habib and Saha **(2013)**	AC fiber Acf-20/ ethanol	14	30	85	12 [kW]	Lumped analysis for adsorption chiller driven by solar energy for hot climate in Malaysia was developed

continued on next page

Table 3. Continued

Author	Pair	Tevap [˚C]	Tcond/ Tads[˚C]	Tdes [˚C]	SCP [W/kg]	Specifications
Chekirou et al. **(2013)**	AC-35/Methanol (Solar cooling)					The thermal performance coefficient COP_{th} increases as long as the internal adsorber radius R2 is less than 60 mm. the solar and the thermal performance coefficient of the system are 0.135 and 0.308, respectively.
Rowe et al. **(2014)**	Ac/Methanol	5	30/25	90	327	4 Copper finned tubes per bed, 4bed chiller with COP 0.33
Jerai et al. **(2015)**	AC KOH-6/Ethanol					Phenol resin based activated carbons, it could adsorb upto 2 kg ref/kg ads
Hassan **(2013)**	AC/Methanol	-5	30	120		Solar making refrigerator with COP of 0.6 was achieved, simulation model was developed
Elsayed et al. **(2014)**	Maxsorb / ethanol Two bed	14	30	85	10 kW	Silica gel has higher capacity at small adsorption time (450 sec) while Maxsorb/ethanol outperforms other materials at large cycle time of 1530 seconds.
Shmroukh et al. **(2013)** **, (2015)**	R-134a,R-407c, R-507A			80		The ACP//R-134a pair is highly recommended to be used as adsorption refrigeration working pair because of its higher maximum adsorption capacity than the other tested pairs,
Umair et al. **(2014)**	ACF/Ethanol				80 [KJ/kg]	Mathematical analysis for continuous solar cooling refrigerator was developed
Habib et al. **(2013)**	AC/R134a	14	30	70	8 kW	The single-stage chiller is superior when the heat source temperature is relatively higher (above 75˚C). However, when the regeneration temperature is at or below 55˚C, the two-stage chiller could still be operate
Ramji et al, **(2014)**	AC/Water AC/Methanol AC/Ammonia			120	0.98[kW]0.65[kW] 0.5[kW]	Activated carbon-water pair simulated to yield a higher COP of 0.58 with Q at 480 kJ mainly due to high heat of evaporation.
Habib et al.**(2014)**	ACF-Ethanol AC-Methanol Silica gel-water	14	30	85	Less than 20 kW	ACF-ethanol had highest adsorption uptake followed by AC-methanol and silica gel-water pairs. Silica gel-water had COP values when regeneration temperature is below 70 °C. However, the COP of ACF-ethanol based adsorption cycle is found to be highest when the regeneration temperature is above 70 °C.
Jribi et al. **(2014)**	Co2/Maxsorb	15	27 to 37	50 to 95	1.7 kW	System is able to produce 1.72 kW of cooling power at driving heat source temperature of 85 °C and with 80 kg of Maxsorb III
Jribi et al. **(2013)**	AC/HFO1234ze(E)	15	30	85	2 kW	System is able to produce 2 kW of cooling power at driving heat source temperature of 85 °C and with 80 kg of Maxsorb III, slightly higher than R134a/ Maxsorb
Uddin et al. **(2014)** El-Sharkawy et al. **(2014)**	Maxsorb III KOH-H2 Maxsorb III H2 treated Maxsorb III	-14 to77 °C Tevap-Tcond		20 to 80 °C Tads-Tdes		Experimental resultsnshow that, adsorption capacity of H2 treated Maxsorb III/ethanol is slightly higher than that of the parent Maxsorb III/ ethanol pair whilst the KOHeH2 treated Maxsorb III has the lowest adsorption capacity
Askalany et al. **(2014)**	Maxsorb III / R410A ACFiber A-20/ R410A			10 to 50°C Tads-Tdes		Maxsorb uptake capacity (Up to 1.8 kg/kgads) was higher by 1.6 times of ACF.
Askalany and Saha **(2015)**	Maxsorb III /R32 ACFiber A-20/R32			25 to 65°C Tads-Tdes		Maxsorb III showed higher capacity

Adsorption Kinetics

Adsorption kinetic is defined as the rate of the adsorption/desorption process with time and is supported by several methods including:

- Gravimetric,
- Calorimetric,
- Volumetric, and
- Micro-flow controller.

In the gravimetric technique, the mass of adsorbent was measured directly using microbalance with ethanol being adsorbed on eleven commercially available activated carbon species (Elsayed, AL-Dadah, Mahmoud, & Kaialy, 2014) and electronic balance used with water/composite adsorbent pair of activated carbon, silica-gel and $CaCl_2$ (Tso & Chao, 2012; Shmroukh, Ali, & Abel-Rahman, 2013). Another method was reported by measuring the weight of the bed by load cell with water/SAPO-34 pair (Ming, 2009; Sapienza, Santamaria, Frazzica, Freni, & Aristov, 2014) with calorimetric method. The concept relates the measured heat rate released due to the adsorption process to the mass adsorbed, when thermoelectric devices was utilized for the water adsorption process on type A silica gel particles bonded to aluminium (Ahamat & Tierney, 2012).

Figure 5. Volumetric techniques for measuring adsorption isotherms
Source: Habib, Saha, Rahman, Chakraborty, Koyama, & Ng, 2010

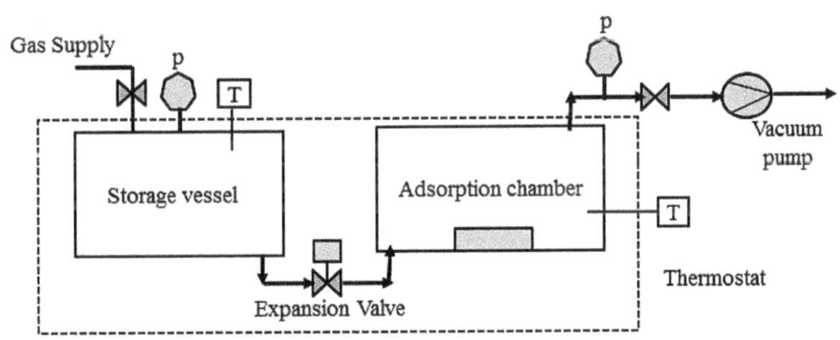

Figure 6. Gravimetric concept of isotherm measurements
Source: Elsayed, AL-Dadah, Mahmoud, & Kaialy, 2014

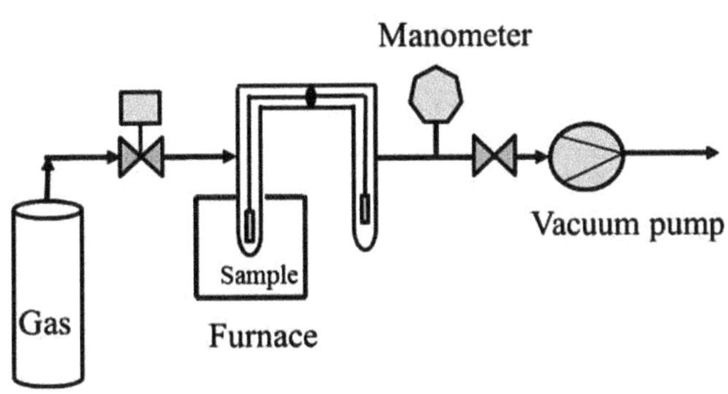

Figure 7 TGA test facility having (1)evaporator; (2)constant temperature water bath; (3) water circulator; (4)TGA reacting chamber; (5)helium cylinder; (6) pressure regulator; (7) valve to (8) the vacuum pump
Source: El-Sharkawy, Saha, Koyama, He, Yap, & Ng, 2008

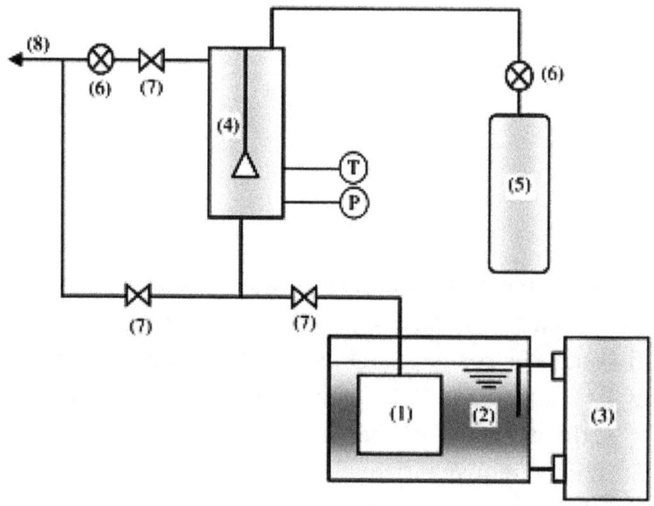

For uses of heat flux sensor, water was adsorbed on a layer of zeolite (UOP DDZ 70) with a thickness of 0.72 mm glued onto an aluminium support (Füldner & Schnable, 2008) and micro-calorimetric sensor using Tian–Calvet-type micro-calorimetry for water vapour adsorbed on the silica gel support (Demir, Mobedi, & Ulku, 2011).

Regarding the volumetric method, the refrigerant volume was tracked by several techniques recording the liquid level by digital microscope as ethanol adsorbed onto Unitika activated carbon fibre (El-Sharkawy, Kuwahara, Saha, Koyama, & Ng, 2006). Other methods of magnetostrictive level sensor (Wang, Wang, Lu, Chen, Wang, & Wu, 2006a; Wang, Wang, Lu, Chen, Wang, & Wu, 2006) and utilized vermiculite matrix of transparent glass vessel for ammonia adsorption on $BaCl_2$ impregnated (Zhong, Critoph, Thorpe, & Tamainot-Telto, 2009; Riffat, Williams, & Stuart, 1997; Ulku, 1986) were reported. Other concept used was to express the liquid level as a function of the pressure difference, when ammonia adsorbed on adsorbents (Zhao, Wang, Wan, Ma, & Jiang, 2013). In constant volume variable pressure method for activated carbon/R134a and activated carbon/R507A pairs, the mass of initial charge was recorded and the mass transferred to the adsorbent chamber at the pressure and temperature measurements is tracked with time (Habib, Saha, Rahman, Chakraborty, Koyama, & Ng, 2010). In large temperature jump method, the adsorption process on pellets of composite sorbent was simulated to real situation of adsorption/desorption process with non-isothermally water vapour (Aristov, Dawoud, Glaznev, & Elyas, 2008). The micro-flow controller technique measured the flow rate of desorbed/adsorbed mass and the integration of the flow rate over a time period gave the refrigerant uptake where composite adsorbents synthesized from zeolite 13X and $CaCl_2$ were used as water vapour adsorbent (Chan, Chao, Sze-To, & Hui, 2012; Saha, Habib, El-Sharkawy, & Koyama, 2009).

In the mentioned techniques with microbalances, only few milligrams of the solid adsorbent (El-sayed, AL-Dadah, Mahmoud, & Kaialy, 2014) or of heat flux sensor (Füldner & Schnable, 2008) were used for suitable to an expensive materials such as metal-organic materials. Other techniques based on volumetric methods required larger quantities ranging from few grams to kilograms as the load cell was charged (Ming, 2009; Sapienza, Santamaria, Frazzica, Freni, & Aristov, 2014), while the micro-flow controller technique was used with packed adsorbent mass (Chan, Chao, Sze-To, & Hui, 2012; Saha, Habib, El-Sharkawy, & Koyama, 2009).

ADSORPTION COOLING CONFIGURATIONS AS SUSTAINABLE SYSTEM

Heat Cycle Configurations of Adsorption Beds

A number of cycle configurations is presented in Figure 8 for classification of adsorption refrigeration strategies. A map of the cycles reported showed the schematic diagrams of the main cycles. Their=cycle configurations are classified in two categories:

1. Conventional and
2. Advanced cycles.

The former cycle is carried by either single stage or multistage with either single bed or multi beds. The latter advanced cycle incorporates more complex techniques to enhance the thermal performance of the conventional cycles like approaches of heat recovery, mass recovery, thermal wave and cascading. In a two beds system, heat recovery was incorporated during the switching process between the two beds where the adsorption bed was pre-heated for the desorption process. But, the desorption bed also needed to be pre-cooled to be ready for the adsorption process. The heat recovery was carried out through allowing the cooling water to pass through the desorption bed at high temperature level before passing on the adsorption bed with medium temperature one (Wang, Wang, & Wu, 2014). This recovered some of the thermal energy and hence improved the overall thermal performance of the system. In cascading adsorption refrigeration system using zeolite/water and activated carbon/methanol as working pairs, mass recovery was achieved during the switching time between the desorption and the adsorption processes by achieving the mechanical equilibrium meaning pressure swing (Wang, Wang, & Wu, 2014). Mass recovery technique was implemented with heat recovery (Riffel, Belo, & Leite, 2009; Wang, Wu, Wang, Xu, Wang, & Li, 2003; Qu, Wang, & Wang, 2001), and with three effects on the heat pipe (cooling, heating and heat recovery) for the adsorption refrigeration system using a composite adsorbent ($CaCl_2$/activated carbon) and ammonia as working pair (Lu, Wang, & Wang, 2011). In the thermal wave cycle, heating/cooling of the adsorption bed =was taken in one single heat transfer fluid that was cooled and heated in separated two heat exchangers. The high temperature heat exchanger (heater) was used to increase the temperature of the heat transfer fluid and the low temperature heat exchanger (cooler) was used to cool the heat transfer fluid (Miles & Shelton, 1996; Burchell, 1999). The heat transfer fluid in thermal wave cycle was oil applying convective thermal wave cycle to other refrigerant (Raymond, 2010; Critoph, 1999; Critoph, 1998).

Cascading the cycle was achieved by integrating two adsorption cycles operating at two different temperature levels so that the condenser of the low temperature cycle acted as the evaporator of the high temperature cycle. Such configuration increases the temperature difference between the heat source and heat sink thus achieving cooling effect at low temperatures, resulting from double or triple cascade arrangements (Burchell, 1999).

Figure 8. Classification of adsorption refrigeration strategies based on reviewed literature

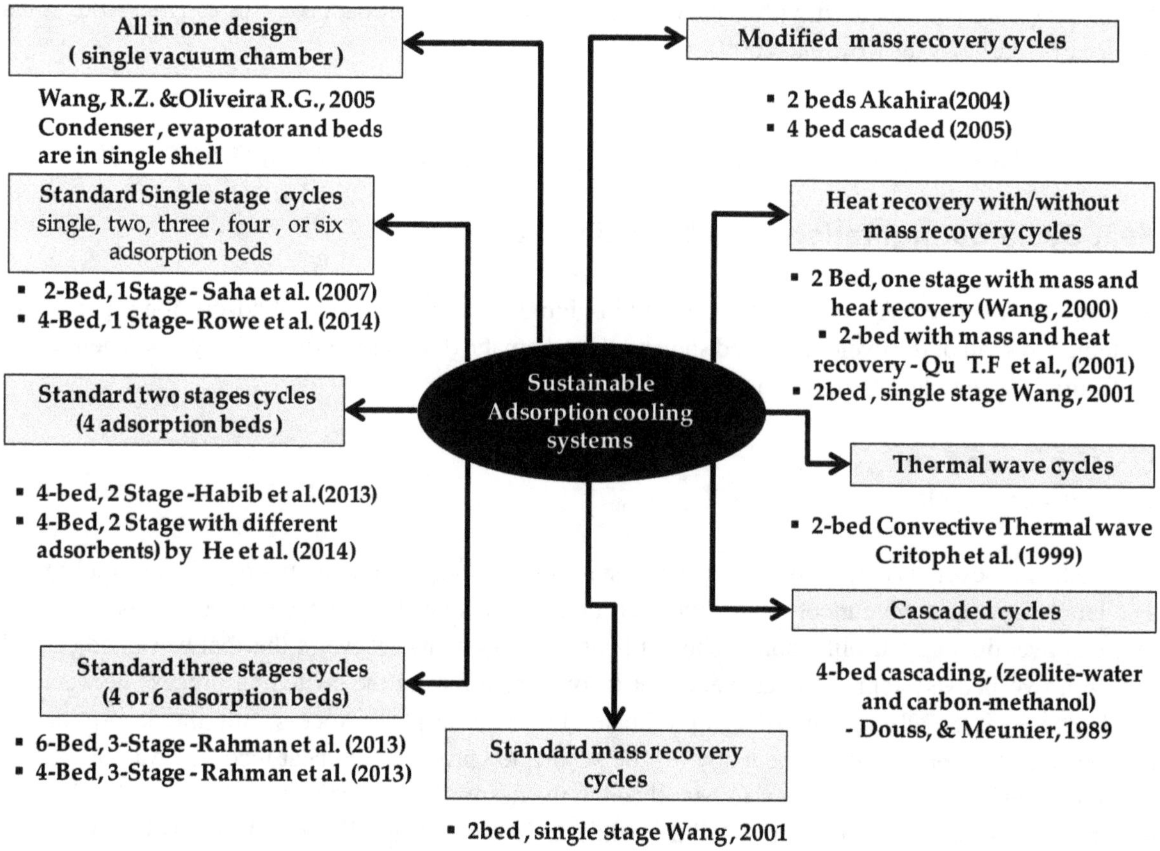

In multiple beds configuration, two or more beds were applied with the same or different modes of adsorption refrigeration system at the same time. At least one adsorption (evaporation) or one desorption (condensation) process was operated during the normal working time (Raymond, 2010). Four beds were numerically simulated to the multi-bed adsorption refrigeration system using activated carbon/methanol as the working pair (Rowe, 2014). Adsorption refrigeration system could be divided in to a single (Saha, El-Sharkawy, Chakraborty, & Koyama, 2007; Khalifa, Hussein, & Hadi, 2011), two (Habib, Majid, & Sulaiman, 2013; He, Hu ang, Deng, Yuan, Kobayashi, Kubota, Huhetaoli, & Zhao, 2014) or three stages (Tseng, R.L., & Wu, 2009) according to the arrangement of the adsorption beds. The single stage was found as one, two, three, four or… etc. for the adsorption beds. The vacuum chamber systems utilised low pressure refrigerants like water as the use of a vacuum chamber offered the advantage of reducing pressure losses between all the components and of improving the overall efficiency (Wang, Wang & Wu, 2014).

Table 4 compares the performance of the various cycle configurations with respect to the operating temperatures, cycle times, COP, SCP and applications. Such multi-stage system has lower COP as compared to single stage ones. However, they could run at a heat source temperature as low as 50°C, thus the exploitation of wide range of low temperature heat sources enabled the heat generation from flat plate collectors and low temperature geothermal heat sources. Higher COP was obtained by cascading (Douss & Meunier, 1989). All-in-one heat pump was successfully implemented, when the condenser, evaporator and adsorption beds were housed in one vacuum chamber for improving the reliability of operation and reducing the number of vacuum valves (Wang, Wang, & Wu, 2014). Modified mass recovery having higher SCP and lower COP compared to conventional mass recovery included flow of secondary heating/cooling fluids allowed during mass recovery process (Akahira, Alam, Hamamoto, Akisawa, & Kashiwagi, 2004, 2005). For other cycle configurations reported, these suffered from serious practical limitations. For example, thermal wave cycle was attractive in terms of performance, but efficient materials with high thermal conductivity were not currently available (Shelton, Wepfer, & Miles, 1990). Additionally, although convective thermal wave cycle had good performance in terms of COP, the bulky and expensive properties to build was problem due to the large gas/gas heat exchangers (Burchell, 1999; Critoph, 1998). Activated carbon adsorption systems were developed for various applications like:

- Solar powered air conditioning (Wang, Wu, Xu, & Wang, 2001),
- Ice making for storing vaccines and drugs (Li, Wang, Wang, Li, & Chen, 2013),
- Food (Wang, Oliveira, 2005; Metcalf, Tamainot-Telto, & Critoph, 2011).

Activated Carbon Adsorption Bed Designs

Various adsorption beds have been designed in order to develop several kinds of activated carbon adsorption systems. For example, solar activated adsorption applied to ice making systems can be classified into two groups (Figure 9). The first group in the adsorption bed with the solar collector was integrated together and the second group in the adsorption bed was separated from the collector (Ji, Li, Fan, Zhang, Luo, & Wang, 2014). The system was packed with activated carbon adsorbent and the solar radiation was collected on the outer surface of collector-adsorbent tubes. Another design was developed for collector-bed (Gonzalez & Rodriguez, 2007). The compound parabolic concentrators (CPC) were used to focus the solar radiation on the bed that was constructed as tubular collectors to receive the radiation from the CPC. Also, for two units of such bed with a surface area of 0.75 m^2/bed (Li, Wang, Xu, Wu, & Dieng, 2002), each flat-plate collector was loaded with 22 kg of activated carbon.

After the black dye was coated on the top surface of the unit, the greenhouse effect was applied by adding glass cover at the top of the flat-plate collector. Hybrid solar adsorption ice making system was used with the inner finned tubes applied with inside of the adsorption unit (Qi, 2005). In that case, the adsorption bed was not integrated with the solar collector but to reduce the size of the adsorption unit, the adsorption bed was divided into units with smaller size. Both adsorption icemaker and four adsorption units were fabricated with each bed having the diameter of 159 mm (Li, Wang, Wang, Li, & Chen, 2013). Seven finned tubes were installed into each shell and the adsorbents were packed in the 28 finned tubes and the COP of the cycle was 0.15 at a regeneration temperature of 105°C and as an evaporation temperature of -5 °C. Similar approach included vapour channels to increase the mass transfer process and stainless steel mesh to increase the heat transfer and seal the adsorbent grains (Wang, Wang, Lu, Chen, Wang, & Wu, 2006; (Oliveira, Silveira, & Wang, 2006). In contrast, another type of adsorption bed configuration for ice production was with application with COP and a specific cooling power (Tamainot-Telto, Metcalf, & Critoph, 2009). The bed was made of aluminium alloy and structured from three passes to circulate different fluids.

Metcalf tested a structure of plate heat exchanger adsorption bed made of nickel-brazed stainless steel and filled out with a layer of 4 mm of activated carbon (Metcalf, Tamainot-Telto, & Critoph, 2011). The secondary fluid was circulated within etched channels (Figure 10). The adsorption system was driven by

Table 4. Performance comparison of the various cycle configurations in terms of operating temperatures, cycle times, Coefficient of Performance (COP) and Specific Cooling Power (SCP)

Author	System Arrangement	COP	T_{chill} (°C)	$T_{cooling}$ (°C)	T_{hot} (°C)	Cycle Time (Sec)	SCP or ACP (W/kg)	Specifications
Riffel et al. **(2009)**	2 bed, one stage with mass and heat recovery, one condenser and one evaporator (activated carbon-methanol)	Improved by 16.23%	7		70	120 sec of heat recovery and 300 sec of mass recovery.	SCP Improved by 8.17%	Regarding this kind of operation procedures, the simulation results shown an increment of both SCP and COP comparing with single cycle.
Wang et al. **(2003)**	2 bed, one stage with mass and heat recovery, one condenser and one evaporator (activated carbon–methanol)	0.125	-10.23	30	77–107	3360 (cycle time) 120 sec of heat recovery and 60 sec of mass recovery.		The performance improved according the mass and heat recovery process than that of adsorber1. The mass and heat recovery processes for the both beds are a slightly different due to a volume difference of the beds.
Li et al. **(2008)**	2 beds one stage heat pipe with mass recovery (CaCl2 /activated carbon-ammonia)	0.20 (after heat recovery) 0.22 (before heat recovery)		30	145	120 sec of heat recovery and 40 sec of mass recovery.	332.2(after heat recovery) 367.8 (before heat recovery)	The influence of mass recovery was investigated on the performance of adsorption cycle in case of after or before the heat recovery process. COP was improved by 46% comparing with no mass recovery cycle.
Meunier **(1986)**	4 bed two stage Cascading cycles (zeolite-water and the active carbon-methanol)	0.65	-10	20	275 and 120			the working temperatures of the both stages were evaluated so the required heat to drive the low temperature stage was provided by the waste heat of high temperature stage.
Rowe et al. **(2014)**	4 bed one stage multi-bed (Activated carbon/methanol)	0.33	3	25	90	37	327	Continuous cooling Desorption and adsorption pressure variation Switching sequence for automatic bed cycling
Habib et al. **(2013)**	two stage. (activated carbon/HFC-134a)	0.3	14	30	50-70	1250		It can be operated with smaller range of regenerating temperature lifts comparing with single-stage chillers. The COP of the two stage cycle is lower than the single stage one.
Tseng et al. **(2009)**	countercurrent two- and three-stage							a 25% adsorbent was saved as equivalent to one extra stage,

continued on next page

Table 4. Continued

Author	System Arrangement	COP	T_{chill} (°C)	$T_{cooling}$ (°C)	T_{hot} (°C)	Cycle Time (Sec)	SCP or ACP (W/kg)	Specifications
Douss et al. **(1989)**	4-bed cascading cycles heat pump (zeolite-water and carbon-methanol)	1.06	2	30	250	9840		the cooling COP of such cycle was experimentally found to be higher than the COP of an intermittent one (about 0.5) and of a two adsorber zeolite water cycle (about 0.75). The COP was very sensitive to any change in the evaporating temperature.
Miles et al. **(1996)**	2-bed thermal wave regeneration cycle (activated carbon/ammonia)	1.19	8.3	26	232	318		The performance of a adsorption heat pump could be significantly improved.
Critoph et al. **(1998)**	2-bed, convective thermal wave (carbon-ammonia)	1.3	0		250	250	1000 to 3000 W/kg	The effectiveness of heat exchanger was anticipated to be 0.8.
Critoph et al. **(1999)**	4-bed, convective thermal wave (granular carbon-ammonia)	0.95	0	42	250	300	404 W/kg	The intensity of heat transfer was achieved to enhance the efficiency and reduce the size of adsorption cycle. The surface area of four beds are very high with cheaper cost. The surface area per volume is (1500 m^2/m^3).
Wang & Oliveira **(2005)**	Vacuum chamber, one condenser and one evaporator (silica gel-methanol/water)	0.45		30	85	2280	199	two vacumm chambers were included in the chiller. The beds, condenser and evaporators were located inside the enclose of vacuum chamber so there is no need of vacuum valves.
Saha et al. **(2007)**	2-bed, one condenser and one evaporator. (activated carbon fiber-ethanol)		13	30	85	600 (cycle time) 60 (switching time)		The main advantage in such system is to utilize from low-temperature waste heat sources.
He et al.**(2014)**	4 -bed two-stage, evaporator and condenser (Zeolite and activated carbon)	0.25-0.35	15	30	45 - 50			The two-stage cycle with different adsorbents could be driven efficiently with temperature range of 45 to 50°C.
Wang et al **(2000)**	2-bed, mass and heat recovery heat pump (activated carbon-methanol)	0.5	5	28.7	100	2400 (cycle time)	143-159	In the COP of such system improved as 25% comparing with basic one. The mass recovery process was applied after the heat recovery process.
Umair et al **(2014)**	2-bed, solar adsorption refrigeration (Activated carbon fiber/ethanol)	0.058 (system)	-3	30	87	172800	80 kJ/kg	Mathematical analysis for continuous solar cooling refrigerator was developed

continued on next page

Table 4. Continued

Author	System Arrangement	COP	T_{chill} (°C)	$T_{cooling}$ (°C)	T_{hot} (°C)	Cycle Time (Sec)	SCP or ACP (W/kg)	Specifications
Hamamoto et al. **(2006)**	4 bed, cascaded (silica gel, methanol top cycle) (activated carbon FX,methanol bottom)	0.062 (overall)	-20 /10	-5/35	80			Equilibrium cycle analysis was used which showed that the COPt of the advanced cycle is 40% higher than that of silica gel based cascade cycles
Wang (2001)	2 beds –single stage Heat and mass recovery system (consolidated activated carbon/ methanol)	0.5	10	24-30	100	40 to 50 min	150 W/kg 4 kW	Summation of ratio of fluid to adsorbent heat capacity (Rf) and metal to adsorbent heat capacity(Rm) (R=Rf+Rm) increases as COP descreases. As a result, oil will be better as it has lower Rf. The mass recovery cycle was found to be very attractive as it increases the cycle uptake difference.
Li et al. (2013)	4 beds- single stage parabolic solar collector-adsorption ice making compound adsorbent (CaCl2/ AC 4:1)/Ammonia	0.15	-5	30	105	5 hr (Desorption) 3:20 (Adsorption)		Ice making
Oliveira & Silveira **(2006)**	4 beds- two stages with mass recovery (AC-Ammonia)	0.1	-27.3	30	85-115			Ice making
Lychnos et al., **(2014)**	1bed-single stage Sorption-vapour compression hybrid system (AC- ammonia)	0.2	-5°C	40	100-120			Ice making and air conditioning applications
Alghoula et al., **(2009)**	2 bed- single stage dual-purpose solar continuous adsorption system (AC / methanol)	0.44	-10	30	85-95	Day time (Desorption) Night time (Adsorption)		Domestic refrigeration and water heating
Metcalf et al, **(2011)**	2 and 4 bed- single stage evacuated tube solar collectors (active carbon-ammonia)	0.38 (2 bed) 0.45 (4 bed)	-5	35	150	3 min	260 (2 bed) 120 (4 bed)	Ice making for food preservation
Li et al., (2002)	flat-plate solar- 1bed -single stage. (AC / methanol)	0.12 (system) 0.428 (cycle)	- 1	23	80 to 90	8 hr (Desorption) 8 hr (Adsorption)		ice maker
Tamainot & Critoph, **(1996)**	1 bed, 2, condenser, 1 evaporator single stage (AC monolithic -ammonia)	0.12	-20 to 0	39	105	1600 sec Heating time 600 sec	32	low temperature purposes

Figure 9. Types of solar collectors used in ice making based on reviewed literature

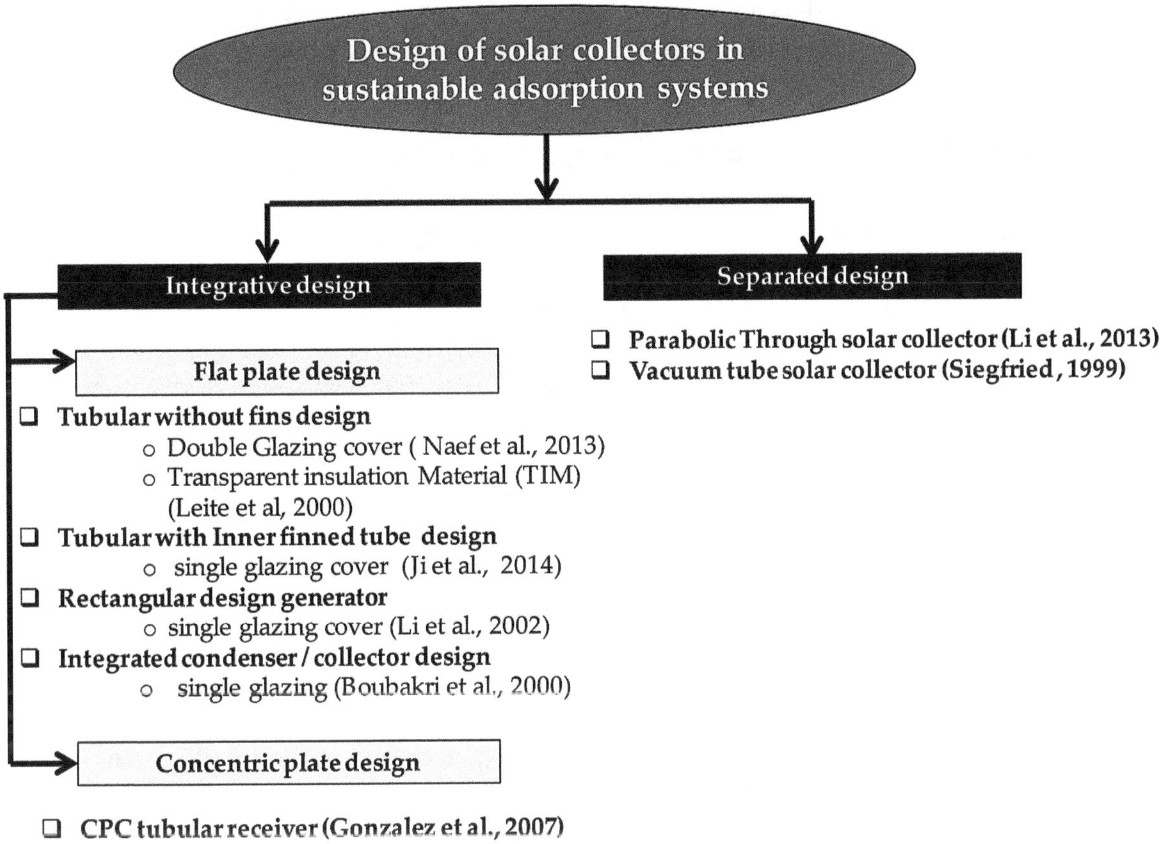

evacuated solar collector. As applied for air conditioning and gas fire heat pumping, the large surface area of the heat transfer provided quick temperature transfer and high SCP. In Figure 11, methanol/ activated carbon system was constructed for heat exchanger having spiral plate shape for ice making application (Wang, Wu, Xu, & Wang, 2001). The secondary fluid was circulated through one channel to transfer the heat through spiral bed area with $2m^2$, causing that the SCP and COP of the system were 150 W/kg and 0.13, respectively. Similarly, shell and tubular shaped bed with two parts was constructed (Nema & Fadiel, 2009). The each part was formed like a hemispherical shell having 22 tubes included in each part and the shell was prepared to contain the secondary fluid. The beds were charged with methanol/ activated carbon as the working pair. A maximum COP of 0.3 at a desorption temperature of 60-70 °C and a condensing temperature of 30°C was achieved. Desk bed with 300mm in diameter and 90 mm in depth with fins was fixed on the circumference of the bed (Rahman, Akhanda, & Islam, 2006). The adsorption bed was located within a water bath in which a solar collector was integrated. Two kind or refrigerants were tested, namely methanol and R-11 based on activated carbon as adsorbent. In Figure 12, adsorption bed made from a glass tube was formed for a circular container in order to enhance the surface area (Khattab, 2004). Two black metallic screens were located on both sides of the circular bed, the adsorbent material was mixed with tiny bodies of blackened steel and the adsorbent material was bonded with these tiny bodies. The cycle COP was 0.505 and 0.464 in summer and winter, respectively. Experimental study on adsorption refrigeration tube (ART) was carried out by amalgamate bed, the condenser, the evaporator, the insulation secondary fluid jacket and the valves (Zhao, 2011). There were some modified conventional ARTs regarding the heat transfer and the pressure control. Such design was in modular form and cooling capacity could be increased by increasing number of modules.

ENVIRONMENTAL CONCERNS OF ADSORPTION COOLING

Eighty percent of the greenhouse gas (GHG) emissions in Europe comes from the energy sector and the higher demand become several problems on 2016 (Livingston, 2016). The extensive use of electrically driven cooling is leading to capacity straining power demand peaks in the summer and increasing of greenhouse gas emissions-either through leakage of cooling fluids or use of non-sustainable energy. Thus, improved building concepts, targeting on reduction of cooling loads by passive and innovative measures, and the use of alternatives in covering the remaining cooling and heating demands of buildings, are of interest (http://www.solair-project.eu/). Adsorption cooling provides innovative technology with almost no GHG emissions. Nevertheless, further research is still necessary not only to improve the performance of the adsorption working pairs in order to build compact, energy efficient, reliable, and long-life adsorption chillers, but also to develop adsorption pairs that have low or no adverse impact on our environment. Refrigerants such as R134a, R-141b and R507a that have been used were found to have a high global warming potential (GWP), also n-butane, hydrogen, methanol, ethanol and diethyl ether are highly flammable gases, while ammonia is a highly toxic refrigerant. Therefore, the importance in the refrigerant is to use water having safe and more environmentally friendly properties. Also a more environmentally friendly approach is developing air conditioning and refrigeration systems that are driven by renewable energy resources such as solar energy. Solar cooling has the potential of significantly reducing the electricity consumption, contribute fossil energy saving and electrical peak load reduction and furthermore, contributes in a positive way with the urban microclimate through absorbing the solar irradiation on the roofs, reducing the carbon emissions and using environmental friendly refrigerants (De Mestrado, 2010).

Adsorption cooling technology in automobile air conditioners provided another application with independent of the conventional energy sources, when the use of the heat from the exhaust gas or the cooling loop was available as the heat sources (Hamdy, Askalany, Harby, & Kora, 2015).

Regarding the cost of the system, more effort is required to make the system cost comparable to other technologies such photovoltaic (PV) cells (https://www.dri.edu). It is evident that recent developments in the adsorption systems have led to a significant reduction in the adsorption system cost. The capital costs of solar cooling technologies are dominated by the solar block (37%), as followed by the thermal chiller (29%). Other cost components are installation (19%), storage (8%), and the heat rejection loop (7%). Total system costs are in the range of USD 4350-5550/kW. Capital costs are in the range of USD 425/m^2 or USD 1915/kW for a 100 kW thermal chiller. However, the costs of the adsorption chiller are highly dependent upon the scale ranging from USD 1560/kW for 20 kW cooling capacity to USD 3600/ kW for 200 kW cooling capacity (http://www.irena.org/). It should be noticed that all the adsorption capacity experimentally were tested on packed beds without improving the bed designs for more efficient heat and mass transfer processes between the adsorbent materials and the refrigerant. Efficient heat and mass transfer processes lead to a more compact and less weight system design. The adsorbent material should be distributed over heat exchanger surface, to allow good heat and mass transfer between the adsorbent and the refrigerant (Shmroukh, Ali, & Ookawara, 2015).

Figure 10. Prototype plate heat exchanger bed
Source: Metcalf, Tamainot-Telto, & Critoph, 2011

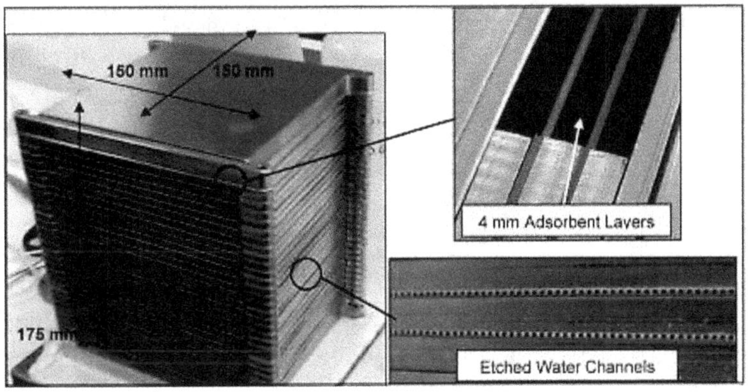

Figure 11. Schematic diagram of spiral plate heat exchanger bed
Source: Wang et al., 2001

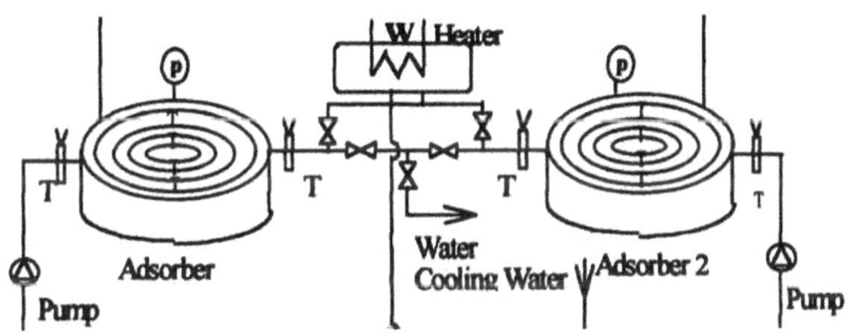

Figure 12. Modified glass tube (a) schematic diagram (b) photograph
Source: Khattab, 2004

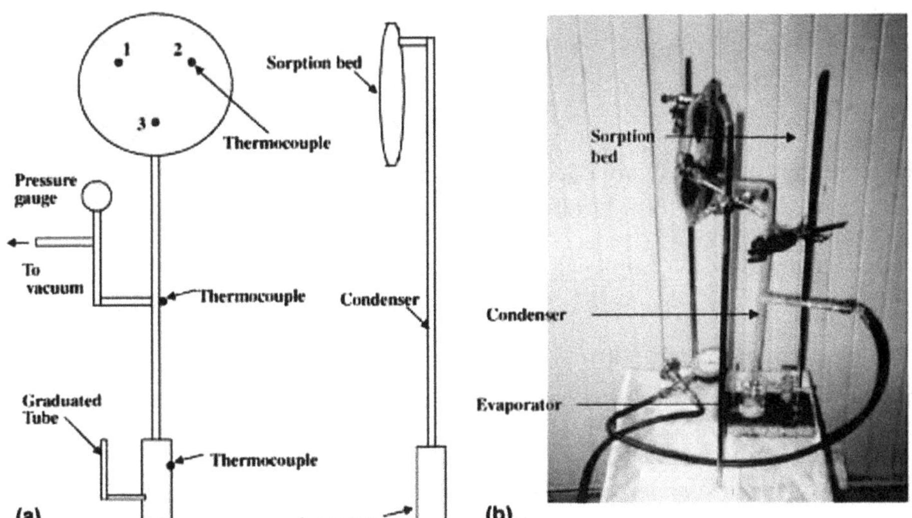

The use of adsorption cooling technology gives us a potential solution for developing air conditioning and refrigeration systems. Such systems offer remarkable advantages such as being environmentally friendly, depend on low temperature heat sources (solar energy, waste heat), the simplicity of the system and long operating life. With such merits, adsorption systems can compete with conventional absorption and compression devices (Shmroukh, Ali, & Ookawara, 2015; Gwadera & Kupiec, 2011).

CONCLUSION

Activated carbon has been an interesting material in the field of adsorption research due to its favourable characteristics including the large uptakes and surface area. This chapter reviews the reported research in adsorption cooling systems by activated carbon. It was focused that the development of efficient materials and sustainable cycle designs approached their environmental and economic impacts. Experimental measurements related to material characterization such particle size, thermal conductivity, packing density and specific heat were identified and illustrated. Finally, the environmental considerations on adsorption cooling systems were discussed. One of the main advantages of any adsorption system was environmentally friendly for almost no GHG emissions. Further research would be required for the adsorption cooling facilities with cost competitiveness.

REFERENCES

Abdullah, M. O., Tan, I. A. W., & Limb, L. S. (2011). Automobile adsorption air-conditioning system using oil palm biomass-based activated carbon: A review. *Renewable & Sustainable Energy Reviews*, *15*(4), 2061–2072. doi:10.1016/j.rser.2011.01.012

Ahamat, M. A., & Tierney, M. J. (2012). Calorimetric assessment of adsorbents bonded to metal surfaces: Application to type A silica gel bonded to aluminium. *Applied Thermal Engineering*, *40*, 258–266. doi:10.1016/j.applthermaleng.2012.02.021

Akahira, A., Alam, K. C. A., & Hamamoto, Y. (2005). Mass recovery four-bed adsorption refrigeration cycle with energy cascading. *Applied Thermal Engineering*, *25*(11-12), 1764–1778. doi:10.1016/j.applthermaleng.2004.10.006

Akahira, A., Alam, K. C. A., Hamamoto, Y., Akisawa, A., & Kashiwagi, T. (2004). Mass recovery adsorption refrigeration cycle-improving cooling capacity. *International Journal of Refrigeration*, *27*(3), 225–234. doi:10.1016/j.ijrefrig.2003.10.004

Alghoula, M. A., Sulaiman, M. Y., Sopian, K., & Azmi, B. Z. (2009). Performance of a dual-purpose solar continuous adsorption system. *Renewable Energy*, *34*(3), 920–927. doi:10.1016/j.renene.2008.05.037

Antonio, P. F. L., & Michel, D. (2000). Performance of a new solid adsorption ice maker with solar energy regeneration. *Energy Conversion and Management*, *4*, 1625–1647.

Anupam, K., Chatterjee, A., Halder, G. N., & Sarkar, S. C. (2011a). Experimental investigation of a single-bed pressure swing adsorption refrigeration system towards replacement of halogenated refrigerants. *Chemical Engineering Journal*, *171*(2), 541–548. doi:10.1016/j.cej.2011.04.025

Anyanwu, E., & Ogueke, N. (2005). Thermodynamic design procedure for solid adsorption solar refrigerator. *Renewable Energy*, *30*(1), 81–96. doi:10.1016/j.renene.2004.05.005

Aristov, Y. (2007). Novel Materials for adsorptive heat pumping and storage: Screening and nanotailoring of sorption properties. *Journal of Chemical Engineering of Japan*, *40*(13), 1242–1251. doi:10.1252/jcej.07WE228

Aristov, Yu. I., Dawoud, B., Glaznev, I. S., & Elyas, A. (2008). A new methodology of studying the dynamics of water sorption/desorption under real operating conditions of adsorption heat pumps: Experiment. *International Journal of Heat and Mass Transfer*, *51*(19–20), 4966–4972. doi:10.1016/j.ijheatmasstransfer.2007.10.042

Askalany, A., Salem, M., Ismail, I. M., Ali, A. H. H., & Morsy, M. G. (2012). Experimental study on adsorptionedesorption characteristics of granular activated carbon/R134a pair. *International Journal of Refrigeration*, *35*(3), 494–498. doi:10.1016/j.ijrefrig.2011.04.002

Askalany, A. A., & Saha, B. B. (2015). Experimental and theoretical study of adsorption kinetics of Difluoromethane onto activated carbons. *International Journal of Refrigeration*, *4*(9), 160–168. doi:10.1016/j.ijrefrig.2014.10.009

Askalany, A. A., Saha, B. B., & Ismail, I. M. (2014). Adsorption isotherms and kinetics of HFC410A onto activated carbons. *Applied Thermal Engineering*, *72*(2), 237–243. doi:10.1016/j.applthermaleng.2014.04.075

Attalla, M., & Sadek, S. (2014). Experimental Investigation of Granular Activated Carbon/R-134a Pair for Adsorption Cooling System Applications. *Journal of Power and Energy Engineering*, *2*(02), 11–20. doi:10.4236/jpee.2014.22002

Banker, N. D., Prasad, M., Dutta, P., & Srinivasan, K. (2010). Development and transient performance results of a single stage activated carbon – HFC 134a closed cycle adsorption cooling system. *Applied Thermal Engineering*, *30*(10), 1126–1132. doi:10.1016/j.applthermaleng.2010.01.027

Bansal, R. C., & Goyal, M. (2005). *Activated Carbon Adsorption*. Taylor & Francis Group. doi:10.1201/9781420028812

Bartell, F.E. & Bower, J.E. (1952). *Adsorption of vapors by silica gels of different structures*. DOI: 10.1016/0095-8522(52)90022-6

Boubakri, A., Guilleminot, J. J., & Meunier, F. (2000). Adsorptive solar powered ice maker: Experiments and model. *Solar Energy*, *69*(3), 249–263. doi:10.1016/S0038-092X(00)00063-3

Burchell, T. D. (1999). *Adsorption refrigerators and heat pumps*. Ltd Carbon Materials for Advanced Technologies.

Çağlar, A. (2012). *Design and experimental testing of an adsorbent bed for a thermal wave adsorption cooling cycle* (Unpublished doctoral dissertation). Middle East Technical University, Ankara, Turkey.

Carson, J. (2002). *Prediction of the thermal conductivity of porous foods* (PhD Thesis). Messay University.

Chan, K. C., Chao, C. Y. H., Sze-To, G. N., & Hui, K. S. (2012). Performance predictions for a new zeolite 13X/CaCl2 composite adsorbent for adsorption cooling systems. *International Journal of Heat and Mass Transfer*, *55*(11-12), 3214–3224. doi:10.1016/j.ijheatmasstransfer.2012.02.054

Chekirou, W., Boukheit, N., & Karaali, A. (2013). Optimal Design Study of Tubular Adsorber for Solid Adsorption Solar Cooling Machine Working with Activated Carbon–Methanol Pair. *16èmes Journées Internationales de Thermique (JITH 2013)*.

Chen, C. J., Wang, R. z., Oliveira, R. G., & Hu, J. (2009). Adsorption Models and Structural Characterization for Activated Carbon Fibers. J. Shanghai Jiaotong Univ. (Sci.), 14(1), 35-39.

Critoph, R. E. (1996). Evaluation of alternative refrigerant-adsorbent pairs for refrigeration cycles. *Applied Thermal Engineering*, *16*(11), 891–900. doi:10.1016/1359-4311(96)00008-7

Critoph, R. E. (1998). Forced Convection Adsorption Cycles. *Applied Thermal Engineering*, *18*(9-10), 799–807. doi:10.1016/S1359-4311(97)00110-5

Critoph, R. E. (1999). Forced Convection Adsorption Cycle with Packed Bed Heat Regeneration. *International Journal of Refrigeration*, *22*(1), 38–46. doi:10.1016/S0140-7007(97)00036-4

Cui, Q., Tao, G., Chen, H., Guo, X., & Yao, H. (2005). Environmentally benign working pairs for adsorption refrigeration. *Energy*, *30*(2-4), 261–271. doi:10.1016/j.energy.2004.05.005

Daou, K., Wang, R. Z., & Xia, Z. Z. (2006). Development of a new synthesized adsorbent for refrigeration and air conditioning applications. *Applied Thermal Engineering*, *26*(1), 56–65. doi:10.1016/j.applthermaleng.2005.04.024

Daou, K., Wang, R. Z., Yang, G. Z., & Xia, Z. Z. (2008). Theoretical comparison of the refrigerating performances of a CaCl2 impregnated composite adsorbent to those of the host silica gel. *International Journal of Thermal Sciences*, *47*(1), 68–75. doi:10.1016/j.ijthermalsci.2007.01.003

Dawoud, B., Sohel, M. I., Freni, A., Vasta, S., & Restuccia, G. (2011). On the effective thermal conductivity of wetted zeolite under the working conditions of an adsorption chiller. *Applied Thermal Engineering*, *31*(14-15), 2241–2246. doi:10.1016/j.applthermaleng.2011.03.016

De Mestrado, D. (2010). *Technical and Economic Assessment of Medium Sized Solar-Assisted Air-Conditioning in Brazil*. Academic Press. http://www.urb.puc-rio.br/dissertacao/dissertacao_till.pdf

Demir, H., Mobedi, M., & Ulku, S. (2011). Microcalorimetric investigation of water vapor adsorption on silica gel. *Journal of Thermal Analysis and Calorimetry*, *105*(1), 375–382. doi:10.1007/s10973-011-1395-y

Douss, N., & Meunier, F. (1989). Experimental study of cascading adsorption cycles. *Chemical Engineering Science*, *44*(2), 225–235. doi:10.1016/0009-2509(89)85060-2

El-Sharkawy, I. I., Hassan, M., Saha, B. B., Koyama, S., & Nasr, M. M. (2009). Study on adsorption of methanol onto carbon based adsorbents. *International Journal of Refrigeration*, *32*(7), 1579–1586. doi:10.1016/j.ijrefrig.2009.06.011

El-Sharkawy, I. I., Kuwahara, K., Saha, B. B., Koyama, S., & Ng, K. C. (2006). Experimental investigation of activated carbon fibers/ethanol pairs for adsorption cooling system application. *Applied Thermal Engineering*, *26*(8-9), 859–865. doi:10.1016/j.applthermaleng.2005.10.010

El-Sharkawy, I. I., Saha, B. B., Koyama, S., He, J., Yap, K. C., & Ng, C. (2008). Experimental investigation on activated carbon–ethanol pair for solar powered adsorption cooling applications. *International Journal of Refrigeration*, *31*(8), 1407–1413. doi:10.1016/j.ijrefrig.2008.03.012

El-Sharkawy, I. I., Uddin, K., Miyazaki, T., Saha, B. B., Koyama, S., Miyawaki, J., & Yoon, S. H. (2014). Adsorption of ethanol onto parent and surface treated activated carbon powders. *International Journal of Heat and Mass Transfer*, *73*, 445–455. doi:10.1016/j.ijheatmasstransfer.2014.02.046

Elsayed, A., Al-Dadah, R. K., Mahmoud, S., & Kaialy, W. (2014). Investigation of activated carbon / Ethanol for low temperature adsorption cooling. *International Journal of Green Energy*.

Elsayed, A., Al-Dadah, R., Mahmoud, S., Elshaer, A., & Bowen, J. (2014). Investigation of efficient adsorbents using transient analysis of adsorption chillers with Simulink. *Proceedings of the International Conference on Heat Transfer and Fluid Flow*.

Elsayed, A., & Elsayed, E. AL-Dadaha, R., Mahmoud, S., Elshaer,A., &Kaialy, W.(2016). Thermal energy storage using metal–organic framework materials. *Applied Energy*. doi:10.1016/j.apenergy.2016.03.113

Eun, T. H., Song, H. K., Han, J. H., Lee, K.-H., & Kim, J. N. (2000). Enhancement of heat and mass transfer in silica gel-expanded graphite composite blocks for adsorption heat pumps: Part I. characterization of the composite blocks. *International Journal of Refrigeration*, *23*, 64–73. doi:10.1016/S0140-7007(99)00035-3

Freni, A., Tokarev, M. M., Restuccia, G., Okunev, A. G., & Aristov, Y. I. (2002). Thermal conductivity of selective water sorbents under the working conditions of a sorption chiller. *Applied Thermal Engineering*, *22*(14), 1631–1642. doi:10.1016/S1359-4311(02)00076-5

Fujioka, K., & Suzuki, H. (2013). Thermophysical properties and reaction rate of composite reactant of calcium chloride and expanded graphite. *Applied Thermal Engineering*, *50*(2), 1627–1632. doi:10.1016/j.applthermaleng.2011.08.024

Füldner, G., & Schnable, L. (2008). Non-isothermal kinetics of water adsorption in compact adsorbent layers on a metal support. *Comsol conference proceeding*. Retrieved from http://www.comsol.com/papers/5118/

Gonzalez, M. I., & Rodriguez, L. R. (2007). Solar powered adsorption refrigerator with CPC collection system: Collector design and experimental test. *Energy Conversion and Management*, *48*(9), 2587–2594. doi:10.1016/j.enconman.2007.03.016

Gurgel, J. M., Oliveira, P. A., Marques Ailson, S., Oliveira, L. G., & Tavares, F. R. M. (2005). Experimental results for thermal conductivity of adsorbed natural gas on activated carbon., *17th European conference on thermophysical properties*.

Gwadera, M., Kupiec, K. (2011). *Adsorption cooling as an effective method of waste heat utilization*. Academic Press. https://suw.biblos.pk.edu.pl/resources/i4/i5/i4/i8/r4548/GwaderaM_AdsorptionCooling.pdf

Habib, K., Majid, M. A. B. A., & Sulaiman, S. A. B. (2013). Study on two stage activated carbon/HFC-134a based adsorption chiller. *4th International Conference on Energy and Environment*. doi:10.1088/1755-1315/16/1/012084

Habib, K., & Saha, B. (2013). Performance evaluation of solar driven activated carbon fiber-ethanol based adsorption cooling system in Malaysia. *Asian Journal of Scientific Research*, *6*, 146–156. doi:10.3923/ajsr.2013.146.156

Habib, K., Saha, B. B., & Koyama, S. (2014). Study of various adsorbent-refrigerant pairs for the application of solar driven adsorption cooling in tropical climates. *Applied Thermal Engineering*, *72*(2), 266–274. doi:10.1016/j.applthermaleng.2014.05.102

Habib, K., Saha, B. B., Rahman, K. A., Chakraborty, A., Koyama, S., & Ng, K. C. (2010). Experimental study on adsorption kinetics of activated carbon/R134a and activated carbon/R507A pairs. *International Journal of Refrigeration*, *33*(4), 706–713. doi:10.1016/j.ijrefrig.2010.01.006

Hamamoto, Y., Alam, K. C. A., Saha, B. B., Koyama, S., Akisawa, A., & Kashiwagi, T. (2006). Study on adsorption refrigeration cycle utilizing activated carbon fibers. Part 2. Cycle Performance evaluation. *International Journal of Refrigeration*, *29*(2), 315–327. doi:10.1016/j.ijrefrig.2005.06.001

Hamdy, M., Askalany, A., Harby, K., & Kora, N. (2015). *An overview on adsorption cooling systems powered by waste heat from internal combustion engine*. Academic Press. Renewable Sustainable Energy Rev., 51,1223–1234.

Hassan, H. Z. (2013). Energy Analysis and Performance Evaluation of the Adsorption refrigeration System. *ISRN Mechanical Engineering*, *2013*, 14. doi:10.1155/2013/704340

He, Z. H., Huang, H.Y., Deng, L.S., Yuan, H.R., Kobayashi, N., Kubota, M., … Zhao, D.D. (2014). Development of novel type of two-stage adsorption chiller with different adsorbents. *The 6th International Conference on Applied Energy – ICAE2014, Energy Procedia, 61*, 1996 – 1999.

Henninger, S. K., Schicktanz, M., Hugenell, P. P. C., Sievers, H., & Henning, H. M. (2012). Evaluation of methanol adsorption on activated carbon for thermally driven chillers part I: Thermophysical characterisation. *International Journal of Refrigeration*, *35*(3), 543–553. doi:10.1016/j.ijrefrig.2011.10.004

Hongyu, H., Zhaohong, H., Haoran, Y., Noriyuki, K., Dandan, Z., Mitsuhiro, K., & Ze-Shao, C. et al. (2009). Analysis for composite zeolite/foam aluminum–water mass recovery adsorption refrigeration system driven by engine exhaust heat. *Energy Conversion and Management*, *50*(2), 255–261. doi:10.1016/j.enconman.2008.09.022

Huafang, G. (2014). Effect of Adsorbent Diameter on the Performance of Adsorption Refrigeration. *Chinese Journal of Chemical Engineering*, *22*(5), 602–606. doi:10.1016/S1004-9541(14)60074-4

Irena. (2015). *Solar Heating and Cooling for Residential Applications Technology Brief*. Retrieved from http://www.irena.org/documentdownloads/publications/irena_etsap_tech_brief_r12_solar_thermal_residential_2015.pdf

Ismail, A. B., Li, A., Thu, K., Ng, K. C., & Chun, W. (2014). Pressurized adsorption cooling cycles driven by solar/waste heat. *Applied Thermal Engineering*, *67*(1-2), 106–113. doi:10.1016/j.applthermaleng.2014.02.063

Ivanova, E. P., Kostova, M. A., & Koumanova, B. K. (2010). Kinetics of water and alcohol vapors adsorption on natural zeolite. *Asia-Pacific Journal of Chemical Engineering*, *5*(6), 869–881. doi:10.1002/apj.412

Jänchen, J., Schumann, K., Thrun, E., Brandt,A., Unger B., & Hellwig, U. (2012). Preparation, hydrothermal stability and thermal adsorption storage properties of binderless zeolite beads. *International Journal of Low-Carbon Technologies*. doi: 10.1093/ijlct/cts037

Jerai, F., Miyazaki, T., Saha, B., & Koyama, S. (2015). Overview of adsorption cooling system based on activated carbon –Alcohol Pair. *Joint Journal of Novel Carbon Resources Sciences & Green Asia Strategy*, 30-40.

Ji, X., Li, M., Fan, J., Zhang, P., Luo, B., & Wang, L. (2014). Structure optimization and performance experiments of a solar-powered finned-tube adsorption refrigeration system. *Applied Energy*, *113*, 1293–1300. doi:10.1016/j.apenergy.2013.08.088

Jiang, L., Wang, L. W., & Wang, R. Z. (2014). Investigation on thermal conductive consolidated composite CaCl2 for adsorption refrigeration. *International Journal of Thermal Sciences*, *81*, 68–75. doi:10.1016/j.ijthermalsci.2014.03.003

Jiayou, Q. (2003). *Characterization of silica gel-water vapor adsorption and its measuring facility* (Unpublished master dissertation). National University of Singapore, Singapore.

Jribi, S., Chakraborty, A., El-Sharkawy, I., Saha, B. B., & Koyama, S. (2009). Thermodynamic Analysis of Activated Carbon CO2 based Adsorption Cooling Cycles. *International Journal of Engineering and Natural Sciences*, *3*(3), 142–145.

Jribi, S., El-Sharkawy, I. I., Koyama, S., & Saha, B. B. (2010b). Study on activated carbon-co2 pair: adsorption characteristics and cycle performance.*International Symposium on Next-generation Air Conditioning and Refrigeration Technology*.

Jribi, S., Koyama, S., & Saha, B. B. (2010a). Performance Investigation of a Novel CO2 Compression-Adsorption Based Hybrid Cooling Cycle. *Engineering Sciences Reports. Kyushu University*, *32*(3), 12–18.

Jribi, S., Saha, B. B., Koyama, S., & Bentaher, H. (2014). Modeling and simulation of an activated carbon–CO2 four bed based adsorption cooling system. *Energy Conversion and Management*, *78*, 985–991. doi:10.1016/j.enconman.2013.06.061

Jribi, S., Saha, B. B., Koyama, S., Chakraborty, A., & Ng, K. C. (2013). Study on activated carbon/HFO-1234ze(E) based adsorption cooling cycle. *Applied Thermal Engineering*, *50*(2), 1570–1575. doi:10.1016/j.applthermaleng.2011.11.066

Khalifa, A. H. N., Hussein, F. M., & Hadi, F. M. (2011). Experimental Study on Two Beds Adsorption Chiller with Regeneration. *Modern Applied Science*, *5*(4), 43–52. doi:10.5539/mas.v5n4p43

Khattab, N. M. (2004). A novel solar-powered adsorption refrigeration module. *Applied Thermal Engineering*, *24*(17-18), 2747–2760. doi:10.1016/j.applthermaleng.2004.04.001

Krause, U. (2009). *Fires in Silos: Hazards, Prevention, and Fire Fighting*. Wiley Publisher, VCH. doi:10.1002/9783527623822

Królicki, Z., Wilk, T., Zajączkowski, B., & Kaczmarczyk, J. (2006). Research on adsorption characteristics of active carbon and silica gel and application in solar refrigeration and heat pump.*61st ATI National Congress –International Session "Solar Heating and Cooling"*.

Kumar, M. (2011). *Adsorption of methane on activated carbon by volumetric method* (Unpublished master dissertation). National Institute of Technology, Rourkela, India.

Lartey, R. B., & Francis, A. (1999). *Developing national capability for manufacture of activated carbon from agricultural wastes.* Institute of Industrial Research, CSIR, Accra, Ghana. Retrieved from http://www.africantechnologyforum.org/GhIE/ActCarbon.htm

Li, C., Wang, R. Z., Wang, L. W., Li, T. X., & Chen, Y. (2013). Experimental study on an adsorption icemaker driven by parabolic trough solar collector. *Renewable Energy*, *57*, 223–233. doi:10.1016/j.renene.2013.01.040

Li, M., Wang, R. Z., Xu, Y. X., Wu, J. Y., & Dieng, A. O. (2002). Experimental study on dynamic performance analysis of a flat-plate solar solid-adsorption refrigeration for ice maker. *Renewable Energy*, *27*(2), 211–221. doi:10.1016/S0960-1481(01)00188-4

Li, T. X., Wang, R. Z., Wang, L. W., Lu, Z. S., & Wu, J. Y. (2008). Influence of mass recovery on the performance of a heat pipe type ammonia sorption refrigeration system using CaCl2 /activated carbon as compound adsorbent. *Applied Thermal Engineering*, *28*(13), 1638–1646. doi:10.1016/j.applthermaleng.2007.10.027

Livingston, W. (2016). *New Liquid Cooled Servers with Adsorption Chillers Improve Efficiency, Reduce Footprint*. Retrieved from http://www.ppiway.com/new-liquid-cooled-servers-with-adsorption-chillers-improve-efficiency-reduce-footprint/

Loh, W., Shah, B., Chakraborty, A., Ng, K., & Chun, W. (2010). Performance analysis of waste heat driven pressurized adsorption chiller. *Journal of Thermal Science and Technology*, *5*(2), 252-265.

Loh, W. S., El-Sharkawy, I. I., Ng, K. C., & Saha, B. B. (2009). Adsorption cooling cycles for alternative adsorbent/adsorbate pairs working at partial vacuum and pressurized conditions. *Applied Thermal Engineering*, *29*(4), 793–798. doi:10.1016/j.applthermaleng.2008.04.014

Lu, Z. S., & Wang, R. Z. (2013). Performance improvement and comparison of mass recovery in CaCl2/activated carbon adsorption refrigerator and silica gel/LiCl adsorption chiller driven by low grade waste heat. *International Journal of Refrigeration*, *36*(5), 1504–1511. doi:10.1016/j.ijrefrig.2013.03.008

Lu, Z. S., Wang, R. Z., & Wang, L. W. (2011). Dynamic characteristics of a novel adsorption refrigerator with compound mass-heat recovery. *International Journal of Energy Research*, *37*(1), 59–68. doi:10.1002/er.1911

Lychnos, G., & Tamainot-Telto, Z. (2014). Performance of hybrid refrigeration system using ammonia. *Applied Thermal Engineering*, *62*(2), 560–565. doi:10.1016/j.applthermaleng.2013.10.013

McGlinchey, D. (2005). *Characterisation of Bulk Solids*. Blackwell Publishing Ltd. doi:10.1002/9781444305456

Metcalf, S.J., Tamainot-Telto, Z., & Critoph, R.E. (2011). Application of a compact sorption generator to solar refrigeration: Case study of Dakar (Senegal). *Applied Thermal Engineering, 31*, 2197-2204.

Meunier, F. (1985). Second law analysis of a solid adsorption heat pump operating on reversible cascade cycles: Application to the zeolite-water pair. *Heat Recovery Systems*, *5*(2), 133–141. doi:10.1016/0198-7593(85)90045-1

Meunier, F. (1986). Theoretical performances of solid adsorbent cascading cycles using the zeolite-water and active carbon-methanol pairs: Four case studies. *Heat Recovery Systems*, *6*(6), 491–498. doi:10.1016/0198-7593(86)90042-1

Meunier, F. (1993). Solid sorption: An alternative to CFCs. *Heat Recovery Systems & CHP*, *13*(4), 289–295. doi:10.1016/0890-4332(93)90051-V

Miles, D. J., & Shelton, S. V. (1996). Design and Testing of a Solid-Sorption Heat-Pump System. *Applied Thermal Engineering*, *16*(5), 389–394. doi:10.1016/1359-4311(95)00021-6

Ming, H. J. (2009). *Adsorption evaporative emission control system for vehicles* (Unpublished doctoral dissertation). National University of Singapore, Singapore.

Miyazaki, T., El-Sharkawy, I. I., Saha, B. B., & Koyama, S. (2014). Optimized Performance of One-Bed Adsorption Cooling System.*International Refrigeration and Air Conditioning Conference*.

Mugnier, D., & Goetz, V. (2001). Energy storage comparison of sorption systems for cooling and refrigeration. *Solar Energy*, *71*(1), 47–55. doi:10.1016/S0038-092X(01)00013-5

Naef, A. A. (2013). Improving ice productivity and performance for an activated carbon/methanol solar adsorption ice-maker. *Solar Energy*, *98*, 523–542. doi:10.1016/j.solener.2013.10.018

Nema, A. H. K., & Fadiel, H. J. (2009). Experimental Study on A Hybrid Adsorption *Refrigeration System. Eng. & Tech. Journal*, *27*(9), 1790–1800.

Oike, T., Huang, H., Watanabe, F., Hayashi, N., Oshima, T., Hasatani, M., & Kobayashi, N. (2007). Development of novel composite adsorbents used for adsorption heat pump.*Proceedings of International Symposium on Eco Topia Science, ISETS07.*

Oliveira, R. G., Silveira, V. Jr, & Wang, R. Z. (2006). Experimental study of mass recovery adsorption cycles for ice making at low generation temperature. *Applied Thermal Engineering*, *26*(2-3), 303–311. doi:10.1016/j.applthermaleng.2005.04.021

Oyarzún, J. (1999). Pigment Processing: Physico-chemical Principles. Vincentz Verlag.

Qi, Z. (2005). Study on hybrid system of solar powered water heater and adsorption ice maker. *International Journal on Architectural Science.*, *6*, 168–172.

Qu, T. F., Wang, R. Z., & Wang, W. (2001). Study on heat and mass recovery in adsorption refrigeration cycles. *Applied Thermal Engineering*, *21*(4), 439–452. doi:10.1016/S1359-4311(00)00050-8

Rahman, A. F., Ueda, Y., Akisawa, A., Miyazaki, T., & Saha, B. B. (2013). Design and Performance of an Innovative Four-Bed, Three-Stage Adsorption Cycle. *Energies*, *6*(3), 1365–1384. doi:10.3390/en6031365

Rahman, M. H., Akhanda, A. R., & Islam, A. K. M. S. (2006). Study, Design and Fabrication of a Solar Powered Adsorption Refrigeration System. *ISESCO*, *2*(1), 10–13.

Ramji, H. R., Leo, S. L., Tan, I. A. W., & Abdullah, M. O. (2014). Comparative study of three different adsorbent-adsorbate working pairs for a waste heat driven adsorption air conditioning system based on simulation. *IJRRAS*, *18*(2), 109–121.

Raymond, A. W. (2010). *Investigation of microparticle to system level phenomena in thermally activated adsorption heat pumps* (Unpublished master dissertation). Georgia Institute of Technology.

Riffat, S. B., Williams, M. D., & Stuart, C. (1997). Adsorption heat pump using HFC refrigerants. *International Journal of Energy Research*, *21*(6), 481–494. doi:10.1002/(SICI)1099-114X(199705)21:6<481::AID-ER277>3.0.CO;2-7

Riffel, D. B., Belo, F. A., & Leite, A. P. F. (2009). Heat and mass recovery in an activated carbon-methanol adsorptive chiller.*20th International Congress of Mechanical Engineering*, Gramado, RS, Brazil.

Robbins, C. (n.d.). *Solar Cooling Technology Review*. Retrieved from https://www.dri.edu/images/stories/editors/receditor/Solar_Cooling_Review.pdf

Rowe, K. C. (2014). Numerical analysis of multi-bed activated carbon/methanol adsorption refrigeration system with allowance for dynamic pressure variation.*3rd IIR Conference on Cold Chain and Sustainability*, London

Saha, B. B., El-Sharkawy, I. I., Chakraborty, A., & Koyama, S. (2007). Study on an activated carbon fiber-ethanol adsorption chiller: Part I-system description and modelling. *International Journal of Refrigeration*, *30*(1), 86–95. doi:10.1016/j.ijrefrig.2006.08.004

Saha, B. B., Habib, K., El-Sharkawy, I. I., & Koyama, S. (2009). Adsorption characteristics and heat of adsorption measurements of R-134a on activated carbon. *International Journal of Refrigeration*, *32*(7), 1563–1569. doi:10.1016/j.ijrefrig.2009.03.010

Sahoo, P. K., John, M., Newalkar, B. L., Choudhary, N. V., & Ayappa, K. G. (2011). Filling Characteristics for an Activated Carbon Based Adsorbed Natural Gas Storage System. *Industrial & Engineering Chemistry Research*, *50*(23), 13000–13011. doi:10.1021/ie200241x

San, J. Y., & Lin, W.-M. (2008). Comparison among three adsorption pairs for using as the working substances in a multi-bed adsorption heat pump. *Applied Thermal Engineering*, *28*(8-9), 988–997. doi:10.1016/j.applthermaleng.2007.06.026

Sapienza, A., Santamaria, S., Frazzica, A., Freni, A., & Aristov, Y. I. (2014). Dynamic study of adsorbers by a new gravimetric version of the Large Temperature Jump method. *Applied Energy*, *113*, 1244–1251. doi:10.1016/j.apenergy.2013.09.005

Seville, J., & Wu, C. (2016). *Particle Technology and Engineering: An Engineer's Guide to Particles and Powders: Fundamentals and Computational Approaches*. Butterworth Heinemann.

Shelton, S. V., Wepfer, W. J., & Miles, D. J. (1990). Ramp Wave Analysis of the Solid/ Vapor Heat Pump. *Journal of Energy Resources Technology*, *112*(1), 69–78. doi:10.1115/1.2905715

Shi, B. (2015). *Development of an MOF based Adsorption air conditioning System for automotive Application* (PhD thesis). University of Birmingham.

Shmroukh, A. N., Ali, A. H. H., & Abel-Rahman, A. K. (2013). Adsorption Refrigeration Working Pairs: The state-of-the-art in the application. *International Journal of Chemical Materials Science and Engineering*, *7*(11b), 793–805.

Shmroukh, A.N., Ali, A.H.H., Abel-Rahman, A.K., & Ookwara, S. (2015). Experimental Investigation on Adsorption Capacity of a Variety of Activated Carbon/Refrigerant Pairs. *Int. Journal of Engineering Research and Applications, 5*(4), 66-76.

Shmroukh, A. N., Ali, A. H. H., & Ookawara, S. (2015). Adsorption working pairs for adsorption cooling chillers: A review based on adsorption capacity and environmental impact. *Renewable & Sustainable Energy Reviews*, *50*, 445–456. doi:10.1016/j.rser.2015.05.035

Siegfried, K. D. B. (1999). Experiments on solar adsorption refrigeration using zeolite and water. University of Applied Sciences Luebeck.

Solair. (n.d.). *Solar Cooling; an Intelligent Option.* Retrieved from http://www.solair-project.eu/fileadmin/SOLAIR_uploads/DOCS/Guidelines/SOLAIR_Brochure_EN.pdf

Spahis, N., Addoun, A., & Mahmoudi, H. (2007). Study on solar adsorption refrigeration cycle utilizing activated carbon prepared from olive stones. *Revue des Energies Renouvelables*, *10*(3), 415–420.

Tamainot-Telto, Z., & Critoph, R.E. (1996). Adsorption refrigerator using monolithic carbon-ammonia pair. *International Journal of Refrigeration*, *20*(2), 146–155.

Tamainot-Telto, Z., Metcalf, S. J., & Critoph, R. E. (2009). Novel compact sorption generators for car air conditioning. *International Journal of Refrigeration*, *32*(4), 727–733. doi:10.1016/j.ijrefrig.2008.11.010

Tseng, R. L., & Wu, F. C. (2009). Analyzing a liquid–solid phase counter current two- and three-stage adsorption process with the Freundlich equation. *Journal of Hazardous Materials*, *162*(1), 237–248. doi:10.1016/j.jhazmat.2008.05.031 PMID:18556116

Tso, C. Y., & Chao, C. Y. H. (2012). Activated carbon, silica-gel and calcium chloride composite adsorbents for energy efficient solar adsorption cooling and dehumidification systems. *International Journal of Refrigeration*, *35*(6), 1626–1638. doi:10.1016/j.ijrefrig.2012.05.007

Tso, C. Y., Chao, C. Y. H., & Fu, S. C. (2012). Performance analysis of a waste heat driven activated carbon based composite adsorbent – Water adsorption chiller using simulation model. *International Journal of Heat and Mass Transfer*, *55*(25-26), 7596–7610. doi:10.1016/j.ijheatmasstransfer.2012.07.064

Uddin, K., El-Sharkawy, I. I., Miyazaki, T., Saha, B. B., Koyama, S., Kil, H. S., & Yoon, S. H. et al. (2014). Adsorption characteristics of ethanol onto functional activated carbons with controlled oxygen content. *Applied Thermal Engineering*, *72*(2), 211–218. doi:10.1016/j.applthermaleng.2014.03.062

Ulku, S. (1986). Adsorption heat pumps. *Heat Recovery Systems*, *6*(4), 277–284. doi:10.1016/0198-7593(86)90113-X

Ullah, K. R., Saidur, R., Ping, H. W., Akikur, R. K., & Shuvo, N. H. (2013). A review of solar thermal refrigeration and cooling methods. *Renewable & Sustainable Energy Reviews*, *24*, 499–513. doi:10.1016/j.rser.2013.03.024

Umair, M., Akisawa, A., & Ueda, Y. (2014). Simulation Study of Continuous Solar Adsorption Refrigeration System Driven by Compound Parabolic Concentrator. *The Open Renewable Energy Journal, 7*(1), 1–12. doi:10.2174/1876387101407010001

Wang, K., Wu, J. Y., Wang, R. Z., & Wang, L. W. (2006a). Composite adsorbent of CaCl2 and expanded graphite for adsorption ice maker on fishing boats. *International Journal of Refrigeration, 29*(2), 199–210. doi:10.1016/j.ijrefrig.2005.06.004

Wang, K., Wu, J. Y., Wang, R. Z., & Wang, L. W. (2006b). Effective thermal conductivity of expanded graphite–CaCl2 composite adsorbent for chemical adsorption chillers. *Energy Conversion and Management, 47*(13–14), 1902–1912. doi:10.1016/j.enconman.2005.09.005

Wang, L. W., Metcalf, S. J., Critoph, R. E., Thorpe, R., & Tamainot-Telto, Z.. (2012). Development of thermal conductive consolidated activated carbon for adsorption refrigeration. *Carbon, 50*(3), 977–986. doi:10.1016/j.carbon.2011.09.061

Wang, L. W., Metcalf, S. J., Wang, R., Wang, L., & Wu, J. (2014). *Adsorption Refrigeration Technology*. Wiley. doi:10.1002/9781118197448

Wang, L. W., Tamainot-Telto, Z., Thrope, R., Critoph, R. E., & Metcalf, S. J. (2011). Study of thermal conductivity permeability and adsorption performance of solidified composite activated carbon adsorbent for refrigeration. *Renewable Energy, 36*(8), 2062–2066. doi:10.1016/j.renene.2011.01.005

Wang, L. W., Wang, R. Z., Lu, Z. S., Chen, C. J., Wang, K., & Wu, J. Y. (2006). The performance of two adsorption ice making test units using activated carbon and a carbon composite as adsorbents. *Carbon, 44*(13), 2671–2680. doi:10.1016/j.carbon.2006.04.013

Wang, L. W., Wang, R. Z., & Oliveira, R. G. (2009a). A review on adsorption working pairs for refrigeration. *Renewable & Sustainable Energy Reviews, 13*(3), 518–534. doi:10.1016/j.rser.2007.12.002

Wang, L. W., Wu, J. Y., Wang, R. Z., Xu, Y. X., Wang, S. G., & Li, X. R. (2003). Study of the performance of activated carbon–methanol adsorption systems concerning heat and mass transfer. *Applied Thermal Engineering, 23*(13), 1605–1617. doi:10.1016/S1359-4311(03)00104-2

Wang, R. (2000). Performance improvement of adsorption heat pump by heat and mass recovery operations. *IV Minsk International Seminar "Heat Pipes, Heat Pumps, Refrigerators"*.

Wang, R. (2001). Performance improvement of adsorption cooling by heat and mass recovery operation. *International Journal of Refrigeration, 24*(7), 602–611. doi:10.1016/S0140-7007(01)00004-4

Wang, R. Z., & Oliveira, R. G. (2005). Adsorption refrigeration- an efficient way to make good use of waste heat and solar energy.*International Sorption Heat Pump Conference*.

Wang, R. Z., Wu, J. Y., Xu, Y. X., & Wang, W. (2001). Performance researches and improvements on heat regenerative adsorption refrigerator and heat pump. *Energy Conversion and Management, 42*(2), 233–249. doi:10.1016/S0196-8904(99)00189-2

Wu, H., Wan, S., & Zhu, D. (2007). Effects of impregnating variables on dynamic sorption characteristics and storage properties of composite sorbnet for solar heat storage. *Solar Energy, 81*(7), 864–871. doi:10.1016/j.solener.2006.11.013

Yang, R. T. (2003). *Adsorbents: Fundamentals and applications*. John Wiley & Sons. doi:10.1002/047144409X

Yeo, T. H. C., Tan, I. A. W., & Abdullah, M. O. (2012). Development of adsorption air-conditioning technology using modified activated carbon – A review. *Renewable & Sustainable Energy Reviews*, *16*(5), 3355–3363. doi:10.1016/j.rser.2012.02.073

Zhao, Y. (2011). *Study of activated carbon/methanol adsorption refrigeration tube and system integration* (Master Science thesis). Adelaide University.

Zhao, Y. J., Wang, L. W., Wan, R. Z., Ma, K. Q., & Jiang, L. (2013). Study on consolidated activated carbon: Choice of optimal adsorbent for refrigeration application. *International Journal of Heat and Mass Transfer*, *67*, 867–876. doi:10.1016/j.ijheatmasstransfer.2013.08.052

Zhao, Y. J., Wang, R. Z., Wang, L. W., & Yu, N. (2014). Development of highly conductive KNO3/NaNO3 composite for TES (thermal energy storage). *Energy*, *70*, 272–27. doi:10.1016/j.energy.2014.03.127

Zhong, Y., Critoph, R. E., & Thorpe, R. (2006). Evaluation of the performance of solid sorption refrigeration systems using carbon dioxide as refrigerant. *Applied Thermal Engineering*, *26*(16), 1807–1811. doi:10.1016/j.applthermaleng.2006.02.006

Zhong, Y., Critoph, R. E., Thorpe, R. N., & Tamainot-Telto, Z. (2009). Dynamics of BaCl2–NH3 adsorption pair. *Applied Thermal Engineering*, *29*(5-6), 1180–1186. doi:10.1016/j.applthermaleng.2008.06.015

KEY TERMS AND DEFINITIONS

Calorimetric Adsorption Technique: A method of measuring adsorption uptake using direct measurements of heat released from adsorption process.

Consolidated Adsorbent: The powder/grain adsorbent material mixed with binders are exposed to high pressures to form consolidated adsorbent. The process could be done with or without heat treatment.

Magnetostrictive Level Sensor: Sensor probe with magnetic float(s) is inserted into a tank and as the float rises or falls with the fluid, the transmitter provides level output.

Permeability: A property of adsorbent material defines the ability of gas to diffuse through the adsorbent. High permeability means uniform pressure distribution of gas through adsorbent bed.

Porosity: It is a measure of the void (i.e., "empty") spaces in a material, and is a fraction of the volume of voids over the total volume.

Swelling and Agglomeration: Instability of adsorbent samples during adsorption process cause the material to expand and cracked.

Chapter 10
Analysis of Lipids Produced by Microalgae Isolated from the Area around Okinawa, Japan

Shinya Ikematsu
National Institute of Technology – Okinawa, Japan

Ipputa Tada
National Institute of Technology – Okinawa, Japan

Yasuma Nagasaki
National Institute of Technology – Okinawa, Japan

ABSTRACT

Petroleum reserves have been decreasing in recent years and microalgae are attractive as a potential source of new biomass petroleum. Microalgae are unicellar microscopic algae and most species microalgae produce lipids. In particular, Botryococcus braunii produces large amount of lipids found with nearly 70% on the basis of the dry weight. This chapter reviews high lipid-producing microalgae found from Okinawa area around National Institute of Technology, Okinawa College (NIT, Okinawa). The microalgae collected were isolated on an AF-6 agar plates, and incubated in AF-6 medium. The fatty acids were extracted from the algae, converted into fatty acid methyl esters, and analysed by GC/MS. As a result, two microalgae strains were identified that the produced fatty acids was loaded in the algae with nearly 20% in the dry weight base. In addition, these two microalgae strains produced palmitic acid as nearly 40% of the total produced lipids. Therefore, the two microalga strains isolated are potentially and highly efficient for the organisms applied for the production of biodiesel fuel.

DOI: 10.4018/978-1-5225-1971-3.ch010

INTRODUCTION

Purpose and Background

Fossil fuels play an extraordinary role in the modern world; however, such fuels are being rapidly consumed as developing countries with further development of their industrial production and also as advanced countries with their expanded technologies. According to the estimation made in 2012, the life of an oil well for drilling will be about 54.2 years (Serive, Kaas, Bérard, Pasquet, Picot, & Cadoret, 2012). Recently, in response to the decline in oil availability, biological production of fossil oils is well paid attention. Among them, microalgae, particularly producing a large amount of lipid components, have attracted as a new energy source alternative to fossil fuels (Agency for Natural Resources and Energy, 2014; Sheehan, Dunahay, Benemann, & Roessler, 1998). Therefore, the present chapter aimed to collect and isolate microalgae producing industrially valuable lipids through sampling several water from a natural environment around National Institute of Technology, Okinawa College (hereinafter referred to as "NIT, Okinawa"), where the northern part of Okinawa's main island is called "Yambaru". This area has abundant exotic biological resources derived from the subtropical climate. For the purpose of discovery of new species of microalgae having lipid component, it was hypothesized that microalgae found in this area would be useful as a biomass. In the present chapter, our efforts in NIT, Okinawa group were described in analysis of lipids produced by microalgae harvested and isolated from the area around Okinawa, Japan.

Microalgae for Lipid Producing

Microalgae comprise a group of very small algae that are invisible to the naked eye. Some microalgae accumulate lipids in their cells. It has been estimated that some algae strains can provide about 21% of the annual consumption of petroleum in Japan (Ministry of Agriculture, Forestry and Fisheries, Japan, 2014), if this cultured in an area of 396,000 ha, which corresponds to the entire area of land not cultivated in Japan (Sheehan, Dunahay, Benemann & Roessler, 1998). Among microalgae, it is known that species of both genus *Botryococcus* and *Aurantiochitrium* have high lipid production efficiency. Since these microalgae have been proposed as potential sources of an algal biomass (Agency for Natural Resources and Energy, 2014; Sheehan, Dunahay, Benemann & Roessler, 1998; Oya, Kanno, Watanabe, Tamura, Nakagawa & Tomishige, 2015), furthermore, then many groups are searching for new lipid-producing all over the world. Therefore, it is expected that additional valuable strains are found for lipid-producing microalgae.

In lipids produced by microalgae, triacylglycerol, a neutral lipid, consists of three long-chain fatty acids bound to one glycerol through an ester bond (Figure 1 (a)). Triacylglycerol is a lipid component, which exists most abundantly within a living organism and functions as a storage form of metabolic energy (BP Statistical Review of World Energy, 2015). There are mainly two types of fatty acids consisting of triacylglycerol chemical structures with saturated fatty acids and unsaturated fatty acids (Yoshimoto, Kaneko & Onodera, 2009).

Saturated and Unsaturated Fatty Acids

Saturated fatty acids form a group of fatty acids that contain no double bonds in their hydrocarbon chain structure (Figure 1(b)). The carbon number of most of natural saturated fatty acids is 4-30 (BP Statistical Review of World Energy, 2015). Saturated fatty acids are generally solid at ambient temperatures. On the other hand, unsaturated fatty acids form a group of fatty acids that contain double carbon bonds in the hydrocarbon chain structure (Figure 1(c)). Unsaturated fatty acids with a single double bond are called monounsaturated fatty acids (monoenoic fatty acids), and unsaturated fatty acids with two or more double bonds are called polyunsaturated fatty acids (polyenoic fatty acids) (BP Statistical Review of World Energy, 2015).

METHODS

Sampling of Microalgae from the Field

Microalgae inhabit a broad range of both land water and seawater. *Botryococcus braunii*, which is projected as a future energy source, is often found in a land water environment (National Institute for Environmental Studies, 2004; Oya, Kanno, Watanabe, Tamura, Nakagawa, & Tomishige, 2015). New several microalgae to prepare some oil into were discovered in Okinawa. On the basis of this information, it was conducted that the sampling from land water environments, was carried out in including mainly rivers and puddles of water at Nago area in Okinawa island for mainly the locations around NIT, Okinawa, where is addressed at (Henoko district, Nago-shi to Gesaji district, Higashi-son in Okinawa Prefecture). The sampling locations are indicated as shown in Figure 2.

Figure 1. Structural formulae of (a) triacylglycerol, (b) stearic acid, and (c) oleic acid (n-9)

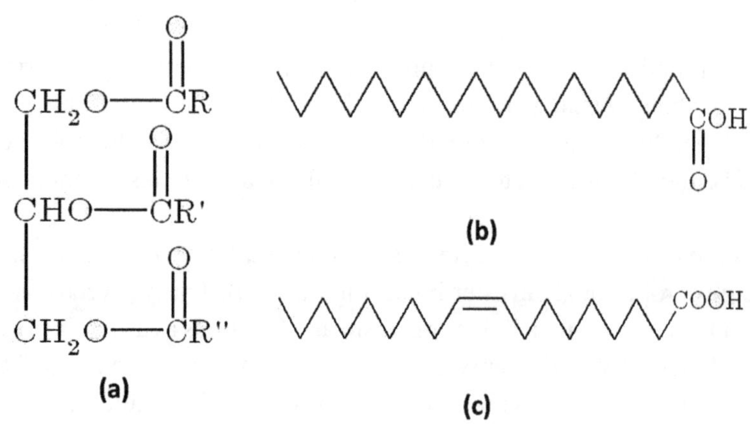

Figure 2. Map showing sampling locations
Numbers on the map correspond to sampling locations of (1) to (5) as shown below:(1) Wansaka Oura Park, (2) Kushi post office, (3) Middle course of Teima River, (4) Puddles in Teniya and (5) Puddles in Arume.

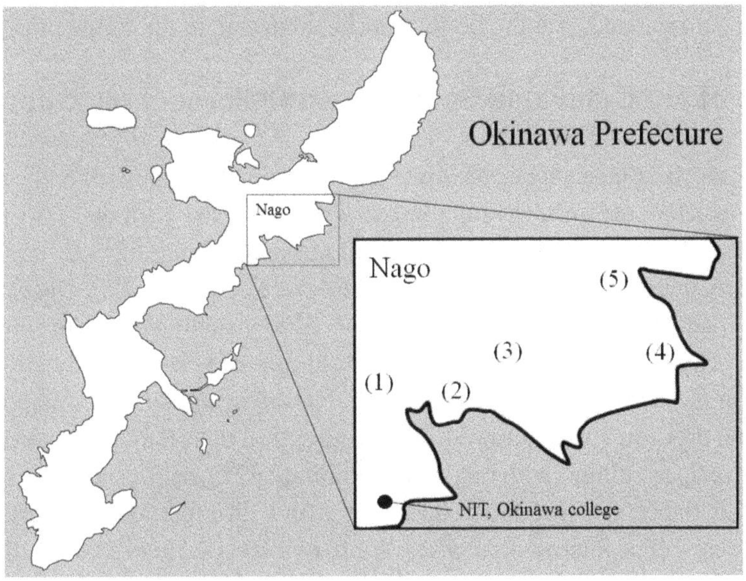

Isolation and Culture of Microalgae from Water Samples and Their Sample Preparation for the Lipid Analysis

It was expected that water samples, which were randomly collected from natural water, might contain a wide variety of algae. Therefore, microalgae sampled were firstly isolated from the water samples. An agar dilution method was used for the isolation of microalgae in this study (Mutanda, Ramesh, Karthikeyan, Kumari, Anandraj, & Bux, 2011). AF-6 medium was chosen as a culture medium, and used in the agar plates by adding 1% agar as the final concentration (Kato, 1982).

Collected water samples were streaked on the AF-6 agar plates with sterile cotton swabs. The plates were cultured under fluorescent lights (light period, 8 am-10 pm) at 25°C. After two weeks, greenish yellow and green colonies were observed on the plates. The greenish yellow colonies were picked with bamboo skewers, transferred and suspended into screw-top test tubes containing 10 ml of the AF-6 medium. Sampling locations and corresponding water sample numbers are shown in Table 1. As seen, total of 25 algae samples were isolated from the water samples collected from the field.

Table 1. Sampling locations and water sample numerical assignments

Location	Sample #	Location	Sample #
(4) Wansaka Oura Park	1 - 7	(4) Puddles in Teniya	18 - 21
(5) Kushi post office	8 - 11	(5) Puddles in Arume	23 - 25
(6) Middle course of Teima River	12 -16		

For lipid analysis, 5 mL of each cultured microalgae from the screw-top test tubes was added to 15 mL centrifuge tubes containing 2 mL of AF-6 medium and cultured under fluorescent lights (light period, 8 am - 10 pm) at 25°C for 6 or 7 days. Though the cultivation, the tubes were stirred every morning because the algae tended to clump together at the bottom of the tubes due to the shape of the centrifuge tube.

Lipid analyses of the Cultivated Samples with Microscopic Observation

Lipid analyses were carried out mainly by measuring microscopic observation for the microalgae sampled at 1)-5) area. The isolated and cultured microalgae were observed with an Axio Imager Z2 (ZEISS) microscope with both differential interference and fluorescence capabilities, to capture morphological features of the microalgae. At the same time, the cells were stained with a fluorescent dye, nile red, which stains lipids. then fluorescence microscope enabled to visually detect the cultured algae and also the producing lipid. Here, nile red is a dye with an excitation wavelength at around 450-600 nm and a fluorescence peak at 520-700 nm (Mutanda, Ramesh, Karthikeyan, Kumari, Anandraj, & Bux, 2011).

The experimental procedure was followed. For staining, 5 μl of a 50 μg/ml stock nile red solution was added to 5 μl of microalgae culture (with final concentration of 25 μg/ml), and the mixture was incubated for 10 minutes. Then, the cells were observed at 400x with a differential interference and fluorescence microscope. For fluorescence microscopic observation, two sets of filters, B excitation filter (excitation wavelength 470 nm, fluorescence wavelength 525 nm) and G excitation filter (excitation wavelength 470 nm, fluorescence wavelength 525 nm), were used. In addition, the influence of intrinsic fluorescence from chlorophyll was examined as a control by observing the microalgae without nile red staining.

Microscopic observation was conducted using a differential interference microscope. Images obtained by the observation are shown in Figure 3. Scale bars represent 50 μm.

The samples harvested at 1) Wansaka Oura Park for samples 2-7 and at 3) Middle course of Teima River for the samples 13-16 were spherical with fine cells having about 1μm diameter. In the cases of 2) at Kushi post office for the samples 9-11, the spherical diameter in the observed cells was somewhat larger than those of 1). In the images of sample 17-19 for the 4) area at Puddles in Teniya, sample 19 showed the presence of the filamentous microalgae and also looked like mixture of larger and fine spheric sizes of cells. At 5) Puddles in Arume similar tendency was observed with that of the 4).

Results of the fluorescence microscopy to the corresponding microscopic views in Figure 3 are shown in Figure 4. Here, lipids stained with nile red produce yellow fluorescence and the microalgae without nile red stain emit red fluorescence (Phillip & Stanly, 1985). This meant that the intrinsic fluorescence from chlorophyll was observed, even if the nile red stain was not used. Through the observation fluorescence views of the microalgae harvested in the 1)-5) areas, it was indicated that samples 3,4 and 6 and for the 1) and samples 9 and 11 for the 2) included much lipids in the cells. In the samples 19-24 at the 4) and 5) the red images suggested that no lipid fluorescence was emited from the cell or very less contents in the cells.

Figure 3. Differential interference microscopic images of cultured microalgae

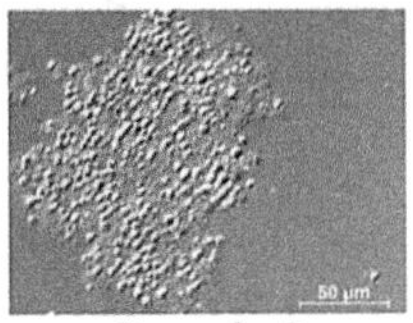
Sample 2
Spherical fine cells

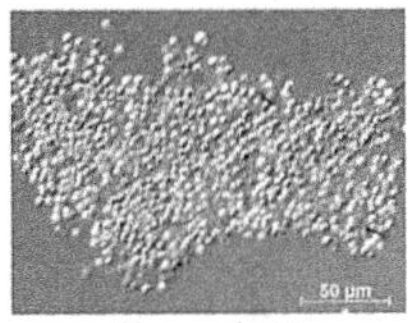
Sample 3
Spherical fine cells

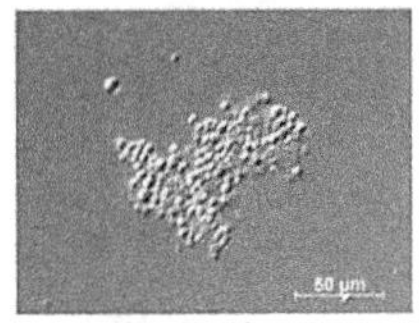
Sample 4
Spherical fine cells

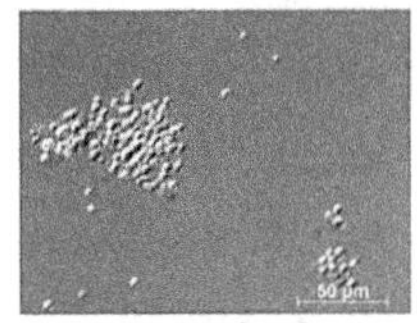
Sample 6
Spherical fine cells

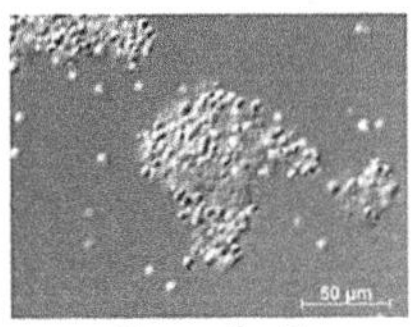
Sample 7
Spherical fine cells

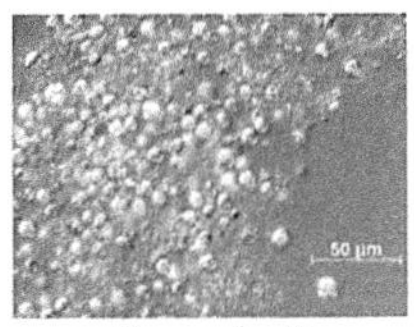
Sample 9
Spherical cells

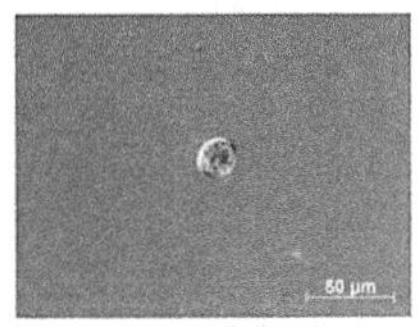
Sample 10
Spherical. Relatively large sized cells

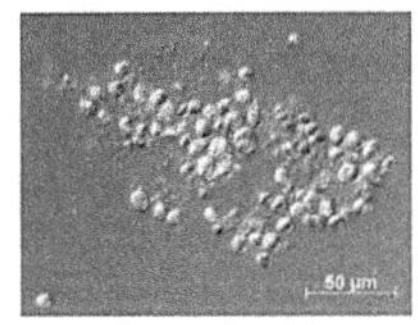
Sample 11
Spherical cells

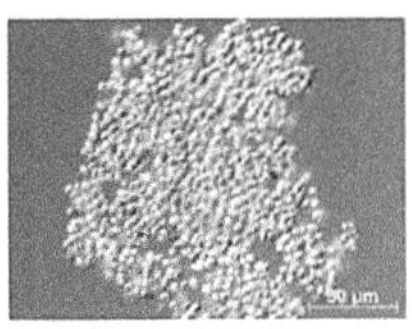
Sample 13
Spherical fine cells

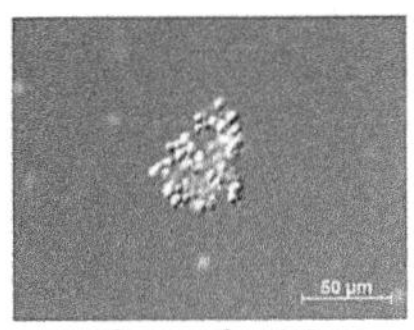
Sample 15
Spherical fine cells

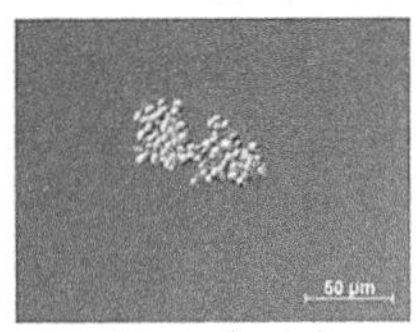
Sample 16
Spherical fine cells

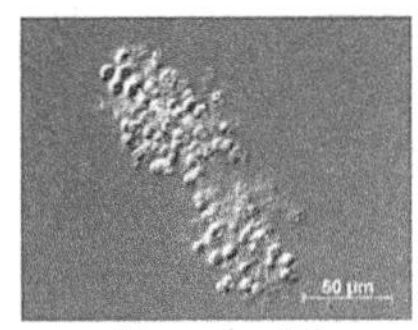
Sample 17
Spherical cells

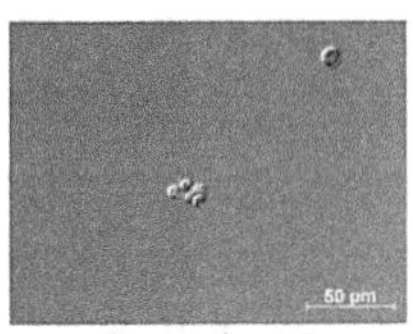
Sample 18
Spherical fine cells

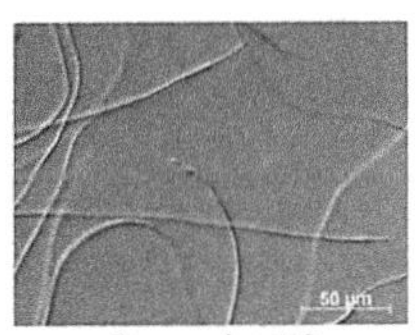
Sample 19
Filamentous microalgae

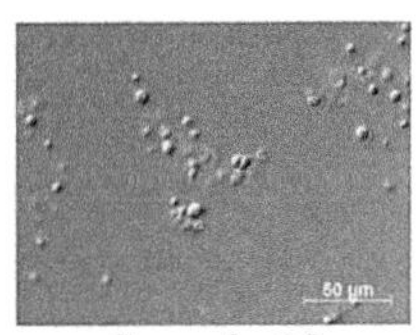
Sample 20
Spherical fine cells

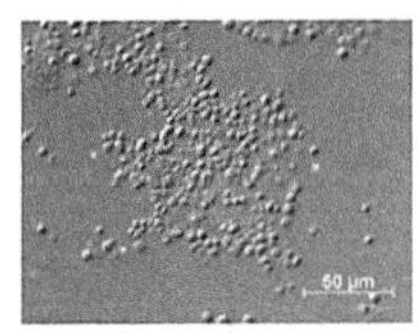
Sample 21
Spherical fine cells

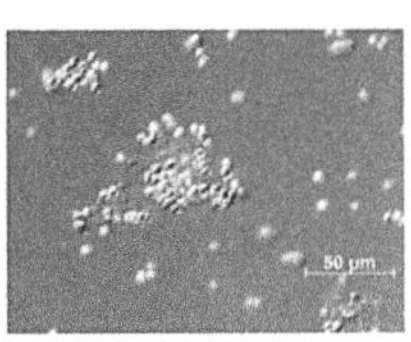
Sample 22
Spherical fine cells

Lipid Analyses of the Cultivated Samples with Gc/Ms for the Lipid Components Produced by Cultured Microalgae

Preparation of the Gc/Ms Samples from the Crude Microalgae Extract

For GC/MS analysis of fatty acids derived from microalgae, crude extract was prepared by an extraction method using a *tert*-butyl methyl ether and methanol solution. After this, the lipid components were methyl-esterified and then applied for GC/MS determination.

First, microalgae cultures were centrifuged with 13,000 rpm for 20 minutes to remove the culture medium. Then, the microalgae were freeze-dried for four hours using a lyophilizer. The freeze-dried microalgae were transferred into 4 ml pre-weighed glass vials, and the dry weight of the microalgae was calculated by subtracting the weight of the glass vial from the total weight of the vial plus the dried sample. Then, 1.6 ml of *tert*-butyl methyl ether and 0.8 ml of methanol were added to each vial containing the freeze-dried microalgae, sonicated for 30 minutes, and left standing overnight. The resultant 1.3 ml of the crude extract of the cultured microalgae was pipetted into 1.5 ml tubes.

The methyl esterification of the fatty acids in the crude extract was carried out as followed in details.

First, 1.3 ml of the crude extracts dispensed in 1.5 ml tubes was transferred into weighed 4 ml glass vials, and each sample was nitrogen dried. Fatty acids in the crude extracts were methyl esterified using a fatty acid methylation (Nakarai e-kit, 2015) according to the manufacturer's protocol to increase the volatility of the fatty acids in the crude extracts. Then, 100 ml of 7-heptadecanoic acid (0.5 μmol/ml), which is an internal standard, 0.5 ml of methylating agent, and 0.5 ml of methylating agent were added to the nitrogen-dried samples. The mixture was heated with a heat block at 37°C for an hour to saponify the fatty acids. Then, 0.5 ml of methylating agent was added to the mixture, and incubated at 37°C for 20 minutes for methylation of the free fatty acids. Then, 1.0 ml of extracting agent was added and the vial contents were mixed with a vortex. After separating into two layers, the upper layer was transferred into a new tube. The collected upper layer was mixed and washed with 1.0 ml of ion-exchanged water, and then the upper layer was collected into another new tube.

The fatty acids that underwent methyl esterification through the process described above were then purified with a methylated fatty acid purification according to above mentioned procedure. More specifically, first, 3 ml of pre-wash solution was injected into and passed through a silica gel cartridge to wash the column. Then, 0.5 ml of the crude extract that was methyl esterified with the fatty acid methylation kit was injected into the silica gel cartridge to absorb fatty acid methyl esters into the silica gel. Then 3 ml of wash solution was injected to elute impurities. Finally, 3 ml of elution solution was injected into the cartridge, and eluted fatty acid methyl esters were collected into a tube yielding purified fatty acid methyl esters.

GC/MS Analysis of Fatty Acid Methyl Esters

To examine the ability of the cultured microalgae to produce lipids, the 25 algae samples were analyzed for fatty acid methyl esters using GC/MS. The conditions used for the GC/MS analysis are followed with column: Agilent J&W Column DB-23, (L X D X B; 30 m X 0.25 mm X 0.25 μm) at column oven temperature of 50 °C, 175 °C (20 °C/min), and 230 °C (4 °C/min), vaporizing chamber temperature at 250 °C, ion source temperature at 200 °C, interface temperature at 250 °C. The pressure was 52.3 kPa, total flow 13.8 ml/min, Column flow rate 0.98 ml/min, linear velocity 36.0 cm/sec, purge flow rate 30 ml/min and split ratio at 10.0.

Furthermore, internal reference controls were employed for the analyses. C16:0, C18:0 (octadecanoic acid), C20:0, C22:0, and C24:0 were used as standard reference materials for saturated fatty acids, and C18:1 (oleic acid), C18:2 (oleic acid), and C18:3 were used as standard reference materials for unsaturated fatty acids. A library of the National Institute of Standard and Technology (NIST) was used for the structural determination by mass spectrometer.

Figure 4. Fluorescence microscopic images of nile red stained cultured microalgae

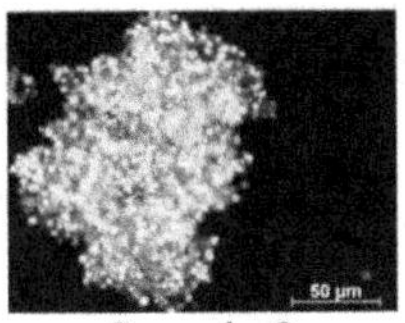

Sample 2
Fine granules of lipid fluorescence

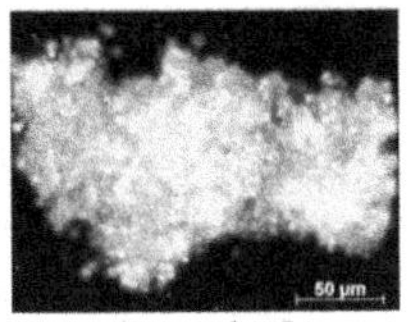

Sample 3
Large granules of lipid fluorescence

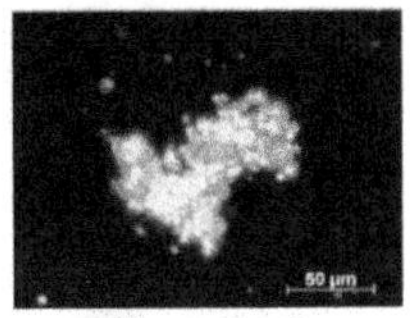

Sample 4
Concentrated fluorescence

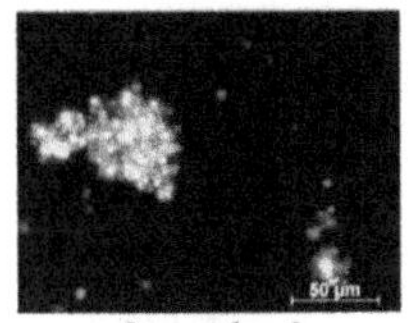

Sample 6
Scattered lipid fluorescence

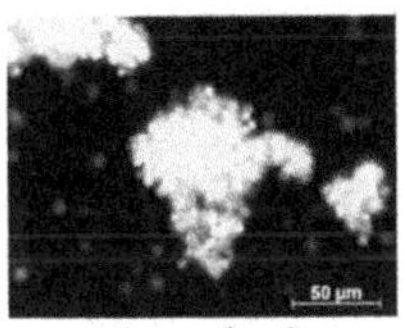

Sample 6
Scattered lipid fluorescence

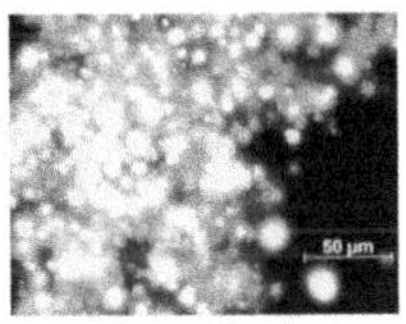

Sample 9
Large fluorescent granules

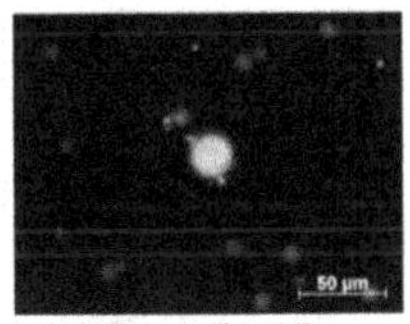

Sample 10
No lipid fluorescence

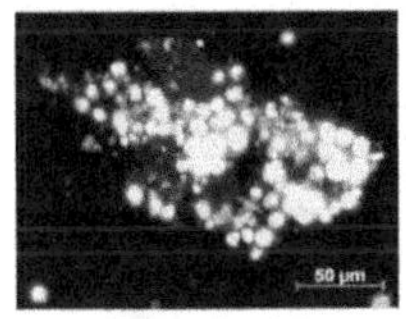

Sample 11
Large fluorescent granules

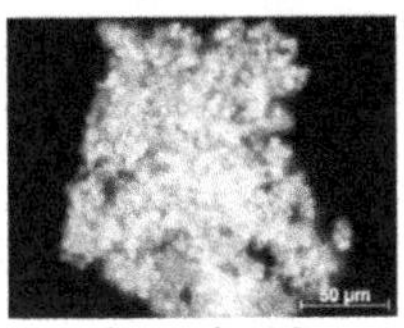

Sample 13
Evenly distributed fluorescence

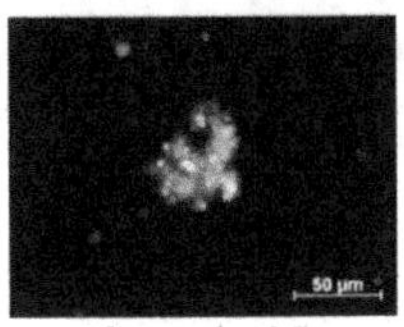

Sample 15
No lipid fluorescence

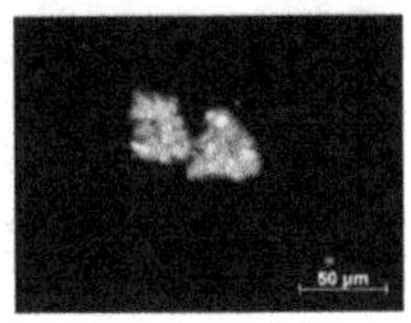

Sample 16
No lipid fluorescence

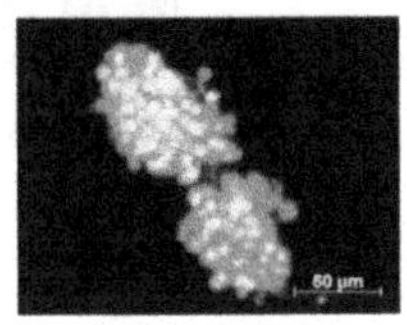

Sample 17
Fine granules of lipid fluorescence

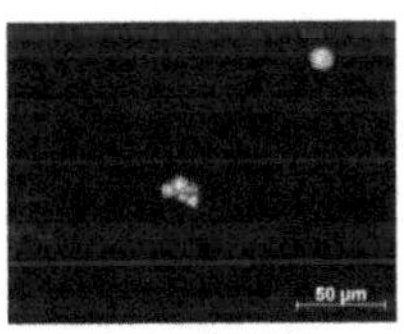

Sample 18
No lipid fluorescence

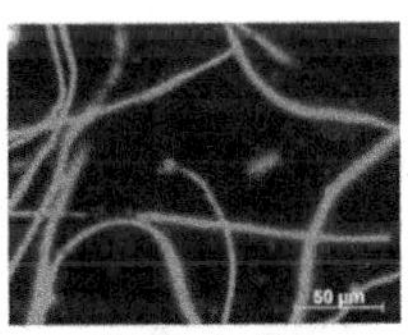

Sample 19
No lipid fluorescence

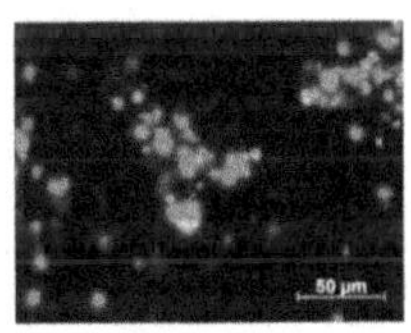

Sample 20
No lipid fluorescence

Sample 21
No lipid fluorescence

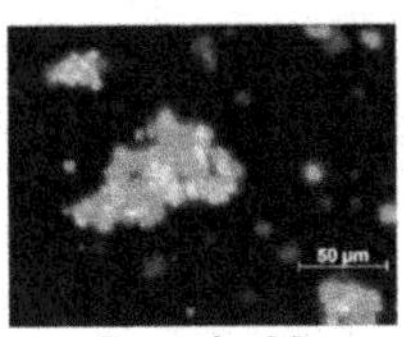

Sample 22
Partial lipid fluorescence

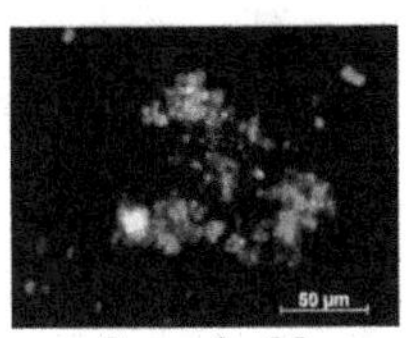

Sample 23
Partial lipid fluorescence

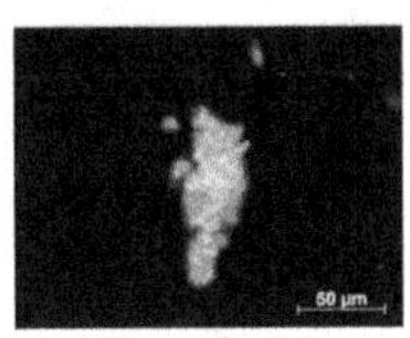

Sample 24
Lipid fluorescence

Figure 5. Quantities of lipid produced per 1 mg freeze-dried microalgae as measured by GC/MS analysis

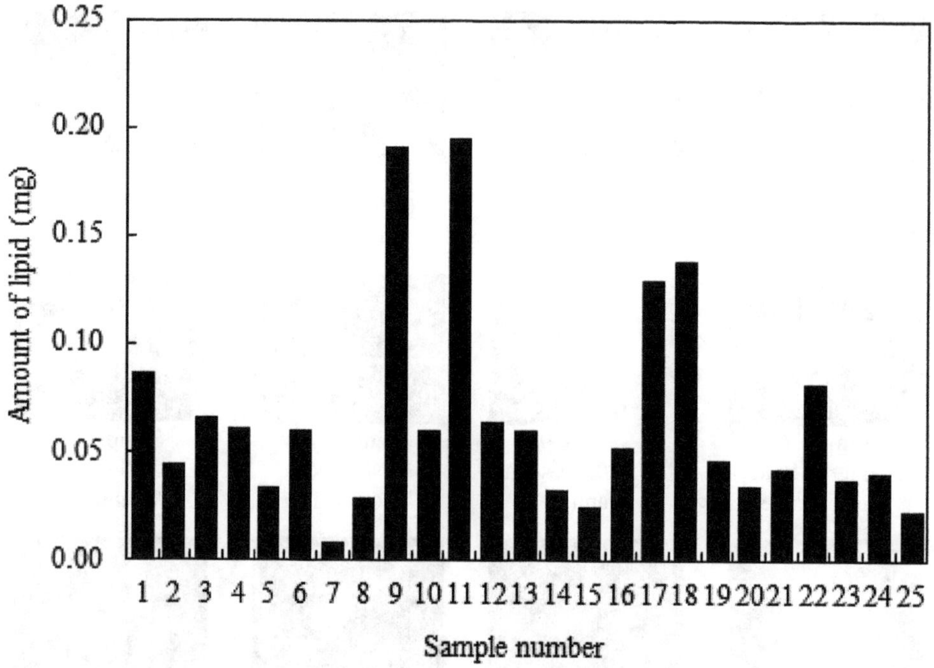

Figure 6. The lipid composition of (a) Sample 9 and (b) Sample 11. C18:0 (oc), octadecanoic acid ; C18:1 (ol), oleic acid ; C18:1 (10), 10-octadecanoic acid; C18:3 (n3), n3-linolenic acid ; C18:3 (n6), n6-linolenic acid

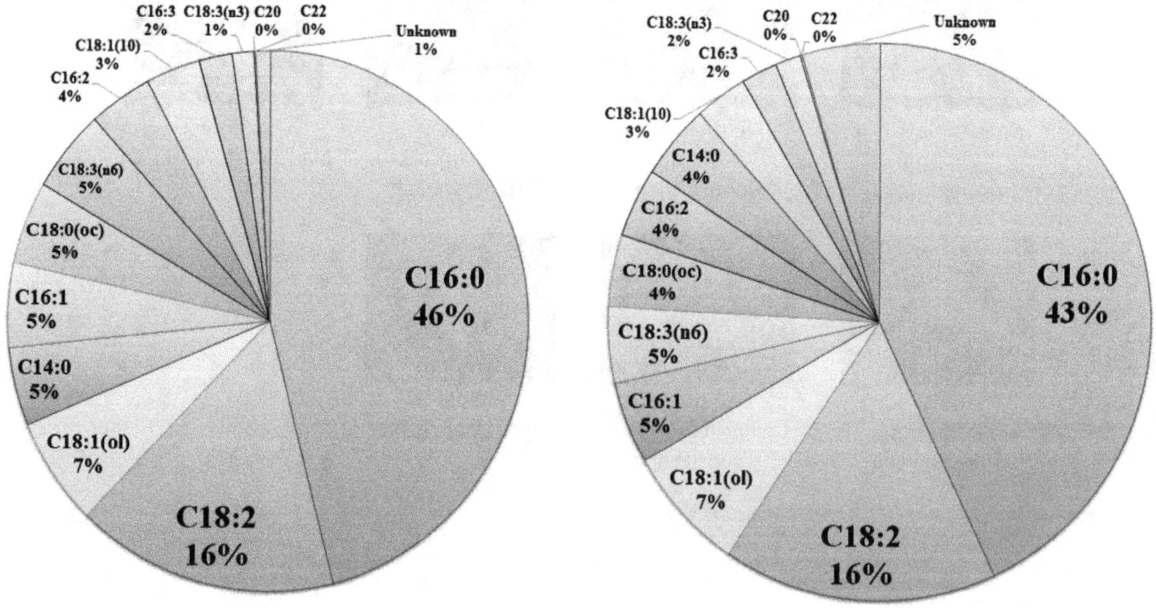

On based on the results of GC/MS analysis, the production quantity of lipid per 1 mg freeze-dried microalgae was calculated. Figure 5 shows the GC/MS analysis for the isolated 25 algae samples. The results of GC/MS analysis indicated that samples 9 and 11 included a high quantity of lipids for the samples at the 2) area of Kushi post office. Then, the composition of the produced lipids was calculated for samples 9 and 11, respectively, as shown in Figure 6. The most abundant lipids contented in the samples 9 and 11 were C16, C18 and C18 (ol). In addition, C16:1 and C18:3 (n6) were included in large quantity. In the sample 9, saturated fatty acids accounted for 56% of the entire fatty acids, and unsaturated fatty acids accounted for 43%. Both samples had more saturated fatty acids than unsaturated fatty acids. Finally, the quantities of each lipid produced in both samples 9 and 11 per 1 mg freeze-dried microalgae were calculated and compared (Figure 7).

EVALUATION RESULTS OF LIPID MICROALGAE ISOLATED FROM OKINAWA

In the present study, 25 microalgae samples were isolated from water samples collected from the locations of Okinawa Nago area. Two of the microalgae sampled at the 1) and 2) areas produced more lipid as compared with the other samples. For the 25 samples of microalgae, the sampling was conducted in shallow waters where the sunlight had full penetration in Nago area, Okinawa. Therefore, the algae were cultured under a limited growth condition with AF-6 medium and the culture was exposed to fluorescent light with a fixed light period length.

Figure 7. Quantities of each lipid produced per 1 mg freeze-dried microalgae by samples 9 and 11

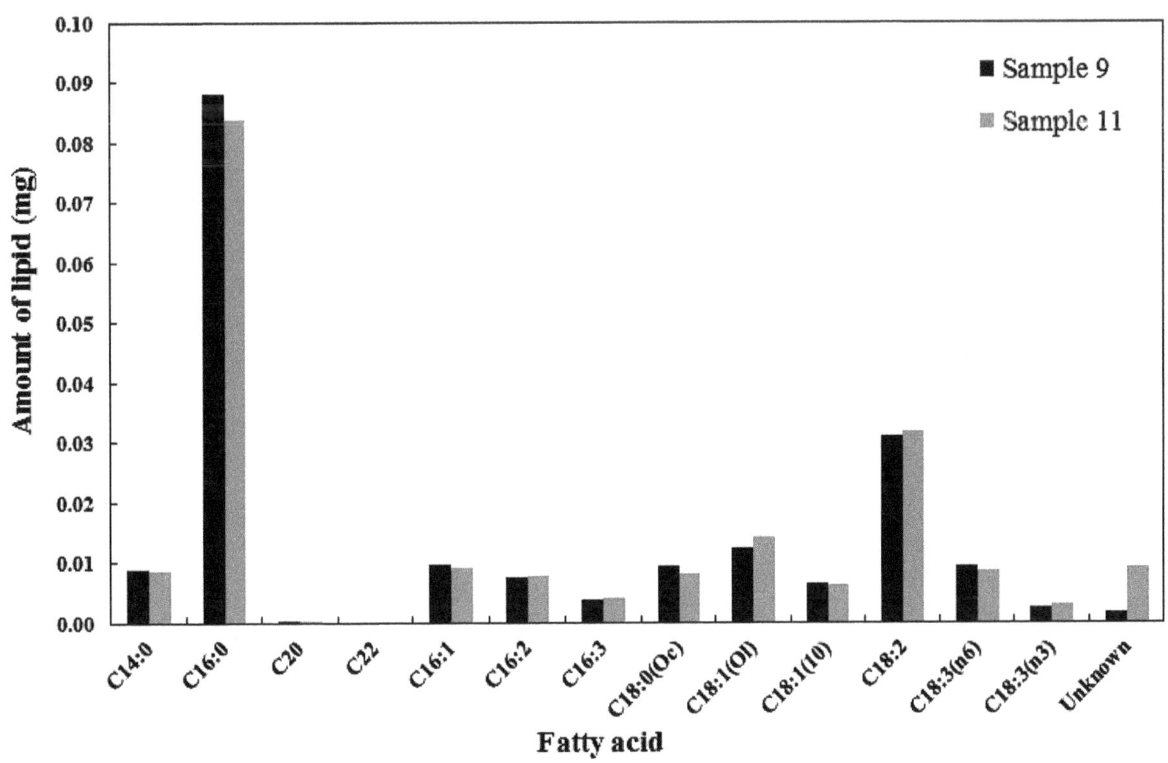

In microscopic differential interference microscopy, it was shown that most of the algae samples were spherical fine cells. Furthermore, it was confirmed that the isolated microalgae had producing lipid components by observing with fluorescence microscope after the nile red staining.

Using GC/MS, the quantity of saturated fatty acids including C16 were determined. It is reported that the fluidity of the microalgae cell membrane is kept in constant by increasing the amount of saturated fatty acids, which are solid at ambient temperatures, and that the fluidity increases with increases in the culture temperature. The 25 samples used for GC/MS analysis were cultured at 25°C but were kept near a window resulting in a temperature increase over 25°C during the daytime. It is seemed that this culture condition may be one of the reasons for the predominance of saturated fatty acids among the fatty acids. It is noted that their lipid components are similar with that of canola oil and palm oil containing with palmitic alkyl group as the major lipid (Lee & Logan, 2011). It was determined by GC/MS analysis that among microalgae isolated as samples 9 and 11 in particular mainly produce C16 component. Therefore, it was speculated that these lipids produced by microalgae might be useful, because of similar to canola oil and palm oil, for application to biodiesel production (Biller & Ross, 2011).

REFERENCES

Agency for Natural Resources and Energy. (2014). *Green Power*. Retrieved from http://www.enecho.meti.go.jp/en/category/others/basic_plan/pdf/4th_strategic_energy_plan.pdf

Biller, P., & Ross, A. B. (2011). Potential yields and properties of oil from the hydrothermal liquefaction of microalgae with different biochemical content. *Bioresource Technology*, *102*(1), 215–225. doi:10.1016/j.biortech.2010.06.028 PMID:20599375

Kato, S. (1982). Laboratory culture and morphology of Colacium vesiculosum Ehrb. (Euglenophyceae). *Japanese Journal of Phycology*, *30*, 63–67.

Lee, D.J., & Logan, B. E. (2011). Preface. *Bioresource Technology*, *102*(1), 1.

Ministry of Agriculture, Forestry and Fisheries, Japan. (2014). *FY2013 Annual Report on Food, Agriculture and Rural Areas in Japan*. Author.

Mutanda, T., Ramesh, D., Karthikeyan, S., Kumari, S., Anandraj, A., & Bux, F. (2011). Bioprospecting for hyper-lipid producing microalgal strains for sustainable biofuel production. *Bioresource Technology*, *102*(1), 57–70. doi:10.1016/j.biortech.2010.06.077 PMID:20624676

Nakarai e-kit. (2015). Retrieved from http://www.nacalai.co.jp/products/recommend/fatty_acid_methylation_kit.html

National Institute for Environmental Studies. (2004). *Collection LIST of STRAINS*. Microalgae and Protozoa.

Oya, S., Kanno, D., Watanabe, H., Tamura, M., Nakagawa, Y., & Tomishige, K. (2015). Catalytic Production on Branched Small Alkanes from Biohydrocarbons. *ChemSusChem*, *8*(15), 2472–2475. doi:10.1002/cssc.201500375 PMID:26097221

Serive, B., Kaas, R., Bérard, J. B., Pasquet, V., Picot, L., & Cadoret, J. P. (2012). Selection and optimisation of a method for efficient metabolites extraction from microalgae. *Bioresource Technology*, *124*, 311–320. doi:10.1016/j.biortech.2012.07.105 PMID:22989659

Sheehan, J., Dunahay, T., Benemann, J., & Roessler, P. (1998). *A Look Back at the U.S. Department of Energy's Aquatic Species Program-Biodiesel from Algae*. National Renewable Energy Laboratory. doi:10.2172/15003040

BP Statistical Review of World Energy. (2015). Retrieved from bp.com/statisticalreview.

Yoshimoto, Y., Kaneko, K., & Onodera, M. (2009). Influence of Fatty Acid Methy Ester Composition on the Combustion Characteristics of Biodiesel Fuels. The Japan Society of Mechanical Engineers, 75(752), 847-854.

KEY TERMS AND DEFINITIONS

Biodiesel: The biodiesel is a generic name of the fuel for diesel engines made from oil derived from a creature by abbreviation of the biodiesel fuel and is one of the biomass energy.

Biomass Petroleum: Biomass is a low-carbon fuel or energy source that comes from plant material such as wood, corn, sugar cane and grasses. Biomass petroleum is oil made using biomass.

Fatty Acid: A fatty acid is univalent carboxylic acid of the long chain hydrocarbon. Generally, we call a thing of the number of carbon which is more than middle chain fatty acid (lower fatty acid), 5-12 things with the thing of 2-4 number of carbon with a long chain fatty acid (higher fatty acid).

Microalgae: The alga refers to bryophyte, pteridophyte, Gymnospermae, the plant except the angio sperm prospering in the land among plants. The alga that a microscope is necessary for the identification of the individual is called a microalga among these algas, and, as for the floating-related thing, all phytoplankton of them is said. The microalga supports the basal production of the hydrosphere and takes an extremely important role in the natural world.

Okinawa: The prefecture where Okinawa assumes Japanese southwest Ryukyu Islands prefecture area.It is the prefecture consisting of 363 islands in the East China Sea and the Pacific. The prefectural seat is Naha-shi of Okinawa Island. In general, it is a climate to be quite considered to be the subtropical zone, and a part belongs to the tropical zone. It is the climate that is warm through the year in Japan, and the differences between gighest temperature and minimum temperature are small.

Yambaru: Yambaru is the area where a lot of mountains of Okinawa, Okinawa Main Island northern part and nature including the forest stay. There is not the clear division, but it include Onna-son and Kin-cho, and north is an approximate range.

Chapter 11
Study on the Use of Biomass Polymer Sheets in Water/Alcohol Pervaporation as a Sustainable Source of Alcohol Energy

Truong Thi Cam Trang
Vietnam National University, Vietnam

ABSTRACT

This chapter presents an overview of the study of cellulose sheets used as a bio-membrane in water/alcohol pervaporation for a variety of purposes in order to offer an environmentally sustainable, renewable, low-energy use and economical alternative. Particular interest is the utilization of readily available biological materials such as cellulose and chitosan to create biomass polymer membranes for the environmentally responsible and sustainable process of reclaiming alcohol from alcohol/water mixtures through the use of cellulose in the pervaporation process. This process is a promising and innovative alternative to the more energy intensive, environmentally destructive methods and materials currently used.

INTRODUCTION

Growing public health and environmental awareness accompany an increasing number of stricter environmental regulations on waste discharge, water purification, air pollution contaminants and the over utilization of natural resources. The ability to use abundant, sustainable and renewable resources in processes and materials related to bio membranes is highly attractive. In particular, the separation of alcohol from an alcohol/water mixture can have important implications in the ability to secure the energy that the reclaimed alcohol can provide, and that may have been lost without this technology. The fuel source provided by the separated alcohol in this ecologically friendly manner can then be utilized in a range of ways, including the automotive, industrial and mechanical fields. Attention has been focused on uses of

DOI: 10.4018/978-1-5225-1971-3.ch011

biopolymers from renewable resources as alternatives to synthetic polymers (Amar, Mohanty, Depan, Tomer & Singh, 2007). Biopolymers produced in nature by living organisms and plants (e.g., microorganisms, plants, crustaceans) have the characteristics of often being biodegradable and non -toxic, being part of the natural bio-cycle and eventually degrading and being reabsorbed into the natural environment. Biopolymers have repeated sequences that vary broadly in chemical composition including a variety of repeating functional groups of such carboxyl, hydroxyl and amino group. This makes them reactive and subject to cross-linking. Therefore, biopolymers having a high molecular weight and containing repeated sequences may have a greater opportunity for chemical interaction with other compounds. Depending on their functional groups, biopolymers can bind metals, organic contaminants, or soil particles and form interpenetrating cross-linking networks with other polymers (Chandra & Rustgi, 1998).

In general, compared to conventional plastics derived from mineral oil, biopolymers have a more diverse chemistry and side chain architecture, giving the material scientist unique possibilities to tailor the properties of the final package (Kumar, Srivastava, Galaev & Mattiasson, 2009). All of these biopolymers are by nature hydrophilic and somewhat crystalline, factors that can cause difficulties, especially in relation to the packaging of moist products. The excess of disulphide cross-linking and thermal degradation of these polymers are also two main problems in their processing and performance. On the other hand, these polymers produce materials with excellent gas barriers (Meyers, 2008). Three interesting materials applicable to this function are whey, gluten and chitosan. The development of their application in food packaging has increased due to the large surpluses of the raw materials which are found in large amounts in the processing of cellulose for paper. Over the last few years, interest in the naturally available class of polymers known as polysaccharides has been increasing rapidly and biopolymers are replacing synthetic polymers in numerous applications (Krochta & Mulder, 1997).

MATERIALS FOR BIOPOLYMER MEMBRANES

Cellulose

Cellulose is the most abundant and major biopolymer on the earth. It is a renewable biopolymer with a long and well-established technological base (Klemm, Fink, & Bohn, 2005). It can be a primary chemical resource of the future as it is renewable, biodegradable, biocompatible and derivatizable. Cellulose is a complex carbohydrate consisting of a chain of linked glucose molecules (Figure 1), a polysaccharide consisting of a linear chain of several hundred to ten thousand β (1→4) linked D-glucose units held together with hydrogen bonds to form the microfibrils that contribute to the strength of cellulose. The configuration of the hydrogen bonds also allows cellulose to be more soluble in both water and aqueous alkaline solutions (Schurz, 1999).

In the past century, cellulose has been produced and used in applications including films and injection-molded products. The development of biomaterial-based cellulose membranes is one of the most exciting fields in relation to many potential applications such as bio-separation, tissue engineering, and food processing. Cellulose membrane and its derivatives have been widely utilized for membranes separation technologies such as dialysis and ultrafiltration and have proved to be important because of the chemical stability, biological compatibility, and high tenacity in the dried state.

Figure 1. Chemical structure of cellulose

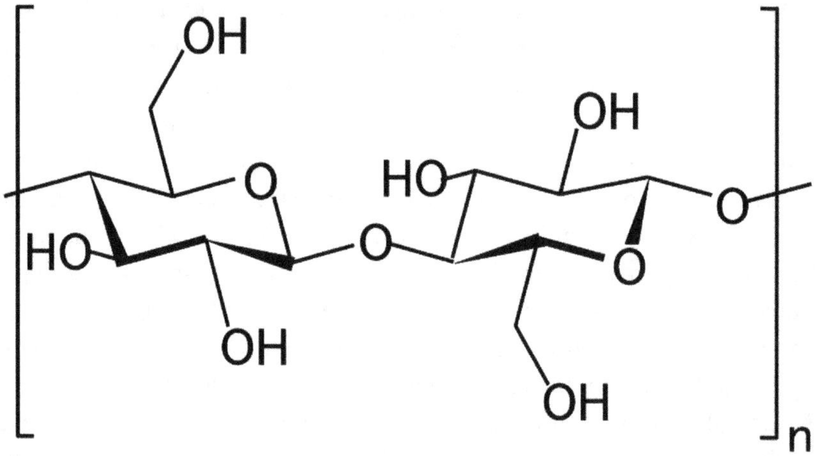

It is widely known that cellulose is the major component of paper, paperboard, and card stock and of textiles made from cotton, linen and other plant fibers. Cellulose can be converted into cellophane, a thin transparent film, and into rayon, an important fiber that has been used for textiles since the beginning of the 20th century. Cellulose is the raw material in the manufacture of nitrocellulose which was historically used in smokeless gunpowder and as the base material for celluloid used for photographic and films until the mid -1930s. In addition, cellulose is used to make water-soluble adhesives and binders such as methyl cellulose and carboxymethyl cellulose which are used in wallpaper paste. Microcrystalline cellulose and powdered cellulose are used as inactive fillers in tablets and as thickeners and stabilizers in processed foods (Guo, Finne-Wistrand, & Albertsson, 2010).

Generally, cellulose constitutes at least 50% of all carbon found in vegetation, due to its importance in the cell walls of plants. Accordingly, it has great potential to be utilized as an alternative and renewable fuel source for the future, when the earth's supplies of fossil fuels decline noticeably and prices are forced upwards. Apart from these economic motivations, other benefits of biomass fuels include the fact that such biomass is composed of carbohydrate polymers, consisting mainly of carbon, hydrogen, oxygen and nitrogen. Thus, such polymers produce very little pollution if the fuel is fully combusted. Therefore, such materials would greatly reduce the current environmental impacts resulting from the use of fossil fuels. Consumers have a lively and vested interest in biopolymers containing cellulose from wood, straw, crop stalks and waste paper, all in plentiful supply, because sustainable biomass materials have the potential to reduce the negative impact that fossil fuels have had on the environment while giving the public an opportunity to contribute positively to a healthier ecosystem. Even with their many advantages such as renewability, low cost, abundance, biodegradability and decreased output of pollutants, natural polymers are being used in the manufacture of consumer goods at a low rate.

Chitosan

Chitosan (CTS) is a commercially available basic polysaccharide (Krajewska, 2000). The basicity of CTS responsible for singular chemical and biological characteristics, biocompatibility, antibacterial properties, heavy metal ion chelation ability, gel-forming properties, hydrophilicity, affinity to proteins and good membrane forming capability.

The chemical structure of chitin and chitosan are shown in Figure 2 (a) and (b). CTS has drawn attention due to its low cost and its activated carbon and high amino and hydroxyl functional group that allow for its high adsorption potential of various aquatic pollutants (Kurita, 2001).

CTS represents an attractive alternative biopolymer relative to other biomaterials because of its physico-chemical characteristics, chemical stability, high reactivity, excellent chelation behavior and high selectivity toward pollutants (Varma, Deshpande, & Kennedy, 2003). Natural chitosan was modified by several methods (either physically or chemically) to enhance the adsorption capacity for various types of pollutants. Derivates from chitosan can be gained using chemical modification which can then be used in various areas (Figure 3). Different shapes of chitosan, e.g. membranes, microspheres, gel beads and films have been prepared and examined for the removal of various pollutants from water and wastewater. CTS has been employed as a pro (Guibal, 2004) missing polymeric matrix for DMFC application considering its low cost, desirable alcohol barrier property and proton conductivity as well as adequate thermal stability after cross-linking (Joel, 2006). Reports are available in which CTS has been used as a membrane forming material in pervaporation for the dehydration of alcohols due to its hydrophilic nature and very high affinity towards water (Uragami, 1989).

The effective and wide application of chitosan based materials or membranes will be possible once the manufacture of chitinous materials according to standardized characteristics accomplished. The distinctive advantages of chitosan include availability, biocompatibility, biodegradability, non-toxicity, antimicrobial properties, gel forming properties, ease of chemical modification, and high affinity to proteins. These characteristics of CTS can be widely exploited in the near future in applications for environmental, bio-technology and energy devices.

A major benefit of the use of bio membranes derived from cellulose and chitosan is the creation of a more environmentally sustainable and renewable process for the reclamation of ethanol from an ethanol/water mixture. This is due to the large supply of both naturally occurring chitosan and the wide availability of cellulose products, both of which can be obtained without undue damage to the environment provided they are harvested responsibly. The reclaimed ethanol can then be utilized and characterized as a more environmentally friendly fuel source for use in such areas as the automotive, mechanical and

Figure 2. Chemical structure of chitin (a) and chitosan (b)

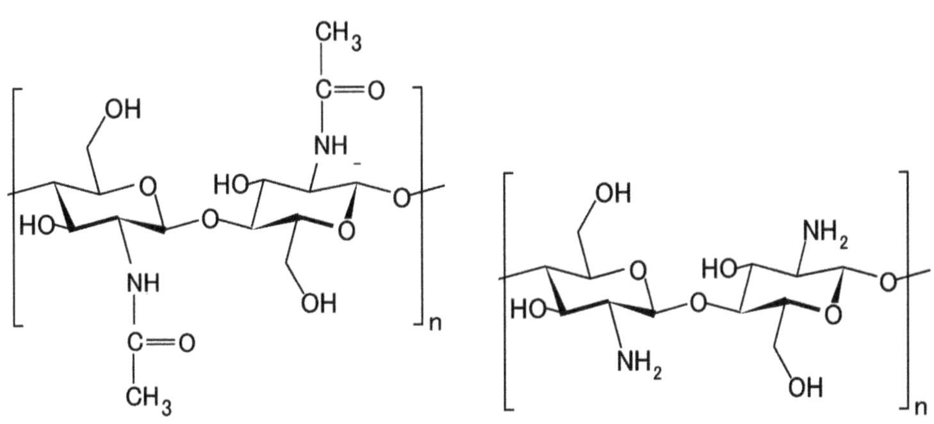

Figure 3. Derivates from chitosan

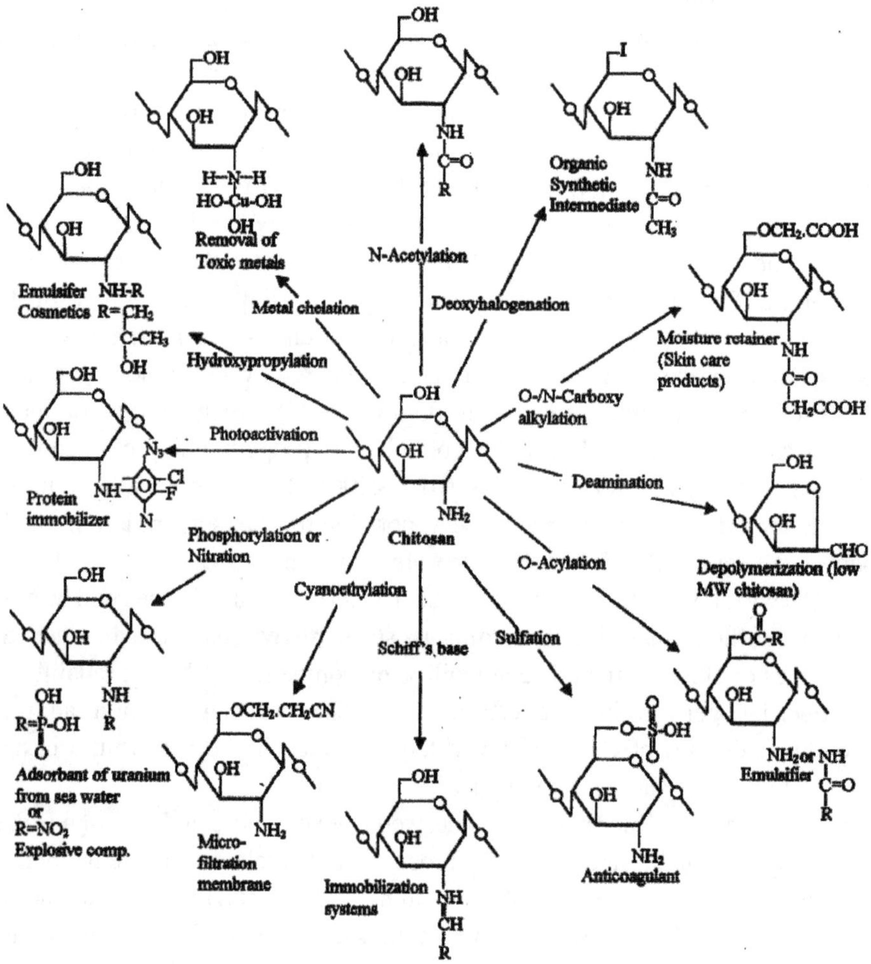

industrial fields. The use of these more sustainable and easily resourced materials to produce the bio membranes used in this alcohol reclamation process is also more economical as they require less heat energy to drive it, thus rendering the process more cost effective (Mukoma, Jooste, & Vosloo 2004; Tripathi & Shahi, 2008).

BIOPOLYMER FILMS AND MEMBRANES MADE OF CELLULOSE

Membranes are a selective barrier between two phases, the term "selective" being inherent to a membrane or a membrane process. A membrane can be thick or thin, its structure can be homogeneous or heterogeneous, transport can be active or passive, natural or synthetic, neutral or charged. Membranes can be made from a large number of different materials and are perhaps the most important biological entity, for it is on membranes that most cellular processes occur. These processes include enzyme activities, active transport of substances for ions, electrons and chemicals across and along membranes, reception and recognition of various stimuli, protein synthesis, nucleic acid synthesis, ATP synthesis, and even membrane synthesis. Membranes are composed mainly of lipids and proteins (in approximately equal amounts) with small amounts of asymmetrically distributed carbohydrates and other prosthetic groups.

Membrane processes were classified by Howell including (Howell, 2002):

- **Cleaner Industrial Processes:** Adsorption, ultrafiltration and electro-ultrafiltration.
- **Energy:** Fuel cell applications
- **Pervaporation:** Separation of organic solvents from their azeotropic mixtures.
- **Water:** Virus–free supply, water reuse and micro-pollutant-free water

Most biopolymer membranes contain appreciable amounts of lipids and proteins (Shahidi, Arachchi & Jeon,1999). These membranes are contributing to cleaner industries, a cleaner environment and safer water and food.

The purpose for developing biopolymer membranes is to offer a commercially valuable replacement for existing synthetic and non -biodegradable products, e.g. in applications for food packaging. Current research in the area covers modified biopolymers and blends, but these approaches still require further research and development before they become competitive with commonly used synthetic polymers (Huang, 1995). An advantage of edible films compared to other polymeric materials is the reduced contribution to environmental pollution, since they are produced from renewable and biodegradable ingredients (Anker, 1996). Edible biopolymers can be consumed with the product which is important since this represents an environmentally ideal package. The film can be used in multi-layer packaging, e.g. with cellulosic material resulting in a totally biodegradable package and can enhance the organoleptic properties of packaged foods, provided that various components are incorporated. Relative to film, membranes can function as carriers for antimicrobial and antioxidant agents and can regulate the transfer of moisture, oxygen, carbon dioxide, lipid, and aroma and flavor compounds in food systems to increase product shelf life and improve quality.

The most common materials used to form edible membranes and coatings are polysaccharides, proteins and lipid compounds. Many of these materials have good membrane-forming properties, as membranes formed from polysaccharides are hydrophilic and provide efficient barriers to oils and lipids, but their moisture barrier properties are poor. Protein based membranes are highly interesting as they confer more potential functional properties. Many lipid compounds such as animal and vegetable fats have been used to make edible membranes and coatings. Lipid membranes have excellent moisture barrier properties but they can cause textural and organoleptical problems due to their oxidation and their waxy taste (Guilbert, Cuq, & Gontard, 1997).

PERVAPORATION PROCESS BY BIOPOLYMER MEMBRANES

Pervaporation (PV) is a method for the separation of mixtures of liquids by partial vaporization through a non-porous or porous membrane. PV is a relatively new membrane separation process that has elements in common with reverse osmosis and membrane gas separation. In PV, the liquid mixture to be separated (feed) is placed in contact with one side of a membrane and the permeated product (permeate) is removed as a low pressure vapor from the other side. The permeate vapor can be condensed and collected or released as desired. The chemical potential gradient across the membrane is the driving force for the mass transport. The commercialization of the PV technique will have a large extent, attributed to the engineering approach of making thin membranes in asymmetric and composite forms.

The separation is based on the selective solution and diffusion, i.e., the physical-chemical interactions between the membrane material and the permeating molecules, not the relative volatility as in distillation. The applications of PV can be classified into three categories:

1. Dehydration of organic solvents,
2. Removal of organic compounds from aqueous solutions, and
3. Separation of anhydrous organic mixtures.

Therefore, PV is commonly considered to be a profitable complement to distillation for the separation of azeo-tropic and close-boiling mixtures, which requires at present the use of energy-intensive processes. PV can be operated at low feed pressures and at ambient temperature or even below this, and no additional chemicals are needed for separation. The early study demonstrated the effectiveness of PV for concentrating dilute protein and enzyme solutions (Farber, 1998).

Unlike reverse osmosis, PV transport is not limited by osmotic pressure because the driving force for mass transfer through the membrane is provided by lowering the chemical potential of the permeate stream on the downstream side, and consequently, the feed pressure is not critical. For example, concentrating ethanol in an aqueous solution is possible to used PV film from 85 to more than 99 wt %. But, an extremely high operating pressure is needed to similarly overcome with the osmotic pressure, if reverse osmosis is used. For a given membrane and a given liquid mixture, both the permeation flux and separation factor in PV are higher than in reverse osmosis (Tanimura, Nakao, & Kimura, 1990).

Since PV involves a phase change of permeating species from the liquid to the vapor state, consequently, energy is needed for the vaporization of the permeation. Thus, from an energy consumption point of view, PV is especially promising when the concentration of the preferentially permeating species in the feed is low. In principle, the heat of vaporization required for permeation can be supplied either in the feed liquid or by a sweeping fluid on the permeate side or directly to the membrane. Direct heating of the membrane has been tested recently (Wnuk & Chmiel,1992). It has not yet been determined which way is more efficient. PV membrane devices are modular in construction. There is no significant economy of scale, so they can be used in either large or small processing capacity. Moreover, it is easy to integrate PV units with other suitable techniques so that the hybrid processes would be more effective than a scheme where the full separation is effected by either technique alone. Compactness, flexibility, simplicity, and versatility are some other strong points of the PV process. In a comprehensive assessment of fluid separation techniques, PV is ranked the third highest among the 31 techniques under evaluation (Bravo, 1986).

For this paragraph on the PV processes on alcohol/water mixtures for methanol, ethanol and isopropanol solutions, the cellulose membranes having pure cellulose frameworks were applied for separation of the mixtures at different concentrations. In observation of the effect of the separation of alcohol aqueous mixtures, the permeation behavior of water and alcohol was mainly examined by means of several methods for with vulcanized cellulose paper (VCP) membranes (Trang & Kobayashi, 2011) and Chitosan/Cellulose (CTS/CL) composite membranes (Trang, Nhung, & Kobayashi, 2011). It was found that the cellulose membranes could enhance the separation of water and alcohol with decrease of values of the permeation flux and increase of values of the separation factor. Furthermore, preparation of cellulose membranes (CM) from novel pulp cellulose/aqueous sodium hydroxide solutions was also studied by a simple method.

RESEARCH ON CELLULOSE SHEETS USED IN WATER/ALCOHOL PERVAPORATION

Vulcanized Paper for Separation of Alcohol Aqueous Solutions by Pervaporation

Vulcanized cellulose paper membranes (VCPM) used in the present work became of interest in the PV process due to its plastic- like properties (Vulcanized fiber Hokuetsu paper Co, 2008). The vulcanization process in cotton can make the paper membranes flexible, tough and durable without the need for resin or bonding materials (Trang & Kobayashi, 2011). It has also an advantage that the VCPM has common cellulose segments without modification of OH side groups. Due to the hydrophilic nature of the VCPM we have noted that the VCPM has the capability for water permeability in the azeotropes as well as other polysaccharide membranes. In addition, until now, no studies have been performed using VCPM for the PV processes. Thus, as a first report, we have studied PV processes for methanol, ethanol and isopropanol solutions using the VCPM for separation of alcohol/ water mixtures at different concentrations respectively.

The SEM morphologies of the VCPM used are shown in Figure 4 for (a) the cross section and (b) the surface of the VCPM. From (a), the VCPM used had a cross section thickness of 250μm and a dense paper layer through the membrane thickness. From the surface image (b) of the VCPM, it could be seen that the minute ravines distributed on the surface of the vulcanized membrane. Also, the surface image showed that the VCPM had dense structure on the surface. For the estimation of the liquid–solid interface, the contact angle method is a technique used to determine the affinity between alcohol/water mixture and the membrane substrate.

In contact angle of the droplets with time, there was a tendency for the value of contact angle on to decrease with the increase in observation time. Also, when the alcohol concentration in the aqueous solutions was high, the value of contact angle on the VCPM decreased, that is, a higher affinity was

Figure 4.SEM pictures of VCPM for (a) cross section and (b) surface

(a) (b)

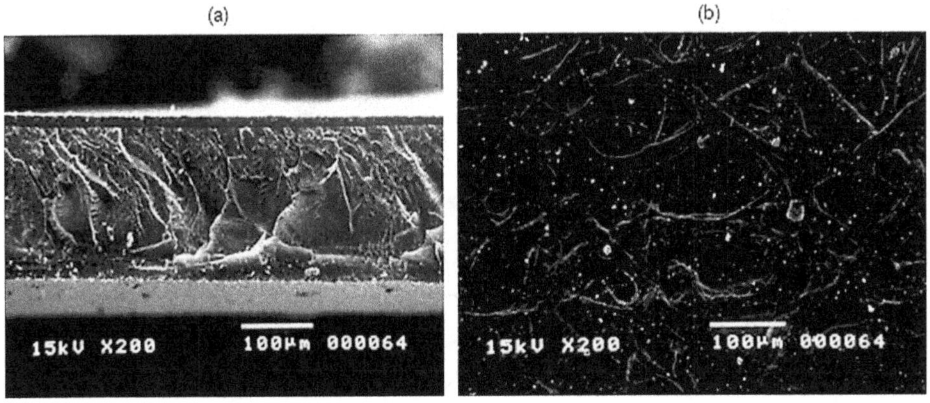

obtained between VCPM and the high aqueous alcohol solution concentration. For example, at 120s, at the alcohol contents of 8 wt% and 65wt%, the values of contact angle of, 25°(±1°) and 7°(±1°) for the methanol/water system, 24°(±1°) and 10°(±1°) for the ethanol/water system, 40°(±1°) and 12°(±1°) for the isopropanol/water system, respectively. This meant that all of alcohol mixtures wet the membrane and their liquids would penetrate spontaneously into the inside membrane. To confirm these phenomena, surface free energy of VCPM was also evaluated. It was measured in water that the value of the air contact angle was 52°(±1°). According to Andrade method (Andrade, 2007), surface free energy γ_{SL} was evaluated as 37.4mN/m by using water contact angle and the air contact angle. This result indicated that the surface of the VCPM had somewhat high surface tension of 22.75mN/m for methanol, 25.5mN/m for paraffin (Grundke, Boerner, & Jacobasch, 1991), and similar surface tension of 44mN/m of polystyrene (Manoj & Chaudhury, 1996). This meant that the hydrophobic segment of cellulose in the VCPM arranged on the membrane surface, indicating that alcohol spread easily on the surface.

The values solvent content (SC) of the VCPM were observed at different alcohol concentrations in the aqueous mixtures (Figure 5). In all cases, the values of the SC decreased with the increase of alcohol concentration in the aqueous solutions. This confirmed that the VCPM made of cellulose preferentially adsorbed water as compared with alcohols. It could be seen that, as the alcohol concentration increased from 0 to 100wt%, the value of the SC decreased from 58.9% to 7.8% for the methanol/water, 56.5% to 4.9% for the ethanol/water and 55.2% to 3.6% for the isopropanol/water, respectively. It was shown that

Figure 5. Solvent content of the VCPM observed at different alcohol concentration alcohol concentration

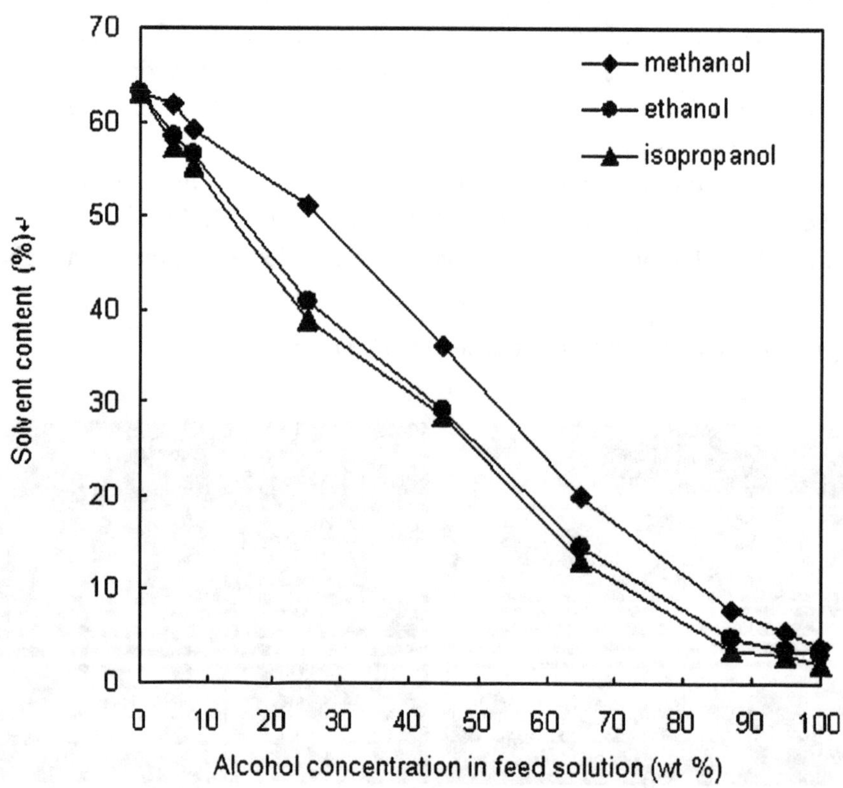

the values of SC in the VCPM for the methanol/water mixtures were larger than those for the ethanol/water and the isopropanol/water mixtures.

In order to determine the PV characteristics of the VCPM, experiments were carried out with a schematic diagram of the PV apparatus and a detailed description of the procedure for performing the PV experiment are shown in Figure 6. The effects of alcohol composition in the feed solution for methanol/water, ethanol/water, and isopropanol/water mixture on total permeation flux are shown in Figure 7. The value of the total permeation flux of VCPM for water was 6.8 kg/ (m^2h), and there was tendency to increase with increasing methanol concentration in the feed. It was observed that the total permeation flux for methanol mixture was 11.0 k g/ (m^2h) at 87wt% concentration, and 12.1 kg/ (m^2h) for methanol meaning that these values were greater than water (6.8kg/m^2h), ethanol (3.9kg/ m^2h) and isopropanol (2.1kg/ m^2h). On the other hand, the total permeation flux of the binary mixtures of ethanol/water and isopropanol/water gradually decreased with increase in alcohol concentration in the feed up to 87wt%, but increased thereafter. The total flux decreased from 6.6 to 2.0 kg/ (m^2h) with ethanol concentration in the feed increasing from 5 to 87wt%. With regards to isopropanol, the total permeation flux decreased from 6.4 to 1.2 kg/ (m^2h) with isopropanol concentration increasing from 5 to 87wt%. While methanol showed an increase in the total permeation flux over the entire concentration range 0-100wt% in the feed, ethanol and isopropanol showed a decrease in the total permeation flux at 87wt%. This is attributable to the occurrence of azeotropic mixture in ethanol/water and isopropanol/water binary mixtures; whereas methanol does form and non-azeatropic mixture with water.

Figure 6. Schematic diagram of the pervaporation apparatus

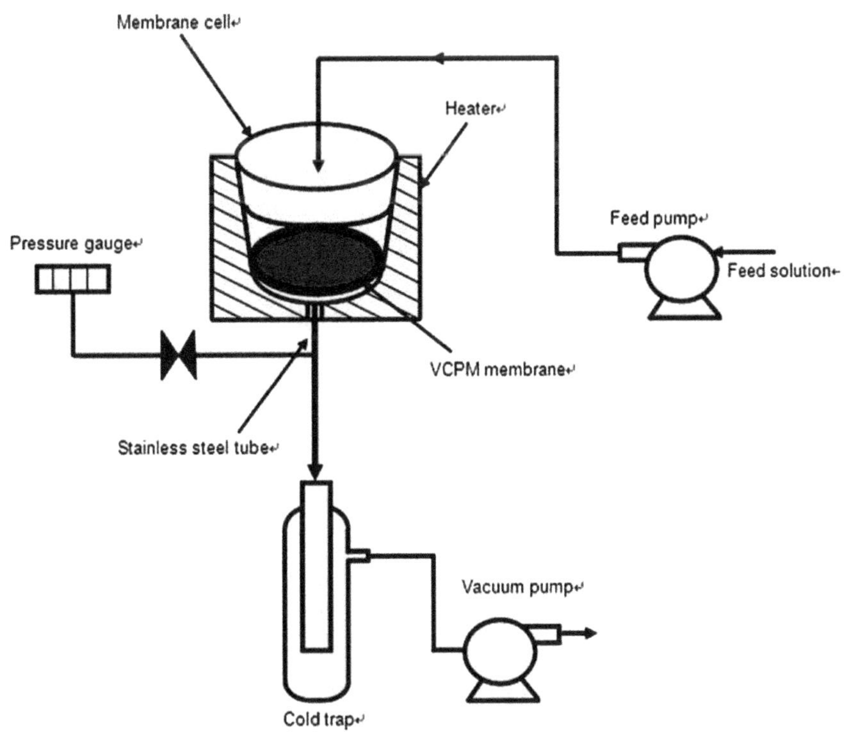

Figure 7. Effect of alcohol feed concentration on total permeation flux

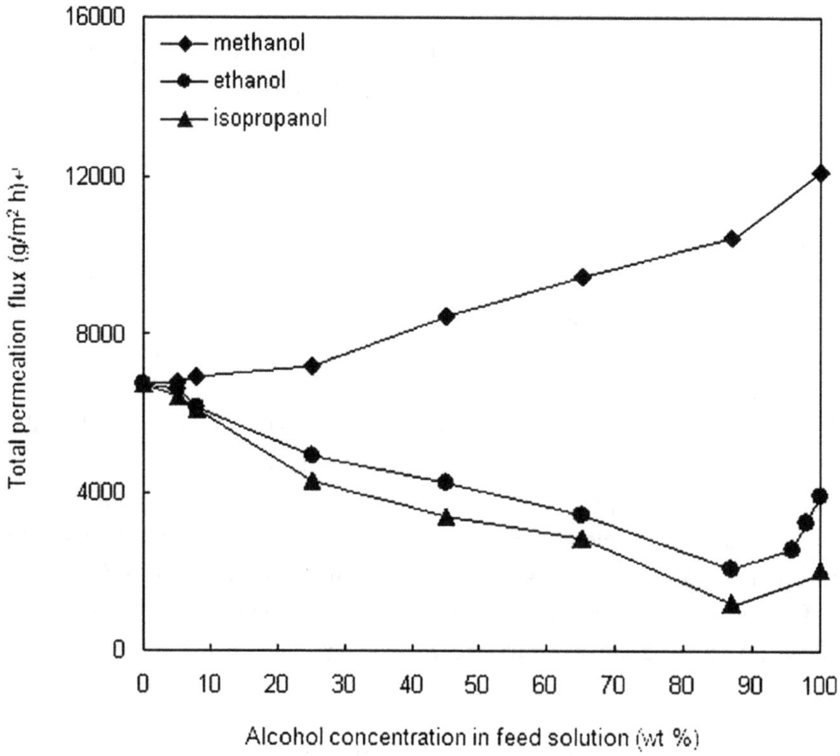

Figures 8 shows the separation factor (α) calculated at each alcohol concentration in (a) and water flux in (b). When the isopropanol concentration increased from 25 to 87wt %, the value of α increased from 4.9 to 16.6, respectively. In the case of ethanol, it was seen that the values of separation factor increase from 2.9 to 6.6 in the range of ethanol concentrations of 25 to 87wt%, respectively. For the methanol/water mixture, the values of separation factor were 2.7 and 4.1 for the methanol concentration of 25 to 87wt%, respectively. While the separation factor increased with alcohol concentration in the feed solution for all alcohol/water mixtures, it increased geometrically with alcohol concentration for isopropanol/water mixture. It was apparent that the separation factor increased with the alkyl chain length of the alcohol used in the mixture, whereas the total permeation flux decreased with the alkyl chain length. Water fluxes were calculated from the permeate compositions, and the value of water flux as a function of alcohol concentration in the feed is shown in Figure 8 (b). The pattern of water flux from alcohol/water mixtures with increasing alcohol concentration was similar to total permeation flux for alcohol/water mixtures as seen in Figure 2.5. However, water flux from methanol/water mixture steadily increased with increasing methanol concentration, and it was higher than from water containing no alcohol. This is presumably due to flow coupling that water permeability is affected by the gradient of methanol.

Figure 9(a) depicts the water permeation results as a function of water content in the feed mixture. It was shown that water content in the permeate increases rapidly with water content in the feed concentration for ethanol/water and isopropanol/water mixtures, but more slowly for methanol/water mixture. This indicated that the VCPM preferentially sorbed water supporting that water could permeate more selectively through VCPM in the alcohol/water mixtures. As a general trend, separation factor decreases

Figure 8. Effect of alcohol feed concentration on (a) separation factor, (b) water flux

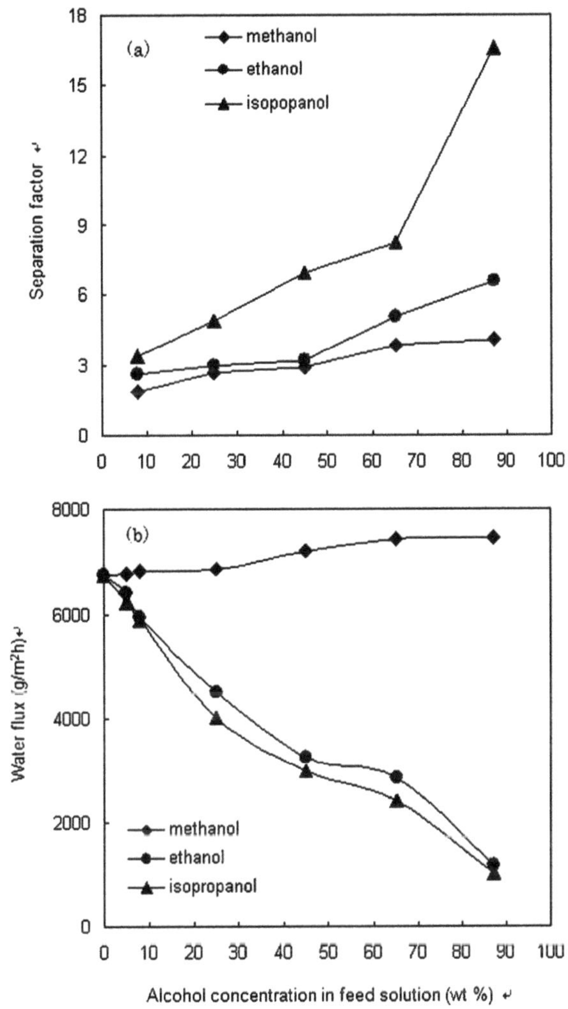

with increase in total permeation flux from 16.6 to 1.5 with total flux increasing from 1.7 to 10.0 kg/ m^2·h. A careful examination of the plot indicated that the inverse relationship was true for ethanol/water and isopropanol/water mixtures, but little for methanol/water mixture. There was a small change in the separation factor with total permeation flux in the methanol/water mixture. Thus, permeation behavior of methanol/water mixture through VCPM was different from that of ethanol/water and isopropanol /water mixtures. The high total permeation flux for methanol/water mixture, which increased with methanol concentration in the feed might be accounted for by the fact that methanol, being a more polar liquid compared with ethanol and isopropanol, could be sorbed by the hydrophilic VCPM along with water leading to greater swelling of the membrane and diffusion through the membrane. However, with increasing methanol concentration in the feed, preferential sorption and permeability of water appeared to increase to a smaller extent, compared with ethanol/water and isopropanol/water mixtures. Thus, this resulted in small increases in separation factor. As considered the separation of aqueous alcohol mixtures where the water molecules are smaller than alcohol molecules and the size of alcohol molecules

Figure 9. Relationship between (a) water feed concentration and water permeate concentration before and after the PV, respectively, (b) separation factor and total permeation flux; ^, methanol; l, ethanol; ~, isopropanol

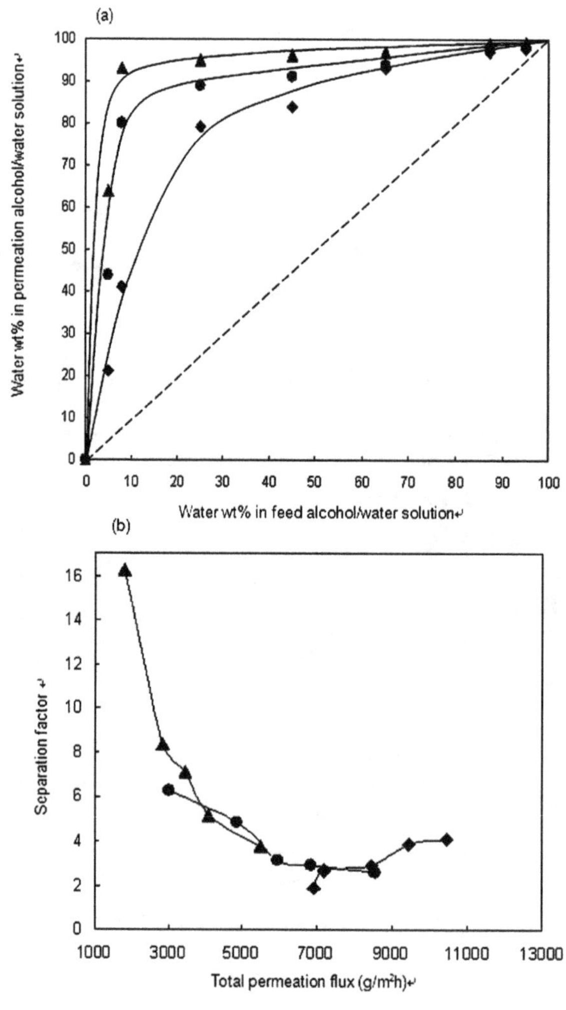

as following methanol< ethanol < isopropanol, the hydrophilic VCPM could favor both solubility and diffusivity for selective permeation of water than alcohol molecules. This was because that the smaller permeating molecules normally exhibited larger diffusivity, while the solubility was often influenced by the chemical affinity of the permeating species to the membrane material.

In Figure 5, there was similar tendency in the permeation flux of methanol/water mixture, but the values of the SC decreased with the increase of the alcohol contents. It was observed that the values of the contact angle became lower in methanol/water mixture. This meant that the methanol/water mixture easily spread on the matrix surface. But the values of the SC in methanol/water mixture were larger than those in ethanol/water and isopropanol/water mixtures. Therefore, the methanol/water mixture was somewhat familiar to contact with the VCPM matrix, indicating that the matrix might be easy to adsorb methanol and water. From the circumstance in the PV process, it was guessed that water could be much familiar to strongly adsorbing to the VCPM but methanol showed a higher accessibility relative to water.

Therefore, this behavior reflected that at a higher permeation flux of methanol and its mixture. As a result, the separation factor became lower in the methanol/water mixture.

FABRICATION AND CHARACTERIZATION OF PULP/CHITOSAN COMPOSITE FOR PERVAPORATION OF ETHANOL/WATER MIXTURE

Characteristics of Chitosan/Cellulose Membranes

Pulp/chitosan composite membranes were studied for alchol PV process (Trang, Nhung, & Kobayashi, 2011). In order to characterize the CTS/CL membranes, scanning electron microscopy (SEM) was used to observe the morphology of the surface and the cross- section of the resultant membranes. Membranes consisting of CTS and CL were prepared with different amounts of CTS and CL for CTS/CL= 100/0, 85/15, 80/20, 75/25, 60/40 and 50/50. Each membrane was cross-linked using 3MGA. Figure 10 shows the resultant SEM images of the CTS/CL membranes with and without the cross-linking treatment. In the pictures (g and h), the cross section of the obtained membranes had about 120μm thickness in the untreated membrane consisting of 80/20 of the CTS/CL. It was clear that the membrane presented many cellulose fibers having about 10μm in diameter. It could be observed that the morphological structure of the surfaces of the four membranes in the pictures (a ~h) was obtained with the cross-linking treatment and had a difference in the membrane thickness and number of CL fibers. With the increasing in the cellulose contents in the CTS/CL membranes, the cross – section which was of lower cellulose content became denser. The FT-IR spectra of CL sample (a), CL/3MGA sample (b), CTS sample (c), CTS/3MGA sample (d), CTS/CL-80/20 without cross-linking treatment, (f) CTS/ CL-80/20 with cross-linking treatment were compared. For the 3MGA treatment (Figure 11), the membranes had IR bands at 1717cm^{-1} for the C=O group of 3MGA. This seemed to be explained due to the cross-linking reaction forming ester groups of 3MGA and OH groups of CL occurred in the CTS/3MGA membranes. Furthermore, (e) shows the FT-IR spectrum of the CTS/CL membrane. When a comparison was made between CL (a) and CTS (c), the IR bands presented a shift in the amide I and amide II of the CTS from 1670 cm^{-1} and 1587 cm^{-1} to 1651 and 1576 cm^{-1}, respectively. Also, the disappearance of band at 1124 cm^{-1} (OH) was observed. Moreover, in the CTS/CL/3MGA membrane (f), both 1651 cm^{-1} (amide I) and 1576 cm^{-1} (amide II) bands were observed in addition with the appearance of 1714 cm^{-1} for the C=O of 3MGA and the disappearance of 1124 cm^{-1} for OH group of CTS and CL.

As characteristic of these CTS/CL membranes, the value of tensile strength and elongation of the membrane were measured and exhibited high tensile strength of 89.2 N/mm^2. There was tendency for the CTS/CL membranes to show a high surface area at CTS/CL= 80/20. The data of the contact angle decreased with the increase in the weight of ratio CTS/CL. This was due to a high affinity of the water molecules with the CTS segments in the CTS/CL membrane.

In the separation process of the mixture, the feed composition of ethanol and water was changed in the range of 10wt% to 90wt% of ethanol. Then the total permeation flux and separation factor was obtained. In order to understand clearly the PV characteristics, the effect of ethanol composition on the total permeation flux was determined in Figure 12(a). As the ethanol concentration in the feed solution was increased, the value of the total permeation flux considerably decreased. Especially, the tendency of the decline was more significant between CTS/CL =80/20 and 85/15. When the CL content was increased in the films, there was a tendency for the values of total permeation flux to increase.

Figure 10. SEM images of CTS/CL blended membranes: (a) surface of membrane CTS/CL-100/0; (b) Cross-section of membrane CTS/CL-100/0; (c) Surface of membrane CTS/CL-80/20; (d) Cross-section of membrane CTS/CL-80/20; (e) Surface of membrane CTS/CL-50/50; (f) Cross-section of membrane CTS/CL-50/50; (g) Surface of CTS/CL-80/20 mem-brane without cross-linking; (h) Cross section of CTS/ CL-80/20 membrane without cross-linking

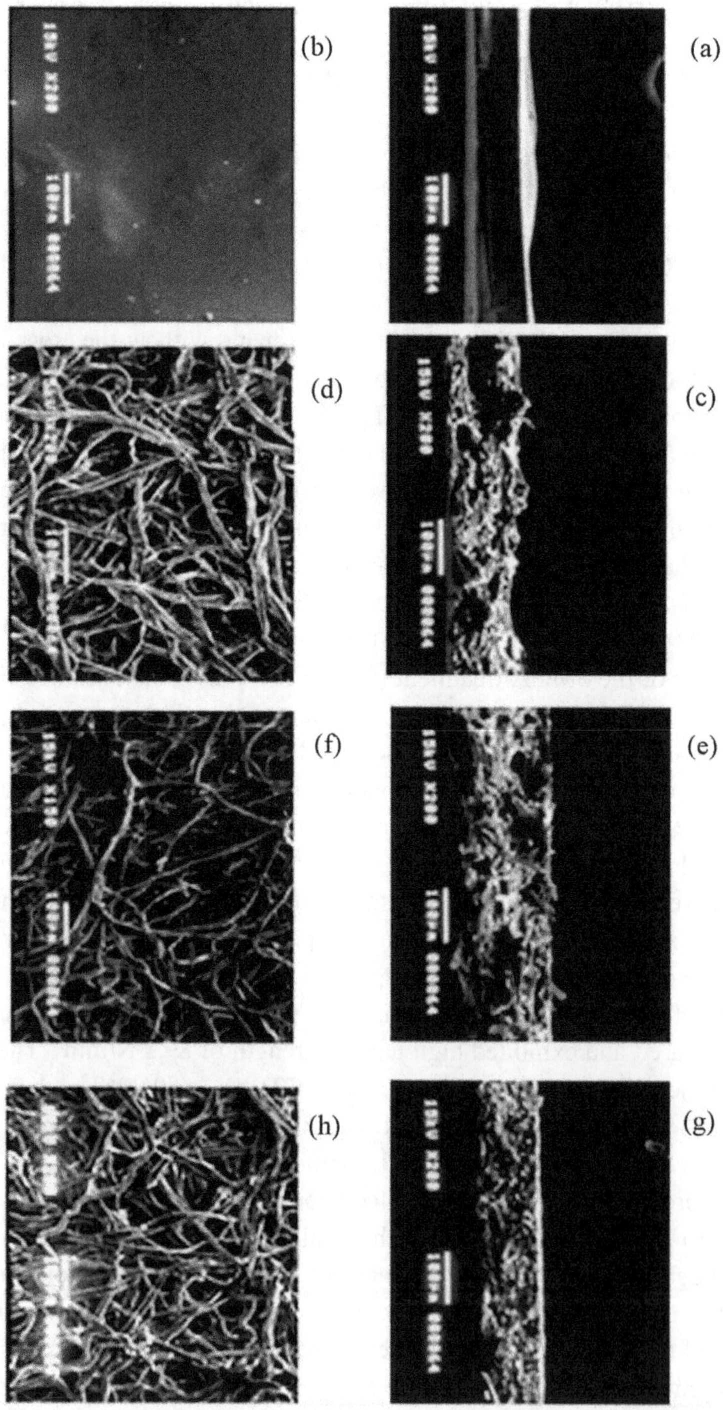

Figure 11. FT-IR spectra of (a) CL, (b) CL/ 3MGA, (c) CTS, (d) CTS/ 3MGA, (e) CTS/ CL-80/20 without cross-linking treatment, (f) CTS/ CL-80/20 with cross-linking treatment

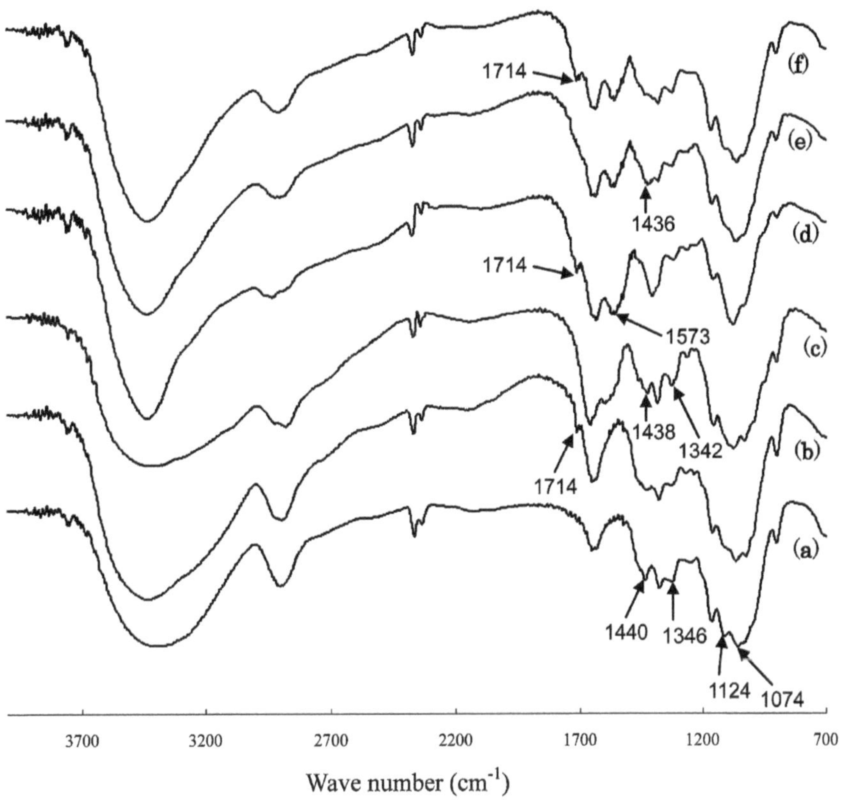

Figure 12(b) shows separation factor (α) for each CTS/CL membrane. When the ethanol concentration increased from 10 to 90wt %, there was a tendency for the value of α to increase from 2.1 to 5.9 for the CTS/CL-75/25, from 5.6 to 17.1 for CTS/CL- 80/20 and from 5.8 to 17.3 for the CTS/CL-85/15. In the CTS/CL-100/0 membrane the values were 5.9 and 17.9 at 10wt% and 90wt% of ethanol solution. It was noted that a reverse tendency could be seen in the data of the permeation flux and α. With the increase in ethanol concentration in the feed solution, the separation factor increased, but also lower permeation flux is obtained. In the case of the CTS/CL without cross-linking treatment, the mechanical properties were not good for the PV experiments, but the value of the faster total permeation flux was observed in the range 90~165g/m^2.h and the separation factor was shown in the range of 2~6.2.

Both the relationship of the total permeation flux and separation factor is clearly indicated (Figure 13). That is, the increase of the total permeation flux led to a decline in the value of separation factor. Among them, it was noted that the CTS/CL-80/20 membrane showed a high PV performance compared to other CTS/CL membranes, although the value of total permeation flux was 500g/m^{2}h and separation factor was about 18. It was also indicated that, the PV performance of untreated cross-linked film and CTS/CL -100/0 film resulted in low total permeation flux and high separation factor.

Figure 12. Effect of ethanol concentration on (a) total permeation flux, (b) separation factor

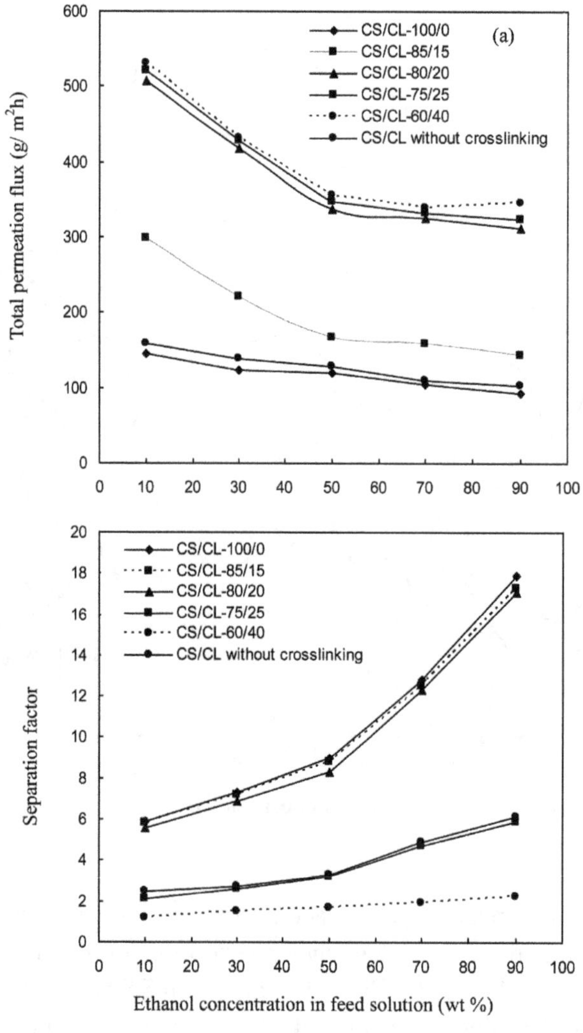

The composite CTS/CL membrane showed permeable and separation performance with 326g/ (m^2h) and α=17.1 for the ethanol/water mixture. The CTS/CL membranes system in the present work therefore could demonstrate a higher permeation flux, even though the separation factor was somewhat lower than that of chitosan- HEC blended film (Hsueh, Kuo, Huang, Wang, & Chen, 2005). This was of advantage in the case of CTS/CL membranes for sustainable materials and firstly reported in the present work. It could be concluded that the film of CTS and CL with 3-MGA treatment would become promising because of the proper alcohol concentration by PV process. Therefore, the CTS/CL biodegradable composite membranes were applicable with high permeation flux and stable operation for separation of ethanol/ water mixtures in food and medical industrial uses.

In the CTS/CL, the membranes showed great performance in the PV process with a total permeation flux, 326g/m^2h by separation factor, 17.1. The CTS/CL membranes also showed a very good stability with benefits in the separation of azeotropic mixtures such as ethanol in aqueous solution by PV process.

Figure 13. Plot of separation factor and total permeation flux

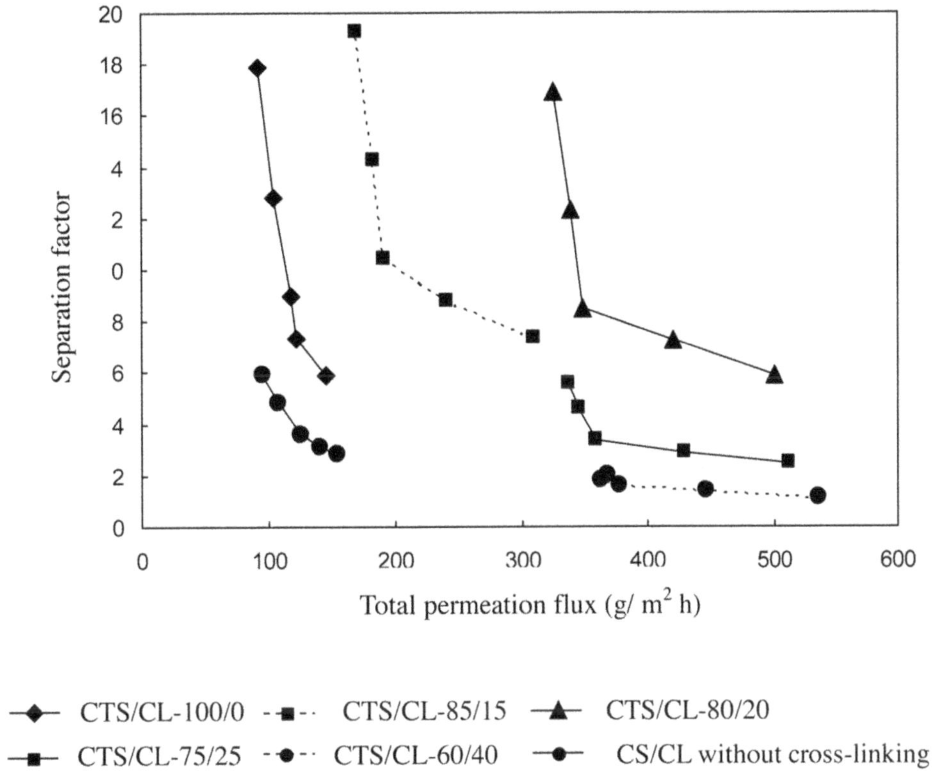

CONCLUSION

In alternative and environmentally conscious and sustainable methods and materials, alcohol separation without heating that will enable us to reduce the negative impact on our energy situation. Utilizing readily available, economical, sustainable and renewable biological materials such as cellulose and chitosan. This chapter supports the moving towards the use of biomass polymer membranes for separating alcohol from alcohol/water mixtures. The further benefit of the reduced necessity of heat energy to drive this process is an added financial incentive to commit to further research on this subject. Our hope is that our research will contribute to a greater enthusiasm and drive towards innovative, sustainable solutions for the challenges we currently face.

REFERENCES

Crini, G. (2005). Recent developments in polysaccharide-based materials used as adsorbents in wastewater treatment. *Progress in Polymer Science*, *30*(1), 38–70. doi:10.1016/j.progpolymsci.2004.11.002

Crini, G., & Badot, P. M. (2008). Application of chitosan, a natural aminopolysaccharide, for dye removal from aqueous solutions by adsorption processes using batch studies: A review of recent literature. *Progress in Polymer Science*, *33*(4), 399–447. doi:10.1016/j.progpolymsci.2007.11.001

Ehrlich, H., Hanke, T., Simon, P., Born, R., Fischer, C., Frolov, A., & Worch, H. et al. (2010). Carboxymethylation of the fibrillar collagen with respect to formation of hydroxyapatite. *Journal of Biomedical Materials Research. Part B, Applied Biomaterials*, *92B*(2), 542–551. PMID:19957363

Farber, L. (1935). Application of pervaporation. *Science*, *82*(2120), 158–158. doi:10.1126/science.82.2120.158 PMID:17811950

Gerente, C., Lee, V. K. C., Cloirec, P. L., & McKay, G. (2007). Application of Chitosan for the Removal of Metals From Wastewaters by Adsorption—Mechanisms and Models Review. *Critical Reviews in Environmental Science and Technology*, *37*(1), 41–127. doi:10.1080/10643380600729089

Grundke, K., Boerner, M., & Jacobasch, H. J. (1991). Characterization of fillers and fibres by wetting and electrokinetic measurements. *Colloids and Surfaces*, *58*(1), 47–59. doi:10.1016/0166-6622(91)80197-V

Guibal, E. (2004). Interactions of metal ions with chitosan-based sorbents: A review. *Separation and Purification Technology*, *38*(1), 43–74. doi:10.1016/j.seppur.2003.10.004

Guilbert, S., Cuq, B., & Gontard, N. (1997). Recent innovations in edible and/or biodegradable packaging materials. *Food Additives and Contaminants*, *14*(6-7), 741–751. doi:10.1080/02652039709374585 PMID:9373537

Guo, B., Finne-Wistrand, A., & Albertsson, A.-C. (2010). Molecular Architecture of Electroactive and Biodegradable Copolymers Composed of Polylactide and Carboxyl-Capped Aniline Trimer. *Biomacromolecules*, *11*(4), 855–863. doi:10.1021/bm9011248 PMID:20201489

Howell, J. A. (2002). Future research and developments in the membrane field. *Desalination*, *144*(1), 127–131. doi:10.1016/S0011-9164(02)00300-4

Hsueh, C. L., Kuo, J. F., Huang, Y. H., Wang, C. C., & Chen, C. Y. (2005). Separation of ethanol–water solution by poly(acrylonitrile-co-acrylic acid) membranes. *Separation and Purification Technology*, *41*(1), 39–47. doi:10.1016/j.seppur.2004.04.002

Huang, S. J. (1995). Polymer Waste Management–Biodegradation, Incineration, and Recycling. *Journal of Macromolecular Science, Part A*, *32*(4), 593–597. doi:10.1080/10601329508010272

Klemm, Fink, & Bohn. (2004). Cellulose: Fascinating Biopolymerand Sustainable Raw Material. *Ang. Chemie (Intl. Edn.)*, *44*, 3358.

Krajewska, B. (2001). Diffusion of metal ions through gel chitosan membranes. *Reactive & Functional Polymers*, *47*(1), 37–47. doi:10.1016/S1381-5148(00)00068-7

Kumar, A., Srivastava, A., Galaev, I. Y., & Mattiasson, B. (2007). Smart polymers: Physical forms and bioengineering applications. *Progress in Polymer Science*, *32*(10), 1205–1237. doi:10.1016/j.progpolymsci.2007.05.003

Kurita, K. (2001). Controlled functionalization of the polysaccharide chitin. *Progress in Polymer Science*, *26*(9), 1921–1971. doi:10.1016/S0079-6700(01)00007-7

Liu, Y.-L., Hsu, C.-Y., Su, Y.-H., & Lai, J.-Y. (2005). Chitosan–Silica Complex Membranes from Sulfonic Acid Functionalized Silica Nanoparticles for Pervaporation Dehydration of Ethanol–Water Solutions. *Biomacromolecules*, *6*(1), 368–373. doi:10.1021/bm049531w PMID:15638541

Meyers, M. A., Chen, P.-Y., Lin, A. Y.-M., & Seki, Y. (2008). Biological materials: Structure and mechanical properties. *Progress in Materials Science*, *53*(1), 1–206. doi:10.1016/j.pmatsci.2007.05.002

Miretzky, P., & Cirelli, A. F. (2009). Hg(II) removal from water by chitosan and chitosan derivatives: A review. *Journal of Hazardous Materials*, *167*(1–3), 10–23. doi:10.1016/j.jhazmat.2009.01.060 PMID:19232467

Mohanty, A. K., Depan, D., Tomer, N. S., & Singh, R. P. (2007). Nanoscale particles for polymer degradation and stabilization—Trends and future perspectives. *Progress in Polymer Science*, *34*, 479–515.

Mohanty. (2005). *Natural Fibers, Biopolymers and Biocomposites*. CRC Press.

Mukoma, P., Jooste, B. R., & Vosloo, H. C. M. (2004). Synthesis and characterization of cross-linked chitosan membranes for application as alternative proton exchange membrane materials in fuel cells. *Journal of Power Sources*, *136*(1), 16–23. doi:10.1016/j.jpowsour.2004.05.027

Nair, L. S., & Laurencin, C. T. (2007). Biodegradable polymers as biomaterials. *Progress in Polymer Science*, *32*(8–9), 762–798. doi:10.1016/j.progpolymsci.2007.05.017

Ramirez-Salgado, J. (2007). Study of basic biopolymer as proton membrane for fuel cell systems. *Electrochimica Acta*, *52*(11), 3766–3778. doi:10.1016/j.electacta.2006.10.051

Ravi Kumar, M. N. V. (2000). A review of chitin and chitosan applications. *Reactive & Functional Polymers*, *46*(1), 1–27. doi:10.1016/S1381-5148(00)00038-9

Sato, K., Kishimoto, T., Morimoto, M., Saimoto, H., & Shigemasa, Y. (2003). Hydrolysis of acetals in water under hydrothermal conditions. *Tetrahedron Letters*, *44*(47), 8623–8625. doi:10.1016/j.tetlet.2003.09.087

Schurz, J. (1999). A bright future for cellulose. *Progress in Polymer Science*, *24*(4), 481–483. doi:10.1016/S0079-6700(99)00011-8

Shahidi, F., Arachchi, J. K. V., & Jeon, Y.-J. (1999). Food applications of chitin and chitosans. *Trends in Food Science & Technology*, *10*(2), 37–51. doi:10.1016/S0924-2244(99)00017-5

She, H., Malotky, D., & Chaudhury, M. K. (1998). Estimation of Adhesion Hysteresis at Polymer/Oxide Interfaces Using Rolling Contact Mechanics. *Langmuir*, *14*(11), 3090–3100. doi:10.1021/la971061m

Tanimura, S., Nakao, S.-i., & Kimura, S. (1990). Ethanol-selective membrane for reverse osmosis of ethanol/water mixture. *AIChE Journal. American Institute of Chemical Engineers*, *36*(7), 1118–1120. doi:10.1002/aic.690360719

Trang, T. C., & Kobayashi, T. (2011). Vulcanized Paper for Separation of Alcohol Aqueous Solutions by Pervaporation. *Journal of Applied Polymer Science*, *121*(2), 639–647. doi:10.1002/app.33258

Trang, T. C., Nhung, N. T., & Kobayashi, T. (2011). Fabrication and Characterization of Pulp/Chitosan Composite Membranes Crosslinked with 3-Methylglutaric Anhydride for Pervaporation of Ethanol/Water Mixture. *Engineering*, *3*, 110–118.

Tripathi, B. P., & Shahi, V. K. (2008). Functionalized Organic–Inorganic Nanostructured N-p-Carboxy Benzyl Chitosan–Silica–PVA Hybrid Polyelectrolyte Complex as Proton Exchange Membrane for DMFC Applications. *The Journal of Physical Chemistry B*, *112*(49), 15678–15690. doi:10.1021/jp806337b PMID:19368033

Uragami, T. (2014). *Pervaporation Membranes. In Encyclopedia of Polymeric Nanomaterials*. Berlin: Springer Berlin Heidelberg. doi:10.1007/978-3-642-36199-9_133-1

Uragami, T., & Takigawa, K. (1990). Permeation and separation characteristics of ethanol-water mixtures through chitosan derivative membranes by pervaporation and evapomeation. *Polymer*, *31*(4), 668–672. doi:10.1016/0032-3861(90)90287-9

Varma, A. J., Deshpande, S. V., & Kennedy, J. F. (2004). Metal complexation by chitosan and its derivatives: A review. *Carbohydrate Polymers*, *55*(1), 77–93. doi:10.1016/j.carbpol.2003.08.005

Varmaa, A. J., Deshpande, S. V., & Kennedy, J. F. (2004). Metal complexation by chitosan and its derivatives: A review. *Carbohydrate Polymers*, *55*(1), 77–93. doi:10.1016/j.carbpol.2003.08.005

Wnuk, R., & Chmiel, H. (1992). Direct heating of composite membranes in pervaporation and gas separation processes. *Journal of Membrane Science*, *68*(3), 293–300. doi:10.1016/0376-7388(92)85030-M

KEY TERMS AND DEFINITIONS

Azeotrope: A mixture of two liquids that has a constant boiling point and composition throughout distillation.

Cellulose: A polysaccharide organic compound composed of a linear chain of several hundred to many thousands of β linked D-glucose units.

Chitosan: A linear polysaccharide organic compound commonly found in nature in the hard outer skeletons of shellfish.

Environmentally Sustainable: The harvesting, extraction or use of a natural resource at a rate that will not exceed the resource's ability to reproduce itself and maintain naturally occurring levels.

Hydrophilic Membrane: A membrane that is easily wetted, allowing for easy passage of water through the membrane. It is the best suited membrane for use with aqueous solutions.

Membrane Permeability: A quality of a membrane that allows for the passage of solutions or solutes through the membrane layer. Permeability can be selective, whereby only certain solvents or solutes may pass through the membrane.

Membrane Permeation Flux: The volume flowing through the membrane per unit area per unit time. In the case of transport of gases and vapors, the volume is strongly dependent on pressure and temperature.

Pervaporation: A method by which the components of a two liquid mixture are separated through a semi-permeable membrane, the component that passes through the membrane being separated by evaporation.

Chapter 12
Bacterial Cellulose:
Biosyntheses, Modifications, and Applications

Siriporn Taokaew
Nagaoka University of Technology, Japan

Muenduen Phisalaphong
Chulalongkorn University, Thailand

Bi-min Zhang Newby
The University of Akron, USA

ABSTRACT

Bacterial cellulose, synthesized by bacteria, is one of the most highly pure cellulose sources. It has gained a great attention due to its unique properties. With molecular similarity to cellulose derived from plants, bacterial cellulose can be modified based on diverse techniques established for the plant-derived cellulose. Modification of cellulose has become one of the major areas of cellulose research to provide cellulose-based materials with novel properties. The progress in cellulose science has also opened up more potentials for bacterial cellulose. This chapter describes an overview of biosyntheses, modifications, and applications of bacterial cellulose.

INTRODUCTION

Cellulose, a straight chained polymer of β-(1,4)-linked D-glucose units, is the most abundant biopolymer on earth. It plays an important role in contributing to the high strength of plant cell wall. Sources of plant-derived cellulose include wood, cotton, tunicate, and algae. To obtain pure cellulose from the plant-derived cellulose, other bio-polymeric components, e.g. lignin and hemicellulose, are required to be removed by bleaching with toxic chemicals (Klemm et al., 2005). One environmentally-friendly approach to obtain pure cellulose is utilizing biosynthesis by bacteria.

Bacterial or microbial cellulose is obtained by extracellular secretion of single chains of cellulose from a certain type of bacteria genera such as:

DOI: 10.4018/978-1-5225-1971-3.ch012

- *Acetobacter,*
- *Aerobacter,*
- *Achromobacter,*
- *Azotobacter,*
- *Escerichia,*
- *Rhizobium,*
- *Sarcina, and*
- *Salmonella* (Shoda et al., 2005).

Among those bacteria, *Acetobacter xylinus* (currently called *Gluconacetobacter xylinus*) appears to be the most-studied cellulose-producing bacterium due to its capability of greater cellulose production from multiple carbon and nitrogen sources (Iguchi et al., 2000a). Single chains of cellulose are polymerized by cellulose synthases using glucose as the monomer, and then extruded as glucan chains through the bacterial cell wall. The glucan chains assemble into a bundle, ribbon, and nanofiber (Ross et al., 1991). The 3-dimensional structure of nanofibers is then formed by inter- and intra- hydrogen bonds with a number of hydroxyl groups bound with water.

The bacterial cellulose exhibits the same chemical structure of linear β-1,4-glucan chains as those found in plant celluloses, but has very different properties from those of the plant-derived cellulose. For instance, it possesses a high crystallinity, a nano-sized fiber network with high porosity and surface area, a high mechanical strength, a great liquid absorption capacity, and an excellent flexibility for fabricating into various sizes and shapes (Czaja et al., 2006). Moreover, it shows outstanding biological characteristics and good biocompatibility owing to the complex network imitating the natural extracellular matrix (ECM). Unlike plant-derived celluloses, bacterial cellulose is free of lignin and hemicellulose, which need to be eliminated by treatment with toxic chemicals. This suggests that some high pollution generations in steps of pulping and bleaching cellulose contained in plants can be removed. Techniques from the genetic and engineering viewpoints have been investigated by researchers to enhance cellulose production from the natural habitats to lab- and mass-scale production (Shoda & Sugano, 2005). In addition, to overcome some limitations and achieve other properties of the native cellulose for perspective applications, modifications such as turning surface properties or introducing additives to alter the bulk properties during or after cellulose biosynthesis can be performed.

BACKGROUND

It has been over a century since bacterial cellulose was discovered by Adrian J. Brown in 1886. He originally observed jelly-like translucent mass formed by *Bacterium aceti* or *Acetobacter xylinus* on the surface of "vinegar plant" or acetic ferment liquid. That polysaccharide mass was found to chemically behave the same as cellulose (Brown, 1886). In 1949, K. Muhlethaler studied the structure of bacterial cellulose by means of electron microscopy (Mühlethaler, 1949). He found that the acetic bacteria could secret cellulose fibers having an approximate diameter of 250 Å within the slime. On the basis of its high purity and unique physicochemical properties that are different from those of plant-derived celluloses, bacterial cellulose becomes attractive to researchers. For the past 20 years, over 4,000 publications relating to the investigation on improving cellulose production from different microbial strains, developing large-scale production, and modifying bacterial cellulose for food, medical, and biotechnological applications have been documented (see Figure 1). A surge in both publication and citation is observed since 2010, with an annual publication rate increase of 50 or more and the annual citation in the thousands. The number of patents subjected to this nanocellulose has also continuously increased (Charreau et al., 2013; Dahman, 2009; Jozala et al., 2016; Lin et al., 2013; Petersen et al., 2011; Yoshinaga et al., 1997).

Publications include commercial products of bacterial cellulose due to the production process of bacterial cellulose poses low-cost, high yield, environmentally friendly, and less waste generation, which is highly desirable in industry. The economical production of bacterial cellulose also obviously accounts for inexpensive raw material, and low operating cost. The first commercial product of bacterial cellulose is food product, which was reported by Martina M. Lapuz and his co-workers in 1969 (Bernardo et al., 1998). This food product has been utilized until now. In recent years, skin care products based on bacterial cellulose have been found in the market worldwide.

Commercial applications of bacterial cellulose have been reported according to its several characteristics. In the point of outstanding mechanical properties and reinforcement capability, applications of bacterial cellulose in the fields of audio component and paper product have been reported. Bacterial cellulose was reported to be a good audio material by Agency of Industrial Science and Technology, Sony Corporation, and Ajinomoto Co., Inc of Japan in 1988 (Iguchi et al., 1988). Ideally, the acoustic diaphragms material needs to be made of a very thin sheet with an extremely high elastic modulus, a low density, a high sonic velocity over a wide range of frequency, and a high internal loss. Bacterial cellulose is found to meet all these requirements, despite other non-composite materials that might have a high sonic velocity generally exhibit a low internal loss. With a light thin dried sheet posing a remarkably high modulus of elastic (>15 GPa), a high sonic velocity (5000 m/s), and a high internal loss (0.03) similar to a good quality diaphragm paper, a bacterial cellulose sheet can be used as diaphragms for acoustic transducer such as loudspeaker units and headsets (Nishi et al., 1990). The speaker diaphragms made from bacterial cellulose was first announced in Japan in the 1990s (Iguchi et al., 2000b). From the perspective

Figure 1. Publications of bacterial cellulose in the latest 20 years (until August 2016) indexed within Web of Science™ Core Collection

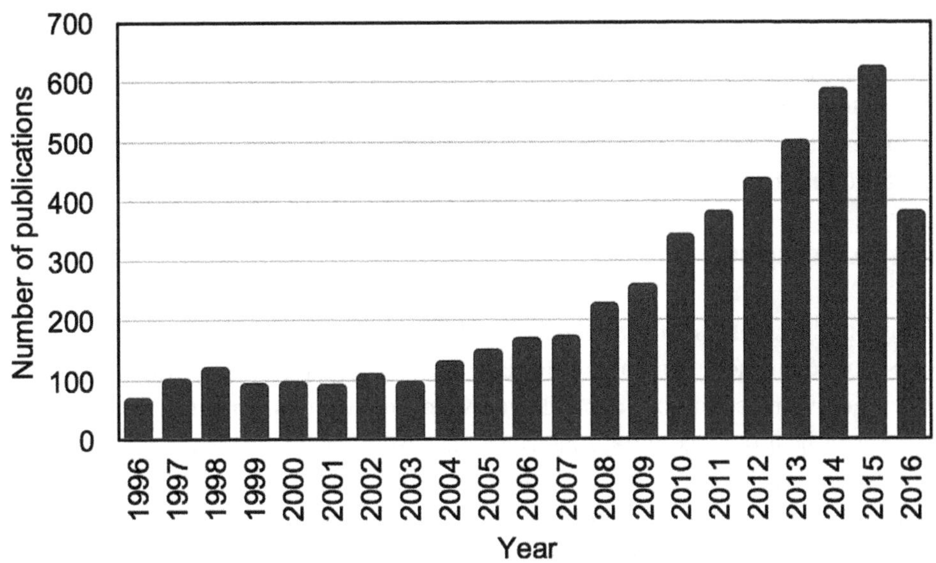

of smaller fibers, higher mechanical strength, and greater emulsion-stabilizing effect as compared to other cellulosic materials, bacterial cellulose (especially the agitated one) could be an effective filling material to improve the durability of traditional integrated pulps for papermaking (Ougiya et al., 1997). Analogous width, curl, and kink indices of bacterial cellulose to those of softwood fibers, an important component to form paper without flocculation, have also been investigated (Gao et al., 2011; Xiu et al., 2005). Incorporation of bacterial cellulose in paper production could provide high performance paper, and this was first patented by Ajinomoto Co. along with Mitsubishi Paper Mills in Japan in the late 1990s (Hioki et al., 1995).

BIOSYNTHESES OF BACTERIAL CELLULOSE HYDROGEL

Culture Media

Bacterial cellulose is synthesized by bacterium genera *Acetobater* in a complex culture medium containing multiple carbon, nitrogen, and vitamin sources. One of the common media is Hestrin and Schramm (HS) medium (Hestrin et al., 1954; Son et al., 2003), which consist of D-glucose, yeast extract, peptone, disodium phosphate, and citric acid. Simple fruit juice-based media, such as pineapple juice, santol, guava, jackfruit, sugarcane, papaya, coconut skim milk, and coconut water (Brown Jr, 1991; Kurosumi et al., 2009; Tecson-Mendoza, 2007), as well as red wine are also used (Brown, 1886). Relatively, mature coconut water and coconut skim milk are cheap promising raw materials, and can be easily found in equatorial area due to the more arable weather for coconuts. Coconut water is the liquid endoderm in the coconut fruit, and coconut skim milk is the layer below the coconut cream made by squeezing ground mature coconut meat with water, filtering, and allowing the cream layer to rise.

As augmented coconut product industries, especially in the tropical zone of Asia, a large amount of mature coconut water, which contains valuable vitamins, carbon, and nitrogen sources, has been discarded as an industrial by-product. Besides, it can be directly utilized as a bacterial culture medium without involving detoxification and hydrolysis steps. The economical coconut-water based medium was first recommended in 1999 by Budhiono and his colleges, and has been used as the basic culture media for bacterial cellulose preparation (Budhiono et al., 1999). However, the growth of cellulose-producing bacteria requires slight additions of a carbon source (e.g. glucose or sucrose) and a nitrogen source (e.g. ammonium sulfate, Bacto peptone, or yeast extract), and a slight acidic-pH adjustment (Kamarudin et al., 2013). Supplementations to the mature-coconut-water based media are therefore essential to achieve a high bacterial cellulose productivity.

Cultivation Conditions

The cultivation condition plays an important role on bacterial cellulose production. After inoculation with cellulose-producing bacteria in the sterile cell culture media, bacterial cellulose could be cultivated in either static or dynamic conditions depended on specific purposes. Under the static condition, a cellulose pellicle forms near the medium-air interface because of an adequate atmospheric oxygen supply. It thickens in the downward direction until the aerobic cellulose producing bacteria are entrapped by their secreted nanofibers and become inactive caused by the deficient diffusions of air and/or liquid medium to the bacteria (Brown, 1886; Brown Jr, 1991). A regular cellulose thick sheet with a visible smooth surface is obtained. This cultivation condition is mostly conducted owing to the most economical condition without a great energy as well as time consumptions. It can also be performed in the warm environment of the tropical countries without needing an incubation oven (Chawla et al., 2009; Taokaew et al., 2014b; Taokaew et al., 2013b).

For the dynamic cultivation condition, the stirred/agitated and aerated cultures have been created according to the kinetic aspect of bacterial-cellulose formation, which is considerably controlled by an oxygen supply (Budhiono et al., 1999). Reactors or fermenters such as cylindrical silicone membrane vessels (Yoshino et al., 1996), rotating disk bioreactors (Serafica et al., 2002), fed-batch aerosol bioreactors (Hornung et al., 2007), and biofilm reactors (Cheng et al., 2009; Cotton et al., 2001) have been designed with the purpose of applying more oxygen to the production system. In addition, the cell immobilization condition has been attempted to increased cell density, thus improving cellulose production (Cotton et al., 2001).

The incubation condition affects some properties of the resulting cellulose. For instance, the varied macroscopic morphology and unshaped massy cellulose could result due to shear stress during agitation or shaking (Watanabe et al., 1998). The microstructural features of agitated-cultured cellulose appear to have lower crystallinity, crystalline polymorphism, degree of polymerization, and Young's modulus. However, a greater water holding capacity, viscosity, filler retention, and emulsion stability index are resulted (Watanabe et al., 1998). Polymerization can also be reduced due to the genetic instability of the cellulose-producing bacteria. Even though a greater productivity of bacterial cellulose has been achieved by dynamic methods, more energy consumption, contamination during fermentation, and membrane fouling are some of the concerns.

Purifications and Storage

After harvesting the cellulose pellicles, to remove the culture medium residues and the bacterial cells debris, the purification procedures are required particularly in medical applications. The purification of bacterial cellulose hydrogel using alkaline treatment, e.g. sodium hydroxide aqueous solution at the concentration ranging 0.1-1 M at 30-100 °C for 15 min to 24 h, is widely adopted to hydrolyze protein substances from the pellicles. The neutralization is then carried out using a dilute acetic acid solution or under a large amount of running water. The light-yellowish bacterial cellulose becomes white after purification (George et al., 2005; Hong et al., 2008; Serafica et al., 2002).

MODIFICATIONS OF BACTERIAL CELLULOSE HYDROGEL

Modifications are necessary for enhancing the capabilities of numerous materials for a wider range of applications. Modification of bacterial cellulose is also aimed to enhance some properties for its more diverse applications. The modification approaches can be broadly classified into physical and chemical modifications. The physical modification is defined as an alteration of structural matrices without the incorporation of any secondary materials. On the other hand, the chemical modification is described as the adjustment of chemical characteristics by introducing other material(s) into the bulk structure or on the surface to generate bacterial cellulose-based composites or surface modified bacterial cellulose, respectively. These modifications (Figure 2) could be carried out during and after biosynthesis.

Figure 2. schematic of modification techniques of bacterial cellulose

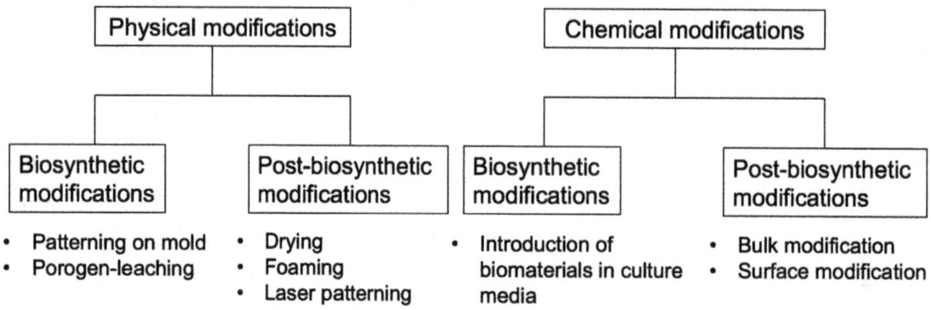

Physical Modifications

Native bacterial cellulose can be tailored physically to have different textures or pore sizes, which refer to the vacant spaces between cellulose fibers, during and after biosynthesis. Physical modifications are referring to the introduction and subsequent removal of foreign particles to the cellulose without compromising the native cellulose chemistry. For physical modification during biosynthesis, cellulose-producing bacteria are grown in the presence of these sterilized particles. After biosynthesis, these particles are removed, allowing the remaining cellulose to have different porous structures. For physical modification after biosynthesis (post-biosynthetic modification), various dehydration techniques and compressing of the native bacterial cellulose are applied. These techniques affect the structure of bacterial cellulose by collapsing its microstructure, resulting in the change of vacant spaces (pore size) between cellulose fibers. To enlarge pore size or reorganize fiber structure, the secondary substances, such as polymeric template mold and removable porogens, can be utilized. The effect of physical techniques on pore sizes of bacterial cellulose is summarized in Table 1.

Biosynthetic Modifications

The major purpose of these techniques is to modify the pore size of the native cellulose to obtain the appropriate size for specific applications. The optimal porous structure facilitates the adsorption capability and mass transport, which can be applied in tissue engineering and separation/filtration applications. In

Table 1. Physical modification techniques during and after biosynthesis of bacterial cellulose hydrogel

Physical Modification Techniques	Pore size (μm)
Biosynthetic modification • Honeycomb-patterning on agarose film (Uraki et al., 2007). • Porogen-leaching using paraffin wax microspheres and potato starch particles (Bäckdahl et al., 2008; Zaborowska et al., 2010).	20-25 30-250
Post-biosynthetic modification • Air drying (Taokaew et al., 2014b). • Freeze drying (Taokaew et al., 2014b). • Supercritical point drying using liquid carbon dioxide (Haimer et al., 2010). • Surfactant-assisted foaming (Yin et al., 2012). • Laser patterning (Ahrem et al., 2014).	0.1-1 1-3 2-10 10-20 200-300

the field of tissue engineering, pore size of a scaffold is very tissue-type specific (Murphy et al., 2010). For bacterial cellulose, due to its packed fibrous structure, the pore size is smaller than 10 microns on the lower surface and even smaller on the upper surface or the air-medium interface. These pores are too small to support ingrowth of skin cell lines in vitro (Sanchavanakit et al., 2006). To regulate or increase the native pore size of bacterial cellulose, various techniques have been performed. Some of the methods are molding, culturing the cellulose-producing bacteria on the mold, and porogen-leaching, which are described below.

Honeycomb-patterned bacterial cellulose can be fabricated by cultivating cellulose-producing bacteria on agarose mold with honeycomb-pattern in micron size. Movement of cellulose-producing bacteria is then controlled within the provided small space. Concave- and convex-typed honeycomb-patterned microgrooves are employed to obtain the highly oriented cellulose fibers with pore arrays. The culture on concave type has been found to yield the honeycomb-patterned bacterial cellulose with well-organized cellulose fibers in the narrow space of groove as compared to the convex one. The pore size of the patterned bacterial cellulose is observed to be 20-25 μm according to the designed honeycomb mold (Uraki et al., 2007).

Bigger pores inside bacterial cellulose can be obtained by the porogen-leaching method; the resulting size depends on the porogen used. Microporous bacterial cellulose scaffolds with controlled microporosity can be prepared by incorporating desired sizes of porogen such as potato starch particles or paraffin wax microspheres into the biosynthesis process of cellulose-producing bacteria. After the biosynthesis, the porogens are removed, leaving behind a network of interconnected pores within the bacterial cellulose. By comparing the two types of porogens, the paraffin wax microspheres is found to provide a more uniform distribution of pores throughout bacterial cellulose scaffold (Bäckdahl et al., 2008; Zaborowska et al., 2010). Besides those porogens mentioned above, the porogen-leaching techniques have been utilized to fabricate other porous materials by numerous research groups using other porogen particles including salt, ice, sugar, and gelatin particles (Weigel et al., 2006). However, these particles can be dissolved in the cell media during biosynthesis, meaning that the chemical composition of the obtained bacterial cellulose could be changed.

Post-Biosynthetic Modifications

The structural network of bacterial cellulose is able to reorganize based on different drying approaches. Air dying is a simple evaporative drying, but results in a large thickness reduction as compared to the never dried bacterial cellulose (Taokaew et al., 2014b). This is the results of capillary pressure in the small pores of bacterial cellulose and the high surface tension of water inside those pores. When replacing water by an organic solvent, such as dichloromethane (DCM), having a lower boiling point (39.6°C for DCM), a higher vapor pressure (~ 0.56 atm for DCM at 25°C), and a lower surface tension (~ 26 mN/m for DCM at 25°C), to allow a faster drying rate, the thickness of dried bacterial cellulose is slightly higher than that without the replacement (Taokaew et al., 2013a).

The large dimensional changes of bacterial cellulose are caused by liquid involvement in the drying systems (Bashaiwoldu et al., 2004). In the case of air drying, the structure shrinks in the thickness direction, and hardly recovers after rehydrating due to the new hydrogen bond formation occurring between fibers in the dry state. This bonding is irreversible between fiber and water. Thus, the air-dried bacterial cellulose cannot uptake the equal amount of initial water. For the freeze drying, liquid can be removed from material by directly transforming into vapor without passing through the liquid state (Mazur, 1970).

It consists of formation of ice crystal from water inside the bacterial cellulose hydrogel and sublimation/drying those ice crystals. It results in retaining dimension of bacterial cellulose as compared to those from air drying.

In addition to thickness, the pore structure of bacterial cellulose can also be modified by the drying techniques. The air dried bacterial cellulose with and without solvent replacement exhibit packed fibers and small pores. The looser packing of fibers and remained highly porous structure are shown on the freeze-dried bacterial cellulose. With regards to the greatly porous structure, the freeze-dried bacterial cellulose allows a good liquid absorption capacity. In the rehydrate state, the thickness of those dried bacterial cellulose appears barely change (Taokaew et al., 2014b). Furthermore, another drying technique reported for physical modification without causing the chemical reactions and changing in the crystalline structure is drying using a supercritical liquid. Carbon dioxide with critical temperature and pressure of 31.3 °C and 72.9 atm, respectively, is the most common fluid in the supercritical drying due to low side reactions, non-toxic, non-pollutant and inexpensive (Zimmermann et al., 2016). In this drying technique, water existing in the bacterial cellulose hydrogel is replaced by ascending grades of ethanol and subsequently the supercritical carbon dioxide fluid. Under the condition above the critical point, liquid and gas phrase of carbon dioxide cannot distinguish, which prevent the liquid-vapor interface. Thus, it effectively prevents the collapse of bacterial cellulose gel network and transforms the cellulose hydrogel into an aerogel with an identical thickness superior to that obtained by the freeze drying (Leitch et al., 2016; Zimmermann et al., 2016). Liebner et. al. (2010) found that after drying with supercritical carbon dioxide at 40 ºC and 100 bar, the average density of the dried bacterial cellulose aerogel is approximately 8 mg/cm^3, which is considerably lighter than the plant cellulose aerogel (Liebner et al., 2010).

An alternative method to the supercritical carbon dioxide drying for retaining the discrete fibril microsctructure of bacterial cellulose is drying using hexamethyldisilazane (HMDS) (Lee et al., 2015). HMDS is an organic solvent that has a surface tension of 18.16 mN/m at 25 °C (Myers et al., 1969), much smaller than that of water (72 dynes/cm at 25 °C), and a low flash point of 11°C. The drying involved a subsequent submersion of bacterial cellulose hydrogel in a series of ascending ethanol in ethanol/water mixture followed with submersion in a series of HDMS/ethanol mixture with assending HDMS. When the HMDS dried bacterial cellulose sheet is placed in ethanol or water, the sheet swells and almost returns back to its original never-dried thickness.

The structural matrices altered by the different drying techniques lead to the deformation of the bulk cellulose 3-dimentional structure, thus the mechanical properties. The variation in thickness results in the varied compressive modulus values, which are proportional to the softness/stiffness of materials. Since the air drying results in denser fiber packing in the thickness direction, the air-dried bacterial cellulose becomes stiffer than the freeze-dried bacterial cellulose. The air-dried (without solvent replacement) and freeze-dried bacterial celluloses was found to provide the compressive moduli in the range of those of bone (90-100 kPa) and muscle (10-20 kPa), respectively (Taokaew et al., 2014b).

The other technique to vary the microstructure of bacterial cellulose hydrogel is compression. After biosynthesis and purification, bacterial cellulose hydrogels can be compressed between two metal plates. Retegi et. al. (2010) reported that under the compression force (10-100 MPa), interfibrillar spaces and porosity of bacterial cellulose became linearly decreased with the compression force, leading to an increased interfibrillar bonding, tensile strength, and modulus (Retegi et al., 2010).

Enlarging the typical pore size of bacterial cellulose hydrogel after biosynthesis has also been proposed. Yin et al. prepared porous bacterial cellulose using the facile Tween80-assisted foaming method in the azodicarbonamide-sodium hydroxide aqueous solution under a mild condition (Yin et al., 2012).

The resulting bacterial cellulose scaffold with interconnected macropores of 10-20 μm in diameter can be generated. However, this technique could not control uniformity of pores. In addition, the pore width of bacterial cellulose using the carbon dioxide laser perforation could effectively increase pore size to 200-300 μm with a greater homogeneity. Pore size of bacterial cellulose could be regulated by the channel diameter of the laser. (Ahrem et al., 2014).

Chemical Modifications

Bacterial cellulose dried differently that leads to physical changes is less useful when applying bacterial cellulose towards *in vitro* cell studies in terms of cell responsive behaviors. The chemical changes, e.g. chemical composition, charges, especially the changes of surface chemistry of bacterial cellulose are more crucial (Taokaew et al., 2015). Many researchers have therefore attempted to modify the typical bacterial cellulose into composites with other secondary materials using chemical modifications. Like physical modifications, chemical modifications can also be carried out during or after biosynthesis.

Biosynthetic Modifications

These modifications are performed by supplementing the secondary substance into the culture media of cellulose-producing bacteria. The main possibly mechanism is the physico-chemical interactions or the hydrogen bonding between fibrillary network of bacterial cellulose or the cellulose fiber and the secondary material. These interactions are aimed to provide the additional stability of bacterial cellulose-based composites (Taokaew et al., 2013b). By supplementing a particular material in the culture media, this material appears to be well incorporated into the cellulose fibril network. Pores of bacterial cellulose can be filled by the material, resulting in the bacterial cellulose-based composites to have smaller pore size as compared to the native cellulose. Moreover, the concentration of secondary material also affects other properties, such as optical transparency, crystallinity, mechanical property, water absorption capacity, and gas transmission rate, of the resulting bacterial cellulose (Saibuatong et al., 2010; Suratago et al., 2015; Taokaew et al., 2013b). In the case of using gelatin as the supplement in the culture medium, the maximum concentration of the gelatin in the culture media is found to be 10% w/v. The higher concentration strongly affects the bacterial cellulose hydrogel formation. At a concentration of > 5%, the cultivation period is found to be 3-7 days longer than the conventional cultivation without gelatin. Moreover, some gelatin gels are observed on the surface of the composite hydrogel and completely filled the space of the bacterial cellulose (Solway et al., 2011). This effect is similar to the introduction of other water-soluble polymers during the biosynthesis such as the introduction of collagen, alginate, aloe vera, and hyaluronic acid. The improved properties resulted from incorporating these secondary biopolymers are summarized in Table 2.

Post-Biosynthetic Modifications

Bulk Modification

The biosynthetic modification technique is majorly limited by such properties of active component supplemented as possessing anti-bacterial activity or providing the cell media with more viscose. The viscose growth medium obviously prevents free mobility of bacteria and dissolution of oxygen neces-

Table 2. Chemical modification during and after biosynthesis of bacterial cellulose

Modification Techniques	Introduced Secondary Biomaterials	Improved Properties
Biosynthetic modification	• Gelatin (Taokaew et al., 2013b). • Collagen (Wiegand et al., 2006). • Alginate (Kanjanamosit et al., 2010). • Aloe vera (Saibuatong & Phisalaphong, 2010). • Hyaluronic acid (Lopes et al., 2014). • Potato starch (Lv et al., 2016).	• Transparency, elongation in rehydrated state, water absorption. • Efficiency in wound healing, reduction of protease, interleukin, and reactive oxygen species. • Water absorption capacity, reduced oxygen transmission. • Water absorption capacity, mechanical properties, crystallinity, water vapor transmission. • Hydrophobic/hydrophobic-tailored character, roughness, thermal stability. • Controllable Pore size between 10-40 μm, muscle cells infiltration, efficiency in wound healing.
Post-biosynthetic modifications • Bulk modification	• Gelatin (Cai et al., 2010). • Collagen (Zhijiang et al., 2011). • Mangosteen extract (Taokaew et al., 2014a). • Alginate (Suratago et al., 2015). • Silk fibroin (Oliveira Barud et al., 2015). • Polyhydroxyalkonoates (Chiulan et al., 2016). • Nisin polypeptide (Nguyen et al., 2008).	• Young's modulus, biocompatibility. • Thermal stability, tensile strength, Young's modulus, skin cell affinity. • Anticancer property, bioactive compound release features. • Hydrophillicity, surface wettability, efficiency in separation of water-ethanol mixture. • Biocompatibility, cell affinity. • Thermal stability, cytocompatibility. • Antimicrobial property, efficiency in packaging films for meat products.

sary for the cellulose-producing aerobic bacteria. Moreover, other parameters affecting the formation of composites during the biosynthesis include temperature and pH (Chawla et al., 2009). Fabrication of bacterial cellulose-based composites by direct immersion mainly can overcome this limitation of the biosynthetic technique, since the higher loading of active component can also be available. With a great liquid absorbency and holding ability of bacterial cellulose, bioactive compounds including drugs can be loaded into its structural matrices.

Bacterial cellulose hydrogel can be immersed in a polymer solution and subsequently dried to entrap the active component/additive. The composites are formed by the penetration of the additive material into the network matrices of synthesized bacterial cellulose hydrogel. The physicochemical binding subsequently occurs between the cellulose fiber and the additives. After drying the bacterial cellulose-based composite hydrogel, its structure becomes dense. When the dried composite is rehydrated, the looser structure is formed, resulting release of the active component (Suratago et al., 2015; Taokaew et al., 2014a). Not only this bulk modification approach allows the change of fiber composition, but it also alters shape, morphology, pore structure, mechanical property of the modified bacterial cellulose. The loading mostly happens through the penetration of liquid containing target biomolecules into structural matrix of bacterial cellulose. Those bioactive compounds reported to be loaded into bacterial cellulose include gelatin, collagen, mangosteen extract, alginate, silk fibroin, polyhydroxyalkonoates, and nisin to provide the improved properties (Table 2).

The bulk modification can be performed by blending bacterial cellulose solution or slurry with other biopolymers, such as alginate, chitosan, and polyvinyl alcohol. Due to the strong hydrogen bonding between cellulose molecules, cellulose is insoluble in water. However, a cellulose solution could be prepared using other solvents, for example, Dimethylacetamide (DMAc)/LiCl, N-methylmorpholine N-oxide (NMMO), NaOH/urea, and ionic liquids (Gericke et al., 2013). In the case of slurry, bacterial cellulose hydrogels blended with other biopolymers can be simply homogenized at room temperature.

The readers are referred to the details of these approaches in a separated publication (Kirdponpattara et al., 2015; Muenduen et al., 2015). Briefly, such bulk modifications allow a higher loading of compatible biopolymers into bacterial cellulose, to achieve homogeneous microstructure with different pore architectures for various applications.

Surface Modification

This chemical modification after biosynthesis bases on termination of different functional groups on bacterial cellulose fibers. According to three hydroxyl groups of the glucose monomer unit resulting in abundance of reactive hydroxyl groups on the polymer chains of cellulose, the secondary functional groups can be covalently bonded to other chemicals. In the biomedical applications for example, the specific functional groups on the large surface area of bacterial cellulose give rise to improvement of biocompatibility due to greater affinity to macromolecules such as proteins in the cell media or cell lines (Sanchavanakit et al., 2006). As compared to expensive protein-based materials, bacterial cellulose exhibits low attachment and proliferation of cell lines according to low generation of contractile force between cells and cellulose. As a result, the cells become round up and are unable to reach confluent on the native bacterial cellulose. These techniques, including the use of gaseous plasma, irradiation, and chemical grafting, have been purposed to improve the cellulose responses of cell lines to bacterial cellulose.

The gaseous plasma generates photons, which strongly and uniformly interact with surface of bacterial cellulose without bulk alteration. The plasmas affect the surface by changing hydrophobicity, wettability, and subsequently cell-surface interactions. New functional groups, such as oxygenated function and amino group, created on the surface depend on the type of processing gas, such as air, oxygen, or nitrogen. The nitrogen plasma produces a lower water contact angle and a higher content of amino group as compared to the air plasma, leading to an increased wettability or imparting the hydrophilic character of bacterial cellulose (Charpentier et al., 2006) and improving cell-surface interaction, respectively (Pertile et al., 2010).

Electron-beam irradiation can also be used to induce charges in bacterial cellulose. The higher dose of irradiation shows a decrease in water absorption, pore size, and mechanical properties (becoming more brittleness, less elasticity) due to the higher crosslink density. However, these irradiated bacterial cellulose membranes were biocompatible with fibroblast cells, and promoted wound healing in an *in vivo* test (Mohamad et al., 2014). Also this surface modification technique could alter the microstructure and porosity of bacterial cellulose, when served as drug carriers, the irradiation dose between 5-25 kGy was found to lower the drug release rate from bacterial cellulose matrix (Stoica-Guzun et al., 2007).

On the other hand, chemical grafting is one of promising approaches without greatly affecting the physical and mechanical properties of the bulk materials. A group of most widely used chemicals to form the grafting on the material surface is organosilanes, which can be chemically grafted on the hydroxyl-rich surface of bacterial cellulose using their head groups (Pertile et al., 2010; Taokaew et al., 2015). As a result, the desired functional groups are pointing towards the surrounding to interact with cells or target proteins. The modification on the surface, especially improving the hydrophobicity of the surface, enhanced physisorption of extracellular matrix proteins through hydrophobic and electrostatic interactions (Faucheux et al., 2004; Toworfe et al., 2009). Positive charges of amine terminated organosilanes, e.g. 3-aminopropyltriethoxysilane (APTES), can be introduced on the surface of bacterial cellulose sheet. The APTES surface-modified bacterial cellulose resulted in cell spreading to cover the majority of the surface, as oppose to little to no cells on the un-modified bacterial cellulose (Taokaew et al., 2015). The positively-charged surface of bacterial cellulose was found to enhance affinity to cells as compared to its unmodified surface because of electrostatic interactions (Figure 3). In other words, the attachment and distribution of cell lines occurred due to the interaction between the positive charge of amine terminated surface and the negatively-charged surface of cells. Moreover, the cells were found to be over-crowded, formed multilayers, and then slightly detached on the surface of the APTES-modified bacterial cellulose. The rapid cell growth is triggered as a result of good cell adhesion, particularly good interaction between cells and the underlying substrate in the first phase of cellular response. This would be beneficial for tissue replacement or wound dressing applications (Taokaew et al., 2015) .

Furthermore, surface modification to create different degrees of hydrophilicity on the surface of bacterial cellulose appears to affect the cells affinity. After grafting methyl group on surface of bacterial cellulose using octadecyltrichlorosilane (OTS) to increase the surface hydrophobicity, non-spread rounded individual cells were mainly observed. The cells numbers on this hydrophobic-surface of bacterial cellulose were much less than that on unmodified bacterial cellulose possessing hydroxyl groups in nature. The faster absorption of proteins in the cell media, in particular bovine serum albumin, on the hydrophobic OTS modified surface of bacterial cellulose occurred owing to an enhanced hydrophobic interaction with proteins (Taokaew et al., 2015). Although the protein adsorption is guided by hydrophobic

Figure 3. Illustration of modifying bacterial cellulose using 3-aminopropyltriethoxysilane (APTES) (a), and responses of normal human dermal fibroblast (b1 and b2) and rat mesenchymal stem cells (b3 and b4) on unmodified (b1 and b3) and APTES-modified bacterial cellulose (b2 and b4). Each scale bars in b1-b4 represents 100 μm

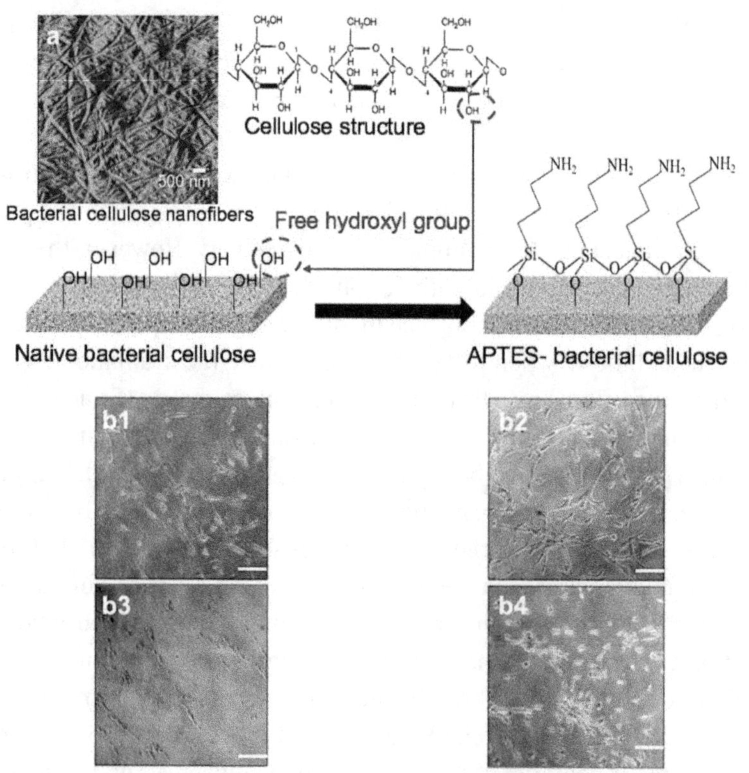

interactions, the proteins adsorbed on the surface are denatured or do not appeared in the appropriate configuration (Webb et al., 1998). Accordingly, the cell attachment and spreading cannot be promoted leading to cell death. For most cell types, survival of cells, including growth and function of cells, is induced by integrins, which are proteins that mechanically link cells to a substrate. With the absence of this focal adhesion through a protein layer, cells undergo programmed cell death.

The other different grafting materials, for instance, monofunctional methyl ether-polyethyl glycol (Kurniawan et al., 2013), epichlorohydrin, diethylenetriamine (Shen et al., 2009), carbon nanotube (O-Rak et al., 2013), and 2-acrylamido-2-methyl-1-propanesulfonic acid (C. W. Lin et al., 2013) were efficiently modified on the surface of bacterial cellulose to achieve the superior properties based on desired applications. Those improved properties such as biofouling, heavy metal adsorption capacity, dielectric property, and efficiency in solid electrolyte (Table 3).

APPLICATIONS OF BACTERIAL CELLULOSE

Since bacterial cellulose has many unique superior properties, such as purity, porosity, mechanical strength, crystallinity, and biocompatibility, based on the insights gained through research studies conducted to date, these properties are being utilized for various applications such as food, cosmetics, biomedical materials, separation technology, and the green products industry.

Food and Cosmetic Products

Bacterial cellulose or *Nata de Coco* has been reported to be the most popular product among the other bacterial cellulose products. In the nutritional consideration, it presents as no cholesterol, low fat, and low calories without any reported toxic reactions from consumption. The translucent and chewy texture of this kind of cellulose contains ~90% of water, and remains are pure cellulose fibers (Stephens et al., 1990). *Nata* is a dessert in syrup and frequently eaten with ice cream, fruit salad, and fruit cocktail (Figure 4). Two types of *Nata* i.e. *Nata de Pina* and *Nata de Coco* produced by pineapple juice and coconut juice, respectively are well known, but *Nata de Coco* is more commonly found in the food store worldwide. This unconventional product of bacterial cellulose is native to the Philippines and locally first developed in 1949. The export of processed *Nata de Coco* from the Philippines reported by Asian

Table 3. Chemical modification based on surface modification of other grafting materials on bacterial cellulose fibers

Grafting Materials	Improved Properties
• 3-aminopropyltriethoxysilane (Rouabhia et al., 2014; Taokaew et al., 2015). • Octadecyltrichlorosilane (Taokaew et al., 2015). • Monofunctional methyl ether- polyethylene glycol (Kurniawan et al., 2013). • Epichlorohydrin and diethylenetriamine (Shen et al., 2009). • Carbon nanotube (O-Rak et al., 2013). • 2-acrylamido-2-methyl-1-propanesulfonic acid (C. W. Lin et al., 2013).	• Skin cell attachment and spreading, antimicrobial property. • Surface hydrophobicity. • Biofouling resistance. • Adsorption of metal ions (Cu (II) and Pb (II)). • Polarozation effect, dielectric property, efficiency in flexible touch screen application. • Methanol anti-sorption, efficiency in solid electrolyte for direct methanol fuel cells.

and Pacific Coconut community (APCC) has had traded more than 5 thousand MT with the revenue of over US$ 5 million in 2009-2011, while the raw *Nata* has exported in much smaller volume ("Market and trade of coconut products," 2013). In 2012, shipment of total *Nata de Coco* slightly dropped by 18.6% from 2011. United Coconut Associations of the Philippines (UCAP) reported earnings from *Nata de Coco* exports that the Philippines contributed US$ 5.030 million from the volume of 5,164 MT. Japan was a principal market responsible for 3,556 MT (68.9%) followed distance behind by USA at 683 MT (13.2%). Thirty-seven other countries including Korea and Canada took in the remaining 926 MT in such year. In 2013, the extensive outlet for ten months was also reported as 3,767 MT. The industry news of UCAP reported that *Nata de Coco* products generated export income more than US$ 100,000 each month. The major volume has gone primarily to Japan (60-90%) until 2014 ("Performance of Philippines' Top Non-Traditional Coco Products Export,"). Japan has been a major destination since 1990s (Mae, 2007) and it controlled over 2,202 MT or 82% of total business in 2014, while the other countries i.e. china, Malaysia, Singapore, Hong Kong, Bahrain, UAE, USA, Italy, Germany, and Qatar shared the other 18%.

For cosmetic application, a clinical study in 2011 reported a good satisfaction of users after applied bacterial cellulose mask trial product (Amnuaikit et al., 2011). The pure bacterial cellulose mask was able to fit well and hydrate skin (Figure 5). On the basis of remarkable absorptive properties, permeability for liquids and gases, low irritation of skin, and high mechanical strength in wet state, bacterial cellulose has been launched as facial mask with incorporated active component. These products made in Asia such as Thailand and South Korea have been found in cosmetic stores.

Figure 4. Nata de Coco products: Nata de Coco in yogurt, heavy syrup, and light syrup with pineapple dices (left to right)

Figure 5. Commercial bacterial cellulose mask products (a) and appearance of bacterial cellulose mask (b1) during application as compared to that of the common cloth mask (b2)

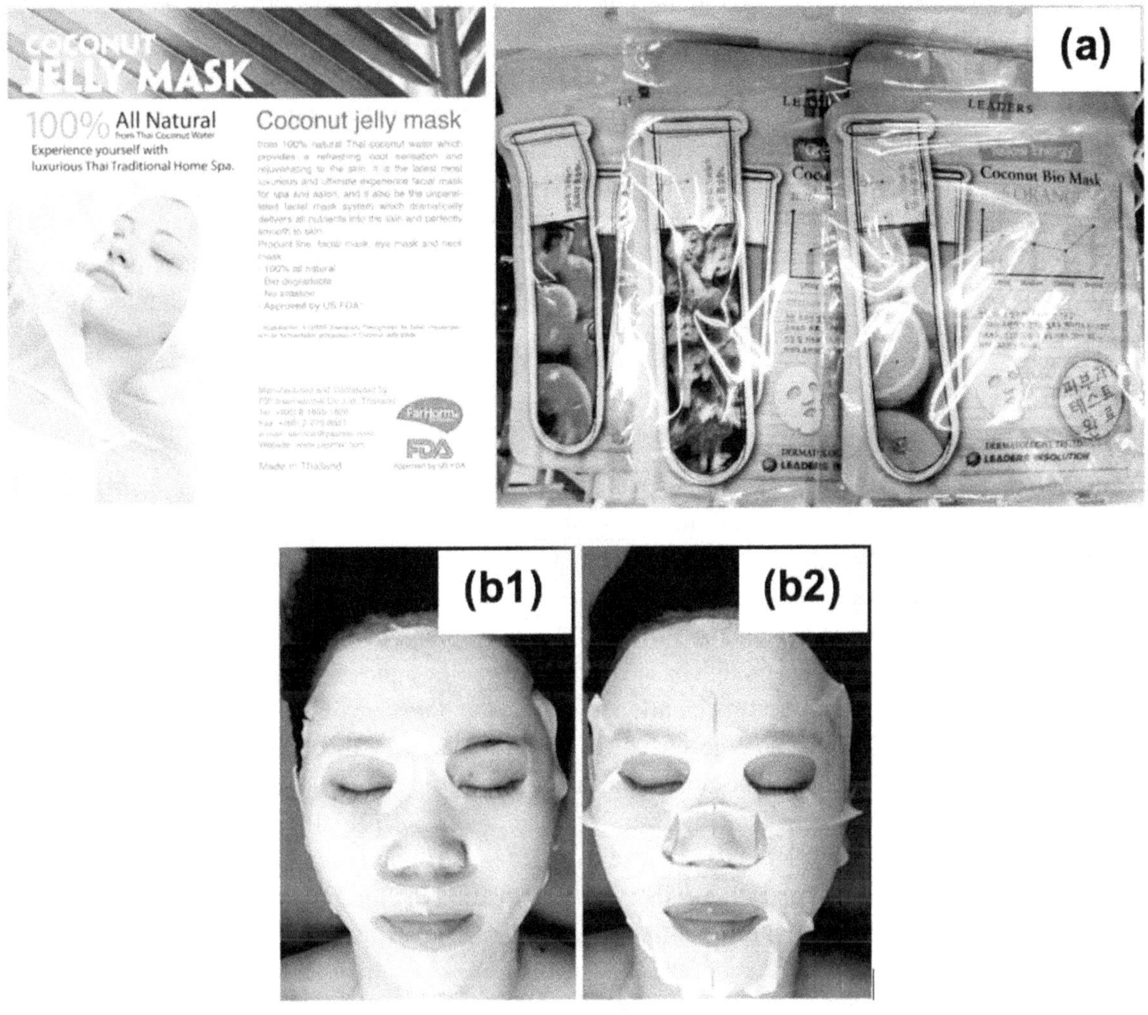

Biomedical Applications

Wound Dressing

According to the high purity and the biocompatibility towards living cells/tissues, bacterial cellulose has become an interesting candidate for medical applications; such as artificial skin, artificial blood vessels, vascular grafts, scaffolds for tissue engineering, and wound dressing. The bacterial cellulose membrane used as a wound dressing material has been reported to shorten healing time in the diabetic foot ulcers and burns as compared to the commercial gauze dressing (Solway et al., 2011). Bacterial cellulose can prevent itself from drying by retaining moisture from surroundings. This feature is beneficial for wound dressing application, especially for burn patients, since when a dressing dries after absorption of the exudates and other wound tears would lead to the sticking of dressing material to the wound, thus causing pain. Although bacterial cellulose is an effective wound dressing material, it has no antimicrobial activities to prevent the infection. Bacterial cellulose-based composites have been synthesized by incorporating an anti-microbial agent to impart the antimicrobial activity (Kucińska-Lipka et al., 2015).

The first wound care product as a liquid-loaded pad made from this kind of cellulose was patented by Johnson & Johnson Products, Inc. in 1987 (Ring et al., 1987). Bacterial cellulose was then commercially developed as wound dressing under the name of Biofill in 1990. It was proved that bacterial cellulose which was obtained from static culture and dried into biofilm could be applied as a temporary skin substitute in the therapy of skin wounds such as burns, ulcers, grafts, and as an adjuvant in dermal abrasions (Fontana et al., 1990; Fontana et al., 1991). Gengiflex and Bioprocess was also lunched for bacterial cellulose product in the family of wound care product for surgery and dental implant (Jonas et al., 1998) Xcell, antimicrobial wound dressing, made from bacterial cellulose was patented by Xylos Corporation in 2010 (Serafica et al., 2010).

Stem Cells Therapy

The moduli of bacterial cellulose can be easily tailored to promote the differentiation of stem cells. Microenvironment with stiffness of underlying substrate close to the native tissue plays a key role on the differentiation of stem cells into specific cell types. After cell attachment, which occurs independently on the stiffness of a substrate, the attached cells sense the surface of substrate by exhibiting morphologic changes (Engler et al., 2006; Rowlands et al., 2008). Early response of stem cells could be observed on bacterial cellulose having different moduli based on the morphology and expression of the cell phenotype. Mesenchymal stem cells, which are sufficiently larger than the pore size of bacterial cellulose, appeared to only deposit, attach, and spread on the surface of bacterial cellulose during 4 weeks of investigation. The stem cells anchored and survived on the surface of bacterial cellulose, and the cells distribution was relatively uniform. However, the cells were not well spread, especially on the air dried bacterial cellulose. The stem cells sense bacterial cellulose with different rigidity mimicking the elasticity of each cell type. Fates of the mesenchymal stem cells affected by varied moduli of bacterial cellulose under the culture condition with and without differentiation agents were investigated. It was found that the stiffness of underlying substrates alone is deficient to regulate the cells differentiation. The differentiation agents are required to increase programming of mesenchymal stem cells into functional cells (Favi et al., 2013; Taokaew et al., 2013a). After induction of differentiation in chondrogenic and osterogenic media, the stem cells differentiate considerably on air-dried and freeze-dried bacterial celluloses (Figure 6). Chondrogenesis (differentiation into chrondrocytes) did not significantly appear on air-dried bacterial cellulose, but the chondrogenic differentiation was observed on the freeze dried bacterial cellulose. The stem cells largely remained undifferentiated on the stiffer air-dried bacterial cellulose. The chondrogenetic differentiation is difficult on the non-protein substrate, and the long period of incubation (more than 3 weeks) is required. In fact, collagen is a protein substrate maintaining the chrondrocyte function (Ikenoue et al., 2003). For the osteogenic differentiation, the differential commitment occurred on the air-dried bacterial cellulose since the aggregate cells mostly observed on the stiffer air-dried bacterial cellulose as compared to the softer freeze-fried bacterial cellulose. These clusters of stem cells directly impact the osteogenesis by increasing hydroxylproline, calcium, and gene expression as well as the cell-cell interactions, but do not influent chondrogenesis (Wang et al., 2010; Xue et al., 2013). These findings open-up the application of bacterial cellulose in the field of stem cells (Acasigua et al., 2014).

Figure 6. Mesenchymal stem cells differentiated into different cells type on bacterial cellulose possessing different compressive moduli. Bacterial cellulose hydrogel (a) was dried by means of air and freeze drying. Air- and freeze-dried bacterial cellulose (b1 and b2) exhibit the compact and loose cross-sectional morphology (c1 and c2) leading to the moduli of ~ 20 and 90 kPa, respectively. The stem cells sensed the dried bacterial cellulose by differentiation into osteoblasts (red) and chondrocytes (blue) indicated by arrows. Each scale bars in SEM and microscopic images represent 2 and 100 μm, respectively

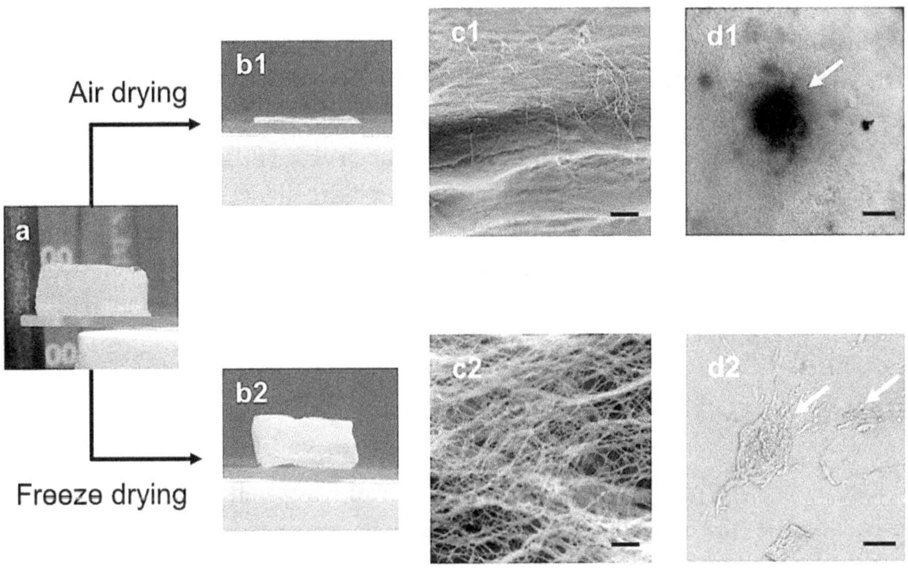

Drug Delivery and Controlled Release of Active Components

Other characteristics of bacterial cellulose, such as small pore size, high water holding ability, allow it to serve as a carrier or matrix for delivery active component to the target area. Bacterial cellulose allowed for gradual release of active component, and the release characteristics could be controlled by the surface modification such as electron beam irradiation to lengthen the release time (Klechkovskaya et al., 2008; Stoica-Guzun et al., 2007; Taokaew et al., 2014a). Drugs can be incorporated after biosynthesis by sorption technique using the advantage of mild condition to avoid the drug deformation. Those drugs available for delivery system and controlled release application from bacterial cellulose matrices includes lidocaine (Trovatti et al., 2011), vancomycin (Mori et al., 2011), gentamicin (Mori et al., 2011), Ibuprofen (Trovatti et al., 2012), Amoxicillin (Pavaloiu et al., 2014), octenidine (Moritz et al., 2014),and doxorubicin (Cacicedo et al., 2016).

Membrane Separation

Highly porous bacterial cellulose thin sheets with a large specific surface area, pore interconnectivity, tunable pore size and extremely low thermal conductivity are excellent in membrane separation technology. Bulk modification of bacterial cellulose hydrogels with alginate and then air-dried to obtain thin dense membranes with small pores (~30-100 μm) would result in the enhanced pervaporation efficiency for separating azeotropic mixture of ethanol/water (Suratago et al., 2015). With a permeate pressure of

10 mmHg, a total permeate flux of 33 g/(m^2h) and a water/ethanol selectivity of 429.9 has been resulted using such membrane. This means that less ethanol (4.2%) was in the permeate. The total permeate flux and the selectivity can be further improved by decreasing the permeate pressure (Suratago et al., 2015).

In addition, bacterial cellulose with surface modified to improve its hydrophobicity, these modified bacterial cellulose membranes can be applied for other separation applications, such as membrane distillation and oil/water separation (Leitch et al., 2016; Sai et al., 2015). The hydrophobic bacterial cellulose obtained from chemical deposition using (tridecafluoro-1,1,2,2-tetrahydrooctyl)-trichlorosilane is applicable in membrane distillation for desalination due to its low thermal conductivity, i.e., 0.027 W/(m·K), at the ambient pressure and a high void fraction in the thin membrane. As a result, it reduced the heat flux from the feed to the permeate stream, and produced 2.4 times higher fresh water permeate as compared to the commercial benchmark membrane (Leitch et al., 2016). Furthermore, surface modification of bacterial cellulose using hydrophobic trimethylchlorosilane leads to hydrophobic/oleophilic character. This is applicable for oil absorption in removing oil spills in the marine environment owing to its selective separation of water and oil (such as gasoline, diesel, and plant oil). Moreover, the trimethylchlorosilane-modified bacterial cellulose is also able to absorb various organic solvents (e.g. n-hexane, acetone, paraffin liquid, toluene, chlorobenzene, dichloromethane, and chloroform) (Sai et al., 2015).

Green Products

Based on a great number of publications for green material, it is well known that bacterial cellulose possesses the peculiarities to contribute to a multidisciplinary material for numerous applications. It has received a lot of attention and possibly will be introduced to the green product market in the near future. Using bacterial cellulose to replace plastic product has been considered.

For example, plant oils have been used for decades in plastic resin applications. These plant-oil-based natural resins need to be reinforced with ultrafine fibers usually (Yano et al., 2005). Cellulose nanofibers would be an alternative to use as reinforcing agent. With biodegradability, transparency, and low toxicity for human, an oxidized-soy bean oil (ESO)-based bacterial cellulose was developed for food packing purpose (Retegi et al., 2012).

In addition, other bacterial cellulose-based composites, such as bacterial cellulose/nisin using as antimicrobial film to preserve processed meats and bacterial cellulose/polylysine for use as sausage packaging film, have been reported. Moreover, this kind of cellulose could be a good emulsifier in food. In traditional Chinese-style fish balls or Mahi-mahi Surimi product, lard is typically added to enhance gelation. It has been investigated that replacing fat and dietary fiber source with bacterial cellulose appears to be feasible (Lin et al., 2011). On the other hand, other potential applications of bacterial cellulose composites containing ESO include airplane parts and car bodies (Retegi et al., 2012), and epoxy resin/bacterial cellulose composite can be used to fabricate organic light-emitting diodes. On the basis of the remarkable properties of bacterial cellulose and the ability to be easily modified, applications of bacterial cellulose seem to be endless.

CONCLUSION

Bacterial cellulose is an alternative cellulose source obtained from environmentally-friendly synthesis by a certain type of bacteria. It is one of functional environmental materials, which is interesting in the area of material science for sustainability. According to outstanding characteristics and inexpensive material, bacterial cellulose and its composites can be prepared to have different properties and then achieve particular applications. The preparation is referred to cellulose-producing bacteria culture media, cultivation conditions, and purification of the resultant bacterial cellulose hydrogel. The optional modification techniques towards the native cellulose hydrogel, including physical and chemical modifications are highlighted in the subsequent sections. Biosynthetic and post-synthetic modifications are classified as subsections of those techniques. The properties of the bacterial cellulose-based composites are deviated from the native cellulose, for example the porous structure, mechanical properties, chemical characteristics, and biological properties. This imparts the wider range of utilization in the fields of biomedical as well as other green applications. Moreover, this kind of environmental material has been launched as commercial products in terms of food and skin care products to date.

REFFERENCES

Acasigua, G. A. X., de Olyveira, G. M., Costa, L. M. M., Braghirolli, D. I., Fossati, A. C. M., Guastaldi, A. C., & Basmaji, P. et al. (2014). Novel chemically modified bacterial cellulose nanocomposite as potential biomaterial for stem cell therapy applications. *Current Stem Cell Research & Therapy*, *9*(2), 117–123. doi:10.2174/1574888X08666131124135654 PMID:24266631

Ahrem, H., Pretzel, D., Endres, M., Conrad, D., Courseau, J., Müller, H., & Kinne, R. W. et al. (2014). Laser-structured bacterial nanocellulose hydrogels support ingrowth and differentiation of chondrocytes and show potential as cartilage implants. *Acta Biomaterialia*, *10*(3), 1341–1353. doi:10.1016/j.actbio.2013.12.004 PMID:24334147

Amnuaikit, T., Chusuit, T., Raknam, P., & Boonme, P. (2011). Effects of a cellulose mask synthesized by a bacterium on facial skin characteristics and user satisfaction. *Medical Devices: Evidence and Research*, *4*(1), 77–81. PMID:22915933

Bäckdahl, H., Esguerra, M., Delbro, D., Risberg, B., & Gatenholm, P. (2008). Engineering microporosity in bacterial cellulose scaffolds. *Journal of Tissue Engineering and Regenerative Medicine*, *2*(6), 320–330. doi:10.1002/term.97 PMID:18615821

Bashaiwoldu, A. B., Podczeck, F., & Newton, J. M. (2004). A study on the effect of drying techniques on the mechanical properties of pellets and compacted pellets. *European Journal of Pharmaceutical Sciences*, *21*(2-3), 119–129. doi:10.1016/j.ejps.2003.09.013 PMID:14757483

Bernardo, E. B., Neilan, B. A., & Couperwhite, I. (1998). Characterization, Differentiation and Identification of Wild-type Cellulose-synthesizing Acetobacter strains Involved in Nata de Coco Production. *Systematic and Applied Microbiology*, *21*(4), 599–608. doi:10.1016/S0723-2020(98)80073-8

Brown, A. J. (1886). XLIII.-On an acetic ferment which forms cellulose. *Journal of the Chemical Society Transactions*, *49*(0), 432–439. doi:10.1039/CT8864900432

Brown, M. R. Jr. (1991). *Advances in Cellulose Biosynthesis Polymers from Biobased Materials*. Prentice hall.

Budhiono, A., Rosidi, B., Taher, H., & Iguchi, M. (1999). Kinetic aspects of bacterial cellulose formation in nata-de-coco culture system. *Carbohydrate Polymers*, *40*(2), 137–143. doi:10.1016/S0144-8617(99)00050-8

Cacicedo, , M., León, I., Gonzalez, J., Porto, L., Alvarez, V., & Castro, G. R. (. (2016). Modified bacterial cellulose scaffolds for localized doxorubicin release in human colorectal HT-29 cells. *Colloids and Surfaces. B, Biointerfaces*, *140*, 421–429. doi:10.1016/j.colsurfb.2016.01.007 PMID:26784658

Cai, Z., & Kim, J. (2010). Preparation and Characterization of Novel Bacterial Cellulose/Gelatin Scaffold for Tissue Regeneration Using Bacterial Cellulose Hydrogel. *Journal of Nanotechnology in Engineering and Medicine*, *1*(2), 1–6. doi:10.1115/1.4000858

Charpentier, P. A., Maguire, A., & Wan, W. (2006). Surface modification of polyester to produce a bacterial cellulose-based vascular prosthetic device. *Applied Surface Science*, *252*(18), 6360–6367. doi:10.1016/j.apsusc.2005.09.064

Charreau, H., Foresti, M. L., & Vázquez, A. (2013). Nanocellulose patents trends: A comprehensive review on patents on cellulose nanocrystals, microfibrillated and bacterial cellulose. *Recent Patents on Nanotechnology*, *7*(1), 56–80. doi:10.2174/187221013804484854 PMID:22747719

Chawla, P. R., Bajaj, I. B., Survase, S. A., & Singhal, R. S. (2009). Microbial cellulose: Fermentative production and applications. *Food Technology and Biotechnology*, *47*(2), 107–124.

Cheng, K.-C., Catchmark, J., & Demirci, A. (2009). Enhanced production of bacterial cellulose by using a biofilm reactor and its material property analysis. *Journal of Biological Engineering*, *3*(1), 1–12. doi:10.1186/1754-1611-3-12 PMID:19630969

Chiulan, I., Mihaela Panaitescu, D., Nicoleta Frone, A., Teodorescu, M., Andi Nicolae, C., Căşărică, A., & Sălăgeanu, A. et al. (2016). Biocompatible polyhydroxyalkanoates/bacterial cellulose composites: Preparation, characterization, and in vitro evaluation. *Journal of Biomedical Materials Research. Part A*, *104*(10), 2576–2584. doi:10.1002/jbm.a.35800 PMID:27242044

Cotton, J., Pometto, A., & Gvozdenovic-Jeremic, J. (2001). Continuous lactic acid fermentation using a plastic composite support biofilm reactor. *Applied Microbiology and Biotechnology*, *57*(5-6), 626–630. doi:10.1007/s002530100820 PMID:11778870

Czaja, W., Krystynowicz, A., Bielecki, S., & Brown, R. M. Jr. (2006). Microbial cellulose - The natural power to heal wounds. *Biomaterials*, *27*(2), 145–151. doi:10.1016/j.biomaterials.2005.07.035 PMID:16099034

Dahman, Y. (2009). Nanostructured biomaterials and biocomposites from bacterial Cellulose nanofibers. *Journal of Nanoscience and Nanotechnology*, *9*(9), 5105–5122. doi:10.1166/jnn.2009.1466 PMID:19928189

Engler, A. J., Sen, S., Sweeney, H. L., & Discher, D. E. (2006). Matrix Elasticity Directs Stem Cell Lineage Specification. *Cell*, *126*(4), 677–689. doi:10.1016/j.cell.2006.06.044 PMID:16923388

Faucheux, N., Schweiss, R., Lützow, K., Werner, C., & Groth, T. (2004). Self-assembled monolayers with different terminating groups as model substrates for cell adhesion studies. *Biomaterials*, *25*(14), 2721–2730. doi:10.1016/j.biomaterials.2003.09.069 PMID:14962551

Favi, P. M., Benson, R. S., Neilsen, N. R., Hammonds, R. L., Bates, C. C., Stephens, C. P., & Dhar, M. S. (2013). Cell proliferation, viability, and in vitro differentiation of equine mesenchymal stem cells seeded on bacterial cellulose hydrogel scaffolds. *Materials Science and Engineering C*, *33*(4), 1935–1944. doi:10.1016/j.msec.2012.12.100 PMID:23498215

Fontana, J., De Souza, A., Fontana, C., Torriani, I., Moreschi, J., Gallotti, B., & Farah, L. et al. (1990). Acetobacter cellulose pellicle as a temporary skin substitute. *Applied Biochemistry and Biotechnology*, *24-25*(1), 253–264. doi:10.1007/BF02920250 PMID:2353811

Fontana, J. D., Franco, V. C., De Souza, S. J., Lyra, I. N., & De Souza, A. M. (1991). Nature of plant stimulators in the production of Acetobacter xylinum (tea fungus) biofilm used in skin therapy. *Applied Biochemistry and Biotechnology*, *28-29*(1), 341–351. doi:10.1007/BF02922613 PMID:1929372

Gao, W. H., Chen, K. F., Yang, R. D., Yang, F., & Han, W. J. (2011). Properties of bacterial cellulose and its influence on the physical properties of paper. *BioResources*, *6*(1), 144–153.

George, J., Ramana, K. V., Sabapathy, S. N., & Bawa, A. S. (2005). Physico-Mechanical Properties of Chemically Treated Bacterial (Acetobacter xylinum) Cellulose Membrane. *World Journal of Microbiology & Biotechnology*, *21*(8), 1323–1327. doi:10.1007/s11274-005-3574-0

Gericke, M., Trygg, J., & Fardim, P. (2013). Functional cellulose beads: Preparation, characterization, and applications. *Chemical Reviews*, *113*(7), 4812–4836. doi:10.1021/cr300242j PMID:23540980

Haimer, E., Wendland, M., Schlufter, K., Frankenfeld, K., Miethe, P., Potthast, A., & Liebner, F. et al. (2010). Loading of Bacterial Cellulose Aerogels with Bioactive Compounds by Antisolvent Precipitation with Supercritical Carbon Dioxide. *Macromolecular Symposia*, *294*(2), 64–74. doi:10.1002/masy.201000008

Hestrin, S., & Schramm, M. (1954). Synthesis of cellulose by Acetobacter xylinum. 2. Preparation of freeze-dried cells capable of polymerizing glucose to cellulose. *The Biochemical Journal*, *58*(2), 345–352. doi:10.1042/bj0580345 PMID:13208601

Hioki, N., Hori, Y., Watanabe, K., Morinaga, Y., Yoshinaga, F., Hibino, Y., & Ogura, T. (1995). Bacterial Cellulose; as a New Material for Papermaking. *Japan Tappi Journal*, *49*(4), 718–723. doi:10.2524/jtappij.49.718

Hong, F., & Qiu, K. (2008). An alternative carbon source from konjac powder for enhancing production of bacterial cellulose in static cultures by a model strain Acetobacter aceti subsp. xylinus ATCC 23770. *Carbohydrate Polymers*, *72*(3), 545–549. doi:10.1016/j.carbpol.2007.09.015

Hornung, M., Ludwig, M., & Schmauder, H. P. (2007). Optimizing the Production of Bacterial Cellulose in Surface Culture: A Novel Aerosol Bioreactor Working on a Fed Batch Principle (Part 3). *Engineering in Life Sciences*, *7*(1), 35–41. doi:10.1002/elsc.200620164

Iguchi, M., Mitsuhashi, S., Ichimura, K., Nishi, Y., Uryu, M., Yamanaka, S., & Watanabe, K. (1988). *Bacterial cellulose-containing molding material having high dynamic strength*. Google Patents.

Iguchi, M., Yamanaka, S., & Budhiono, A. (2000a). Bacterial cellulose - a masterpiece of natures arts. *Journal of Materials Science*, *35*(2), 261–270. doi:10.1023/A:1004775229149

Iguchi, M., Yamanaka, S., & Budhiono, A. (2000b). Bacterial cellulose-a masterpiece of natures arts. *Journal of Materials Science*, *35*(2), 261–270. doi:10.1023/A:1004775229149

Ikenoue, T., Trindade, M. C. D., Lee, M. S., Lin, E. Y., Schurman, D. J., Goodman, S. B., & Smith, R. L. (2003). Mechanoregulation of human articular chondrocyte aggrecan and type II collagen expression by intermittent hydrostatic pressure in vitro. *Journal of Orthopaedic Research*, *21*(1), 110–116. doi:10.1016/S0736-0266(02)00091-8 PMID:12507587

Jonas, R., & Farah, L. F. (1998). Production and application of microbial cellulose. *Polymer Degradation & Stability*, *59*(1–3), 101–106. doi:10.1016/S0141-3910(97)00197-3

Jozala, A. F., de Lencastre-Novaes, L. C., Lopes, A. M., de Carvalho Santos-Ebinuma, V., Mazzola, P. G., Pessoa-Jr, A., & Chaud, M. V. et al. (2016). Bacterial nanocellulose production and application: A 10-year overview. *Applied Microbiology and Biotechnology*, *100*(5), 2063–2072. doi:10.1007/s00253-015-7243-4 PMID:26743657

Kamarudin, S., Mohd Sahaid, K., Mohd Sobri, T., Wan Mohtar, W. Y., Dayang Radiah, A. B., & Norhasliza, H. (2013). Different media formulation on biocellulose production by Acetobacter xylinum (0416). *Pertanika Journal of Science and Technology*, *21*(1), 29–36.

Kanjanamosit, N., Muangnapoh, C., & Phisalaphong, M. (2010). Biosynthesis and characterization of bacteria cellulose–alginate film. *Journal of Applied Polymer Science*, *115*(3), 1581–1588. doi:10.1002/app.31138

Kirdponpattara, S., Khamkeaw, A., Sanchavanakit, N., Pavasant, P., & Phisalaphong, M. (2015). Structural modification and characterization of bacterial cellulose–alginate composite scaffolds for tissue engineering. *Carbohydrate Polymers*, *132*, 146–155. doi:10.1016/j.carbpol.2015.06.059 PMID:26256335

Klechkovskaya, V. V., Volkov, V. V., Shtykova, E. V., Arkharova, N. A., Baklagina, Y. G., Khripunov, A. K., & Tkachenko, A. A. (2008). Network Model of Acetobacter Xylinum Cellulose Intercalated by Drug Nanoparticles. In M. Giersig & G. Khomutov (Eds.), *Nanomaterials for Application in Medicine and Biology* (pp. 165–177). Springer Netherlands. doi:10.1007/978-1-4020-6829-4_14

Klemm, D., Heublein, B., Fink, H. P., & Bohn, A. (2005). Cellulose: Fascinating biopolymer and sustainable raw material. *Angewandte Chemie International Edition, 44*(22), 3358–3393. doi:10.1002/anie.200460587 PMID:15861454

Kucińska-Lipka, J., Gubanska, I., & Janik, H. (2015). Bacterial cellulose in the field of wound healing and regenerative medicine of skin: Recent trends and future prospectives. *Polymer Bulletin (Berlin), 72*(9), 2399–2419. doi:10.1007/s00289-015-1407-3

Kurniawan, H., Ye, Y.-S., Kuo, W.-H., Lai, J.-T., Wang, M.-J., & Liu, H.-S. (2013). Improvement of biofouling resistance on bacterial cellulose membranes. *Biochemical Engineering Journal, 78*, 138–145. doi:10.1016/j.bej.2013.03.021

Kurosumi, A., Sasaki, C., Yamashita, Y., & Nakamura, Y. (2009). Utilization of various fruit juices as carbon source for production of bacterial cellulose by Acetobacter xylinum NBRC 13693. *Carbohydrate Polymers, 76*(2), 333–335. doi:10.1016/j.carbpol.2008.11.009

Lee, S. H., Lim, Y. M., Jeong, S. I., An, S. J., Kang, S. S., Jeong, C. M., & Huh, J. B. (2015). The effect of bacterial cellulose membrane compared with collagen membrane on guided bone regeneration. *Journal of Advanced Prosthodontics, 7*(6), 484–495. doi:10.4047/jap.2015.7.6.484 PMID:26816579

Leitch, M. E., Li, C., Ikkala, O., Mauter, M. S., & Lowry, G. V. (2016). Bacterial Nanocellulose Aerogel Membranes: Novel High-Porosity Materials for Membrane Distillation. *Environmental Science and Technology Letters, 3*(3), 85–91. doi:10.1021/acs.estlett.6b00030

Liebner, F., Haimer, E., Wendland, M., Neouze, M. A., Schlufter, K., Miethe, P., & Rosenau, T. et al. (2010). Aerogels from unaltered bacterial cellulose: Application of scCO <inf>2</inf> drying for the preparation of shaped, ultra-lightweight cellulosic aerogels. *Macromolecular Bioscience, 10*(4), 349–352. doi:10.1002/mabi.200900371 PMID:20166232

Lin, C. W., Liang, S. S., Chen, S. W., & Lai, J. T. (2013). Sorption and transport properties of 2-acrylamido-2-methyl-1- propanesulfonic acid-grafted bacterial cellulose membranes for fuel cell application. *Journal of Power Sources, 232*, 297–305. doi:10.1016/j.jpowsour.2013.01.047

Lin, S. B., Chen, L. C., & Chen, H. H. (2011). Physical characteristics of surimi and bacterial cellulose composite gel. *Journal of Food Process Engineering, 34*(4), 1363–1379. doi:10.1111/j.1745-4530.2009.00533.x

Lin, S. P., Loira Calvar, I., Catchmark, J. M., Liu, J. R., Demirci, A., & Cheng, K. C. (2013). Biosynthesis, production and applications of bacterial cellulose. *Cellulose (London, England), 20*(5), 2191–2219. doi:10.1007/s10570-013-9994-3

Lopes, T. D., Riegel-Vidotti, I. C., Grein, A., Tischer, C. A., & Faria-Tischer, P. C. S. (2014). Bacterial cellulose and hyaluronic acid hybrid membranes: Production and characterization. *International Journal of Biological Macromolecules, 67*, 401–408. doi:10.1016/j.ijbiomac.2014.03.047 PMID:24704166

Lv, X., Yang, J., Feng, C., Li, Z., Chen, S., Xie, M., & Xu, Y. et al. (2016). Bacterial Cellulose-Based Biomimetic Nanofibrous Scaffold with Muscle Cells for Hollow Organ Tissue Engineering. *ACS Biomaterial Science and Engineering, 2*(1), 19–29. doi:10.1021/acsbiomaterials.5b00259

Mae, T.-M. E. (2007). Development of Functional Foods in the Philippines. *Food Science and Technology Research, 13*(3), 179–186. doi:10.3136/fstr.13.179

Market and trade of coconut products. (2013). Retrieved March 2015 from http://www.fao.org/fileadmin/templates/rap/files/meetings/2013/131030-market.pdf

Mazur, P. (1970). Cryobiology: The freezing of biological systems. *Science, 168*(3934), 939–949. doi:10.1126/science.168.3934.939 PMID:5462399

Mohamad, N., Mohd Amin, M. C. I., Pandey, M., Ahmad, N., & Rajab, N. F. (2014). Bacterial cellulose/acrylic acid hydrogel synthesized via electron beam irradiation: Accelerated burn wound healing in an animal model. *Carbohydrate Polymers, 114*, 312–320. doi:10.1016/j.carbpol.2014.08.025 PMID:25263896

Mori, R., Nakai, T., Enomoto, K., Uchio, Y., & Yoshino, K. (2011). Increased antibiotic release from a bone cement containing bacterial cellulose. *Clinical Orthopaedics and Related Research, 469*(2), 600–606. doi:10.1007/s11999-010-1626-8 PMID:20945120

Moritz, S., Wiegand, C., Wesarg, F., Hessler, N., Müller, F. A., Kralisch, D., & Fischer, D. et al. (2014). Active wound dressings based on bacterial nanocellulose as drug delivery system for octenidine. *International Journal of Pharmaceutics, 471*(1–2), 45–55. doi:10.1016/j.ijpharm.2014.04.062 PMID:24792978

Muenduen, P., & Suchata, K. (2015). *Synthesis and Characterization of Bacterial Cellulose-Based Composites and Their Applications. In Handbook of Sustainable Polymers* (pp. 495–532). Pan Stanford.

Mühlethaler, K. (1949). The structure of bacterial cellulose. *Biochimica et Biophysica Acta, 3*(0), 527–535. doi:10.1016/0006-3002(49)90126-2

Murphy, C. M., & OBrien, F. J. (2010). Understanding the effect of mean pore size on cell activity in collagen-glycosaminoglycan scaffolds. *Cell Adhesion & Migration, 4*(3), 377–381. doi:10.4161/cam.4.3.11747 PMID:20421733

Myers, R. S., & Clever, H. L. (1969). Surface tension of octamethylcyclotetrasiloxane and hexamethyldisilazane and their solutions with carbon tetrachloride and n-hexadecane. *Journal of Chemical & Engineering Data, 14*(2), 161–164. doi:10.1021/je60041a014

Nguyen, V. T., Gidley, M. J., & Dykes, G. A. (2008). Potential of a nisin-containing bacterial cellulose film to inhibit Listeria monocytogenes on processed meats. *Food Microbiology, 25*(3), 471–478. doi:10.1016/j.fm.2008.01.004 PMID:18355672

Nishi, Y., Uryu, M., Yamanaka, S., Watanabe, K., Kitamura, N., Iguchi, M., & Mitsuhashi, S. (1990). The structure and mechanical properties of sheets prepared from bacterial cellulose. *Journal of Materials Science, 25*(6), 2997–3001. doi:10.1007/BF00584917

O-Rak, K., Ummartyotin, S., Sain, M., & Manuspiya, H. (2013). Covalently grafted carbon nanotube on bacterial cellulose composite for flexible touch screen application. *Materials Letters, 107*, 247–250. doi:10.1016/j.matlet.2013.05.126

Oliveira Barud, H. G., Barud, H. D. S., Cavicchioli, M., Do Amaral, T. S., De Oliveira Junior, O. B., Santos, D. M., & Ribeiro, S. J. L. et al. (2015). Preparation and characterization of a bacterial cellulose/silk fibroin sponge scaffold for tissue regeneration. *Carbohydrate Polymers*, *128*, 41–51. doi:10.1016/j.carbpol.2015.04.007 PMID:26005138

Ougiya, H., Watanabe, K., Morinaga, Y., & Yoshinaga, F. (1997). Emulsion-stabilizing Effect of Bacterial Cellulose. *Bioscience, Biotechnology, and Biochemistry*, *61*(9), 1541–1545. doi:10.1271/bbb.61.1541

Pavaloiu, R.-D., Stoica, A., Stroescu, M., & Dobre, T. (2014). Controlled release of amoxicillin from bacterial cellulose membranes. *Central European Journal of Chemistry*, *12*(9), 962–967. doi:10.2478/s11532-014-0541-3

Performance of Philippines' Top Non-Traditional Coco Products Export. (n.d.). Retrieved March 2015 from http://ucap.org.ph/news-and-events/category/industry-news

Pertile, R. A. N., Andrade, F. K., Alves, C. Jr, & Gama, M. (2010). Surface modification of bacterial cellulose by nitrogen-containing plasma for improved interaction with cells. *Carbohydrate Polymers*, *82*(3), 692–698. doi:10.1016/j.carbpol.2010.05.037

Petersen, N., & Gatenholm, P. (2011). Bacterial cellulose-based materials and medical devices: Current state and perspectives. *Applied Microbiology and Biotechnology*, *91*(5), 1277–1286. doi:10.1007/s00253-011-3432-y PMID:21744133

Retegi, A., Algar, I., Martin, L., Altuna, F., Stefani, P., Zuluaga, R., & Mondragon, I. et al. (2012). Sustainable optically transparent composites based on epoxidized soy-bean oil (ESO) matrix and high contents of bacterial cellulose (BC). *Cellulose (London, England)*, *19*(1), 103–109. doi:10.1007/s10570-011-9598-8

Retegi, A., Gabilondo, N., Peña, C., Zuluaga, R., Castro, C., Gañan, P., & Mondragon, I. et al. (2010). Bacterial cellulose films with controlled microstructure-mechanical property relationships. *Cellulose (London, England)*, *17*(3), 661–669. doi:10.1007/s10570-009-9389-7

Ring, D. F., Nashed, W., & Dow, T. (1987). *Pressing, releasing polysaccharide pellicles to rearrange into lamellas*. Google Patents.

Ross, P., Mayer, R., & Benziman, M. (1991). Cellulose biosynthesis and function in bacteria. *Microbiological Reviews*, *55*(1), 35–58. PMID:2030672

Rouabhia, M., Asselin, J., Tazi, N., Messaddeq, Y., Levinson, D., & Zhang, Z. (2014). Production of biocompatible and antimicrobial bacterial cellulose polymers functionalized by RGDC grafting groups and gentamicin. *ACS Applied Materials & Interfaces*, *6*(3), 1439–1446. doi:10.1021/am4027983 PMID:24422537

Rowlands, A. S., George, P. A., & Cooper-White, J. J. (2008). Directing osteogenic and myogenic differentiation of MSCs: Interplay of stiffness and adhesive ligand presentation. *American Journal of Physiology. Cell Physiology*, *295*(4), C1037–C1044. doi:10.1152/ajpcell.67.2008 PMID:18753317

Sai, H., Fu, R., Xing, L., Xiang, J., Li, Z., Li, F., & Zhang, T. (2015). Surface Modification of Bacterial Cellulose Aerogels Web-like Skeleton for Oil/Water Separation. *ACS Applied Materials & Interfaces*, *7*(13), 7373–7381. doi:10.1021/acsami.5b00846 PMID:25799389

Saibuatong, O., & Phisalaphong, M. (2010). Novo aloe vera–bacterial cellulose composite film from biosynthesis. *Carbohydrate Polymers*, *79*(2), 455–460. doi:10.1016/j.carbpol.2009.08.039

Sanchavanakit, N., Sangrungraungroj, W., Kaomongkolgit, R., Banaprasert, T., Pavasant, P., & Phisalaphong, M. (2006). Growth of human keratinocytes and fibroblasts on bacterial cellulose film. *Biotechnology Progress*, *22*(4), 1194–1199. doi:10.1021/bp060035o PMID:16889398

Serafica, G., Mormino, R., & Bungay, H. (2002). Inclusion of solid particles in bacterial cellulose. *Applied Microbiology and Biotechnology*, *58*(6), 756–760. doi:10.1007/s00253-002-0978-8 PMID:12021795

Serafica, G., Mormino, R., Oster, G. A., Lentz, K. E., & Koehler, K. P. (2010). *Microbial cellulose wound dressing for treating chronic wounds*. Google Patents.

Shen, W., Chen, S., Shi, S., Li, X., Zhang, X., Hu, W., & Wang, H. (2009). Adsorption of Cu(II) and Pb(II) onto diethylenetriamine-bacterial cellulose. *Carbohydrate Polymers*, *75*(1), 110–114. doi:10.1016/j.carbpol.2008.07.006

Shoda, M., & Sugano, Y. (2005). Recent advances in bacterial cellulose production. *Biotechnology and Bioprocess Engineering*, *10*(1), 1–8. doi:10.1007/BF02931175

Solway, D. R., Clark, W. A., & Levinson, D. J. (2011). A parallel open-label trial to evaluate microbial cellulose wound dressing in the treatment of diabetic foot ulcers. *International Wound Journal*, *8*(1), 69–73. doi:10.1111/j.1742-481X.2010.00750.x PMID:21159127

Son, H.-J., Kim, H.-G., Kim, K.-K., Kim, H.-S., Kim, Y.-G., & Lee, S.-J. (2003). Increased production of bacterial cellulose by Acetobacter sp. V6 in synthetic media under shaking culture conditions. *Bioresource Technology*, *86*(3), 215–219. doi:10.1016/S0960-8524(02)00176-1 PMID:12688462

Stephens, R. S., Westland, J. A., & Neogi, A. N. (1990). *Method of using bacterial cellulose as a dietary fiber component*. Google Patents.

Stoica-Guzun, A., Stroescu, M., Tache, F., Zaharescu, T., & Grosu, E. (2007). Effect of electron beam irradiation on bacterial cellulose membranes used as transdermal drug delivery systems. *Nuclear Instruments & Methods in Physics Research. Section B, Beam Interactions with Materials and Atoms*, *265*(1), 434–438. doi:10.1016/j.nimb.2007.09.036

Suratago, T., Taokaew, S., Kanjanamosit, N., Kanjanaprapakul, K., Burapatana, V., & Phisalaphong, M. (2015). Development of bacterial cellulose/alginate nanocomposite membrane for separation of ethanol–water mixtures. *Journal of Industrial and Engineering Chemistry*, *32*, 305–312. doi:10.1016/j.jiec.2015.09.004

Taokaew, S., Nunkaew, N., Siripong, P., & Phisalaphong, M. (2014a). Characteristics and anticancer properties of bacterial cellulose films containing ethanolic extract of mangosteen peel. *Journal of Biomaterials Science. Polymer Edition*, *25*(9), 907–922. doi:10.1080/09205063.2014.913464 PMID:24802115

Taokaew, S., Phisalaphong, M., & Newby, B. M. Z. (2015). Modification of bacterial cellulose with organosilanes to improve attachment and spreading of human fibroblasts. *Cellulose (London, England)*, *22*(4), 2311–2324. doi:10.1007/s10570-015-0651-x PMID:26478661

Taokaew, S., Phisalaphong, M., Yin, L. Y., & Newby, B. M. Z. (2013a). Effects of drying methods of bacterial cellulose membrane on differentiation and proliferation of induced pluripotent stem cells. *Abstracts of Papers of the American Chemical Society, 245*.

Taokaew, S., Phisalaphong, M., & Zhang Newby, B. M. (2014b). In vitro behaviors of rat mesenchymal stem cells on bacterial celluloses with different moduli. *Materials Science and Engineering C*, *38*(1), 263–271. doi:10.1016/j.msec.2014.02.005 PMID:24656377

Taokaew, S., Seetabhawang, S., Siripong, P., & Phisalaphong, M. (2013b). Biosynthesis and characterization of nanocellulose-gelatin films. *Materials (Basel)*, *6*(3), 782–794. doi:10.3390/ma6030782

Tecson-Mendoza, E. M. (2007). Development of Functional Foods in the Philippines. *Food Science and Technology Research*, *13*(3), 179–186. doi:10.3136/fstr.13.179

Toworfe, G. K., Bhattacharyya, S., Composto, R. J., Adams, C. S., Shapiro, I. M., & Ducheyne, P. (2009). Effect of functional end groups of silane self-assembled monolayer surfaces on apatite formation, fibronectin adsorption and osteoblast cell function. *Journal of Tissue Engineering and Regenerative Medicine*, *3*(1), 26–36. doi:10.1002/term.131 PMID:19012271

Trovatti, E., Freire, C. S. R., Pinto, P. C., Almeida, I. F., Costa, P., Silvestre, A. J. D., & Rosado, C. et al. (2012). Bacterial cellulose membranes applied in topical and transdermal delivery of lidocaine hydrochloride and ibuprofen: In vitro diffusion studies. *International Journal of Pharmaceutics*, *435*(1), 83–87. doi:10.1016/j.ijpharm.2012.01.002 PMID:22266531

Trovatti, E., Silva, N. H. C. S., Duarte, I. F., Rosado, C. F., Almeida, I. F., Costa, P., & Neto, C. P. et al. (2011). Biocellulose Membranes as Supports for Dermal Release of Lidocaine. *Biomacromolecules*, *12*(11), 4162–4168. doi:10.1021/bm201303r PMID:21999108

Uraki, Y., Nemoto, J., Otsuka, H., Tamai, Y., Sugiyama, J., Kishimoto, T., & Shimomura, M. et al. (2007). Honeycomb-like architecture produced by living bacteria, Gluconacetobacter xylinus. *Carbohydrate Polymers*, *69*(1), 1–6. doi:10.1016/j.carbpol.2006.08.021

Wang, L., Dormer, N. H., Bonewald, L. F., & Detamore, M. S. (2010). Osteogenic differentiation of human umbilical cord mesenchymal stromal cells in polyglycolic acid scaffolds. *Tissue Engineering. Part A*, *16*(6), 1937–1948. doi:10.1089/ten.tea.2009.0706 PMID:20070186

Watanabe, K., Tabuchi, M., Morinaga, Y., & Yoshinaga, F. (1998). Structural Features and Properties of Bacterial Cellulose Produced in Agitated Culture. *Cellulose (London, England)*, *5*(3), 187–200. doi:10.1023/A:1009272904582

Webb, K., Hlady, V., & Tresco, P. A. (1998). Relative importance of surface wettability and charged functional groups on NIH 3T3 fibroblast attachment, spreading, and cytoskeletal organization. *Journal of Biomedical Materials Research*, *41*(3), 422–430. doi:10.1002/(SICI)1097-4636(19980905)41:3<422::AID-JBM12>3.0.CO;2-K PMID:9659612

Weigel, T., Schinkel, G., & Lendlein, A. (2006). Design and preparation of polymeric scaffolds for tissue engineering. *Expert Review of Medical Devices*, *3*(6), 835–851. doi:10.1586/17434440.3.6.835 PMID:17280547

Wiegand, C., Elsner, P., Hipler, U.-C., & Klemm, D. (2006). Protease and ROS activities influenced by a composite of bacterial cellulose and collagen type I in vitro. *Cellulose (London, England)*, *13*(6), 689–696. doi:10.1007/s10570-006-9073-0

Xiu, H. J., Wang, Z. J., & Li, J. B. (2005). Effect of bacterial cellulose on paper property. *Chung-kuo Tsao Chih/China Pulp and Paper, 24*(3), 14-17.

Xue, R., Li, J. Y.-S., Yeh, Y., Yang, L., & Chien, S. (2013). Effects of matrix elasticity and cell density on human mesenchymal stem cells differentiation. *Journal of Orthopaedic Research*, *31*(9), 1360–1365. doi:10.1002/jor.22374 PMID:23606500

Yano, H., Sugiyama, J., Nakagaito, A. N., Nogi, M., Matsuura, T., Hikita, M., & Handa, K. (2005). Optically transparent composites reinforced with networks of bacterial nanofibers. *Advanced Materials*, *17*(2), 153–155. doi:10.1002/adma.200400597

Yin, N., Chen, S., Li, Z., Ouyang, Y., Hu, W., Tang, L., & Wang, H. et al. (2012). Porous bacterial cellulose prepared by a facile surfactant-assisted foaming method in azodicarbonamide-NaOH aqueous solution. *Materials Letters*, *81*, 131–134. doi:10.1016/j.matlet.2012.04.133

Yoshinaga, F., Tonouchi, N., & Watanabe, K. (1997). Research progress in production of bacterial cellulose by aeration and agitation culture and its application as a new industrial material. *Bioscience, Biotechnology, and Biochemistry*, *61*(2), 219–224. doi:10.1271/bbb.61.219

Yoshino, T., Asakura, T., & Toda, K. (1996). Cellulose production by Acetobacter pasteurianus on silicone membrane. *Journal of Fermentation and Bioengineering*, *81*(1), 32–36. doi:10.1016/0922-338X(96)83116-3

Zaborowska, M., Bodin, A., Bäckdahl, H., Popp, J., Goldstein, A., & Gatenholm, P. (2010). Microporous bacterial cellulose as a potential scaffold for bone regeneration. *Acta Biomaterialia*, *6*(7), 2540–2547. doi:10.1016/j.actbio.2010.01.004 PMID:20060935

Zhijiang, C., & Guang, Y. (2011). Bacterial cellulose/collagen composite: Characterization and first evaluation of cytocompatibility. *Journal of Applied Polymer Science*, *120*(5), 2938–2944. doi:10.1002/app.33318

Zimmermann, M. V. G., Borsoi, C., Lavoratti, A., Zanini, M., Zattera, A. J., & Santana, R. M. C. (2016). Drying techniques applied to cellulose nanofibers. *Journal of Reinforced Plastics and Composites*, *35*(8), 682–697. doi:10.1177/0731684415626286

KEY TERMS AND DEFINITIONS

Bacterial Cellulose: Cellulose hydrogel produced from certain types of bacteria, predominantly of genera *Acetobacter* (or *Gluconacetobacter*).

Biosynthesis: Production of cellulose single chain by cellulose synthases within bacteria cells. The cellulose chains are secreted through bacterial cell wall and assemble into a bundle, ribbon, and nanofiber.

Biosynthetic Modification: Change of bacterial cellulose characteristics by modifying during biosynthesis to obtain physical, chemical, and/or biological properties that are different from those of the original.

Bulk Modification: Total change of bacterial cellulose characteristics by chemically or physically modifying it during or after biosynthesis to obtain physical, chemical, and/or biological properties that are different from those of the original.

Chemical Modification: Change of bulk or surface properties of bacterial cellulose by modifying it during or after biosynthesis to obtain chemical and/or biological characteristics different from those of the original.

Physical Modification: Change of the properties of bacterial cellulose by modifying it during or after biosynthesis to obtain physical and/or biological characteristics different from those of the original without alteration its chemical characteristics.

Post-Biosynthetic Modification: Change of bacterial cellulose characteristics by modifying it after biosynthesis to obtain physical, chemical, and/or biological properties that are different from those of the original.

Surface Modification: Characteristic changes of bacterial cellulose interface to the surrounding environment or surface of cellulose fibers by chemically modifying it after biosynthesis to obtain physical, chemical, and/or biological properties that are different from those of the original.

Chapter 13
Metal Ion Separation with Functional Adsorbents and Phytoremediation Used as Sustainable Technologies

Minoru Satoh
National Institute of Technology, Ibaraki College, Japan

ABSTRACT

The amounts of "Electronic wastes" including heavy metals are increasing day by day. Such waste is in the rich resource of various metals having the precious metals. Therefore, these wastes are considered as urban mine, if people successfully can separate them to each source. In sustainable viewpoints, separation technologies applied for such electronics waste are essential and important to efficiently recover various metals at a low concentration from these sources. This chapter reviews functional adsorbents made of polymers, ionic liquid, and dendrimer. Also, membrane technology is introduced as separation toll for heavy metals. Among them, topics of phytoremediation are made as an effective sustainable method, utilizing certain plants to clean up the environmental contaminants. Here, plants are able to remove harmful chemicals such as metals, which are present in the soil, when their roots absorb water and nutrients from the contaminated soil, sediment and surface, or ground water. The contaminants are removed by trapping them into harvestable plant biomass. Furthermore, cleanup methods of environments and recovery of precious and rare metals are mentioned for sources of urban and submarine mines with low cost and high recovery efficiency.

INTRODUCTION

In Brundtland's Commission, World Commission on Environment and Development in United Nations following has submitted, "sustainable development is development that which meets the needs of the present without compromising the ability of future generations to meet their own needs" (Brundtland, 1987). There are several environmental problems in the sustainable society in coming our future. Among

DOI: 10.4018/978-1-5225-1971-3.ch013

them, the metal elements are heterogeneously distributed all over the earth. Humans have used Au, Ag, Cu, and Fe since the earliest our history. Precious metals, such as Au, Ag, Pt, Pd and so on, are widely used in several fields, for example, in catalysts in various chemical processes, electrical and electronic materials, medicine and jewelry (Ramesh, Hasegawa, Sugimoto, Maki, & Ueda, 2008). Recently, precious metals become very important because various high tech industries require and consume them for advanced technology. Under the circumstance of technological innovation and market expansion, these products strongly are concerned with economic growth. However, the increase in waste has caused environmental problems (He, Li, Ma, Wang, Huang, Xu, & Huang, WEEE, 2006).

The separation and purification of the metals are key technologies needed with reducing the energy consuming to effect the separation processes. In addition, much attention has been paid to the decontamination technologies in the environmental conservation aspects. For instance, adsorption of metal ions using low cost natural adsorbents such as agricultural waste, clay materials, biomasses and seafood processing waste is one of the technologies to maintain the sustainability of the human society (Yang & Zall, 1984). As shown in Figure 1, cycling process of metals in our human society composed of producing and wasting in industries. In these most of things, our life faces many kinds of metals. The metals mainly come from mine and the products have been spread to human societies from industries. In the cycle, "electronic wastes (e-waste)" can be found in discarded computers, office electronic equipment, entertainment electronic devices, mobile phones, television sets, and refrigerators, which are increasing the quantity of disposal day by day. It is important to note that such electronic waste becomes the rich resource of various metals including the precious metals. Therefore, these wastes are considered as urban mine (Sthiannopkaoa & Wongb, 2013). The recycling process of urban ore is different from the smelting process of natural ore, that is, in the points of view of the required energy and material inputs (Yamasue, Minamino, Numata, Nakajima, Murakami, Daigo, Hashimoto, Okumura, & Ishihara, 2009). For example, the amount of the rare metals in urban mines in Japan corresponds approximately to leading natural mines. An estimated 50 million tons of e-waste are produced each year globally as indicated by International Environmental Technology Center (URL2013). The estimation of the accumulated

Figure 1. Cycling drawing of metals in producing, wasting and removing processes against metal pollutings

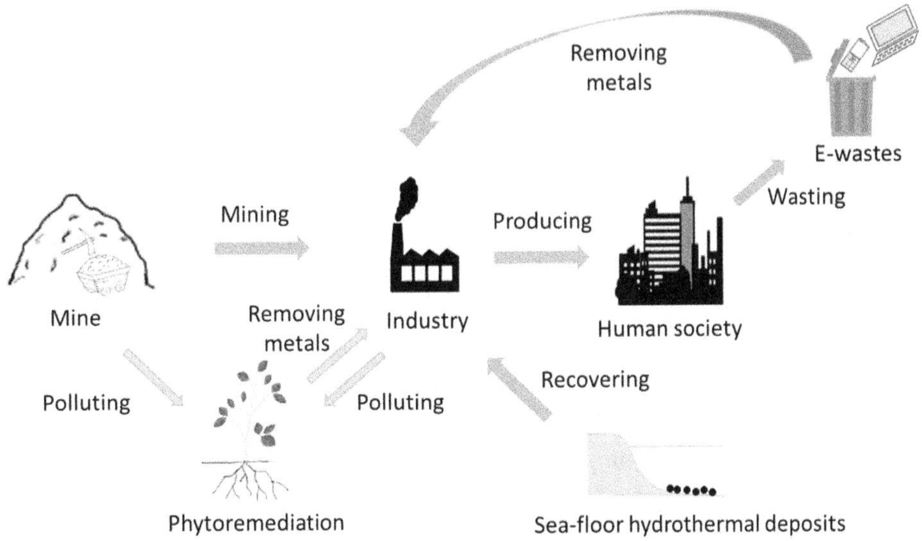

amounts of several elements in Japanese urban mines was Au 16%, Ag 22%, In 61%, Ti 42% and Ta 10% as percentages of the world's known deposite (Harada, 2008).

On the other hand, oceans cover 71% of the earth. Thus, there are enormous amounts of precious and rare metals untapped in submarine mine. Recently, sea-floor hydrothermal deposits have been expected to be new natural resources under the seabed. This is due to that water heated by magma near submarine volcano is at near 300°C without boiling under higher pressure and higher temperature to become supercritical water. Many metal elements are dissolved in the hot water and are deposited, when the temperature of the hot water containing many metal elements is cooled by mixing with cold water. Such hydrothermal deposition forms the sea-floor. Rare metals are widely at low concentration in sea-floor hydrothermal deposits. Thus separation technologies are essential to efficiently recover various metals at a low concentration from these sources.

On the other hand, the increase in polluting by metals in soil has become a serious problem near mines or industrial areas in developed countries. Phytoremediation is an effective method in which certain plants are utilized to clean up the environmental contaminants. So, plants are separation materials to remove harmful chemicals, such as metals, pesticides, organic pollutants, and oil, from the ground, when their roots absorb water and nutrients from the contaminated soil, sediment and surface, or ground water. Through the plant glowing the contaminants in soil are trapped into harvestable plant biomass. Then plants store the contaminants in their roots and/or transport them into the stems and/or leaves and continue to absorb them allowing the environment to be contaminant free gradually. Thus, phytoremediation is very important in separation method of metal recovers from contaminated environments.

ADVANCED TOPICS ON SEPARATION TOOLS OF METAL IONS

Human developed many kinds of separation technologies. In practical processes, chemical precipitation, chemical oxidation and reduction, ion exchange and activated carbon adsorption were used as conventional methods for removing dissolved metal ions from wastewater. For instance, ion exchange method has been often used with following chemical precipitation process to lower the metal concentration (Robert & Young, 1985). Ion exchange includes a reversible chemical reaction wherein metal ions from a wastewater solution are exchanged for charged ions attached to ion exchange supports of polymeric resin and inorganic mine. These ion exchange resins are either naturally obtaining inorganic zeolites or synthetically produced organic resins. Especially, the synthetic organic resins are predominant because their characteristics can be controlled to specific applications (EPA, 1981). Chelating with imino diacetic acid and amino phosphoric acid-types is one of most important example typically using ion exchange resins. The resin method has advantages in terms of high regeneration of materials and metal selectivity. In contrast, disadvantages are high cost and less number of metal ions removed (Farooq, Kozinski, Khan, & Athar, 2010).

More recently, ionic liquids (ILs) consist of an organic cation and an organic or inorganic anion only have found the chelating effect with metal ions. Here, ILs have particularly significant characters that are high thermal stability and the low vapor pressures in most instances which contrast the environmental problems of volatile organic solvents (Welton, 1999; Hallett & Welton, 2011). Then, ILs are known as environmental-friendly solvents, as compared to classical organic solvents. Since it is easy to perpetrate ILs with specific properties, ILs have possibility to be excellent extracting solvents as alternatives to the classical organic solvents as separation media.

Separation of Metal Ions with Chemical Materials

Separation process contains the progresses inexorably toward entropy maximization in nature principle. Input of energy is required to the system. The process is now positioned as the basic process in chemistry, chemical engineering and the comparable industry. For example, coagulation-sedimentation method, solvent extraction, ion exchange resin method, method for electrolytic deposition, crystallization method, membrane separation process, and cementation as separation methods of metal ions from waste solutions are important in their fields. For the adequate method, the most of the requirements are concerned with metal separation. It is known that the processes are dependent on metal concentration, volume of waste solution, and coexistence ion. In hydroxide coagulation method, the one of sedimentation method is used mostly in the industrial processes (Patterson & Minear, 1975). The sulfide coagulation method is excellent separation process for metals ($M^{2+} + H_2S \rightarrow MS + 2H^+$). But it has the two problems are concerned in bad odor of H_2S and formation of colloid for the excess sluge. Coagulation and flocculation are essential processes in industrial wastewater treatment. Basically, coagulation is divided among four categories (Sahu & Chaudhari, 2013; Bolto & Gregory, 2007) as seen in Table 1.

In recent, two-phase system including the coagulation method and another separation process can be avoided by application based on membrane process. In their coagulant substances, a number of soluble and hydrophilic polymers are suitable for the sedimentation tools for the enrichment of metal ions in conjunction with membrane filtration (Rivas, Pereira, & Moreno-Villoslada, 2003). After coagulation with metal ions, the filtration easily allows to separate metal ions bound to soluble polymers from non-bound metals (Figure 2). Usually, the polyelectrolytes have charged groups, or easily ionizable groups in aqueous solution. These are called as chelating polymers (polychelatogens) bearing functional groups with the ability to form coordination bonds. The most representative ligands in the polychelatogens are amines, carboxylic acids, amides, alcohols, amino acids, pyridines, thioureas, iminos, etc. Among them, polymers containing amino groups were extensively studied in connection with ultrafiltration process,

Table 1. Coagulants and flocculants

Types of Coagulants and Flocculants	Examples	References
Inorganic salt	Aluminum salt Ferric salt Ferrous salt	Benschoten, & Edzwald, 1990; Tripathy& De, 2006.
Cationic polymers	Poly (diallyldimethyl ammonium chloride) Epichlorohydrin/dimethylamine polymers Cationic Polyacrylamides Chitosan Partially deacetylated chitin	Wandrey & Jaeger, 1985; Vorchheimer, 1981; Hoover, 1970; Roussy, Van Vooren, Dempsey & Guibal, 2005.
Anionic polymers	Carboxylic acid polymers Anionic PAMs Heparin Dextran sulphate Mannan sulphate Chondroitin sulphate Tannins	Vorchheimer, 1981; Wassmer, Schroeder & Horn, 1991; Voycheck, Tan, 1993; Rice, Denysschen& Stander, 1964
Non-ionic polymers	Non-ionic Polyacrylamide	Tripathy & De, 2006.

Figure 2. Sedimentation-ultrafiltration process

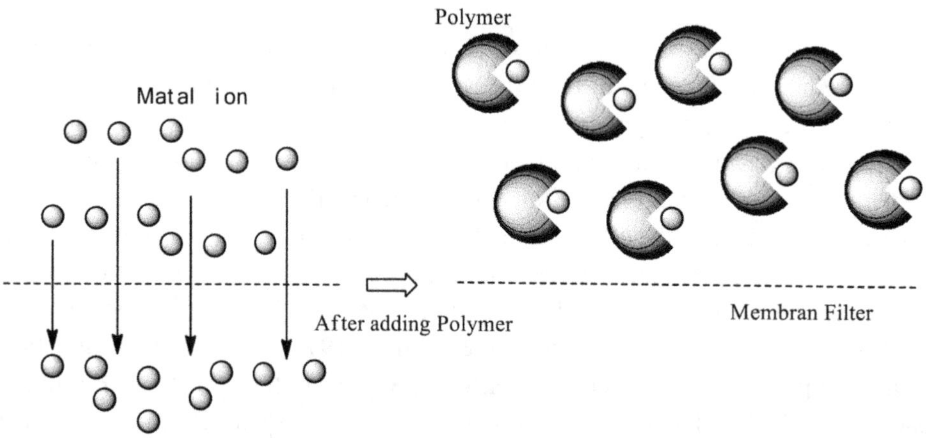

which including cut-off behavior depending on pore size of membrane, solute–membrane interaction, and shape and size of the solute macromolecule at a given pH and ionic strength (Rivas & Geckeler, 1992).

Membrane process is also composed of two-phase system. In separation technology, the advanced methods are now familiar as membrane filtration, which is one of the most frequently studied and widely applied for the treatment of wastewater (Mulder, 1991; Noble & Stern, 1995). Ultrafiltration has its high removal efficiency, low costs and large throughput but is often limited by membrane fouling which severely decreases membrane throughput over time (Kurniawan, Chan, Lo, & Babel, 2006). These advanced membranes are made of polymeric materials like poly(sulfone), cellulose acetate, poly(ethersulfone), poly(ethylene), poly(vinyldine fluoride) (Rivas, Pereira, & Moreno-Villoslada, 2003). However, the membrane process strongly depends on the chemical properties of the materials. Although cellulose acetate-based membranes are stain-resistant and typically with higher flux rather than poly(sulfone) membranes at equivalent rejection. It is characteristics for poly(sulfone) membranes that have many applications because of higher stabilities in chemicals like acid and base and heat.

The polymer-enhanced ultrafiltration (PEUF) is paid attention as a promising process for treatment of water contaminated by toxic metal ions. In the PEUF, a water-soluble polymer like dendrimer with strong binding affinity for toxic metal ions is added to contaminated water (Geckeler, & Volchek, 1996). The polymeric dendrimers are highly branched polymers with nanometer-scale dimensions defined by three components: a central core, an interior dendritic structure (the branches), and an exterior surface with functional surface groups. The properties of dendrimers such as size, shape, core chemistry, surface chemistry and polydispersity can be controlled by the dendomeric reactions. Poly(amidoamine) (PAMAM) dendrimer is nanoscale polymer with ethylenediamine (EDA) core and terminal NH_2 groups and are hyperbranched polymers with unparalleled molecular uniformity, narrow molecular weight distribution, defined size and shape characteristics and a multifunctional terminal surface (Figure 3, Tomalia, 2004).

It was reported that the Cu(II) binding capacities of PAMAM were much larger and more sensitive to solution pH as compared with those of linear polymers with amine groups (Diallo, Christie, Swaminathan, Johnson, & Goddard, 2005). Then it was noted that both chemical structure and the pore size of membrane affected the separation behavior of Cu(II) ion, because strongly bound to large molecules PAMAM were easily retained by the membrane. It is found that the Metal Products and Machinery (MP&M, URL) industry is selected as the first target for the application of dendrimer enhanced ultra-

Figure 3. Chemical structures of Poly(amidoamine) dendrimer PAMAM

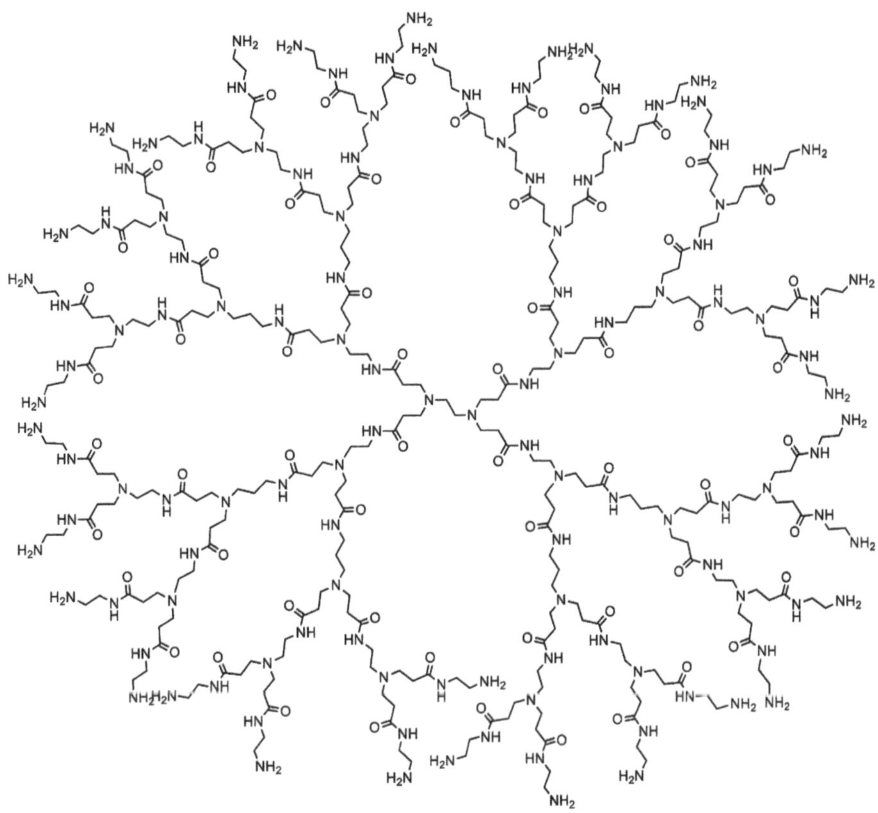

filtration (Diallo, Johnson, & Balogh, 2000) for wastewater contaminated by toxic metal ions. Under current regulations in USA, the MP& M industry is required to treat its contaminated wastewater prior to the discharge with lower costs than chemical precipitation, promising for recovering metal ions such as Cu(II) from aqueous solutions.

Ionic Liquids for Green Solvent Used as Separation Chemicals

As mentioned, the Brundtland's Commission, the point of view well meets with properties of ionic liquids, which called sustainable solvents. It is noted that ionic liquids (ILs) have attracted considerable attention because they refer to a kind of organic salt composed of a cation and anion and possessing a melting point below 100°C and have some unique properties of such as low vapor pressure, low flammability, wide electrochemical windows, tunable properties, and the ability to solubilize for many organic/ inorganic compounds (Welton, 1999; Plechkova & Seddon, 2008). Thus, ILs have been widely used in organic reaction, separation process, electrochemistry, etc (Petkovic, Seddon, Rebelo, & Pereira, 2011; Hubbard, Illner, & van Eldik, 2011). It was found that ILs with the bis(trifluoromethanesulfonyl) imide counter anion made them easy to separate from water (Nakashima & Kawai, 2005; Nakamura, Saiwaki & Fukao, 2010). In these cases, ILs were utilized as an ideal media in analytical chemistry.

Figure 4. Chemical structures of ILs and $TMPyPH_2$

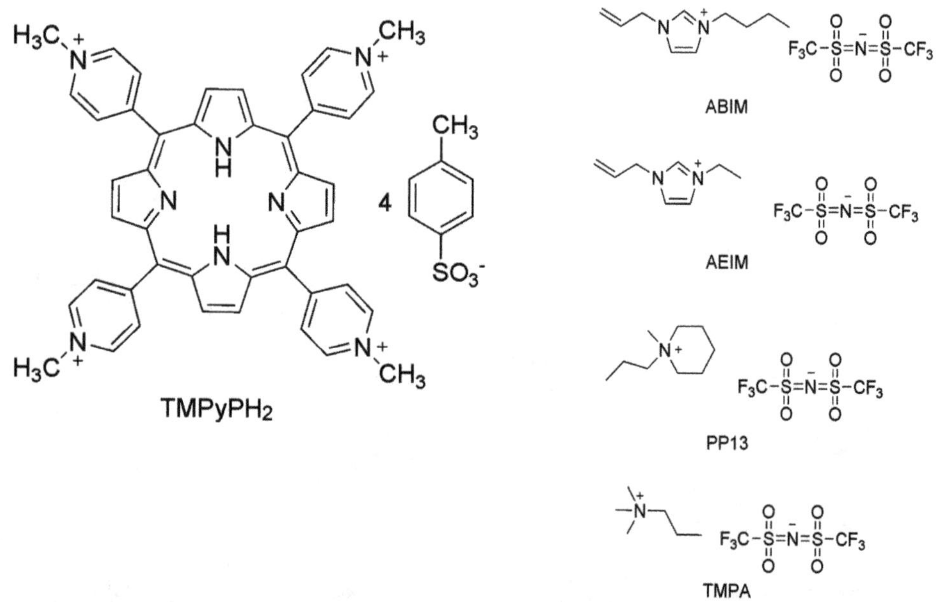

In particular, imidazolium ILs are solvents of main interest. The use of ILs as extraction solvents was reported for metal cations (Dai, Ju & Barnes, 1999). Some ammonium and phosphonium ILs were also studied in solvent extraction (Bradaric, Downard, Kennedy, Robertson, & Zhou, 2003; Bradaric, Downard, Kennedy, Robertson, & Zhou 2003; Del Sesto, Corley, Robertson, & Wilkes, 2005; Pernak, Stefaniak, & Weglewski, 2005). Many studies of the metal extraction with ILs were carried out as extractants such as crown ethers, organophosphorus compounds, and nitrogen-donor compounds.

Table 2 shows separation of rare earth metals by ILs.

Table 2. Removal by using ILs

Ils	Extractants	Metals	References
[1-R^1-2-R^2-3-min][Tf_2N] [1-R^1-2-R^2-3-min][PF_6] R^1 = ethyl, propyl, or butyl, R^2 = H, or methyl	DCH18C6	Sr	Dai, Ju, & Barnes, 1999.
[$C_n$mim][PF_6] n=4, 6, 8	DC18C6 dicyclohexyl-18-crown-6	Na, Cs, Sr	Visser, Swatloski, Reichert,Griffin & Rogers, 2000.
[C_4mim][PF_6] [C_6mim][PF_6]	PAN, TAN, CN−, OCN−, SCN−, F−, Cl−, Br−, I−.	Cd, Co, Ni, Fe, Hg	Visser, Swatloski, Griffin, Hartman & Rogers, 2001.
[$C_n$mim][PF_6] n=4, 8	imidazoles with urea-, thiourea-, and thioether-substituted alkyl groups	Hg, Cd	Visser, Swatloski, R. P., Reichert, W. M., Mayton, R., Sheff, S., Wierzbicki, Davis& Rogers, 2002.
[C_4mim][PF_6]	Dithizone	Pb, Zn, Cd, Hg, Ag	Wei, Yang & Chen, (2003).
[C_4mim][PF_6]	CMPO	Ln, Y	Nakashima, Kubota, Maruyama & Goto, 2003.

continued on next page

Table 2. Continued

Ils	Extractants	Metals	References
$[C_4mim][Tf_2N]$	Htta	Nd, Eu	Jensen, Neuefeind, Beitz, Skanthakumar & Soderholm, 2003.
$[C_nmim][PF_6]$ n=4, 6, 8	Pyridinocalix[4]arene	Ag	Shimojo& Goto, 2004.
$[C_4mim][PF_6]$	tri-n-butylphosphate	U	Giridhar, Venkatesan, Srinivasan & Vasudewa Rao, 2005
$[C_nmim][Tf_2N]$ n=5-10	DCH18C6	Na	Dietz & Stepinski, 2005.
$[C_{10}mim][Tf_2N]$	D2EHPA, Cyanex272	U, Am, Nd, Eu	Cocalia, Jensen, Holbrey, Spear, Stepinski, Rogers, 2005.
$[C_4mim][PF_6]$ $[C_4mim][Tf_2N]$	CMPO	Ln, Y	Nakashima, Kubota, Maruyama & Goto, 2005.
$[C_4mim][Tf_2N]$	Htta/supercritical-CO_2	Eu, La	Mekki, Wai, Billard, Moutiers, Burt, Yen, Wang, Gaillard, Ouadi, & Hesemann, 2006.
$[C_4mim][Tf_2N]$	TPEN	Eu	Shimojo, Naganawa, Kubota & Goto, 2006.
$[C_4mim][NfO]$	-	La, Sr, Ca, Li, Na, Cs	Kozonoi & Ikeda, 2007.
$[C_8mim][PF_6]$	Cyanex925	Sc, Y, La, Yb	Sun, Wu, Chen & Li, 2007.
$[C_4mim][PF_6]$ $[C_4mim][Tf_2N]$	TPMDPO, Ph_2Et_2 Bu_2Et_2	Ln, Y	Turanov, Karandashev & Baulin, 2008.
$[C_nmim][Tf_2N]$ (n=8,12)	PC-88A	Y, Eu, Ce, La	Kubota, Koyanagi, Nakashima, Shimojo, Kamiya& Goto, 2008.
$[C_4mim][Tf_2N]$	Htta-18C6/synergistic ex.	La, Eu, Lu	Hirayama, Okamura, Kidani & Imura, 2008.
$[C_nmim][Tf_2N]$ (n=2, 4, 6)	TODGA	Ln	Shimojo, Kurahashi& Naganawa, 2008.
$[C_8mim][PF_6]$	-	Ce, Th, Gd, Yb	Zuo Liu, Chen, & Li, 2008.
$[C_8mim][PF_6]$	primary amine N1923	Th/Rare earths	Zuo, Chen & Li, 2008.
$[C_8im][PF_6]$	D2EHPA	Ln,Sc, Y, Th	Zuo, Liu, Chen, & Li2009.
$[C_nmim][Tf_2N]$ (n= 4, 6)	HYD	Sm, Eu, Gd, Dy	Mallah, Shemirani & Maragheh, 2009.
$[C_nmim][PF_6]$ (n=4,8), Cyphos IL 104	Cyphos IL 104/ TSIL	Y, Dy, Ho, Er, Yb	Liu, Zhu, Sun, Chen & Luo,2009.
$[C_nmim][PF_6]$ n=4, 8 $[C_nmim][Tf_2N]$ $[C_nmim][BF_4]$ [MTOA][Cl],	-	Zn, Fe	Perez de los Rios, Hernandez-Fernandez, Sanchez-Segado, Lozano, Moreno, & Godinez, 2010.
[A336][CA-100] (TSIL)	[A336][CA-100]/TSIL	Sc, Y, Eu, Ce	Sun, Ji, Hu, He, Chen & Li, 2010.
$[C_6mim][Tf_2N]$	PAN-HYD/ syner.ex.	Sm, Eu, Gd, Dy	Mallah, Shemirani, Maragheh, & Jamali, 2010.
$[C_nmim][Tf_2N]$ (n=4, 8, 12)	DODGAA/ SLM	Y, Eu, Zn (fluorescent powder)	Kubota, Shimobori, Koyanagi, Shimojo, Kamiya & Goto, 2010.
CMPO-modified IL (TSIL)	CMPO-modified IL /TSIL	U, Pu, Eu, Am	Odinets, Sharova, Artyshin, Lyssenko, Nelyubina, Myasoedova, Molochnikova, & Zakharchenro, 2010.
$[C_nmim][PF_6]$ (n=2, 4), $[C_4mpy][PF_6]$	D2EHPA	Ce, Nd, Sm, Dy, Yb	Yoon, Lee, Tajima, Yamasaki, Kiyono, Nakazato, Tao, 2010.

continued on next page

Table 2. Continued

Ils	Extractants	Metals	References
[C_4mim][Tf_2N]	Htta, Htta-18C6, -DC18C6, -DB18C6, -15C5/synergistic. ex.	Ln	Okamura, Hirayama, Morita, Shimojo, Naganawa & Imura, 2010.
[C_8mim][Tf_2N]	D2EHPA, HDEHDGA / masking DTPA	Eu, Am	Rout, Karmakar, Venkatesan, Srinivasan & Rao, 2011.
[$C_n$mim][Tf_2N] [$C_n$mim][BETI] (n=4, 8,12) [C_4mPy][Tf_2N]	D2EHPA/aq. Complex agents	Ln	Sun, Bell, Luo & Dai, 2011.
[A336][NO3] (TSIL)	Cyanex925	Sc, Y	Sun, Ji, Guo, Chen, & Li, 2011.
[A336][CA-100] (TSIL)	[A336][CA-100] / Tas-specific ionic liquid	Ln, Y	Wang, Yang, Cui, Zhang, Liu, & Chen, 2011.
Ac-functionalised IL	Ac-functionalised IL /Task-specific ionic liquid	Eu, Tb	Olivier, Camerel, & Ziessel, 2011.
[C_8mim][Tf_2N]	DODGAA	Nd, Dy, Fe(magnet)	Baba, Kubota, Kamiya& Goto, 2011.
[$C_n$mim][Tf_2N] (n=4, 8, 12)	DODGAA	Y, Eu, Zn (fluorescent powder)	Kubota, Shimobori, Baba, Koyanagi, Shimojo, Kamiya, & Goto, 2011.
[A336][P204], [A336][P507]	[A336][P204], [A336][P507] /n-heptane	Y, Eu, Gd, Ce, Tb(fluorescent powder)	Yang, Wang, Cui, Zhang, Liu, Chen, 2012.
[C_8mim][Tf_2N]	DODGAA	Ln	Yang, Baba, Kubota, Kamiya, & Goto, 2012.
Carboxyl-functionalized IL betainium bis (trifluoromethylsulfonyl)imide, [Hbet][Tf2N]	zwitterionic betaine	La, Pr, Nd, Dy, Ho, Er, Y, Sc, Ga, In, Mn, Ni, Cu, Zn, Ag	Hoogerstraete, Onghena & Binnemans, 2013
[C_6mim][DEHP] [C_6mpyr][DEHP] [N4444][DEHP] [C_6mim][Tf_2N] [C_6mpyr][Tf_2N] [N4444][Tf_2N]	DEHPA	La, Ce, Pr, Nd, Yb, Y	Rout, Kotlarska, Dehaen, & Binnemans, 2013.
[P66614][MA] [P66614][NO3] [P66614][Cl]	TEHMA	Eu, La, Ce, Nd, Sm, Ho, Yb, Ni, Co, Zn	Rout & Binnemans, 2014.
[C_2mim][Cl] [P66614][R2POO]	-	Y, La, Nd, Eu, Dy, Ho, Yb	Rout, Wellens & Binnemans, 2014
[Hbet][Tf_2N]	-	Sc	Onghena & Binnemans, 2015.

$C_n$mim: 1-alkyl-3-methylimidazolium (n=carbon number), PF_6: hexafuorophosphate, Tf_2N: bis[(trifluoromethyl)-sulfonyl]imide, CMPO: oct(phenyl)-*N*, *N*-diisobutylcarbamoylmethyl phosphine oxide, Htta: 2-thenoyltrifluoroacetone, D2EHPA: bis(2-ethylhexyl) phosphoric acid, HDEHDGA: bis(2-ethylhexyl)diglycolamic acid, TPEN: *N*, *N*, *N'*, *N'*-tetrakis(2-pyridylmethyl)-ethylenediamine, NfO: Nonafluorobutanesulfonate, Cyanex 925: bis(2,4,4-trimethylpentyl)octylphosphine oxide, TPMDPO: tetraphenylmethylenediphosphine dioxide, Ph_2Et_2: diphenyl(diethylcarbamoylmethyl)phosphine oxide, Bu_2Et_2: dibutyl(diethylcarbamoylmethyl)phosphine oxide, PC-88A: 2-ethylhexylphosponic acid mono-2-ethylhexyl ester, HYD: 1-hydroxy-2, 5-pyrroliidinedione, DCH18C6: dicyclohexyl-18-crown-6, 18C6: 18-crown-6 ether, PAN: 1-(2-pyridylazo)-2-naphthol, TAN: 1- (2-thiazolyl)-2-naphthol, Dithizone, [MTOA] [Cl]: methyltrioctylammonium chloride, DEHPA: bis(2-ethylhexyl)phosphoric acid, TEHMA: N,N,N',N'-tetra(2-ethylhexyl)malonamide, [A336][CA-100]: [trialkymethylammonium][sec-nonyphenoxyacetate], Cyphos IL 104: Trihexyl(tetradecyl)phosphonium bis(2,4.4-trimethylpentyl) phosphinate, DODGAA: N, N, N', N'-dioctyldiglycol amic acid, mpy: 4-methylpyridinium, TSIL: Task-specific ion liquids, DEHP: bis(2-ethylhexyl)phosphate, C6mpyr: 1-hexyl-1-methylpyrrolidinium, [N4444]: tetrabutylammonium, [P66614][MA]: trihexyl(tetradecyl)phosphonium *N*, *N*, *N'*, *N'*-tetra(2-ethylhexyl)malonate, [P66614][NO3]: trihexyl(tetradecyl)phosphonium nitrate, [Hbet]: betaine, [P66614][R2POO]: trihexyl(tetradecyl)phosphonium bis(2,4,4-trimethylpentyl)phosphinate, [P204]: tetradecyl(trihexyl) phosphonium, [P507]: di-(2-ethlhexyl)orthphosphinate.

Following is explanation example for an IL-porphyrin concentrator developed for the detection of Cu(II) ions in aqueous solution (Li, Satoh & Kobayashi, 2016). As shown in Figure 4, when an aqueous solution of 5,10,15,20-Tetrakis(*N*-methylpyridinium-4-yl)-21H, 23H-porphyrin, tetrakis- (*p*-toluenesulfonate), $TMPyPH_2$ and 1-allyl-3-butylimidazolium bis(trifluoromethanesulfonyl) imide (ABIM) were mixed together and kept for some time, it was found that $TMPyPH_2$ effectively moved to the ABIM IL layer, indicating that the ILs extracted $TMPyPH_2$ from water medium. Figure 5 shows the absorption spectra of the porphyrin soret band in ABIM layer containing different Cu(II) concentration. The UV-vis results presented that the absorption intensity increased with the increasing of Cu(II) concentration in water. The ILs behavior was like Cu(II) concentrator because $TMPyPH_2$ bound Cu(II) ions in solution and became a facile method to detect Cu(II) ions. A trace amount, 4 ppb, of copper could be extracted from aqueous solution. The ABIM IL-$TMPyPH_2$ was effective for Cu(II) ion concentrator, since nterference studies showed that other metals (Cd(II), Zn(II), Fe(III), Mn(II), and Ni(II)) did not interfere significantly with the determination of Cu(II) ion concentration using this method.

The separation and recovery of precious metals such as Au(III), Pt(IV) and Pd(II) ions were carried out by using 1-butyl-3-methylimidazolium bis(trifluoromethylsulfonyl)imide ([Bmim][Tf_2N] (Figure 6) (Yang, Kubota, Baba, Kamiya, & Goto, 2014). Here, Au(III) was efficiently extracted from the aqueous phase to undiluted IL and was concentrated almost 8 times. The extraction percentage of each metal ion was evaluated as a function of the HCl concentration (0.01M – 2M) but, Pt(IV) was partly extracted, and Pd(II) was hardly extracted. In the presence of HCl, Au(III) existed as $[AuCl_4]^-$, and Pt(IV) and Pd(II) were $[PtCl_6]^{2-}$ and $[PdCl_4]^{2-}$, respectively as main species. The extraction order of the chloro-complexes of precious metals $[AuCl_4]^- > [PtCl_6]^{2-} > [PdCl_4]^{2-}$, was consistent with the extraction via anion exchange mechanism related with counter ions of the ILs.

Figure 5. Cu^{2+} concentraton dependence of absorption spectra of TMPyPCu in ABIM IL

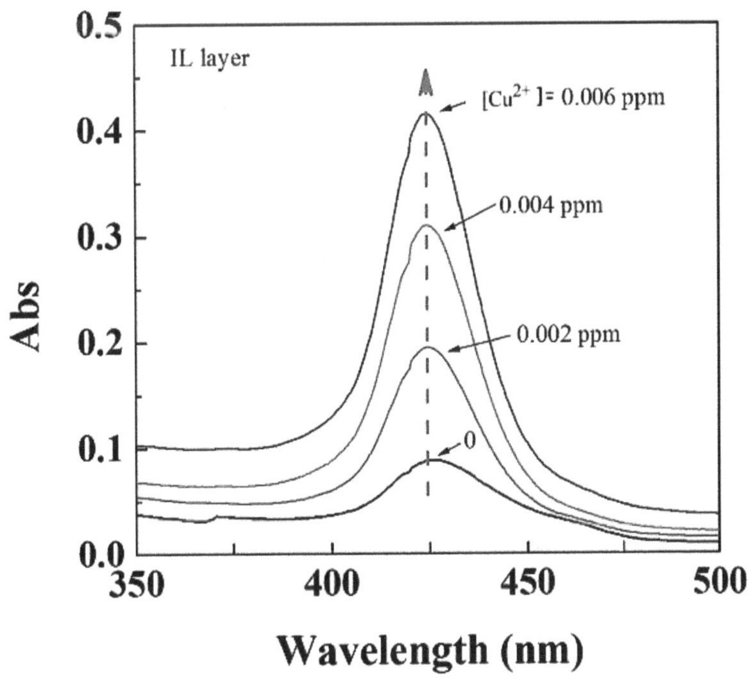

Figure 6. Chemical structures of Imidazolium-based ILs having different alkyl chain length

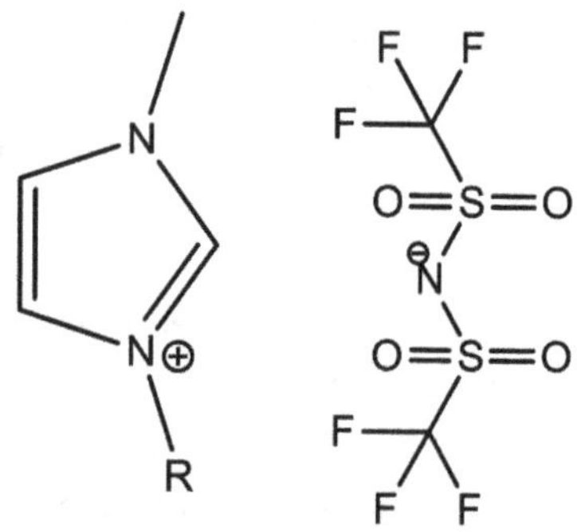

R=butyl [Bmin][Tf2N]
hexyl [Hmin][Tf2N]
octyl [Omin][Tf2N]

In the case of IL extractants, Imidazolium-based ILs with hexyl, and octyl, instead of butyl of [Bmim] [Tf_2N], were studied without dilution for the extractive separation of Pt and Pd in an HCl solution (Yang, J., Kubota, F., Baba, Y., Kamiya, N., & Goto, M., 2014). The effect of the alkyl chain length of the imidazolium ring on the extraction performance by using three ILs (Figure 6). With increasing the alkyl chain length, the extraction efficiency of Pt(IV) was significantly increased, while that of Pd(III) increased slightly. In the IL, 1-methyl-3-octylimidazolium bis(trifluoromethylsulfonyl)imide ([Omim] [Tf_2N]), the separation of Pt(IV) and Pd(II) was highly effective. The highest separation factor between Pt(IV) and Pd(II) was 312 at the optimum conditions in a binary component system, suggesting that the extraction ability of ILs for the precious metals ions was controllable.

SEPARATION BY USING BIOMASS WASTES

Plant and Wooden Wastes for Biosorbents

In the popular methods for recovery and separation of metal ions, these artificial processes have some disadvantages such as incomplete removal of metal ions, high amounts of toxic sludge, and high cost. In contrast, agricultural biomass sourced from forestry and fisheries waste recently attract attention in the field of wastewater remediation (Table 3). This is because of their abundance and sustainable nature.

In wooden wastes sourced from plants, biosorbents are main approaching to decontaminateon of metal ions. For the binding metal ions from aqueous solutions, the moss peat contains in its structure different polar functional groups (ex. –COOH, –OH, –NH_2 etc.) (Caramalău, Bulgariu, & Macoveanu, 2009). Thus adsorption behaviour of Co (II) from aqueous solutions on peat moss sampled from Poiana Stampei, Romania influenced initial pH, peat moss dose, initial Co (II) concentration and contact time, in batch experiments. The optimum pH was 6.0 (acetate buffer) with an adsorbent dose of 5 g/l and an initial concentration of 240 mg Co(II)/L The sorption of Pb(II) and Hg(II) ions to moss peat was also studied in a batch system (Bulgariu, Răţoi, Bulgariu, & Macoveanu, 2008). Analyzing by the Langmuir model, the maximum uptake capacities of moss peat for Pb(II) and Hg(II) were obtained as 117.58 mg/g and 81.97 mg/g, respectively. In Canadian Sphagnum peat moss, humic acid and humin were extracted with the bound Cu (II) in batch experiments (Gardea-Torresdey, Tang, & Salvador, 1995). The adsorption of Cu(II) also depended on pH dependent. For example, at pH 4.0 and 5.0, about 99% of the Cu(II) was bound by all three biomasses with very rapid binding. Then, more than 90% of the Cu(II) adsorpted on the biomasses was recovered by treatment with 0.1 M HCl.

Table 3. Adsorption by using biomasses

Materials	Metals	Maximum Adsorption Capacity / mg/g	References
Peat moss	Co	34.2 (treated with NaOH) 29.75(un-treated) 20.85(treated with HNO_3)	Caramalău, Bulgariu & Macoveanu., 2009
Peat moss	Pb Hg	117.58 81.97	Bulgariu, Răţoi, Bulgariu & Macoveanu, 2008
Peat moss	Cu	16.1	Gardea-Torresdey, Tang & Salvador, 1995
Peat moss	Pb	-	Lee, Lee, Chung, Park, Huh & Jun, 2013
Biochar of peat moss	Pb Cd Cu	81.3 39.8 18.2	Lee, Park, Ahn & Chung, 2015
Okra	Cu Zn Cd Pb	72.72 57.11 121.51 273.97	Singha & Guleria, 2015
Coffee beans	Cd Cu Zn Pb Fe	6.72 - - - -	Kaikake, Hoaki, Sunada, Dhakal, & Baba, 2007
Cicer arientinum (chickpea var. black gram)	Pb Cd Zn Cu Ni	49.97 39.99 33.81 25.73 19.56	Saeed, Iqbal & Akhtar, 2005
Wheat shell	Cu	8.34	Basci, Kocadagistan, & Kocadagistan, 2004
Papaya (Carica papaya)	Cu Cd Zn	19.88 17.22 13.45	Saeed, Akhter, & Iqbal, 2005
Treated sawdust	Cr Pb Hg Cu	111.61 52.38 20.62 5.64	Meena, Kadirvelu, Mishra, Rajagopal & Nagar, 2008
Crab cell	Cu Co	243.9 322.6	Vijayaraghavan, Palanivelu & Velan, 2006
Salmon milt	Nd La, Ce, Pr, Pm, Sm, Eu, Gd, Tb, Dy, Y, Ho, Er, Tm, Yb, Lu	50.1 - -	Takahashi, Kondo, Miyaji,Watanabe, & Fan, 2014
DNA	Nd La, Ce, Pr, Sm, Eu, Gd, Tb, Dy, Ho, Er, Tm, Yb, Lu	0.182 - -	Takahashi, Kondo, Miyaji, Umeo & Honma, 2012
Eggshell	Cr	21-160	Chojnacka, 2005

The immobilization of Pb on peat moss, were evaluated in three types of the column leaching test, microcosm test, and batch incubation test (Lee, Lee, Chung, Park, Huh, & Jun, 2013) for the Pb leaching. Thus, the peat moss was effective in Pb immobilization in contaminated soil. The peat moss immobilization reduced the mobility and toxicity of Pb more effectively in the column leaching and microcosm tests than bioavailability in the batch incubation test. In recent research interest as a metal adsorbent. biochar was produced by the pyrolysis of peat moss over a temperature range of 400–1000 °C for 30–90 min adsorpted heavy metals (Pb, Cu, and Cd) under different carbonization conditions (Lee, Park, Ahn, & Chung, 2015). Biochar produced at 800 °C for 90 min was the most effective for the removal of Pb and Cu. The highest maximum adsorption capacity for Pb (81.3 mg/g) on the biochar, followed by Cd and Cu, which were 39.8 and 18.2 mg/g, respectively.

Although about 8 million tons of coffee were produced around the world each year most of the coffee beans become residue after being dripped and degreased (Potts, Lynch, Wilkings, Huppé, Cunningham, & Voora, 2014). The residue was used as an adsorbent for base metals such as Cu(II), Zn(II), Pb(II), Fe(III) and Cd(II) (Kaikake, Hoaki, Sunada, Dhakal, & Baba, 2007). In the beans, the plant cell wall had porous structure with specific surface area, 1.2 m^2/g. The adsorption of Gu(II), Zn(II), Pb(II), Fe(III) and Cd(II) depended on pH in batch adsorption experiments, but was observed to bind approximately 100% at the region over pH 4, while not adsorbed in the lower pH than 2. The reaction mechanism was cation exchange one between metal ions and the beans. It was supported that this had possibility to remove harmful metal ions from drinking water, underground water and waste water.

By chemically succinylated okra biomass, the adsorbent used could be removed of Cu(II), Zn(II), Cd(II) and Pb(II) for such toxic metal ion from aqueous solution (Singha & Guleria, 2015).

As biosorbents, the milling agrowaste of *Cicer arientinum* (chickpea var. black gram) was useful for biological adsorbent that occured naturally in certain biomass (Saeed, Iqbal, & Akhtar, 2005). The removal of the highly toxic Pb, Cd, and other heavy metals in binary and ternary systems by black gram husk of an agrowaste. The selectivity order of heavy metal ions and increasing the initial metal concentration were observed as well fitted by the Langmuir and Freundlich adsorption isotherms. In addition, complete recovery of these metals in single and multi-metal solutions was achieved with 0.1 M HCl in both shake flask and fixed bed column methods. Wheat shell was also used as a biosorbent of Cu(II) (Basci, Kocadagistan, & Kocadagistan, 2004). The adsorption capacity of wheat shell for Cu (II) was estimated at various pH 2–7 with stirring speeds and initial concentrations of Cu (II) from 10 to 250 mg/ l^{-1}. Wheat shell was an effective biosorbent for removing Cu(II) from aqueous solutions.

It was reported that papaya (Carica papaya) wood obtained from the felled trunk of matured tree was used as a biosorbent of heavy metals (Saeed, Akhter, & Iqbal, 2005). According to the treatment procedure, the trunk was debarked, cut into small pieces, soaked in boiling water for 30 min. Then, it thoroughly washed under tap water, and left for 2–3 h in distilled water. The washed wood pieces were ground into fibers and oven dried at 80 °C. Batch biosorption assays were carried out in 100 ml of solutions contacting 10 mg l^{-1} Cu (II), Cd (II) and Zn (II) solutions with 5 g l^{-1} papaya wood, during shake flask contact time of 60 min. The metal removals were noted to be 97.8, 94.9 and 66.8% for Cu (II), Cd (II) and Zn (II) respectively at pH 5. The biosorption data perfectly fitted the Langmuir adsorption isotherms model for all the metal ions, indicating saturation binding to the sites on the papaya wood.

Similarly, treated sawdust was used for removing heavy metals from wastewater and aqueous solution (Meena, Kadirvelu, Mishra, Rajagopal & Nagar, 2008). The lignin was removed before using adsorption of heavy metals, Cr(VI), Pb(II), Hg(II) and Cu(II), because lignin inhibited the adsorption and color treated water became red. The heavy metal ions adsorbed on treated sawdust by batch technique.

The maximum removal of Hg(II), Cu(II), Pb(II) and Cr(VI) at pH 6 were 99.4, 92.2, 94.6 and 99.3% with adsorption capacities for treated sawdust were Cr(VI) (111.61 mg/g), Pb(II) (52.38 mg/g), Hg(II) (20.62 mg/g), and Cu(II) (5.64 mg/g), respectively. The possible sites on sawdust for complexing metal ions are –COOH, -C_6H_5-OH and amide group. It was suggested that the surface complexation and ion exchange are the major removal mechanisms by the treated sawdust.

Animal Wastes for Biosorbents

In animal wastes, following is chitin, chitosan and crab cell were reported as biosorbents to metal ions. Chitosan adsorbed Cd in batch experiments (Jha, Iyengar, & Rao, 1988). An adsorption capacity was 5.93 mg/g, and pH dependent and increased up to pH 8.3 with decreasing in particle size of chitosan. The presence of chloride ion and ionic strength had no influence, but complexing agents such as ethylenediamine tetracetate (EDTA) decreased the adsorption of Cd. Heavy metals Cu (II) and Co (II) were removed as by crab cell (Vijayaraghavan, Palanivelu, & Velan, 2006). Crab shells were pretreated by washing with 0.1 M HCl for 4 h to remove $CaCO_3$. The treated shells were then washed with water and dried naturally. The biosorption capacities of crab shell for copper and cobalt depended on their particle sizes (0.456–1.117 mm), biosorbent dosages (1–10 g/l), initial metal concentrations (500–2000 mg/l) and solution pH values (3.5–6) in batch mode for the achieved maximum capacity of copper and cobalt of 243.9 and 322.6 mg/g, respectively. The optimum particle size was 0.767 mm with the dosage (5 g/l) and initial solution at pH6. The presence of heavy metal co-ions of Ni(II), Cd(II) and Zn(II), caused de-sorption of the copper and cobalt on the crab shell. The presence of EDTA (pH 3.5, in HCl) performed well for the eluting agents, and had low biosorbent damage. Chitin is distributed in many places throughout the natural world as main component in the exoskeletons of arthropods such as crustaceans of crabs, lobsters and shrimps and also insects for the radulae of molluscs. Chitin $(C_8H_{13}O_5N)_n$ is a long-chain polymer having *N*-acetylglucosamine with which partial derivative of glucose. Chitosan is obtained from chitin by treating shrimp and other crustacean shells with the alkali, sodium hydroxide (No & Meyers, 1995). It is characterized that these polymers are not fear of depletion, are a biodegradable and highly safe, to our body with biocompatibility. Because of less solubility of chitin, the research works of the biosorbents were not enough relative to chitosan. It was reported that chitosan was modified for different polymers and to be formed in such powder, nano-particle, bead, membrane, sponge and fiber for various applications (Table 4).A number of chitosan derivatives for the recovery of metal ions were reported because of their great potentials for the adsorption of metal ions owing to the amino group in tis chemical structure (Table 5). To interesting application in biosorbents, aminophenyl thiomethyl chitosan derivative, APTMC (Figure 7), was used to selectively separate precious metals such as Au(III), Pd(II), and Pt(IV) from base metal solutions (Yamashita, Yoshida, Oshima, & Baba, 2015). The modified chitosan adsorbed Au(III), Pd(II), and Pt(IV) depending on the HCl concentration. With increasing the HCl concentration, Au(III), Pd(II), and Pt(IV) were selectively separated from base metal solutions.

Considering salmon milt is also interesting in regeneration of waste from fishery industries and an environmentally friendly biomaterial. The use of milt can be a cost-effective method to recover rare earth elements (REEs) and has a substantial potential for the extraction to REEs from liquid ore waste (Takahashi, Kondo, Miyaji, Watanabe, & Fan, 2014). The REE adsorption onto the milt was effective by the phosphate in the milt sufficiently high affinity with the adsorption capacity 1.04 mEq/g. Furthermore, milt powder was used for batch-type recovery and separation of REEs containing. Y and lanthanides except for Pm (for the total 15 elements were mixed as each REE with 0.10 mg/L; milt: 2.0 g/L) at pH

Table 4. Examples of application for chitin and chitosan

Forms	Examples of Application	References
Hydrogel	Wound dressings, drug delivery system, coatings	Dutta, Khatua, Duttaz, & Prasadz, 2003; Roughley Hoemann,Desrosiers, Mwale, Antoniou & Alini, 2006.
Powder	Anti-bacteria	Andres, Giraud, Gerente & Le Cloirec, 2007; Mujeeb Rahman, Muraleedaran, & Abdul Mujeeb, 2015
Nano-particle	Gaolinium neutran capture therapy Cytotoxicity and antibacterial activity	Tokumitsu, Ichikawa & Fukumori,1999; Tokumitsu,Hiratsuka, Sakurai, Kobayashi, Ichikawa, Fukumori,2000; Anitha, Divya, Krishna, Sreeja, Selvamurugan, Nair, Tamura, H., Jayakumar, 2009.
Beads	Drug delivery Wastewater treatment	Jayakumar, Reis & Mano 2007; Prabhu & Meenakshi, 2015.
Membrane	Artificial kidney membranes Template for hydroxyapatite crystal growth Osteoblast cell attachment	Dutta, Ravikumar, & Dutta, 2002; Ehrlich, Krajewska, Hanke, Born, Heinemann, Knieb, Worch, 2006; Jayakumar, Divya, Shalumon, Sudheesh Kumar, Nair, Furuike, Tamura, 2009.
Sponge	Biodegradability materals Wound dressings	Muramatsu, Masuda, Yoshihara& Fujisawa, 2003; Dai, Zheng, Xu, Kong, Li, Guo, Luo, Zhao, Wei, & Qian, 2009.
Fiber	Wastewater treatment	Lim & Hudson, 2003; Rodrigues, Laranjeira, e Fávere, & Stadler, 1998; Haider, & Park, 2009. Ngah, Endud, C. S., Mayanar, 2002; Huang, Chung, Liou, 1996.
	Braided fiber scaffold for tendon or ligaments tissue	Albanna, Bou-Akl, Blowytsky, Walters III, & Matthew, 2013.
	Filtration Biosensors Wound dressing Tissue engineering Drug delivery	Desai, Kit, Li, Davidson, Zivanovic, & Meyer 2009; Huang, Ge, & Xu, 2007; Chen, Chang, & Chen, 2008; Bhattarai, Edmondson, Veiseh, Matsen & Zhang, 2004; Jiang, Fang, Hsiao, Chu, & Chen, 2004.

Table 5 Adsorption capacities of metals on Chitosan

Materials	Metals	Maximum Adsorption Capacity (mg/g)	References
Chitosan	Cd	5.93	Jha, Iyengar & Rao, 1988.
Chitosan	Cu	4.7	Ngah, & Isa, 1998.
N-(2-pyridyl-methyl)chitosan N-(4-pyridyl-methyl)chitosan	Cu	104 54	Rodrigues, Laranjeira,., de Fávere &, Stadler, 1998.
N-[3-(methylthio)propyl]chitosan	Hg	420	Baba, Matsumura, Shiomori, Kawano, 1998.
Non-cross-linked chitosan	Cr Cu	78 85	Schmuhl, Krieg, & Keizer, 2001.
Cross-linked chitosan	Cr Cu	50 86	Schmuhl, Krieg, & Keizer, 2001.
GLA, ECH, EGDE	Cu	60 62 46	Ngah, Endud & Mayanar, 2002.

continued on next page

Table 5. Continued

Materials	Metals	Maximum Adsorption Capacity (mg/g)	References
N,O-carboxymethyl-chitosan	Cu Co Ni Zn Pb	124 87.2 113 98.1 257	Sun, Wang & Wang, 2006.
Crosslinked N,O-carboxymethyl-chitosan	Cu Co Ni Zn Pb	116 68.9 46 48 392	Sun, Wang & Wang, 2006.
Chitosan–GLA beads (1:1) Chitosan–GLA beads (2:1) Chitosan–alginate beads (2:1)	Cu	64.62 31.20 19.51	Ngah, & Fatinathan, 2008.
APTMC	Pd Pt	304 569	Yamashita, Yoshida, Oshima, & Baba, 2015.

GLA: glutaraldehyde chitosan derivative, ECH: epichlorohydrine chitosan derivative, EGDE: ethylene glycol diglycidyl ether chitosan derivative, APTMC: aminophenyl thiomethyl chitosan

Figure 7. Chemical structure of APTMC

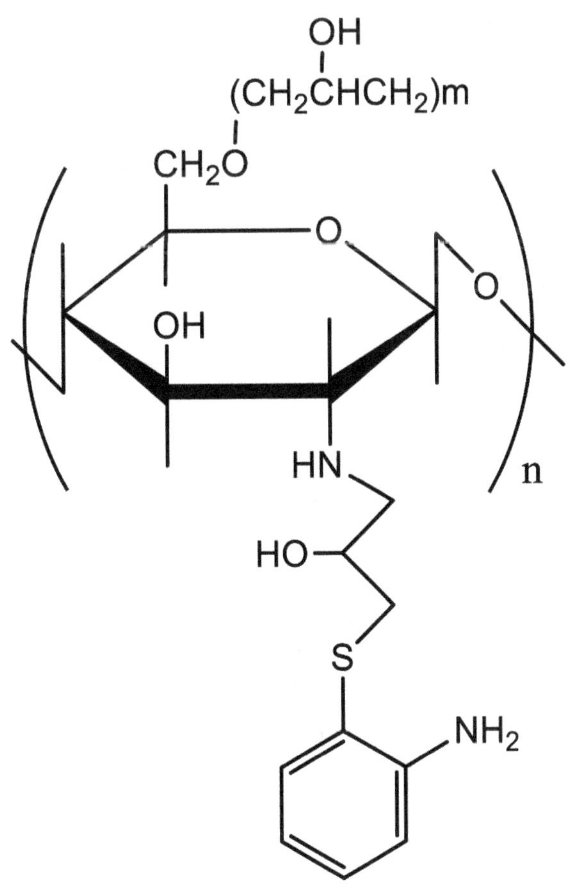

4 and 25°C for 1h. Through the batch experiments, more than 80% of all REEs were adsorbed on the milt at pH 4 especially for such Tm, Yb, and Lu. The lower than 20% of the adsorbed REE remained in the milt by changing of pH values from 2 to 4, except for Yb and Lu.

DNA is very interesting materials extracted from the raw wastes in marine products. Thus, DNA is a relatively low-cost and environmental-friendly biomaterial. The attached cellulose DNA-filter hybrid was successfully utilized for the recovery and separation of rare earth elements by column method (Takahashi, Kondo, Miyaji, Umeo, Honma, 2012). The binding site of REEs was due to the presence of the phosphate within the DNA. By *N,N′*-Disuccinimidyl reaction with amine, two succinimidyl groups in *N,N′*-disuccinimidyl suberate combined two amine groups in filer paper as a cross-linker reagent. For the fixation of DNA onto a fibrous cellulose filter sufficiently high affinity appeared to adsorb REEs with adsorption capacity of 0.182 mg/g for Nd. It was observed that the affinity of REEs for DNA became stronger for REEs with larger atomic numbers.

In the point of view of practical applications, it was useful that Nd and Fe(III) were separated from a synthetic solution of Nd magnet waste using columns loaded with the DNA-filter hybrid.

Crushed eggshell was wasted with 120,000tones/year of eggshells in USA and disposed in landfills. (Anonymous, Food Eng. Int. 2000). The crushed eggshells have relatively high sorption capacity of Cr(III) ions (Chojnacka, K.2005). The dry eggshell contains about 85-95% calcium carbonate, 1.4% magnesium carbonate, 0.76% phosphates, 4% organic matter, traces of sodium, potassium, zinc, manganese, iron and copper (Schaafsma, Pakan, Hofstede, Muskiet, Van Der Veer, & De Vries, 2000; Daengprok, Garnjanagoonchorn, & Mine, 2002). The sorption capacity increased with the increase of Cr(III) concentration, temperature and sorbent concentration, meaning that eggshells were able to remove the concentration of Cr(III) ions below the acceptable level, i.e. at 40 °C, at the initial concentration of metal ions 100 mg/kg with sorbent concentration 15 g/l.

Phytoremediation for Heavy Metal Decontamination

Since humans commence effectively using heavy metals in industries and doing smelted metals for mine development, the contamination of soils, water, and food supply is progressing. It is difficult to restore contaminated soil or water effectively. Some plants can adapt to a high concentration of metals, in which other plant species would not alive and also accumulate toxic metals in their body up to a very high level (Peterson, 1983). Phytoremediation is one of special application of bioremediation and very important as new technology for sustainable development. The term of phytoremediation combines the Greek word “phyton”, plant, with the Latin word “remediare”, to remedy. Phytoremediation is generally classified into five types: phytoextraction (hyperaccumutalor), phytodegradation, rhizofiltration, phytostabilization and phytovolatilization (Salt, Smith, & Raskin, 1998). More than 500 plant species or about 0.2% of all angiosperms are able to accumulate heavy metals (Sarma, 2011). Especially, hyperaccumlator was very important for phytoremediation for heavy metals. Hyperaccumlators were generally defined as plants can accumulate more than 1000mg/kg of Cu, Co, Cr, Ni and Pb, or 10000 mg/kg of Fe, Mn and Zn in their shoot dry matter (Baker & Brooks, 1989: Market, 2003). The phytoremediation process utilizes plants for the removal of contaminants and metals from the soil and water, or to render them harmless. It is basically the decontamination or stabilization of the polluted area using plants. The feature is that the phytoremediation is a low cost and environmentally friendly technology because of only using natural solar energy driven for the cleanup process, and useful for treating a wide variety of environmental contaminants (Brooks, Chambers, Nicks, & Robinson,1998). The plants absorb various inorganic matters from a root in the soil and uses them for their growth as nutrition as shown in:

1. Growing in soils with very high concentrations of metals,
2. Absorbing these metals through their roots, and
3. Concentrating up to extremely high levels of metals in its plant body. As compared to non-hyper-accumulating species, roots of hyperaccumulator can extract the metal from the soil, transfer it to their shoots, and store large amounts in leaves and roots (Rascio & Navari-Izzo, 2011; Hossner, Loeppert, Newton, & Szaniszlo, 1998).

On considerable researchs on hyperaccumulators approximately 75% of these was reported about accumulation of Ni and other metals such as Zn, Cd, Cu, Pb, Co, Tl, Mn, Al, As, and Se were hyperaccumulated (Brooks, 1998; Baker, McGrath, Reeves, & Smith, 2000; Jansen, Broadley, Robbrecht, & Smets,2003). In New Zealand accumulated for gold from ores with high gold yield, 9.27-19.34 μg/g was treated with ammonium thiocyanate (Anderson, Brooks, Stewart, & Simcock, 1998). Also when growing in soil contained nickel/cobalt as well as platinum and palladium metal family members (Chaney, Angle, Baker,, Li, 1998), the soil was adjusted to a low pH to be low Ca concentration by the addition of high-ammonium *N*-fertilizers in the presence of chelating agents such a nitrilo triacetic acid, or any of a variety of amino-acetic acids thereto. Ni concentrations were as high as 2.5 to 5.0% in the above-ground plant tissues. Then the metal recovery was sequestered therein at 10-20 fold higher concentrations than in the soil.

It was reported that copper roofs at Buddhist temples in Asian countries inhibited most of moss growth as the corrosion product, copper oxide by rainy washing down the roof by the rain (Nomura, Itouga, Kojima, Kato, Sakakibara, & Hasezawa, 2015; Shaw, 1994).

Scopelophila cataractae (Mitt.) Broth is one of the copper moss and is often found in Cu-enriched environments, but cannot flourish under normal conditions in nature. The moss, Scopelophila cataractae (Mitt.) Broth., is a typical hyperaccumulator for the phytoremediation of soils contaminated with heavy metals due to usually copper-rich environments and accumulation of copper with high concentrations in its tissues. In the quantitative studies on the concentrations of Cu and 15 other elements (Na, Mg, Ai, P, K, Ca, Ti, V, Cr, Mn, Fe, Co, Ni, Zn, and Pb) in the copper-rich samples the concentration of Cu in the shoots was a level as high as 1–3% in dry weight (Satake, Shibata, Nishikawa, & Fuwa, 1988), although the Cu concentration in rain water was in the order of ppm. The cell wall was a particularly important site of Cu accumulation in comparison with other cell components. Therefore, this moss species may be recognized as a hyperaccumulator (Rascio & Navari-Izzo, 2011).

Water spinach (Ipomoea aquatica) and okra (Abelmoschus esculentus) are tropical vegetables found in area located across the equator region.

The growth response, metal tolerance and phyto-accumulation properties of water spinach and okra were studied under different contaminated soils: control, 50 mg Pb/kg soil, 50 mg Zn/kg soil and 50 mg Cu/kg soil (Ng, Rahman, Boyce & Abas, 2016). The concentration and accumulation of Pb, Zn and Cu metals from soil to roots and shoots (edible parts) were estimated in terms of translocation factor, accumulation factor and tolerance index. It was interesting to note that Okra showed the highest accumulation of Pb (80.20 mg/kg) in its root followed by Zn in roots (35.70 mg/kg) and shoots (34.80 mg/kg) of water spinach, respectively. Different orders of accumulation between okra and water spinach were observed

Pb > Zn > Cu for okra and Zn > Pb > Cu for water spinach. Both water spinach and okra have good potential as Pb and Zn phytoremediators.

Athyrium yokoscense is plant for a species of fern in the family *Athyriaceae*, called *Hebino-negoza* in Japanese whose has mine sites for tolerant of metals and thrive even in soils contaminated with high concentrations of heavy metals such zinc, cadmium, lead, and copper (Morishita & Boratynski, 1992). The *Athyrium yokoscense* exhibited the highest level of Cd in the range of 451-996 mg/kg in the leaflets. Also, Arabis geminifera was the second highest accumulator for cadmium, their concentration was 94.8mg/kg in leaves and *Miscanthus ssp.* on polluted soil accumulated the lowest cadmium level of 1.75mg/kg in leaves. *Chengiopanax sciadophylloides* hyperaccumulates Mn as reported with:

1. Mn uptake ability relative to soil-Mn availability,
2. The potential use of this species in phytomining, and
3. Potential key physiological factors associated with Mn hyperaccumulation ability (Mizunoa, Emorib, & Itoa, 2013).

Among plant tests sampled from 17 locations in their native Japan provenance, the maximum concentration of Mn in foliar was 23,000 mg / kg dry weight. There was a strong positive correlation between foliar Mn and calcium (Ca) concentrations.

Three wild aquatic macrophyte species, *Ipomea* sp., *Eclipta* sp. and *Marsilea* sp., were accumulated multiple heavy metals such as Fe, Cr, Mn, Zn, Cu, Cd in roots and shoots (Gupta, Nayek, Saha, & Satpati, 2008). The accumulation of metals was in the order of Fe > Cr > Mn > Zn > Cu > Cd. In the metal accumulation ability of *Marsilea* sp., was inferior remediation was seen as compared to those of *Eclipta,* sp. and *Ipomea*.sp. The concentrations of heavy metals in different parts, root, stem, leaves, and fruits, of these three species was resulted that the ratio of the concentrations between soil and plant parts (TF_S) important to select plant species for the phytoremediation of soils contaminated with highly heavy metals. Thus, the value was at TFs >1, meaning higher accumulation of metals in plant parts than soil (Barman, Sahu, Bhargava, Chatterjee, 2000). For each plat, the values were founded as followed;

***Ipomea* sp.:** Fe (2.38) > Cu (1.9) > Cd (1.8) > Mn (1.39) > Zn (1.0) > Cr (0.75)
***Eclipta* sp.:** Fe (2.15) > Cu (1.35) > Cd (1.24) > Mn (0.9) > Zn (0.74) > Cr (0.63)
***Marsilea* sp.:** Fe (2.17) > Cu (1.19) > Cd (1.1) > Mn (0.88) > Zn (0.79) > Cr (0.53)

Here, the *Ipomea* sp. showed $TF_S \geq 1$ for Cd, Cu, Mn, and Zn, while *Eclipta* and *Marsilea* sp. were at $TF_S \geq 1$ for Fe, Cu, and Cd, resulting in especially very useful for phytoremediation in the former plant.

CONCLUSION

In the nature, metals are localized at mines and thermal deposits. The metals have been coming into our lives as products with developing economies. Unused or harmful metals are in E-wastes after being used the products, industrial wastes, or by-products which are produced from producing process. In some case, Soil or water around mines or factories have been polluted with metals. For example, the Ashio Copper Mine Mineral Pollution Incident occurred in Japan is the environmental disaster that occurred as a result of the Ashio mining operations in the late 19th and early 20th centuries. In emerging countries, such a problem has been reported even now with developing. Scientists and engineers have to proceed with technological development with considering the problems. This chapter introduced advanced methods of separation and removal of metals with using ILs, dendrimers, and biomasses. The removal of metals with environmental friendly methods and remediation of polluted soil and water by phytoremediation are sustainable and would be increasing importance.

REFERENCES

Anderson, C. W. N., Brooks, R. R., Stewart, R. B., & Simcock, R. (1998). Harvesting a crop of gold in plants. *Nature*, *395*(6702), 553–554. doi:10.1038/26875

Anonymous, . (2000). Valuable raw materials extracted from egg shells. *Food Eng. Int.*, *25*(2), 45–48.

Baba, Y., Kubota, F., Kamiya, N., & Goto, M. (2011). Selective recovery of dysprosium and neodymium ions by a supported liquid membrane based on ionic liquids. *Solv. Extr. Res. Dev. Jpn.*, *18*(0), 193–198. doi:10.15261/serdj.18.193

Baker, A. J. M., & Brooks, R. R. (1989). Terrestrial higher plants which hyperaccumulate metallic elements—a review of their distribution, ecology and phytochemistry. *Biorecovery*, *1*, 81–126.

Baker, A. J. M., McGrath, S. P., Reeves, R. D., & Smith, J. A. C. (2000). Metal hyperaccumulator plants: A review of the ecology and physiology of a biological resource for phytoremediation of metal-polluted soils. In N. Terry & G. Bañuelos (Eds.), *Phytoremediation of Contaminated Soil and Water* (pp. 85–108). Boca Raton, FL: CRC Press.

Barman, S. C., Sahu, R. K., Bhargava, S. K., & Chatterjee, C. (2000). Distribution of heavy metals in wheat, mustard, and weed grown in fields irrigated with industrial effluents. *Bulletin of Environmental Contamination and Toxicology*, *64*(4), 489–496. doi:10.1007/s001280000030 PMID:10754044

Basci, N., Kocadagistan, E., & Kocadagistan, B. (2004). Biosorption of copper (II) from aqueous solutions by wheat shell. *Desalination*, *164*(2), 1, 135–1400. doi:10.1016/S0011-9164(04)00172-9

Bolto, B., & Gregory, J. (2007). Organic polyelectrolytes in water treatment. *Water Research*, *41*(11), 2301–2324. doi:10.1016/j.watres.2007.03.012 PMID:17462699

Bradaric, C. J., Downard, A., Kennedy, C., Robertson, A. J., & Zhou, Y. (2003a). Phosphonium Ionic Liquids. *The Strem Chemiker*, *XX*(1), 2–11.

Bradaric, C. J., Downard, A., Kennedy, C., Robertson, A. J., & Zhou, Y. (2003b). Industrial Preparation of Phosphonium Ionic Liquids. *Green Chemistry*, *5*(2), 143–152. doi:10.1039/b209734f

Brooks, R. R. (1998). *Plants That Hyperaccumulate Heavy Metals*. Oxford, UK: CAB International.

Brooks, R. R., Chambers, M. F., Nicks, L. J., & Robinson, B. H. (1998). Phytomining. *Trends in Plant Science*, *3*(9), 359–362. doi:10.1016/S1360-1385(98)01283-7

Brundtland, G. H. (1987). *Towards Sustainable Development.* A/42/427. Our Common Future: Report of the World Commission on Environment and Development.

Bulgariu, L., Răţoi, M., Bulgariu, D., & Macoveanu, M. (2008). Equilibrium Study of Pb(II) and Hg(II) Sorption from Aqueous Solutions by Moss Peat. *Environmental Engineering and Management Journal.*, *7*(5), 511–516.

Caramalău, C., Bulgariu, L., & Macoveanu, M. (2009). Cobalt (II) Removal from Aqueous Solutions by Adsorption on. Modified Peat Moss. *Chem. Bull. "POLITEHNICA"Univ. (Timisoara)*, *54*(68), 1.

Chaney, R. L., Angle, R. S., Baker, A. J. M., & Li, Y. M. (1998). *Method for phytomining of nickel cobalt and other metals from soil.* U.S. Patent No. 5711784.

Chojnacka, K. (2005). Biosorption of Cr(III) ions by eggshellsOriginal Research Article. *Journal of Hazardous Materials*, *121*(1-3), 167–173. doi:10.1016/j.jhazmat.2005.02.004 PMID:15885418

Cocalia, V. A., Jensen, M. P., Holbrey, J. D., Spear, S. K., Stepinski, D. C., & Rogers, R. D. (2005). Identical extraction behavior and coordination of trivalent or hexavalent f-element cations using ionic liquid and molecular solvents. *Dalton Transactions (Cambridge, England)*, (11): 1966–1971. doi:10.1039/b502016f PMID:15909044

Daengprok, W., Garnjanagoonchorn, W., & Mine, Y. (2002). Fermented pork sausage fortified with commercial or hen eggshell calcium lactate. *Meat Science*, *62*(2), 199–204. doi:10.1016/S0309-1740(01)00247-9 PMID:22061412

Dai, S., Ju, Y. H., & Barnes, C. E. (1999). Solvent extraction of strontium nitrate by a crown ether using room-temperature ionic liquids. *Journal of the Chemical Society, Dalton Transactions: Inorganic Chemistry*, (8): 1201–1202. doi:10.1039/a809672d

Del Sesto, R. E., Corley, C., Robertson, A., & Wilkes, J. S. (2005). Tetraalkylphosphonium-Based Ionic Liquids. *Journal of Organometallic Chemistry*, *690*(10), 2536–2542. doi:10.1016/j.jorganchem.2004.09.060

Diallo, M. S., Christie, S., Swaminathan, P., Johnson, J. H. Jr, & Goddard, W. A. III. (2005). Dendrimer Enhanced Ultrafiltration. 1. Recovery of Cu(II) from Aqueous Solutions Using PAMAM Dendrimers with Ethylene Diamine Core and Terminal NH_2 Groups. *Environmental Science & Technology*, *39*(5), 1366–1377. doi:10.1021/es048961r PMID:15787379

Diallo, M.S., Johnson, J., & Balogh, L. (2000). *Removal of Toxic Metal Ions from Contaminated Water by Dendrimer Enhanced Ultrafiltration.* National Science Foundation.

Dietz, M. L., & Stepinski, D. C. (2005). A Ternary Mechanism for the Facilitated Transfer of Metal Ions into Room-Temperature Ionic Liquids (RTILs): Implications for the Greenness of RTILs as Extraction Solvents. *Green Chemistry*, *7*(10), 747–750. doi:10.1039/b508604c

EPA. (1981). *Summary Report: Control and Treatment Technology for the Metal Finishing Industry; Ion Exchange.* EPA 625/8-81-007.

Farooq, J., Kozinski, A., Khan, M. A., & Athar, M. (2010). Biosorption of heavy metal ions using wheat based biosorbents. A review of the recent literature. *Bioresource Technology*, *101*(14), 5043–5053. doi:10.1016/j.biortech.2010.02.030 PMID:20223652

Gardea-Torresdey, J. L., Tang, L., & Salvador, J. M. (1995). Copper Adosoption by Sphagnum Peat Moss and Its Different Humic Fractions.*Proceedings of the 10th Annual Conference on Hazardous Waste Research*, (pp. 249-260).

Geckeler, K. E., & Volchek, K. (1996). Removal of Hazardous Substances from Water Using Ultrafiltration in Conjunction with Soluble Polymers. *Environmental Science & Technology*, *30*(3), 725–734. doi:10.1021/es950326l

Giridhar, P., Venkatesan, K. A., Srinivasan, T. G., & Vasudewa Rao, P. R. (2005). Extraction of Uranium(VI) from Nitric Acid Medium by 1.1 M Tri-n-butylphosphonate in Ionic Liquid Diluent. *Journal of Radioanalytical and Nuclear Chemistry*, *265*(1), 31–38. doi:10.1007/s10967-005-0785-7

Gupta, S., Nayek, S., Saha, R. N., & Satpati, S. (2008). Assessment of heavy metal accumulation in macrophyte, agricultural soil, and crop plants adjacent to discharge zone of sponge iron factory. *Environmental Geology*, *55*(4), 731–739. doi:10.1007/s00254-007-1025-y

Hallett, J. P., & Welton, T. (2011). Room-temperature ionic liquids: Solvents for synthesisand catalysis. 2. *Chemical Reviews*, *111*(5), 3508–3576. doi:10.1021/cr1003248 PMID:21469639

Harada, K. (2008, January 11). *Press Release.* National Institute for Materials Science.

He, W., Li, G., Ma, X., Wang, H., Huang, J., Xu, M., & Huang, C. (2006). Recovery strategies and the WEEE treatment status in China. *Journal of Hazardous Materials*, *136*(3), 502–512. doi:10.1016/j.jhazmat.2006.04.060 PMID:16820262

Hirayama, N., Okamura, H., Kidani, K., & Imura, H. (2008). Ionic Liquid Synergistic Cation-Exchange System for the Selective Extraction of Lanthanum(III) Using 2-Thenoyltrifluoroacetone and 18-Crown-6. *Analytical Sciences*, *24*(6), 697–699. doi:10.2116/analsci.24.697 PMID:18544855

Hoogerstraete, T. V., Onghena, B., & Binnemans, K. (2013). Homogeneous Liquid–Liquid Extraction of Metal Ions with a Functionalized Ionic Liquid. *The Journal of Physical Chemistry Letters*, *4*(10), 1659–1663. doi:10.1021/jz4005366 PMID:26282975

Hossner, L. R., Loeppert, R. H., Newton, R. J., & Szaniszlo, P. J. (1998). Literature review: Phytoaccumulation of chromium, uranium, and plutonium in plant systems. Amarillo National Resource Center for Plutonium, TX (United States) Technical Report.

Hubbard, C. D., Illner, P., & van Eldik, R. (2011). Understanding Chemical Reaction Mechanisms in Ionic Liquids: Successes and Challenges. *Chemical Society Reviews*, *40*(1), 272–290. doi:10.1039/C0CS00043D PMID:21079861

International Environmental Technology Center. (2013, May 13). Policy Brief on E-waste. International Environmental Technology Center.

Jansen, S., Broadley, M. R., Robbrecht, E., & Smets, E. (2003). Aluminium hyperaccumulation in angiosperms: A review of its phylogenetic significance. *Botanical Review*, *68*(2), 235–269. doi:10.1663/0006-8101(2002)068[0235:AHIAAR]2.0.CO;2

Jensen, M. P., Neuefeind, J., Beitz, J. V., Skanthakumar, S., & Soderholm, L. (2003). Mechanisms of metal ion transfer into room-temperature ionic liquids: The role of anion exchange. *Journal of the American Chemical Society*, *125*(50), 15466–15473. doi:10.1021/ja037577b PMID:14664592

Kaikake, K., Hoaki, K., Sunada, H., Dhakal, R. P., & Baba, Y. (2007). Removal characteristics of metal ions using degreased coffee beans: Adsorption equilibrium of cadmium(II). *Bioresource Technology*, *98*(15), 2787–2791. doi:10.1016/j.biortech.2006.02.040 PMID:17400448

Kozonoi, N., & Ikeda, Y. (2007). Extraction Mechanism of Metal Ion from Aqueous Solution to the Hydrophobic Ionic Liquid, 1-Butyl-3-methylimidazolium Nonafluorobutanesulfonate. *Monatshefte fur Chemie*, *138*(11), 1145–1151. doi:10.1007/s00706-007-0727-x

Kubota, F., Koyanagi, Y., Nakashima, K., Shimojo, K., Kamiya, N., & Goto, M. (2008). Extraction of Lanthanide Ions with an Organophosphorous Extractant into Ionic Liquids. *Solv. Extr. Res. Dev. Jpn.*, *15*, 81–87.

Kubota, F., Shimobori, Y., Baba, Y., Koyanagi, Y., Shimojo, K., Kamiya, N., & Goto, M. (2011). Application of ionic liquids to extraction separation of rare earth metals with an effective diglycol amic acid extractant. *Journal of Chemical Engineering of Japan*, *44*(5), 307–312. doi:10.1252/jcej.11we005

Kubota, F., Shimobori, Y., Koyanagi, Y., Shimojo, K., Kamiya, N., & Goto, M. (2010). Uphill transport of rare-earth metals through a highly stable supported liquid membrane based on an ionic liquid. *Analytical Sciences*, *26*(3), 289–290. doi:10.2116/analsci.26.289 PMID:20215675

Kurniawan, T. A., Chan, G. Y. S., Lo, W., & Babel, S. (2006). Physico-chemical treatment techniques for wastewater laden with heavy metals. *Chemical Engineering Journal*, *118*(1-2), 83–98. doi:10.1016/j.cej.2006.01.015

Lee, S-J., Lee, M-E., Chung, J. W., Park J. H., Huh, K. Y., & Jun, G-I. (2013). Immobilization of Lead from Pb-Contaminated Soil Amended with Peat Moss. *Journal of Chemistry*, 1-6.

Lee, S.-J., Park, J. H., Ahn, Y.-T., & Chung, J. W. (2015, February). Comparison of Heavy Metal Adsorption by Peat Moss and Peat Moss-Derived Biochar Produced Under Different Carbonization Conditions. *Water, Air, and Soil Pollution*, *226*(2), 9. doi:10.1007/s11270-014-2275-4

Li, K., Satoh, M., & Kobayashi, T. (2016). A Facile Method to Determine the Cu^{2+} Concentration Using Ionic Liquids as Concentrated Reagent. *Journal of Chemical Engineering of Japan*, *49*(1), 16–21. doi:10.1252/jcej.15we137

Liu, Y., Zhu, L., Sun, X., Chen, J., & Luo, F. (2009). Silica materials doped with bifunctional ionic liquid extractant for yttrium extraction. *Industrial & Engineering Chemistry Research*, *48*(15), 7308–7313. doi:10.1021/ie900468c

Mallah, M.H., Shemirani, F., & Maragheh, M. G. (2009). Ionic liquids for simultaneous preconcentration of some lanthanoids using dispersive liquid-liquid microextraction technique in uranium dioxide powder. *Environ Sci Technol.*, *43*(6), 1947-1951.

Mallah, M. H., Shemirani, F., Maragheh, M. G., & Jamali, M. R. (2010). Evaluation of synergism in dispersive liquid–liquid microextraction for simultaneous preconcentration of some lanthanoids. *Journal of Molecular Liquids*, *151*(2-3), 122–124. doi:10.1016/j.molliq.2009.12.003

Market, B. (2003). *Element concentration in ecosystems*. Zittau, Germany: International Institute of Advanced Ecological and Economic Studies.

Meena, A. K., Kadirvelu, K., Mishra, G. K., Rajagopal, C., & Nagar, P. N. (2008). Adsorptive Removal of Heavy Metals from Aqueous Solution by Treated Sawdust (Acacia arabica). *Journal of Hazardous Materials*, *150*(3), 604–611. doi:10.1016/j.jhazmat.2007.05.030 PMID:17600619

Mekki, S., Wai, C. M., Billard, I., Moutiers, G., Burt, J., Yen, C. H., & Hesemann, P. et al. (2006). Extraction of Lanthanides from Aqueous Solution by Using Room-Temperature Ionic Liquid and Supercritical Carbon Dioxide in Conjunction. *Chemistry (Weinheim an der Bergstrasse, Germany)*, *12*(6), 1760–1766. doi:10.1002/chem.200500559 PMID:16311989

Mizunoa, T., Emorib, K., & Itoa, S. (2013). Manganese hyperaccumulation from non-contaminated soil in Chengiopanax sciadophylloides Franch. et Sav. and its correlation with calcium accumulation. *Soil Science and Plant Nutrition*, *59*(4), 591–602. doi:10.1080/00380768.2013.807213

Morishita, T., & Boratynski, J. K. (1992). Accumulation of cadmium and other metals in organs of plants growing around metal smelters in Japan. *Soil Science and Plant Nutrition*, *38*(4), 781–785. doi: 10.1080/00380768.1992.10416712

Mulder, M. (1991). *Basic Principles of Membrane Technology*. Dordrecht, The Netherland: Kluwer. doi:10.1007/978-94-017-0835-7

Nakamura, K., Saiwaki, T., & Fukao, K. (2010). Dielectric Relaxation Behavior of Polymerized Ionic Liquid. *Macromolecules*, *43*(14), 6092–6098. doi:10.1021/ma100918e

Nakashima, K., Kubota, F., Maruyama, T., & Goto, M. (2003). Ionic liquids as a novel solvent for lanthanide extraction. *Analytical Sciences*, *19*(8), 1097–1098. doi:10.2116/analsci.19.1097 PMID:12945658

Nakashima, K., Kubota, F., Maruyama, T., & Goto, M. (2005). Feasibility of ionic liquidsas alternative separation mediaforindustrial solvent extraction processes. *Industrial & Engineering Chemistry Research*, *44*(12), 4368–4372. doi:10.1021/ie049050t

Nakashima, T., & Kawai, T. (2005). Quantum Dots-Ionic Liquid Hybrids: Efficient Extraction of Cationic Cdte Nanocrystals into An Ionic Liquid. *Chemical Communications (Cambridge)*, (12): 1643–1645. doi:10.1039/B418001A PMID:15770287

Ng, C. C., Rahman, M. M., Boyce, A. N., & Abas, M. R. (2016). Heavy metals phyto-assessment in commonly grown vegetables: Water spinach (I. aquatica) and okra (A. esculentus). *SpringerPlus*, *5*(1), 469. doi:10.1186/s40064-016-2125-5 PMID:27119073

No, H. K., & Meyers, S. P. (1995). Preparation and Characterization of Chitin and Chitosan – A review. *Journal of Aquatic Food Product Technology*, *4*(2), 27–52. doi:10.1300/J030v04n02_03

Noble R. D., & Stern, S. A. (Eds.). (1995). *Membrane Separation Technology, Principles and Applications*. Elsevier.

Nomura, T., Itouga, M., Kojima, M., Kato, Y., Sakakibara, H., & Hasezawa, S. (2015). Copper mediates auxin signalling to control cell differentiation in the copper moss Scopelophila cataractae. *Journal of Experimental Botany*, *66*(5), 1205–1213. doi:10.1093/jxb/eru470 PMID:25428998

Odinets, I. L., Sharova, E. V., Artyshin, O. I., Lyssenko, K. A., Nelyubina, Y. V., Myasoedova, G. V., & Zakharchenro, E. A. et al. (2010). Novel class of functionalized ionic liquids with grafted CMPO-moieties for actinides and rare-earth elements recovery. *Dalton Transactions (Cambridge, England)*, *39*(17), 4170–4178. doi:10.1039/b926377b PMID:20390180

Okamura, H., Hirayama, N., Morita, K., Shimojo, K., Naganawa, H., & Imura, H. (2010). Synergistic effect of 18-crown-6 derivatives on chelate extraction of lanthanoids(III) into an ionic liquid with 2-thenoyltrifluoroacetone. *Analytical Sciences*, *26*(5), 607–611. doi:10.2116/analsci.26.607 PMID:20467139

Olivier, J. H., Camerel, F., & Ziessel, R. (2011). Lanthanide ion extraction by trifluoromethyl-1,3diketonate-functionalised ionic liquids adsorbed on silica. *Chemistry (Weinheim an der Bergstrasse, Germany)*, *17*(33), 9113–9122. doi:10.1002/chem.201100296 PMID:21739493

Onghena, B., & Binnemans, K. (2015). Recovery of Scandium(III) from Aqueous Solutions by Solvent Extraction with the Functionalized Ionic Liquid Betainium Bis(trifluoromethylsulfonyl)imide. *Industrial & Engineering Chemistry Research*, *54*(6), 1887–1898. doi:10.1021/ie504765v

Patterson, J. W., & Minear, R. A. (1975). Physical-Chemical Methods of Heavy Metal Removal. In P. A. Kenkel (Ed.), *Heavy Metals in the Aquatic Environment* (pp. 261–276). Oxford, UK: Pergamon Press. doi:10.1016/B978-0-08-018068-7.50040-X

Perez de los Rios, A., Hernandez-Fernandez, F. J., Sanchez-Segado, S., Lozano, L. J., Moreno, J. I., & Godinez, C. (2010). Selective Separation of Zn(II) and Fe(III) from Acidic Media Using Ionic Liquids as Sole Extraction Agents. *Chem. Eng. Transactions*, *21*, 625–630.

Pernak, J., Stefaniak, F., & Weglewski, J. (2005). Phosphonium Acesulfamate Based Ionic Liquids. *European Journal of Organic Chemistry*, (4): 650–652. doi:10.1002/ejoc.200400658

Peterson, P. J. (1983). CHAPTER 4 Adaptation to Toxic Metals. In D. A. Robb (Ed.), *Metals and Micronutrients* (pp. 51–69). doi:10.1016/B978-0-12-589580-4.50010-9

Petkovic, M., Seddon, K. R., Rebelo, L. P., & Pereira, C. S. (2011). Ionic Liquids: A Pathway to Environmental Acceptability. *Chemical Society Reviews*, *40*(3), 1383–1403. doi:10.1039/C004968A PMID:21116514

Plechkova, N. V., & Seddon, K. R. (2008). Applications of Ionic Liquids in The Chemical Industry. *Chemical Society Reviews*, *37*(1), 123–150. doi:10.1039/B006677J PMID:18197338

Potts, J., Lynch, M., Wilkings, A., Huppé, G., Cunningham, M., & Voora, V. (2014). Chaper 8 Coffee Market. In R. Ilnyckyj, D. Holmes, & E. Rickert (Eds.), *SSI Review* (pp. 155–186). London: International Institute for Environment and Development.

Ramesh, A., Hasegawa, H., Sugimoto, W., Maki, T., & Ueda, K. (2008). Adsorption of gold(III), platinum(IV) and palladium(II) onto glycine modified crosslinked chitosan resin. *Bioresource Technology*, *99*(9), 3801–3809. doi:10.1016/j.biortech.2007.07.008 PMID:17698356

Rascio, N., & Navari-Izzo, F. (2011). Heavy metal hyperaccumulating plants: How and why do they do it? And what makes them so interesting? *Plant Science*, *180*(2), 169–181. doi:10.1016/j.plantsci.2010.08.016 PMID:21421358

Rivas, B. L., & Geckeler, K. E. (1992). Synthesis and metal complexation of poly(ethylenimine) and derivatives. *Advances in Polymer Science*, *102*, 171–188. doi:10.1007/3-540-55090-9_6

Rivas, B. L., Pereira, E. D., & Moreno-Villoslada, I. (2003). Water-soluble polymer–metal ion interactions Review Article. *Progress in Polymer Science*, *28*(2), 173–208. doi:10.1016/S0079-6700(02)00028-X

Robert, W. P., & Young, K. H. (1985). Evaluation of recent treatment techniques for removal of heavy metals from industrial wastewaters.*American Industrial Chemical Engineering Symposium Series, No. 243,*Vol. 81, 165-203.

Rout, A., & Binnemans, K. (2014). Liquid–liquid extraction of europium(III) and other trivalent rare-earth ions using a non-fluorinated functionalized ionic liquid. *Dalton Transactions (Cambridge, England)*, *43*(4), 1862–1872. doi:10.1039/C3DT52285G PMID:24257814

Rout, A., Karmakar, S., Venkatesan, K. A., Srinivasan, T. G., & Rao, P. R. V. (2011). Room temperature ionic liquid diluent for the mutual separation of europium (III) from americium (III). *Separation and Purification Technology*, *81*(2), 109–115. doi:10.1016/j.seppur.2011.04.033

Rout, A., Kotlarska, J., Dehaen, W., & Binnemans, K. (2013). Liquid–liquid extraction of neodymium(III) by dialkylphosphate ionic liquids from acidic medium: The importance of the ionic liquid cation. *Physical Chemistry Chemical Physics*, *15*(39), 16533–16541. doi:10.1039/c3cp52218k PMID:23949284

Rout, A., Wellens, S., & Binnemans, S. (2014). Separation of rare earths and nickel by solvent extraction with two mutually immiscible ionic liquids. *RSC Advances*, *4*(11), 5753–5758. doi:10.1039/c3ra46261g

Saeed, A., Akhter, M. W., & Iqbal, M. (2005). Removal and recovery of heavy metals from aqueous solution using papaya wood as a new biosorbent. *Separation and Purification Technology*, *45*(1), 25–31. doi:10.1016/j.seppur.2005.02.004

Saeed, A., Iqbal, M., & Akhtar, M. W. (2005). Removal and recovery of lead (II) from single and multimetal (Cd, Cu, Ni, Zn) solutions by crop milling waste (black gram husk). *Journal of Hazardous Materials*, *117*(1), 65–73. doi:10.1016/j.jhazmat.2004.09.008 PMID:15621354

Sahu, O. P., & Chaudhari, P. K. (2013). Review on Chemical treatment of Industrial Waste Water. *J. Appl. Sci. Environ. Manage.*, *17*(2), 241–257.

Salt, D. E., Smith, R. D., & Raskin, I. (1998). Phytoremediation. *Annual Review of Plant Physiology and Plant Molecular Biology*, *49*(1), 643–668. doi:10.1146/annurev.arplant.49.1.643 PMID:15012249

Sarma, H. (2011). Metal hyperaccumuulation in plants: A Review focusing on phytoremediation technology. *Journal of Environmental Science and Technology.*, *4*(2), 118–138. doi:10.3923/jest.2011.118.138

Satake, K., Shibata, K., Nishikawa, M., & Fuwa, K. (1988). Copper accumulation and location in the moss Scopelophila cataractae. *Journal of Bryology*, *15*(2), 353–376. doi:10.1179/jbr.1988.15.2.353

Schaafsma, A., Pakan, I., Hofstede, G. J. H., Muskiet, F. A. J., Van Der Veer, E., & De Vries, P. J. F. (2000). Mineral, amino acid, and hormonal composition of chicken eggshell powder and the evaluation of its use in human nutrition. *Poultry Science*, *79*(12), 1833–1838. doi:10.1093/ps/79.12.1833 PMID:11194049

Shaw, A. J. (1994). Adaptation to metals in widespread and endemic plants. *Environmental Health Perspectives*, *102*(Suppl 12), 105–108. doi:10.1289/ehp.94102s12105 PMID:7713025

Shimojo, K., & Goto, M. (2004). Solvent Extraction and Stripping of Silver Ions in Room-Temperature Ionic Liquids Containing Calixarenes. *Analytical Chemistry*, *76*(17), 5039–5044. doi:10.1021/ac049549x PMID:15373439

Shimojo, K., Kurahashi, K., & Naganawa, H. (2008). Extraction behavior of lanthanides using a diglycolamide derivative TODGA in ionic liquids. *Dalton Transactions (Cambridge, England)*, (37): 5083–5088. doi:10.1039/b810277p PMID:18802624

Shimojo, K., Naganawa, H., Kubota, F., & Goto, M. (2006). Solvent Extraction of Lanthanides into an Ionic Liquid Containing N, N, N, N-tetrakis(2-yridylmethyl)ethylenediamine. *Chemistry Letters*, *35*(5), 484–485. doi:10.1246/cl.2006.484

Singha, A. S., & Guleria, A. (2015). Utility of chemically modified agricultural waste okra biomass for removal of toxic heavy metal ions from aqueous solution. *Engineering in Agriculture. Environment Folio*, *8*(1), 52–60.

Sthiannopkaoa, S., & Wongb, M. H. (2013). Handling e-waste in developed and developing countries: Initiatives, practices, and consequences. *The Science of the Total Environment*, *463–464*, 1147–1153. doi:10.1016/j.scitotenv.2012.06.088 PMID:22858354

Sun, X., Bell, J. R., Luo, H., & Dai, S. (2011). Extraction separation of rare-earth ions via competitive ligand complexations between aqueous and ionic-liquid phases. *Dalton Transactions (Cambridge, England)*, *40*(31), 8019–8023. doi:10.1039/c1dt10873e PMID:21735029

Sun, X., Ji, Y., Guo, L., Chen, J., & Li, D. (2011). A novel ammonium ionic liquid based extraction strategy for separating scandium from yttrium and lanthanides. *Separation and Purification Technology*, *81*(1), 25–30.

Sun, X., Ji, Y., Hu, F., He, B., Chen, J., & Li, D. (2010). The inner synergistic effect of bifunctional ionic liquid extractant for solvent extraction. *Talanta*, *81*(4-5), 1877–1883. doi:10.1016/j.talanta.2010.03.041 PMID:20441990

Sun, X., Wu, D., Chen, J., & Li, D. (2007). Separation of scandium(III) from lanthanides(III) with room temperature ionic liquid based extraction containing. Cyanex 925. *Journal of Chemical Technology and Biotechnology (Oxford, Oxfordshire)*, *82*(3), 267–272. doi:10.1002/jctb.1664

Takahashi, Y., Kondo, K., Miyaji, A., Umeo, M., Honma, T., & Asaoka, S. (2012). Recovery and separation of rare earth elements using columns loaded with DNA-filter hybrid. *Analytical Sciences*, *28*(10), 985–992. doi:10.2116/analsci.28.985 PMID:23059995

Takahashi, Y., Kondo, K., Miyaji, A., Watanabe, Y., Fan, Q., Honma, T., & Tanaka, K. (2014). Recovery and Separation of Rare *Earth Elements Using Salmon Milt. PLoS ONE*, *9*(12), e114858. doi:10.1371/journal.pone.0114858 PMID:25490035

Tomalia, D. A. (2004). Birth of a New Macromolecular Architecture: Dendrimers as Quantized Building Blocks for Nanoscale Synthetic Organic Chemistry. *Aldrichimica Acta*, *37*(2), 39.

Turanov, A. N., Karandashev, V. K., & Baulin, V. E. (2008). Effect of ionic liquids on the extraction of rare-earth elements by bidentate neutral organophosphorus compounds from chloride solutions. *Russian Journal of Inorganic Chemistry*, *53*(6), 970–975. doi:10.1134/S0036023608060272

Vijayaraghavan, K., Palanivelu, K., & Velan, M. (2006). Biosorption of copper(II) and cobalt(II) from aqueous solutions by crab shell particles. *Bioresource Technology*, *97*(12), 1411–1419. doi:10.1016/j.biortech.2005.07.001 PMID:16112568

Visser, A. E., Swatloski, R. P., Griffin, S. T., Hartman, D. H., & Rogers, R. D. (2001). Liquid/liquid extraction of metal ions in room temperature ionic liquids. *Separation Science and Technology*, *36*(5-6), 785–804. doi:10.1081/SS-100103620

Visser, A. E., Swatloski, R. P., Reichert, W. M., Griffin, S. T., & Rogers, R. D. (2000). Traditional Extractants in Nontraditional Solvents: Groups 1 and 2 Extraction by Crown Ethers in Room-Temperature Ionic Liquids. *Industrial & Engineering Chemistry Research*, *39*(10), 3596–3604. doi:10.1021/ie000426m

Visser, A. E., Swatloski, R. P., Reichert, W. M., Mayton, R., Sheff, S., Wierzbicki, A., & Rogers, R. D. et al. (2002). Task Specific Ionic Liquids Incorporating Novel Cations for the Coordination and Extraction of Hg^{2+} and Cd^{2+}: Synthesis, Characterization and Extraction Studies. *Environmental Science & Technology*, *36*(11), 2523–2529. doi:10.1021/es0158004 PMID:12075816

Wang, W., Yang, H., Cui, H., Zhang, D., Liu, Y., & Chen, J. (2011). Application of Bifunctional Ionic Liquid Extractants [A336][CA-12] and [A336][CA-100] to the Lanthanum Extraction and Separation from Rare Earths in the Chloride Medium. *Industrial & Engineering Chemistry Research*, *50*(12), 7534–7541. doi:10.1021/ie2001633

Wei, G. T., Yang, Z., & Chen, C. J. (2003). Room Temperature Ionic Liquid as a Novel Medium for Liquid/Liquid Extraction of Metal Ions. *Analytica Chimica Acta*, *488*(2), 183–192. doi:10.1016/S0003-2670(03)00660-3

Welton, T. (1999). Room-Temperature Ionic Liquids. Solvents for Synthesis and Catalysis. *Chemical Reviews*, *99*(8), 2071–2084. doi:10.1021/cr980032t PMID:11849019

Welton, T. (1999). Room-temperature ionic liquids. Solvents for synthesis and catalysis. *Chemical Reviews*, *99*(8), 2071–2084. doi:10.1021/cr980032t PMID:11849019

Yamashita, A., Yoshida, K., Oshima, T., & Baba, Y. (2015). Synthesis of an Aminophenylthiomethyl Chitosan Derivative and Its Adsorption of Precious Metals. *Journal of Chemical Engineering of Japan*, *48*(11), 897–902. doi:10.1252/jcej.15we025

Yamasue, E., Minamino, R., Numata, T., Nakajima, K., Murakami, S., Daigo, I., & Ishihara, K. N. et al. (2009). Novel Evaluation Method of Elemental Recyclability from Urban Mine. *Materials Transactions*, *50*(6), 1536–1540. doi:10.2320/matertrans.MBW200816

Yang, F., Baba, Y., Kubota, F., Kamiya, N., & Goto, M. (2012). Extraction and separation of rare earth metal ions with DODGAA in ionic liquids. *Solvent Extr. Res. Dev.*, *19*(0), 69–76. doi:10.15261/serdj.19.69

Yang, H., Wang, W., Cui, H., Zhang, D., Liu, Y., & Chen, J. (2012). Recovery of rare earth elements from simulated fluorescent powder using bifunctional ionic liquid extractants (Bif-ILEs). *Journal of Chemical Technology and Biotechnology (Oxford, Oxfordshire)*, *87*(2), 198–205. doi:10.1002/jctb.2696

Yang, J., Kubota, F., Baba, Y., Kamiya, N., & Goto, M. (2014). Separation of Precious Metals by Using Undiluted Ionic Liquids. *Solvent Extr. Res. Dev.*, *21*(1), 89–94. doi:10.15261/serdj.21.89

Yang, J., Kubota, F., Baba, Y., Kamiya, N., & Goto, M. (2014). One Step Effective Separation of Platinum and Palladium in an Acidic Chloride Solution by Using Undiluted Ionic Liquids. *Solvent Extr. Res. Dev.*, *21*(2), 129–135. doi:10.15261/serdj.21.129

Yang, T. C., & Zall, R. R. (1984). Absorption of metals by natural polymers generated from seafood processing wastes. *Industrial & Engineering Chemistry Product Research and Development*, *23*(1), 168–172. doi:10.1021/i300013a033

Yoon, S. J., Lee, J. G., Tajima, H., Yamasaki, A., Kiyono, F., Nakazato, T., & Tao, H. (2010). Extraction of lanthanide ions from aqueous solution by bis(2-ethylhexyl)phosphoric acid with room-temperature ionic liquids. *Journal of Industrial and Engineering Chemistry*, *16*(3), 350–354. doi:10.1016/j.jiec.2009.09.063

Zuo, Y., Chen, J., & Li, D. (2008). Reversed micellar solubilization extraction and separation of thorium(IV) from rare earth(III) by primary amine N1923 in ionic liquid. *Separation and Purification Technology*, *63*(3), 684–690. doi:10.1016/j.seppur.2008.07.014

Zuo, Y., Liu, Y., Chen, J., & Li, D. Q. (2008). The separation of cerium(IV) from nitric acid solutions containing thorium(IV) and lanthanides(III) using pure [C8mim]PF6 as extracting phase. *Industrial & Engineering Chemistry Research*, *47*(7), 2349–2355. doi:10.1021/ie071486w

Zuo, Y., Liu, Y., Chen, J., & Li, D. Q. (2009). Extraction and recovery of cerium(IV) along with fluorine(I) from bastnasite leaching liquor by DEHEHP in [C_8mim]PF_6. *Journal of Chemical Technology and Biotechnology (Oxford, Oxfordshire)*, *84*(7), 949–956. doi:10.1002/jctb.2116

Chapter 14
Eco-Friendly On-Site Water Analyses for Ultra-Trace Harmful Ions

Yukiko Takahashi
Nagaoka University of Technology, Japan

ABSTRACT

Governments in developed countries specify water quality standards to be applied in their own countries. Recently the allowance values of harmful metals are getting tighter for environmental water and industrial wastewater throughout the world. For example, World Health Organization (WHO) recommends the criterion value of 3 ppb for cadmium for protection of human health, furthermore, governments of some countries provide the wastewater standard for zinc for conservation of aquatic life. High-performance analytical instruments such as AAS, ICP, and ICP-MS have been employed as official analytical methods of ultra-trace elements. Besides the costly initial and running expenses and the material and energy consuming of the instruments, specific technical skills and long analytical time are required for sample pre-treatment. Therefore, simple yet highly sensitive and eco-friendly on-site methods have been demanded for quick judgment of industrial and environmental water in developed countries as well as evaluation of drinking water in developing countries.

INTRODUCTION

Harmful Ions in Public Water

The total volume of water on the earth surface contains liquid, solid, or gaseous is stated in the atmosphere, on the earth surface, and in the crust down to the depth of 2000 meters. By approximate estimates, its volume is an abundant 1,386 million km^3 (Shiklomanov, 2000). However, 97.5 percent of this amount is salty water, and only about 2.5 percent is fresh water. The greater portion of the fresh water is locked up in the form of ice in the Antarctic and the Arctic, permafrost area and mountainous regions (68.7 percent of total fresh water resources). Besides except for fresh groundwater (30.9 percent) and gaseous

DOI: 10.4018/978-1-5225-1971-3.ch014

water, only about 0.3 percent of the total amount of fresh water on the earth is concentrated in lakes, reservoirs, and river systems. The latter water sources are most accessible for human activities such as drinking water, domestic water, agricultural water, and industrial water, as well as very important for all terrestrial animals and water ecosystems. Mankind has utilized freshwater by uptaking mainly from river and lake, and then after use for each purpose domestic wastewater, agricultural and industrial effluents, mine drainages have been discharged to a public water area including river, lake, canal and sea. In this process, public water has been artificially polluted by various toxic materials, which has resulted in serious environmental pollutions all around the world. In addition, naturally originating contaminations based on soil composition have severed damage to water quality. Figure 1 shows examples of contaminated area by inorganic substances worldwide, however they are only a part of historical and geographical water pollution. Especially heavy metals among them are indispensable and industrially utilizable to human being. Such local cases of inorganic contamination are too numerous to be reported in many developing countries as well as advanced countries. As a well-known example, an extensive arsenic (As) contamination in groundwater in south and east Asian countries is derived from strata which highly concentrated arsenic accumulates (The World Bank, 2005), an estimated 60 million people are at risk from high levels of naturally-occurring arsenic in groundwater, and at least 700,000 people in the region have thus far been affected by arsenicosis. Besides, there is a lot of artificially-occurring contamination all over the world; *Minamata disease* that is a type of poisoning caused by industrial methyl mercury (met-Hg) contamination in Japan (Harada, 1995), inorganic mercury (Hg) pollution around gold mines that have been reported in the Amazonian valley in Brazil and around Lake Victoria in Tanzania (Ikingura, Akagi, Mujumba, & Messo, 2006) from the end of the 1970's, a recent lead contamination in tap water in US state of Michigan (Ganim & Tran, 2016), *Itai-Itai disease* caused by cadmium (Cd) poisoning due to mining in Japan (Järup & Åkesson, 2009), etc. These heavy metals have accumulated in living bodies through foods and drinking water, and then cause serious health damages. In environmental water, these inorganic substances are dissolved as particular ionic forms by surrounding of H_2O molecules having high dielectric constant, for instance, Cd^{2+}, Pb^{2+}, and $HAs^{V}O_4^{2-}$, etc., so that they easily spread in public water.

Figure 1. Man-made inorganic contaminations on a world map

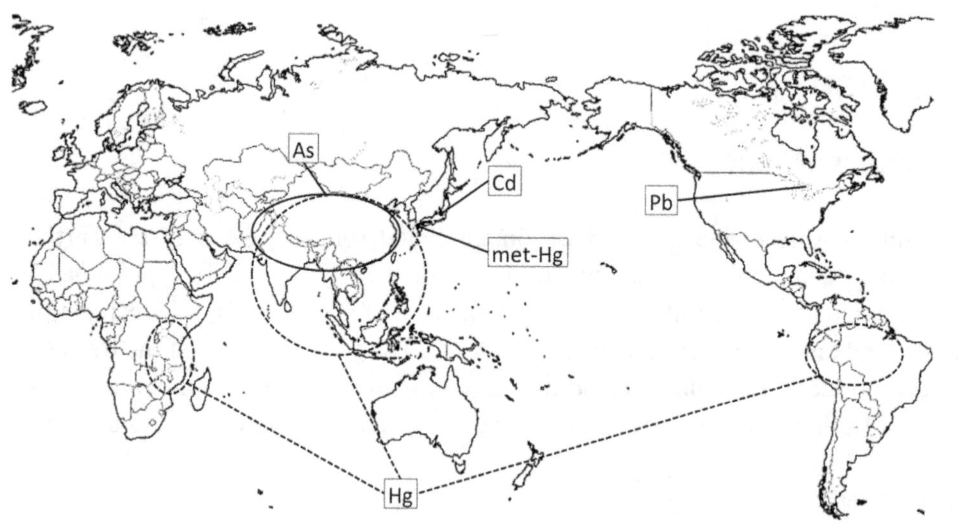

Water Quality Standards

Heavy metal ions are found in public water in some stable chemical forms depending on discharge sources, surrounding pHs, dissolved oxygen (DO), and biological activities, therefore, they change their oxidation states and chemical structures according to an ambient environment. Arsenic (As) is found as both inorganic 'oxo-anion' and organic forms with different oxidation states, highly toxic arsenite As^{3+} ($H_3As^{III}O_3$), arsenate As^{5+} ($H_2As^{V}O_4^-$ or $HAs^{V}O_4^{2-}$), lowly toxic organic forms (arsenobetaine $(CH_3)_3As^+ CH_2COO^-$, dimethylarsinic acid $(CH_3)_2AsO_2H$,) particularly in fish and shellfish. Lead is used principally for production activities of pigments, batteries, solders, alloys, and water pipes. In addition, the organolead compounds (tetramethyl and tetraethyl lead $Pb(CH_3CH_2)_4$) have extensively used as anti-knocking agents in petrol, although their use has been phased out in many countries because of high toxicity, therefore they mainly exist inorganic Pb^{2+} and organic lead forms in water. Mercury is released to the environment from natural sources and human activities including catalysts of production of acetylene and vinyl chloride and globally circulated as various chemical forms such as Hg^0, Hg^{2+}, organic mercury forms (CH_3Hg, $(CH_3)_2Hg$, etc) in air and natural water. The Minamata Convention on Mercury signed by delegates from 140 countries on 19 January 2013 is an international treaty to protect human health and the environment from anthropogenic emissions and releases of mercury and mercury compounds. Thus metal ions have several different forms controlled by the surrounding environment and show unique toxicities. Besides they accumulate human organs and have a long biological half-life time in human bodies, for example, the half-life time of cadmium is 10-35 years, which indicate that chronic toxicity must be taken into consideration for determining quality of water we drink and use everyday as well as acute toxicity. World health organization (WHO) has provided public normative guidelines for drinking water quality (World Health Organization, 2011) as shown in Table 1 based on the risk management from hazards that may compromise the safety of drinking water. The latest regulated values of inorganic substances for drinking water or similar environmental water in three different countries Thailand (Pollution Control Department, Thailand), Japan (Ministry of the Environment, Government of Japan, 2003), and US (United States Environmental Protection Agency, Drinking Water Contaminants) are also compiled in Table 1. It indicates that these legally binding standards are more or less different from WHO guidelines, however roughly similar to them. It is notable that all standard values of thirteen contaminants in Table 1 ranging from 0.003 to 5 mg in 1 L of water are at the level of trace or ultra-trace. The unit milligrams per liter (mg/L) can be expressed as parts per million (ppm) and micrograms per liter (μg/L) also can be converted to parts per billion (ppm). In other words, to prevent chronic toxicity of harmful ions for human, the regulated values of harmful ions in drinkable water restricted down to a low concentration in a ppb level.

A movement in a new direction about water regulation is that some countries expand water regulation from protection of human health to conservation of aquatic environment; there is no heath damage to human by exposing zinc (Zn) and zinc compounds but great harm to aquatic organisms such as *daphnia*, *rotifers*, *earthworms*, *rainbow trouts*, and *chars* according to EU risk assessment report in 2008 (European Commission, 2008). Water Framework Directive formulated on 2000 by European Commission for protecting aquatic ecosystem influenced on Ministry of the Environment (MOE), Japan and the environmental quality standard of total Zn in river and lake is limited to be less than 0.03 mg/L (MOE, 2003) as shown in Table 1.

Table 1. Comparison of water quality standards for drinking water or human health in some countries

				mg / L = ppm
	WHO	Thailand	Japan	US
Contaminant	Drinking Water Guideline (4th ed)[a]	Surface Water Quality Standard[b]	Environmental Quality Standard[c]	Drinking Water Regulation[d]
Arsenic (As)	0.01	0.01	0.01	0.01
Cadmium (Cd)	0.003	0.005	0.003	0.005
Chromium (Cr)	0.05	0.05	0.05 (Cr(VI))	0.1 (total)
Copper (Cu)	2	0.1		1.3
Fluoride (F)	1.5		0.8	4
Iron (Fe)	0.3			0.3
Lead (Pb)	0.01	0.05	0.01	0.015
Manganese (Mn)	0.4	1	0.2	0.05
Mercury (Hg)	0.006	0.002	0.0005	0.002
Molybdenum (Mo)	0.07		0.07	
Nickel (Ni)	0.07	0.1	– (Monitored)	
Selenium (Se)	0.04		0.01	0.05
Zinc (Zn)	3		0.03 (Aquatic organism)	5

[a] World Health Organization, 2011, [b] Pollution Control Department, Thailand, [c] Ministry of the Environment, Government of Japan, 2003, [d] United States Environmental Protection Agency, Drinking Water Contaminants

Moreover, some administrative organizations and union, including Thailand Pollution Control Department (Pollution Control Department, Thailand), US Environmental Protection Agency (EPA) (EPA Effluent Guidelines), and MOE (MOE National Effluent Standards, 2015) also decided the effluent standards about discharge from each of particular factories and industrial plants to public water. The water emission standards disclosed on their homepages, those are almost 10 times as much as the drinking or environmental water standard values shown in Table 1.

Current Analytical Methods

As stated above, a water quality test requires the analytical methodology that enables the accurate detection at the level of ultra trace even if the sample contains unknown and complicated interfering components. International Organization for Standardization (ISO) (ISO ICS 13.060), European Committee for Standardization (CEN) (CEN TC230), US EPA, and Japanese Industrial Standards (JIS) (JIS K 0102:2013) cover all the analytical procedures of inorganic ions in aqueous water and have consistency each other. Especially, USEPA test methods are freely available and easy to find the method you require on the web site National Environmental Methods Index (National Environmental Methods Index). Table 2 shows official analytical methods of the thirteen heavy metals based on the above mentioned standard methods. As you can see in Table 2, except for fluoride (F), inorganic targets are measured by analytical instruments based on atomic vapor generation by heating with flammable gas or with plasma gas and varying magnetic field, such as Atomic Absorption Spectroscopy (AAS), Inductively Coupled Spectroscopy (ICP)

Table 2. Official analytical methods prescribed by ISO, CEN, USEPA, and JIS for harmful heavy metals

	μg / L = ppb	
Contaminant	**WHO Guidelines / ppb**	**Official Analytical Methods**
Arsenic (As)	10	Hydride Generation Atomic Absorption Spectroscopy (HG-AAS), Hydride Generation Inductively Coupled Plasma Spectroscopy (HG-ICP)
Cadmium (Cd)	3	Atomic Absorption Spectroscopy (AAS), Electro Thermal Atomic Absorption Spectroscopy (ETAAS), Inductively Coupled Plasma Mass Spectrometry (ICP-MS)
Chromium (Cr)	50	Spectrometric Method using 1,5-Diphenylcarbazide, AAS, ETAAS, ICP, ICP-MS
Copper (Cu)	2,000	AAS, ETAAS, ICP, ICP-MS
Fluoride (F)	1,500	Lanthanum-alizarin Complexone Method, Ion Chromatography
Iron (Fe)	300	AAS, ETAAS, ICP
Lead (Pb)	10	AAS, ETAAS, ICP, ICP-MS
Manganese (Mn)	400	AAS, ETAAS, ICP, ICP-MS
Mercury (Hg)	6	Cold Vapor Atomic Absorption Spectrophotometry
Molybdenum (Mo)	70	ICP, ICP-MS, ETAAS
Nickel (Ni)	70	ICP, ICP-MS, ETAAS
Selenium (Se)	40	HG-AAS, HG-ICP
Zinc (Zn)	3,000	AAS, ETAAS, ICP, ICP-MS

and Mass Spectroscopy (MS). AAS and ICP methods are spectroanalytical procedures for the quantitative determination of chemical elements based on individual atomic spectrum, using the absorption of light by free atoms in the gaseous state (AAS) and the emission of ionized gas (ICP), respectively. ICP-MS is a mass-pectrometric procedure based on gaseous ions generated by ICP phenomena. Therefore, these analytical instruments require a large quantity of acetylene or argon gas, a great volume of water for cooling, and a large amount of electric power, that is, official analytical methods for trace or ultra-trace inorganic contaminants impose a large environmental burden. Figure 2 shows a brief flowchart of analysis of water quality based on official procedure of public water. Since instrumental analysis must be carried out only in laboratories (in lab) due to cleanness against contamination and stabilization of instrument as well as the above-mentioned infrastructure, water sample must be acquired and transferred from a sampling spot, refilled into a bottle, and immediately added with acid for sample preservation. After that, generally pretreatment process of sample solution including separation and/or enrichment of target ion using by solvent extraction or solid extraction technique should be carried out in lab by a technician having knowledge of sample handling of various kinds of water. The pretreatment is the most important part of analytical procedure of trace analysis of water, strongly influencing on the accuracy of analytical data. That's because sometimes water samples contains not only many kinds of inorganic and organic contaminants other than a target ion in the wide range from low to high concentration but also very low content of a target ion below the detection limit of a instrument. Consequently, it takes at least a few weeks to obtain the analytical results from sampling corresponding to an analytical flowchart in Figure 2. In addition, the costly initial and running expenses of the instruments and waste liquid from the instruments during sample analysis are very worrisome problems.

Figure 2. A typical flowchart of analytical procedure of environmental water by instrumental analysis

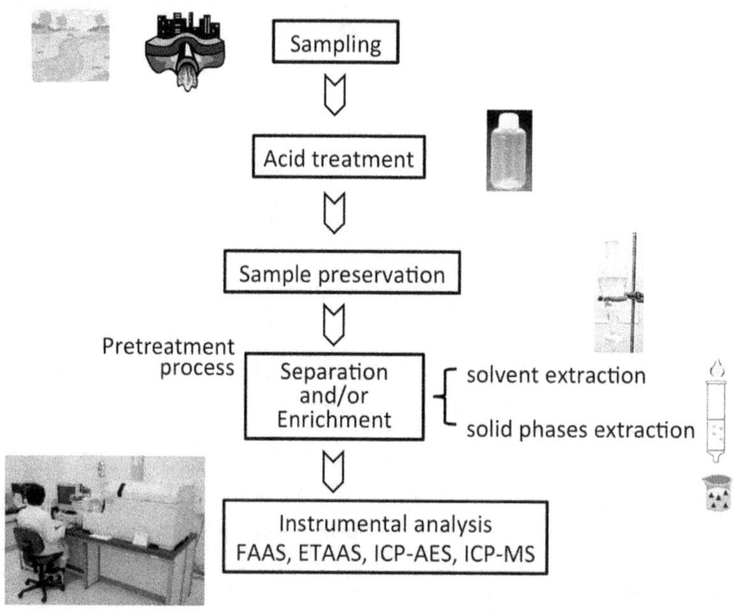

The frequency of water examination depends on various factors such as the purpose of using water, contaminant type, and volume, etc., for example, both quality tests for raw water of tap water and industrial wastewater are generally conducted annual or every half a year. The official analytical methods in Table 2 are suitable for these periodic water inspections from the viewpoint of highly accurate and precise analysis. However, quantities of harmful ions in water vary day by day, the annual or every half-year examination is insufficient to water quality management at a particular time. Especially in effluents from factories and industrial plants quantities of contaminants greatly vary depend on their daily operations, and also in environmental water quantities of contaminants easily fluctuate depend on human activities, crustal movements and climate change, etc. Therefore, analysts in general have desired 'on-site' simple test methods to determine whether the amount of a contaminant exceeds the water quality standard or not at the site where there is a measuring target water. The on-site analytical methods are overwhelmingly advantageous for saving analytical time, expenses, efforts in comparison with 'in lab' instrumental analysis.

To conclude, although having sufficient sensitivity for trace or ultra-trace harmful inorganic pollutants in aqueous solutions, current analytical methods based on instrumental analysis are material consuming, energy consuming, time-consuming, and also very expensive. In addition, specific technical skills are required for complicated pretreatment before instrumental analysis, so that the official methods are exclusively suitable for annual or semiannual water inspection like legally regulated examination with high sensitivity and quantification. However, rapid and on-site daily management of changing at ppb level is essential to understand for controlling water in real time. Simple test methods for the trace or ultra trace harmful ions by the judgment whether a water sample is polluted or not are desired as highly sensitive on-site analysis such as quick monitoring of industrial wastewater, evaluation of drinking water, urgent assessment of well water in times of disaster, as well as in environmental education in schools.

ECO-FRIENDLY ON-SITE ANALYSES FOR WATER QUALITY

Based on the background as described above, various different types of simple on-site analytical methods are being developed for detecting the thirteen harmful ions of the regulated inorganic substances shown in Table 1 at ultra trace levels. Here the leading candidates among them are introduced. There has been a strong demand for the on-site analysis of harmful inorganic ions in water because they enable daily or continuous monitoring and high-density analysis of water quality in a contrasting manner of once-a-year thorough examination by official analytical methods using analytical instruments. Figure 3 shows the comparison of analytical flow charts between on-site analysis and instrumental analysis from the viewpoint of location, operation and time required. Water quality is always fluctuating and quite different depends on sampling points even though within a same water area, therefore, if we truly achieve water quality management of environmental water for human health and aquatic ecosystem, it's absolutely necessary to use rapid and simple on-site analytical methods. As shown in Figure 3, the on-site analysis is based on using portable analytical instruments like handheld X-Ray Fluorescence (XRF) detector, portable Cathode or Anodic Stripping Voltammetry (CSV or ASV) analyzer, compact Flow Injection Analysis (FIA) system, handy colorimeter and spectrophotometer, etc. Since the on-site analytical methods using these portable apparatus require no special gas, no cooling water, undoubtedly low electric power, and low cost, that is to say, they are eco-friendly analytical methods for sustainability. They also have advantages of real time measurements achieved by omission of the complicated processes such as preservation, transportation, and separation/enrichment for instrumental analysis 'in lab'. Moreover, accurate analysis without sample preservation can be carried out, which means no loss of a trace or ultra trace ion by adsorption onto a container. In order to establish the on-site analysis of ultra-trace and trace of heavy metal ions in environmental water and industrial wastewater, it is mandatory to have performance requirements such as high sensitivity that make ppb level of metal ions detectable, high selectivity or specificity that make only trace of target in a sample containing large amounts of various organic and inorganic measurable, and versatility applicable for some toxic inorganic ions.

Figure 3. Comparison between instrumental analysis and on-site analyses for water quality

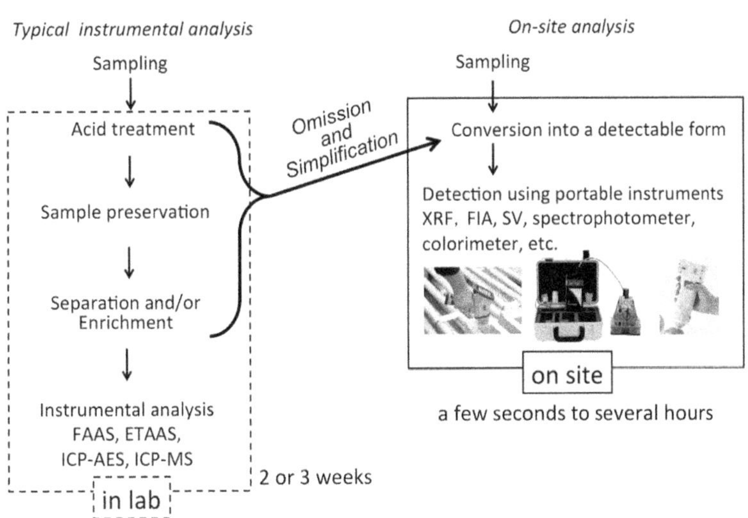

The SV method is an electrochemical assay for detection of ions in the field using a commercially available and inexpensive portable instrument (Lewtas, Fujikawa, & Wajrak, 2015). ASV shown in Figure 4 involves two major steps. First a fraction of the dissolved cationic metal ion is reduced and concentrated onto the surface of electrode (concentration step), and the metal removed from an aqueous solution forms a layer onto the electrode. Subsequently, the metal are oxidized off during the stripping step. The oxidation of metal onto the electrode is registered as peaks in the current signal at the potential at which the metal begins to be oxidized. It is highly sensitive enough to detect ppb level of analyte ion, moreover applicable for discrimination analysis of oxidation state such as As^{3+} / As^{5+} and Cr^{3+} / Cr^{6+}. Approximately 12-15 metal ions can be analyzed by this method. Actually, ASV has been approved by EPA as analytical method of arsenic in solution (SW-846-7063) that is capable of measuring from 0.1 to 300 ppb of free arsenic. SV methods work best for comparatively clean water such as groundwater; because natural organic matters in lake and river such as humic substances and highly concentrated ions significantly disturb the SV measurements by the adsorption on the surface.

Simultaneous determination of unspecified trace heavy metal ions have been achieved by a handheld type XRF instrument based on energy-dispersive XRF (EDXRF) shown in Figure 4 with pre-concentration of metal ions into a disk membrane by the addition of a certain reagent like solid phase extraction (Margui, Zawisza, & Sitko, 2014). Direct measurements of aqueous samples by XRF have low sensitivity; therefore, the pre-concentration process into the thin layer is the essence of the detection in range from ppm to ppb of target ions in water. The handheld XRF instruments are little pricy, however, they are capable of providing multi-element simultaneous detection, simple operations, and rapid measurements. One reported example is simultaneous determination of Ni, Cu, Zn, Pb, and Cd in tap and waste water with the preconcentration using carbon nanotube based on adsorption (Margui, Hidalgo, Queralt, Van Meel, & Fontas, 2012); the other is the multi trace determination of heavy metals Cd, Cr, Cu, Fe, Mn, Ni, Pb, and Zn using a azo dye (Taguchi et al., 2015).

Figure 4. Conceptual illustrations of anodic stripping voltammetry analysis and energy-dispersive X-ray fluorescence analysis combined with pre-concentration of target ions

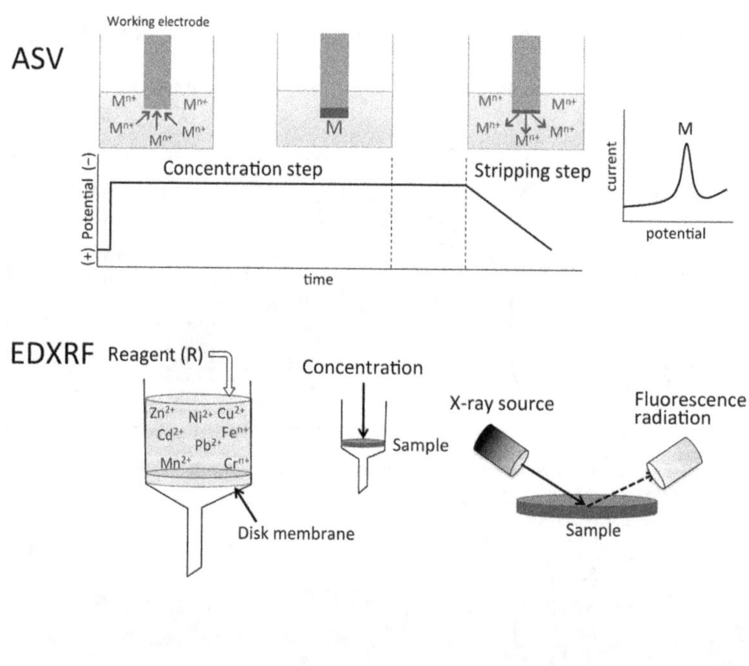

Other eco-friendly methods are based on spectrophotometry or colorimetry using simple and portable apparatus such as handy reflectometer, small sized spectrophotometer equipped with a LED light and CCD array detector. There are some considerably significant pre-concentration and separation processes before the spectrophotometric or colorimetric analysis due to highly sensitive and selective detection of a target ion in a real sample on the sampling site. One of the most typical and effective examples of these processes is solid phase extraction with an assisting agent. A highly effective enrichment and determination of trace components in water were achieved by filtration with a membrane filter and hydrophobic chelating reagent for metal ions combination based on solid phase extraction into membrane filter (Taguchi, Nakayama, Hata, Kasahara, & Goto, 1994; Taguchi, Hui-Feng, Hata, & Kasahara, 2000). A solid phase spectrophotometry have been proposed for phosphate, Cr^{6+}, and Fe^{2+} at µg/l or sub µg/l using ion exchange resin adsorbing its ionic complex with a chromogenic reagent (Saputro, Kai, Satoda, Matsuoka, & Yoshimura, 2013). Recently paper-based devices prepared with photoresist and inkjet printing technology are being presented in literature. Some of them have several effective zones such as sample zone, pretreatment zone, separate zone and detection zone, which enable separation analysis removing contaminants in a sample (Nery & Kubota, 2013). Mercury detection at ppb level using silver nanoparticles was reported (Apilux, Siangroph, Praphairaksit, & Chailapakul, 2012), however, paper-based sensing methods for metal ions in an aqueous sample are few and limited, further functional improvement especially in detecting procedure is necessary to apply for real samples. Recently, by using one of the most accessible devices, spatial mapping of Hg contamination in water has been reported (Wei et al., 2014). This is based on a smart phone integrated with optomechanical attachment as a camera module to quantify Hg concentration using a plasmonic gold nanoparticle (Au NP) and aptamer based colorimetric transmission assay. The attachment contains two different light emitting diodes (LEDs) at 523 nm and 635 nm for two-color radiometric analysis of dispersed and aggregated Au NPs in response to Hg^{2+} ions. The actual measurements of tap water and natural water including revers, lakes and beaches on the California coast was successful with the detection limit of 3.5 ppb and the reaction time within 20 min per one sample.

Among them, a dye-nanoparticle coated test strip (DNTS) is the most promising method of all on-site analysis in the near future with high sensitivity to trace ions at ppb levels (Takahashi, Kasai, Nakanishi, & Suzuki, 2006). Characteristically, the DNTS has dual detection mode including not only normal way like ordinary test papers 'immersion test' but also 'filtration enrichment' in which target metal ions are concentrated by passing sample solutions through it (Figure 5a) (Takahashi, 2014). As shown in Figure 5b, the DNTS is loaded with a thin layer (400 nm ~ 3 µm) of indicator dye nanoparticles on the top surface of a membrane filter and provides a remarkably concentrated color signaling surface (Takahashi, Danwittayakul, & Suzuki, 2009; Takahashi, 2015a). Therefore, almost all signals are available efficiently, but in contrast to conventional test papers such as pH test papers indicator dyes are distributed over the entire support. Two simple fabrication methods of DNTSs were designed; initially the aqueous dispersion of nanoparticle or nanocomposite is prepared and then the dispersion is simply filtered by a fine woven membrane filter having microscopic pores (Takahashi, 2015b). Versatility of the preparative methods means that DNTS is a novel platform technology for producing any highly sensitive test strips for various analysis targets. Figure 5c explains azo dye PAN nanofiber coated DNTS can be applied for the detection of Zn^{2+} ion in an aqueous sample with the dual detection mode. Notably 65 ppb of Zn^{2+} was detected by ordinary immersion test and sub-ppb concentrations of Zn^{2+} were also successfully detected by filtration enrichment of 100 mL of the sample solution. Detections of ppb leveled Zn^{2+} (Takahashi, Kasai, Nakanishi, & Suzuki, 2006), Hg^{2+} (Takahashi, Danwittayakul & Suzuki, 2009), Cd^{2+} (Takahashi,

Figure 5. Explanatory drawings of DNTS. a) dual detection modes for sensing ions with DNTS. (Reproduced from (Takahashi, 2014) by permission of Elsevier) b) Comparison of DNTS and conventional test strip from the viewpoint of the structure of the signaling layers with a cross-sectional SEM image of DNTS (Reproduced from (Takahashi, 2009) by permission of The Royal Society of Chemistry) c) Detection of Zn^{II} by PAN nanofiber DNTS via immersion test and filtration enrichment. Immersion test: Piece of test strip was dip into 10 mL of aqueous solution containing Zn^{II} at pH 8.4 for 15 min. Filtration enrichment: A 100 mL of aqueous solution containing Zn^{II} at pH 8.4 was filtrated through the PAN membrane under ca. 6.9 mL min^{-1} flow rate

Source: Reproduced from Takahashi, 2006, by permission of WILEY-VCH Verlag GmbH & Co. KGaA

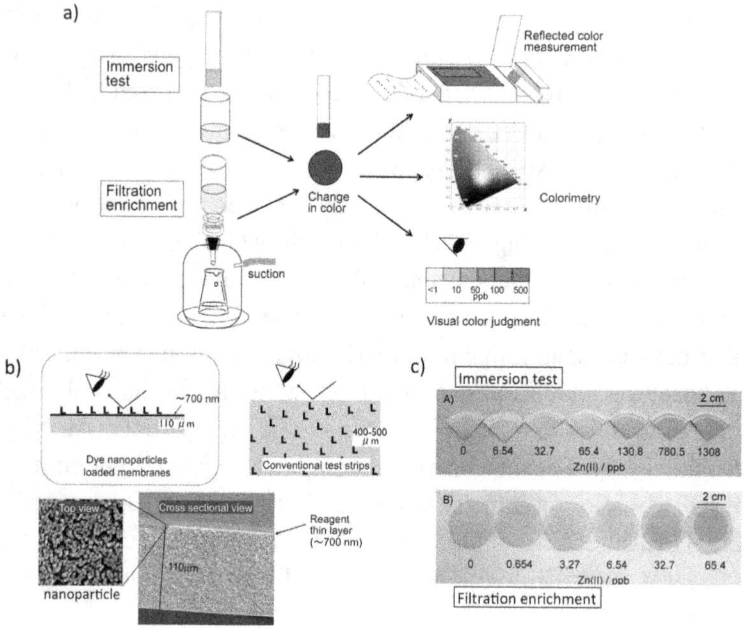

Souma, & Wakui, 2012), Pb^{2+} (Latt & Takahashi, 2011) were reported to be successful with respective different dyes loaded DNTSs. Especially, the DNTS for Hg^{2+} achieved highly selective detection using a masking reagent without pretreatment of real samples including industrial waste water and sea water with the detection limit of 0.057 ppb for 17 minutes (Takahashi, 2012).

CONCLUSION

Mankind can use only 0.3 percent of the total amount of fresh water on the earth that is concentrated in lakes, reservoirs, and river. Now public water has been artificially polluted by various toxic heavy metals, which are dissolved in water as water soluble ionic forms so that easily spread in water environment, resulting in serious environmental pollutions all around the world. WHO as well as governments of some countries pose individual water quality standards for these metals in the range of from ppb to ppm level. Accordingly, official methods of analysis are disclosed by standardization organizations such as ISO, CEN, US EPA, and JIS etc. Most inorganic contaminants have been measured at ppb or ppm level by advanced instruments including ICP, AAS, and ICP-MS. They guarantee detection sensitivity, however, require a large amount of carrier gas, cooling water, electric power having a high environmental load. In addition, specific technical skills are required for complicated pretreatment process before instrumental analysis, because water samples contains a large amount of various inorganic and organic contaminants. In order to realize more frequent water management, eco-friendly 'on-site' analytical methods have been strongly desired all over the world without the bothersome pretreatment process of a sample by using commercially available portable instruments. XRF with a handheld XRF detector and SV with an inexpensive electrochemical apparatus are introduced as associate official methods for comparatively clean water. Spectrophotometry based on-site analyses with a handy reflectometer and LED detector are necessary to be combined with significant pre-concentration and separation processes before the analyses due to achieve highly sensitive and selective detection. Solid phase spectrophotometry, paper-based sensor and smart phone based are interesting enough in terms of simplicity and cost suitable for on-site analysis, however, applicable samples and detectable target ions are few and relatively limited. DNTS has the highest sensitivity test method in test paper history and enables the ultra-trace detection of various harmful ions. The thin dye layer composed of nano-sized dye provides a remarkably concentrated signaling surface, which enables the highly sensitive detection. Additionally, the DNTS has important practical advantages such as material saving, energy saving, timesaving features both in fabrication and detection processes. Future success of on-site analytical methods is greatly anticipated by worldwide users.

REFERENCES

Apilux, A., Siangroph, W., Praphairaksit, N., & Chailapakul, O. (2012). Simple and rapid colorimetric detection of Hg(II) by a paper-based device using silver nanoplates. *Talanta*, *97*, 388–394. doi:10.1016/j.talanta.2012.04.050 PMID:22841097

European Commission. (2008). *European Union Risk Assessment Report, Zinc metal*. DOI 10.2788/40041

Ganim, S. & Tran, L. (2016). *How tap water became toxic in Flint, Michigan*. CNN.

Harada, M. (1995). Minamata disease: Methylmercury poisoning in Japan caused by environmental pollution. *Critical Reviews in Toxicology*, *25*(1), 1–24. doi:10.3109/10408449509089885 PMID:7734058

Ikingura, J. R., Akagi, H., Mujumba, J., & Messo, C. (2006). Environmental assessment of mercury dispersion, transformation and bioavailability in the Lake Victoria Goldfields, Tanzania. *Journal of Environmental Management*, *81*(2), 167–173. doi:10.1016/j.jenvman.2005.09.026 PMID:16782263

Järup, L., & Åkesson, A. (2009). Current status of cadmium as an environmental health problem. *Toxicology and Applied Pharmacology*, *238*(3), 201–208. doi:10.1016/j.taap.2009.04.020 PMID:19409405

Latt, K. K., & Takahashi, Y. (2011). Fabrication and characterization of a α,β,γ,δ-Tetrakis(1-methylpyridinium-4-yl)porphine/silica nanocomposite thin-layer membrane for detection of ppb-level heavy metal ion. *Analytica Chimica Acta*, *689*(1), 103–109. doi:10.1016/j.aca.2011.01.032 PMID:21338764

Lewtas, P., Fujikawa, Y., & Wajrak, M. (2015). Stripping Voltammetry for On-Site Analysis of Heavy Metals: Working Electrode and Automation Options. *Kankyo Gijyutsu*, *144*, 537–544.

Margui, E., Hidalgo, M., Queralt, I., Van Meel, K., & Fontas, C. (2012). Analytical capabilities of laboratory, bench-top and hand-held X-ray fluorescence systems for detection of metals in aqueous samples pre-concentrated with solid phase extraction disks, *Spectrochimica Acta. B*, *67*, 17–23.

Margui, E., Zawisza, B., & Sitko, R. (2014). Trace and ultratrace analysis of liquid samples by X-ray fluorescence spectrometry. *Trends in Analytical Chemistry*, *53*, 73–83. doi:10.1016/j.trac.2013.09.009

Ministry of the Environment, Government of Japan. (2003). *Environmental Quality Standards for Water Pollution*. Retrieved from https://www.env.go.jp/en/water/wq/wp.pdf

Ministry of the Environment, Government of Japan. (2015). *National Effluent Standards*. Retrieved from http://www.env.go.jp/en/water/wq/nes.html

National Environmental Methods Index. (n.d.). Retrieved from https://www.nemi.gov/home/

Nery, W. E., & Kubota, L. T. (2013). Sensing approaches on paper-based devices: A review. *Analytical and Bioanalytical Chemistry*, *405*(24), 7573–7595. doi:10.1007/s00216-013-6911-4 PMID:23604524

Pollution Control Department, Thailand. (n.d.a). *Industrial Effluent Standards*. Retrieved from http://www.pcd.go.th/info_serv/en_reg_std_water04.html

Pollution Control Department, Thailand. (n.d.b). *Surface water quality standards*. Retrieved from http://www.pcd.go.th/info_serv/en_reg_std_water05.html

Saputro, S. S., Kai, S., Satoda, M., Matsuoka, S., & Yoshimura, K. (2013). Solid-phase Spectrophotometry for on-site analysis of trace elements in natural water. *Analytical Sciences*, *29*, 677–680. doi:10.2116/analsci.29.677 PMID:23749138

Shiklomanov, I. A. (2000). Appraisal and Assessment of World Water Resources. *Water International*, *25*(1), 11–32. doi:10.1080/02508060008686794

Taguchi, S., Asaoka, M., Hirokami, E., Hata, N., Kuramitz, H., Kawakami, T., & Miyake, R. (2015). A Simple and Rapid Method for Simultaneous Preconcentration of Eight Trace-heavy-metals in Water Using 1-(2-Pyridylazo)-2-naphthol and Yttrium for X-ray Fluorescence Spectrometry. *Analytical Methods*, *7*(16), 6545–6551. doi:10.1039/C5AY00657K

Taguchi, S., Hui-Feng, S., Hata, N., & Kasahara, I. (2000). Design of simple and sensitive methods for trace analysis by mean of concentration with a membrane filter. *Bunseki Kagaku*, *49*, 941–952. doi:10.2116/bunsekikagaku.49.941

Taguchi, S., Nakayama, K., Hata, N., Kasahara, I., & Goto, K. (1994). Membrane filters for concentration of trace elements in water: Distribution of 2-(2-pyridylazo)-5-diethylaminophenol between the membrane filter and the aqueous phases. *Analyst (London)*, *119*(1), 135–139. doi:10.1039/an9941900135

Takahashi, Y. (2012). Elimination of interferences in the Mercury(II) determination with a dithizone nanofiber-coated test strip based on filtration-enrichment. *Bunseki Kagaku*, *61*(2), 123–126. doi:10.2116/bunsekikagaku.61.123

Takahashi, Y. (2014). Dye nanoparticle-coated test strips for detection of ppb-level ions in water. In A. Street, R. Sustich, J. Duncan, & N. Savage (Eds.), *Nanotechnology Applications for Clean Water* (2nd ed.; pp. 63–72). Oxford, UK: Elsevier. doi:10.1016/B978-1-4557-3116-9.00004-4

Takahashi, Y. (2015a). Dye Nanoparticle or Nanocomposite-Coated Test Papers for Detection at Ppb Levels of Harmful Ions. In M. Aliofkhazraei (Ed.), *Handbook of Nanoparticles* (pp. 1–13). New York: Springer. doi:10.1007/978-3-319-13188-7_38-1

Takahashi, Y. (2015b). Preparation and characterization of dye nanoparticle-coated test strips composed of anionic dyes and basic nano-adsorbents. *Bunseki Kagaku*, *63*, 525–531. doi:10.2116/bunsekikagaku.63.525

Takahashi, Y., Danwittayakul, S., & Suzuki, T. M. (2009). Dithizone nanofiber-coated membrane for filtration-enrichment and colorimetric detection of trace Hg(II) ion. *Analyst (London)*, *134*(7), 1380–1385. doi:10.1039/b816461d PMID:19562205

Takahashi, Y., Kasai, H., Nakanishi, H., & Suzuki, T. M. (2006). Test strips for heavy-metal ions fabricated from nanosized dye compounds. *Angewandte Chemie International Edition*, *45*(6), 913–916. doi:10.1002/anie.200503015 PMID:16382478

Takahashi, Y., Souma, S., & Wakui, Y. (2012). Preparation of 2-(5-bromo-2-pyridylazo)-5-diethylaminophenol nanoparticle-coated test strip and its detection of trace Cd^{II} Ion by immersion test. *Bunseki Kagaku*, *61*, 229–234. doi:10.2116/bunsekikagaku.61.229

The World Bank. (2005). *Towards a More Effective Operational Response: Arsenic Contamination of Groundwater in South and East Asian Countries*. No 71518. United States Environmental Protection Agency. Effluent Guidelines. Retrieved from http://www.epa.gov/eg

United States Environmental Protection Agency. (n.d.). *Table of Regulated Drinking Water Contaminants*. Retrieved from https://www.epa.gov/your-drinking-water/table-regulated-drinking-water-contaminants

Wei, Q., Nagi, R., Sadeghi, K., Feng, S., Yan, E., Ki, S. J., & Ozcan, A. et al. (2014). Detection and spatial mapping of mercury contamination in water samples using a smart-phone. *ACS Nano*, *8*(2), 1121–1129. doi:10.1021/nn406571t PMID:24437470

World Health Organization. (2011). Guidelines for drinking water quality (4th ed.). Author.

KEY TERMS AND DEFINITIONS

Drinking Water Standards: Legally mandated standards of hazardous contaminants in public water for human health in some countries. WHO has provided an authoritative basis for the setting of national regulations and standards for water safety in support of public health.

Dye Nanoparticle-Coated Test Strip: A highly sensitive test strip to having a thin layer composed of indicator dye nanoparticles enables the detection of aqueous harmful ions regulated in the world at ppb levels.

Effluent Standards: Water quality standards for monitoring the water mixed with waste matter in a plant or an office flowing outward and to the public water area.

Harmful Ions: Aqueous ions that are harmful to the environment and the human body, for example, cadmium ion, arsenic ion, mercury ion, lead ion, chromium (VI) ion, etc.

Official Analytical Methods: Analytical procedures that are prescribed by public organization for standardization, ISO, CEN, US EPA, JIS, etc.

On-Site Analysis: Analyses that are performed outdoors, especially at each sampling point such as discharge port, river, pound, well.

Portable Analytical Instruments: Analytical instruments that are greatly reduced in size and weight allow users carrying to the outsides near sampling points by a car or their hands.

Chapter 15
Ionic Liquids and Poly (Ionic Liquid)s Used as Green Solvent and Ultrasound Responded Materials

Kai Li
Nagaoka University of Technology, Japan

Takaomi Kobayashi
Nagaoka University of Technology, Japan

ABSTRACT

Ionic liquids (ILs) have been known as a class of green solvent and played an important role in green chemistry in recently years. Materials based on ILs, such as poly(ionic liquid) and polymer/ILs composites, are also widely investigated and possess many unique properties and applications. In this chapter, green solvent IL is reviewed and the responsive behavior of ILs based materials to external stimulus, like ultrasound (US), is introduced. Especially, it is described that FT-IR spectroscopy was effective for analyzing hydrogen bonding in ILs, polymeric ILs and PVA/IL composites. In addition, shear viscosity and viscoelasticity measurement and techniques of two-dimensional correlation and deconvolution are applied to analyze the change of interaction between IL to different substances and US responded behavior.

INTRODUCTION

Nowadays, ionic liquids (ILs) is widely investigated in different fields since their unique properties. The applications of ILs in green chemistry becomes a promising topic, since ILs have no vapor press. And ILs have been treated as a replacement for the traditional solvents. The interaction between ILs and different compounds strongly influence the properties of ILs. Therefore, the investigation of interaction between IL and different compounds and the change of the interaction under external stimuli are very interesting. In addition, the existence of hydrogen bonding network in ILs systems have been fully understood. While, the stimuli responsive behavior based on the change of hydrogen bonding network of

DOI: 10.4018/978-1-5225-1971-3.ch015

ILs is rare. In additional, recently studies have shown that ultrasound (UA), which is treated as a green technology, could change the hydrogen bonding network (Venegas-Sanchez, Tagaya, & Kobayashi, 2013b). However, researches on the interaction change in ILs systems under ultrasound have not been fully understood. In this chapter, green solvent IL is reviewed and the responsive behavior of ILs based materials to US is introduced. The change of hydrogen bonds in ILs systems was studies with FT-IR spectroscopy, shear viscosity and viscoelasticity measurement and techniques of two-dimensional correlation and deconvolution.

BACKGROUD

The history of ILs can be dated back to 1914 (Walden, 1914) for the first IL, $[EtNH_3]$-$[NO_3]$ (Mp: 12 °C). The IL was reported as a kind of organic salt with melting point below 100 °C and was consisted totally of cations and anions. Since some ILs are liquid in room temperature, they are also called room temperature ionic liquids (RTILs). As a well-known green solvent, ionic liquids (ILs) are actively studied in recent decades. ILs have hardly detectable vapor press as compared with the common used solvent, such as THF, DMF and chloroform and have been treated as a green solvent/medium. ILs also have other unique properties of high conductivity, negligible flammability, high thermal and electrical stability, wide electrochemical window, and the ability to dissolve many organic and inorganic compounds (Hallett & Welton, 2011; Welton, 1999). Therefore, IL was widely used in organic synthesis (Li, Yuan, Zheng, & Fang, 2012), material science (Rogers, 2007), separation process (Sun, Luo, & Dai, 2011), and electrochemical conductive layer (Lu et al., 2002). Figure 1 shows the commonly used cations and anions for the composition of ILs. One of the feature property of ILs is that different ILs could be prepared by together of different cations and anions. Therefore, the IL properties enable to changing hydrophilicity

Figure 1. The structures of common used cation and anion

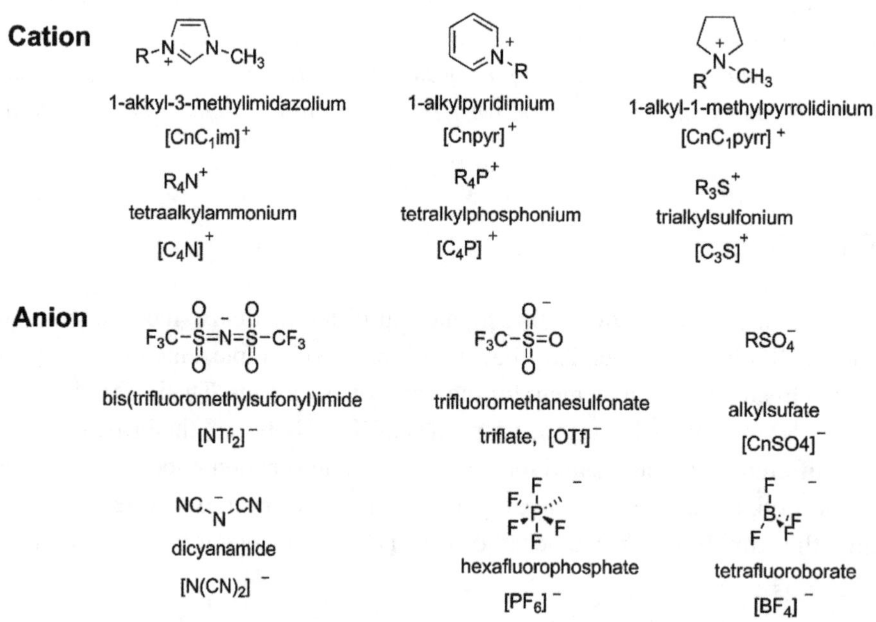

and conductivity by the combination of their ions. ILs are called as "*designer solvent*". This characteristic also enables researchers to design the chemical structure based on their demand. Among the researches of ILs, the properties of ILs and the interaction of ILs with different compounds, especially for the mediums such as water, alcohol and acetonitrile (Bester-Rogac, Stoppa, Hunger, Hefter, & Buchner, 2011; Zheng, Wang, Luo, Zhou, & Yu, 2013) .

Furthermore, there are polymer ionic liquids (PIL) or polymerized ionic liquids, which are a kind of polymers and polymerized monomers having IL properties. The definition of PILs was described as followed (Shaplov, Lozinskaya, & Vygodskii, 2010). PILs contain at least one ionic center and similarity in the chemical structure covalently bonded part of their constitutional repeating unit in the monomer. Relative to ILs, PILs have unique properties of high conductivity (Zhang et al., 2011), catalyst reactivity (Liu et al., 2014), high stability and the process ability (Yuan, Mecerreyes, & Antonietti, 2013). Sometimes PILs are treated as an analogue of ILs. In contrast to the liquid state of common ILs, PILs are commonly solid at room temperature due to their high molecular weight. PILs has several unique properties in high bulk conductivity (Zhang et al., 2011), catalyst activity (Liu et al., 2014), and high CO_2 absorption ability (Tang et al., 2005) and so on (Yuan et al., 2013). Thus, PILs have been used in electrochemical actuator (Zhao et al., 2015), organic synthesis (Yu et al., 2011), material science (Zhou, Wu, Hu, Yang, & Han, 2010) and separation process (Bi, Tang, & Row, 2013).

Polymer/IL Composites

Despite pure ILs and PILs, polymer/IL composite is another class of materials based on ILs and have been widely investigated in recent years. Such polymer/IL composites could be fabricated by dissolving polymer into ILs and combines the unique properties of the ILs and the polymers. The representative polymer/IL composites is iongel. Iongel is developed from the concept of "polymer in salt" by Angell (Angell, Liu, & Sanchez, 1993). This always refers a kind of hybrid materials formed with swelling polymers into ILs. Here, polymers, such as poly(methyl methacrylate) (Susan, Kaneko, Noda, & Watanabe, 2005), poly(*N*-isopropylacrylamide) (Ueki & Watanabe, 2006), poly(*N*-isopropyl acrylamide-*b*-ethylene oxide-*b*-*N*-isopropyl acrylamide) (He & Lodge, 2007) and poly(vinyl alcohol) (Liew, Ramesh, & Arof, 2014) were known readily to dissolve in ILs. The mobilized ions of ILs endeavor iongel conductivity. In general, there are two methods for the preparation of iongels:

1. In situ polymerization of monomer in IL (Susan et al., 2005) and
2. Direct mixing polymer into IL with or without assistant solvent (He, Boswell, Bühlmann, & Lodge, 2007).

Watanabe and co-workers (Susan et al., 2005) firstly reported the formation of iongel by using the in situ polymerization of vinyl monomers in 1-ethyl-3-methyl imidazolium bis(trifluoromethane sulfonyl) imide and applied these iongels as polymer eletrolytes. Timothy group (He et al., 2007; Lee, Panzer, He, Lodge, & Frisbie, 2007) developed different triblock copolymers based iongels through mixing the copolymers and ILs in methylene chloride (He et al., 2007). Till now, iongel has been applied in transistors (Lee et al., 2007), batteries (Osada, de Vries, Scrosati, & Passerini, 2015), solar cells (Liew et al., 2014), electromechanical actuators (Ding et al., 2003) and others (Lodge, 2008). The high ionic conductivity of iongels becomes ideal candidates for electrochemical application.

Chemical Properties of ILs based Systems

As described, IL consists of total cations and anions. Most of anions contain strong electro-withdraw atoms of such as Cl, N, O and F (Figure 1). These groups can form hydrogen bonds with the hydrogen atoms as presented in Figure 2b. Different studies have shown the evidences of the existence of hydrogen bonds networks in the most ILs (Dymek, Grossie, Fratini, & Wade Adams, 1989; Matsumoto & Hagiwara, 2007; Reichert et al., 2007). For the imidazolium based ILs (Figure 2a), the electro-withdrawing nature of N atom in imidazolium ring can result in the larger positive charges on the C_2-H, C_4-H, and C_5-H groups and becomes the main interactive sites (Anderson, Ding, Welton, & Armstrong, 2002; Crowhurst, Mawdsley, Perez-Arlandis, Salter, & Welton, 2003; Hunt, Kirchner, & Welton, 2006). Therefore, the hydrogen bond interaction between imidazolium cation and anion is caused with interactions of X···C_2-H, X···C_4-H, and X···C_5-H. For anions with a single atom anion, such as Cl^-, C_2-H···Cl hydrogen bond forms between cation and anion (Figure 2b). For other anion groups, like the tetrafluoroborate (BF_4^-) and hexafluorophosphate (PF_6^-), the C_2H···F and $C_{4(5)}$-H···F are hydrogen bonds between the cation and the anion (Figure 2b) (Anderson et al., 2002; Dong, Zhang, Wang, & Yao, 2006; Heimer, Del Sesto, Meng, Wilkes, & Carper, 2006).

Despite the pure ILs, hydrogen bonding also very widely investigated between ILs and different substances such as water (Wang, Li, & Han, 2006), acetonitrile (Zheng et al., 2013), methanol (López-Pastor, Ayora-Cañada, Valcárcel, & Lendl, 2006) and polymer (Jiang et al., 2010). Cammarata et al. (Cammarata, Kazarian, Salter, & Welton, 2001) investigated water molecules interacted with ILs and found that anion affected the formation of hydrogen bond with water as a form of anion···H_2O···anion. Lo´pez-Pastor et al (López-Pastor et al., 2006) discovered that water could interact with BF_4^- contained ionic liquids in the forms of 1: 2 (BF_4^-···H_2O···BF_4^-), and a cyclic water dimer interacting with BF_4^-(BF_4^-···H_2O···H_2O ···BF_4^-).

Figure 2. (a) Chemical structure of imidazolium based ILs, (b) hydrogen bonding in imidazolium based ILs

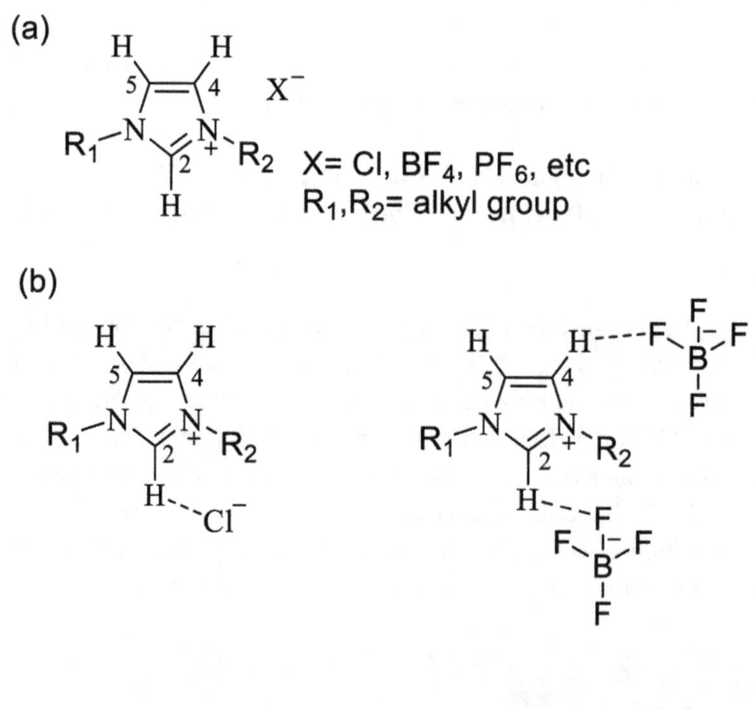

Green ILs Having Different Counter Ions

It's well known that the properties of ILs are strongly influenced by the counter anions having hydrophilicity and conductivity. As reported as the physical and chemical properties of ILs, for instance, viscosity, electrical conductivity, and reactivity (Jacquemin, Husson, Padua, & Majer, 2006; Koddermann, Wertz, Heintz, & Ludwig, 2006; J. A. Widegren, Laesecke, & Magee, 2005; Jason A. Widegren, Saurer, Marsh, & Magee, 2005) could be influenced with the water presence. Therefore, the interaction between water and ILs is an important topic for research. Moreover, the change of hydrogen bonding networks in imidazolium ILs with different counter anions was investigated by measuring the viscosity and FT-IR spectra (Li & Kobayashi, 2016a). In the change of FT-IR spectra as shown in Figure 3, it could be seen that in the BMI-PF_6 (Figure 3a) almost no change of their peaks was observed at different US exposure system. While BMI-Cl (Figure 3b) and BMI-BF_4 (Figure 3c) showed the apparent change in OH stretching (3000-3700 cm^{-1}). Additionally, in the BMI-BF_4, it was much apparent observed as the B-F stretching band in 950-1152 cm^{-1} (Dupont, Consorti, Suarez, & Souza, 2002; Gruzdev, Ramenskaya, Chervonova, & Kumeev, 2009). The different changes of these three ILs were cause by dependence of hydrophobicity, showing in the order of $PF_6^- > BF_4^- > Cl^-$ (D. Holbrey & R. Seddon, 1999; Kakiuchi & Nishi, 2006).

US Responsive ILs Systems

Smart materials always draw a lot of intention of researchers, since they can response to certain stimuli, such as:

- Light (Guang Wang, Tong, & Zhao, 2004),
- pH (Horton, Bao, Bai, Lodge, & Zhao, 2011),
- Temperature (Kotsuchibashi, Yamamoto, & Aoyagi, 2009),
- Magnetic (Meng & Jinlian Hu, 2010) and
- US (Lentacker et al., 2006).

For ILs and their composites, such stimuli response has been studied by Ohno et al (Kohno & Ohno, 2012). The hydrophilicity of ILs can be tuned by changing the counter anions. They conducted a lot investigations on the temperature-responsive behavior of ILs-water interface and found that the some ILs had a lower critical temperature in the phase transition with water. The separated bi-phases became miscible upon cooling (Kohno & Ohno, 2012). Responsive behavior was reported by Hamaguchi et al. (Hayashi & Hamaguchi, 2004). Magnetic IL by mixture by mixing 1-butyl-3-methylimidazolium chloride and $FeCl_3$ showed IL response to the magnetic field (Zhu et al., 2013) by using magnetic IL (*N*-butylpyridinium tetrachloroferrate). ILs have potential uses for stimuli response materials, especially for the ILs/water system, because their tunable miscibility. In PILs with stimuli response, their responsive behavior was reported in 2007. Texter et al (Yan & Texter, 2007) firstly reported the reversible change of PIL gel between porous polymer and gel through varying the solvent. Vijayakrishna et al (Vijayakrishna et al., 2008) reported the ionic responsiveness of hydrophilic PIL copolymer in 2008. Later, some of PIL copolymers showed responsive stimulus such to temperature (Mori, Yahagi, & Endo, 2009). The responsive PIL homopolymer was reported till recently (Deguchi, Kohno, & Ohno, 2015) (Ge Wang, & Wu, 2015) for the thermoresponsive behavior of PIL solution. Furthermore, in a smart iongel system, therefore, the stimuli responsive behavior was widely investigated. Ueki et al (Ueki, Usui, Kitazawa,

Figure 3. FT-IR spectra before and after 23 kHz US exposure at different time for BMI-PF_6 (a), BMI-BF_4 (b) and BMI-Cl (c) with 2.6wt % water content
Source: Li & Kobayashi, 2016a

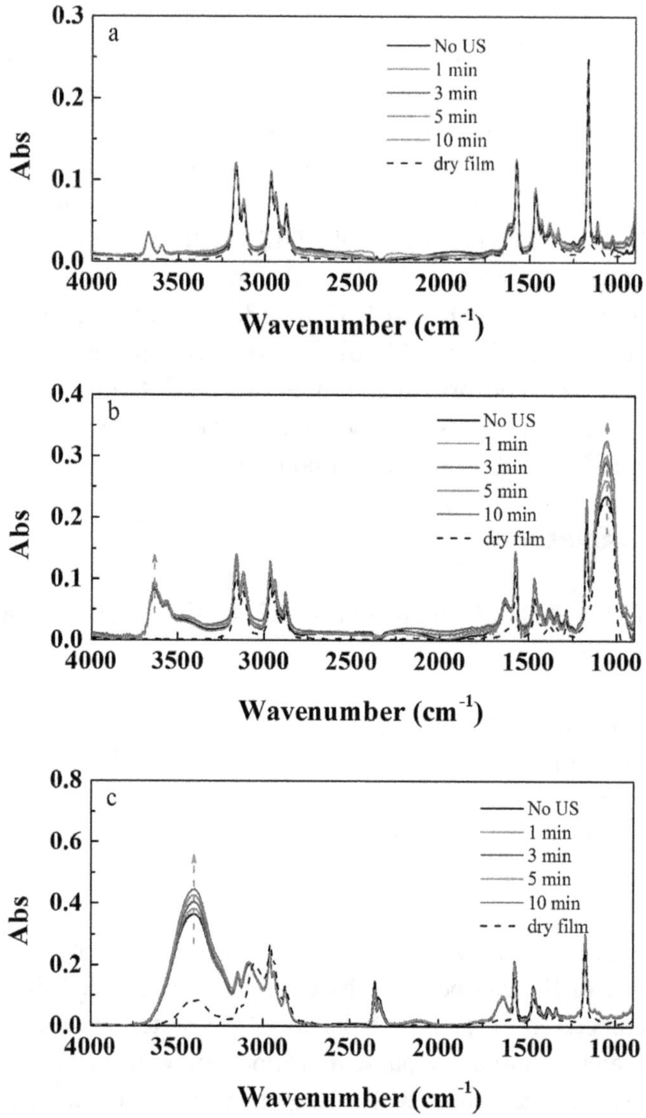

Lodge, & Watanabe, 2015) investigated the iongel based on 1-butyl-3-methylimidazolium hexafluorophosphate and poly(ethylene oxide)/ *N*-isopropylacrylamide random block copolymer for photo-healing ability and thermal reversible behavior. Lodge et al (He & Lodge, 2007) also studied the thermoreversible iongel by triblock copolymer self-assemble with an IL.

In recent years, ultrasound (US) effected aqueous polymer solution was investigated with FT-IR spectroscopy and shear viscosity. Ngoc et al investigated the influence of US on the viscosity of aqueous slurry composing polyacrylic acid (PAA) and alumina with the frequencies of 28, 45, and 100 kHz (Ngoc & Kobayashi, 2010), indicating that US caused the breakage of the hydrogen bonds between PAA and Al_2O_3. Later, different aqueous polymers of poly(*N*-isopropyl acrylamide co acrylic acid), poly(acrylic acid), poly(vinyl alcohol), poly(ethylene glycol) and polysaccharides (Huo & Kobayashi, 2006; Venegas-Sanchez, Kusunoki, Yamamoto, & Kobayashi, 2013; Venegas-Sanchez, Tagaya, & Kobayashi, 2013a; Josue Addiel Venegas-Sanchez et al., 2013b; Venegas-Sanchez, Tagaya, & Kobayashi, 2014) were fully used for the investigation of US effect. When the US shear stress presses the solution, the aqueous viscosity of these polymer solution largely decreased, meaning that US stress broke hydrogen bond network in the solution. In these studies, the US exposure could effectively change the water-polymer interaction.

Since ILs based systems also contain hydrogen bond network, which is similar with the polymer solution. US might be play as an external stimulus to ILs systems and could be used to control the properties and application of ILs.

The responsive behavior of IL systems for the aqueous ILs, PIL solution and PVA/IL composites, to US effect is summarized. It was noted that the FT-IR band of the sample was returned to the original spectra without US exposure. The relative intensity was defined as I_t/I_0, where I_t referred to the intensity of OH stretching region after US exposure for t time and the I_0 was the intensity of OH stretching region before US exposure. The relative intensities returned to the origin value gradually for both ionic liquids after the US exposure, strongly indicating the reforming of hydrogen bond after removing the US. Moreover, shear viscosity results of BMI-Cl solution also showed periodical change because of the breakage of hydrogen bond during the US exposure and the reforming of hydrogen bond, when the US exposure was stopped.

Figure 4 shows the hydrogen bonds for these three ionic liquids. For BMI-Cl (Figure 4c), the e and f were assigned to the structure of $H_2O\cdots Cl^-$. The d was the $Cl^-\cdots H_2O\cdots Cl^-$. The a and b were the hydrogen bonds in $H_2O\cdots C_{4(5)}$-H and $H_2O\cdots C_2$-H, respectively. For BMI-PF_6, the c was assigned to free water. The a and b were a hydrogen bonds formed by water$\cdots C_{4(5)}$-H and water$\cdots C_2$-H, respectively and d was for the hydrogen bonds of $PF_6^-\cdots$water$\cdots PF_6^-$. For BMI-BF_4, a cyclic water dimer interacted with BF_4^- contained two kinds of hydrogen bond shown with f and peak d. The marked a and b were hydrogen bonds in $H_2O\cdots C_{4(5)}$-H and $H_2O\cdots C_2$-H. The c was the O-H stretching of $BF_4^-\cdots$water$\cdots BF_4^-$. Through analyzing the change of the FT-IR deconvolution results before and after US exposure, it could be seen that US could break the hydrogen bond in the BMI-Cl for the $C_{4(5)}$-H...H_2O, Cl^-...H_2O...Cl^- and H_2O...Cl^- (Li & Kobayashi, 2016a). For the BMI-BF_4, the hydrogen bonds in cyclic water dimer, BF_4^-...water...BF_4^- and hydrogen bonds formed between imidazole ring and water were broken by US irradiation. Their counter ions with different hydrophilicity in ILs played an important role for water solvation under US exposure. However, US has no influence on the hydrophobic ILs (BMI-PF_6).

For PIL, US similarly changed the hydrogen bonding network. The PIL solution was probed by shear viscosity and FT-IR spectra (Li & Kobayashi, 2016b). The shear viscosity was tested using the experimental setup shown in Figure 5 (a). The results in Figure 6 showed that the aqueous PIL solution viscosity decreased from 1550 Cp to 1360 Cp under US exposure. It gradually returned to the original values after removing the US exposure. Moreover, the change pattern was observed in the following two cycles and shear viscosity in the aqueous PIL solution was varied periodically in the aqueous PIL solution. Besides, the viscosity change was significantly greater for 43 kHz US exposure. The absorption of US exposure was mostly occured for 43 kHz as compared with 23 and 96 kHz in the PIL solutions with different concentrations of 40 wt%, 50 wt%, and 66 wt%.

As seen Figure 6a for the FT-IR spectra, the US exposure gradually enhanced the intensities of the 3000-3700 cm^{-1} region and the intensities increased with increasing of exposure time of 1 to 10 min. It was interesting to note that the results of the OH band enhancement corresponded well to the shear viscosity change. Two-dimensional (2D) correlation analysis (Noda, 1993; Noda, Dowrey, Marcoli,

Figure 4. Illustration of interaction in these three ionic liquids solution
Source: Li & Kobayashi, 2016a

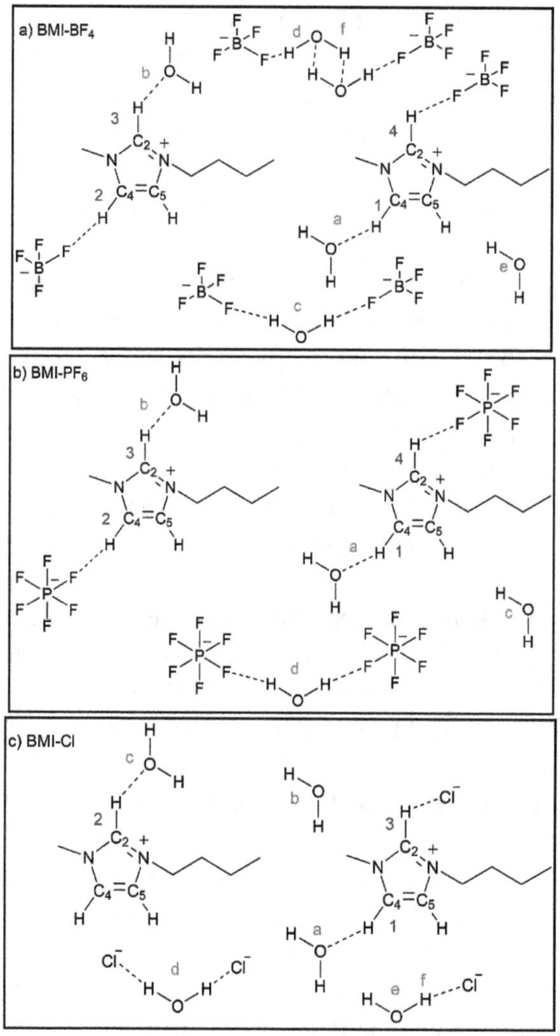

Figure 5. Shear viscosity measurement setup (a) and shear viscosity with and without US exposure at 50 W for PIL solution (b)
Source: Li & Kobayashi, 2016b

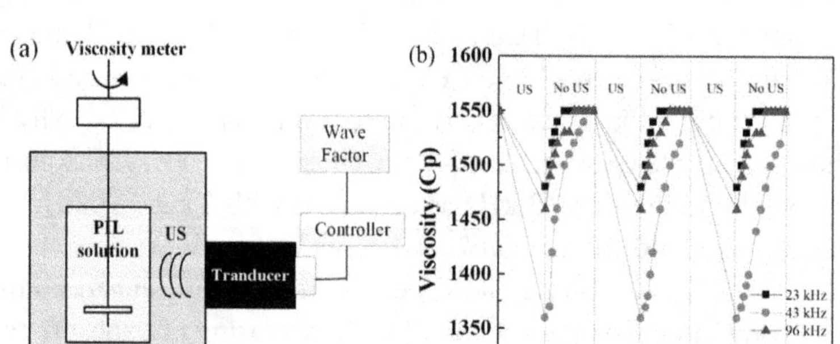

Story, & Ozaki, 2000) and deconvolution (Gemmei-Ide, Motonaga, & Kitano, 2006) were introduced in these studies to reveal the hydrogen bonds change. As shown in Figure 6b for the hydrogen bonds of PILs, one autopeak at 3418 cnm^{-1} and two cross-peaks at (3418, 3136) and (3418, 3093) were found. For asynchronous spectrum (Figure 6c), five cross-peaks are found as following:

- (3093, 3418),
- (3136, 3418),
- (3226, 3418),
- (3524, 3418),
- (3588, 3418) and
- (3648, 3418).

Based on 2D analysis, seven peaks might exist in the region of 3000-3700 cm^{-1}, as 3093, 3136, 3226, 3418, 3524, 3588, and 3648 cm^{-1}. Deconvolution of the 3000-3700 cm^{-1} region was conducted with a gaussian function to elucidate the change of hydrogen bonds in the system. There were seven peaks (Figure 6d) and their assignment were as follows:

- Peaks of c, $Cl^{-}\cdots H_2O\cdots Cl^{-}$;
- Peaks of d, H_2O;
- Peak of e, $H_2O\cdots Cl^{-}$;
- Peak of f, $H_2O\cdots C_2$-H and
- Peak of g, $H_2O\cdots C_4$-H;
- Peak of a and b, C-H stretching of imidazole ring.

The doconvolution results were compared in Figure 6d and e. It was shown that peaks of c and d enhanced after US exposure for 10 min, whereas peak g and f were decreased. This was caused by the breakage of the hydrogen bonds occurred in the PIL solution. Here, the peak of c was assigned to $Cl^{-}\cdots H_2O\cdots Cl^{-}$. The concentration of free water, which possesses the peak of d, increased after US exposure, strongly suggesting that free water was released from the polymer networks during US exposure. These changes in the intensities after the US exposure were followed by the breakage of hydrogen bonding in the cases of water and the PIL (Figure 7). The aqueous PIL solution responded to US exposure through breakage and reformation of hydrogen bonds.

Ionic Liquid-Poly (Vinyl Alcohol) Composites Response to US Stimuli

In order to fully understand the US responsive behavior of IL-polymer composites, ionic liquid-poly(vinyl alcohol) (PVA/BMI-Cl) gelatinous composites were studied for IL-poly(vinyl alcohol) composites (Li, Noguchi, & Kobayashi, In Press). The viscoelasticity was also studied in the presence and absence of US exposure. Here, the gelatinous composites of PVA/BMI-Cl using different PVAs were fabricated as listed in Table 1. The shear viscosity of PVA/BMI-Cl composites was determined in the absence and presence of US exposure (Figure 8). It was seen that the shear viscosity of four PVA/BMI-Cl composites decreased under the US exposure (43 kHz, 12.5 W) and returned to the original values after removing the US exposure within about 5 min. The similar change on shear viscosity was observed in the following two cycles. These meant that US changed the shear viscosity of the PVA/BMI-Cl periodically

Figure 6. (a) FT-IR spectra of 40 wt% PIL solution before and after 43 kHz US exposure, synchronous (b) and asynchronous (c) 2D correlation IR spectra of PIL solution in the 3000-3700 cm^{-1} region, deconvolution results for PIL solution before (d) and after (e) US exposure. Source: Li & Kobayashi, 2016b

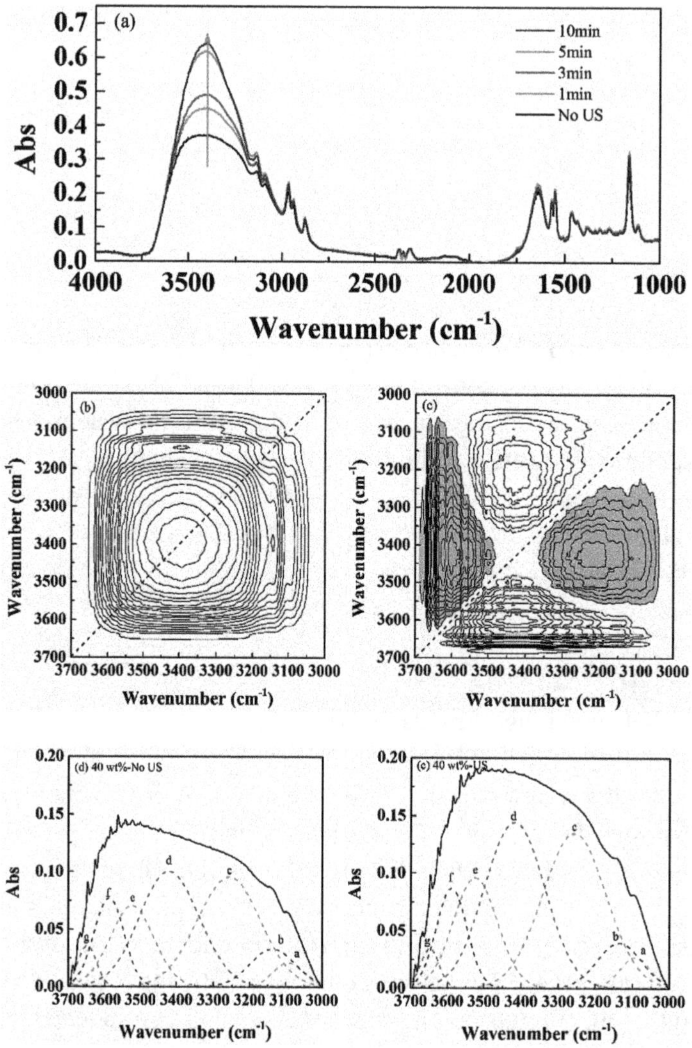

Figure 7. Illustration of hydrogen bond change during US exposure
Source: Li & Kobayashi, 2016b

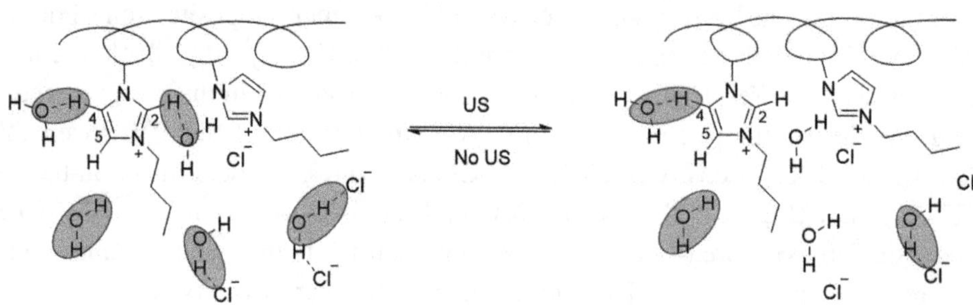

Table 1. PVA/BMI-Cl composites used in the present study

PVA	*Mn*	Alcoholysis Degree
PVA13	13000-23000	87%
PVA31	31000-50000	87%
PVA217	80000	80%
PVA117	146000-180000	98%

Figure 8. Experimental setup for (a) measuring shear viscosity and shear viscosity change with and without US for PVA/BMI-Cl composites
Source: Li et al., In Press

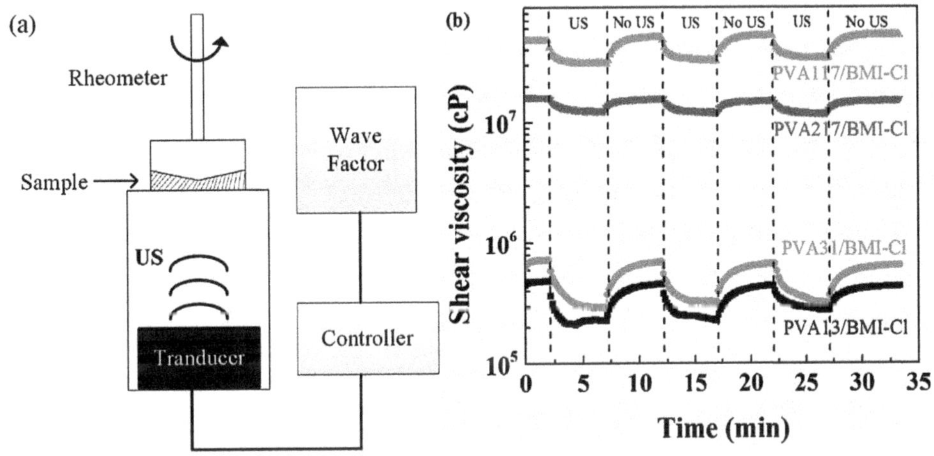

and the PVA/BMI-Cl composites responded to the US exposure. For the viscoelasticity (Figure 9), the storage moduli G' and loss moduli G'' decreased for four PVA/BMI-Cl composites under the US exposure and returned to the original ones after stopping the US exposure. The change of Tan δ in Figure 9b showed that the value of Tan δ increased under the US exposure and returned to the original values after stopping US exposure. This indicated that these four composites behaved much more like a liquid under the US exposure, which could be explained by the breakage of hydrogen bonds in the composites under the US exposure for the gelatinous composites. These results provided a possibility to control the properties of polymer/ILs composites by using US. The FT-IR spectra before and after the US exposure were also recorded and shown in Figure 10a. The intensities of the OH stretching band increased as the increasing of the exposure time from 1 to 10 min. Similar experiments were conducted for other PVA/BMI-Cl composites, namely PVA31/BMI-Cl, PVA217/BMI-Cl and PVA117/BMI-Cl. It was reasonable to consider that the US exposure broke the hydrogen bonds in the PVA/BMI-Cl composites. The results of 2D correlation had the synchronous maps and asynchronous maps (Figure 10b and c) in the range of 3200-3700 cm^{-1}. One autopeaks of (3450, 3450) and two cross-peaks of (3648, 3450) and (3580, 3450) were observed in the synchronous map. In the asynchronous map of PVA13/BMI-Cl (Figure 10c), four cross-peaks lied on:

- (3648, 3450),

Figure 9. (a) Viscoelasticity and (b) tan δ change in the absence and presence US exposure for PVA/ BMI-Cl composites
Source: Li et al., In Press

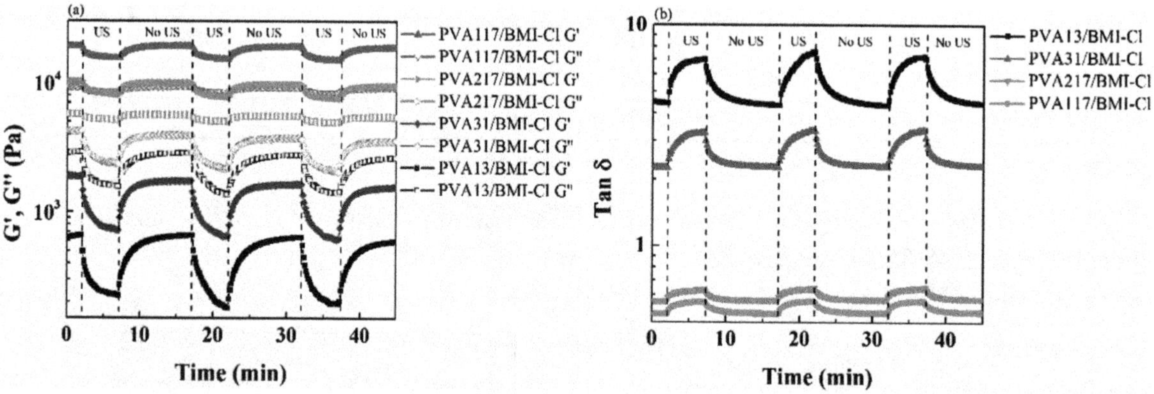

Figure 10. The FT-IR results for the PAV/BMI-Cl composites (a-c) after US exposure for different time
Source: Li et al., In Press

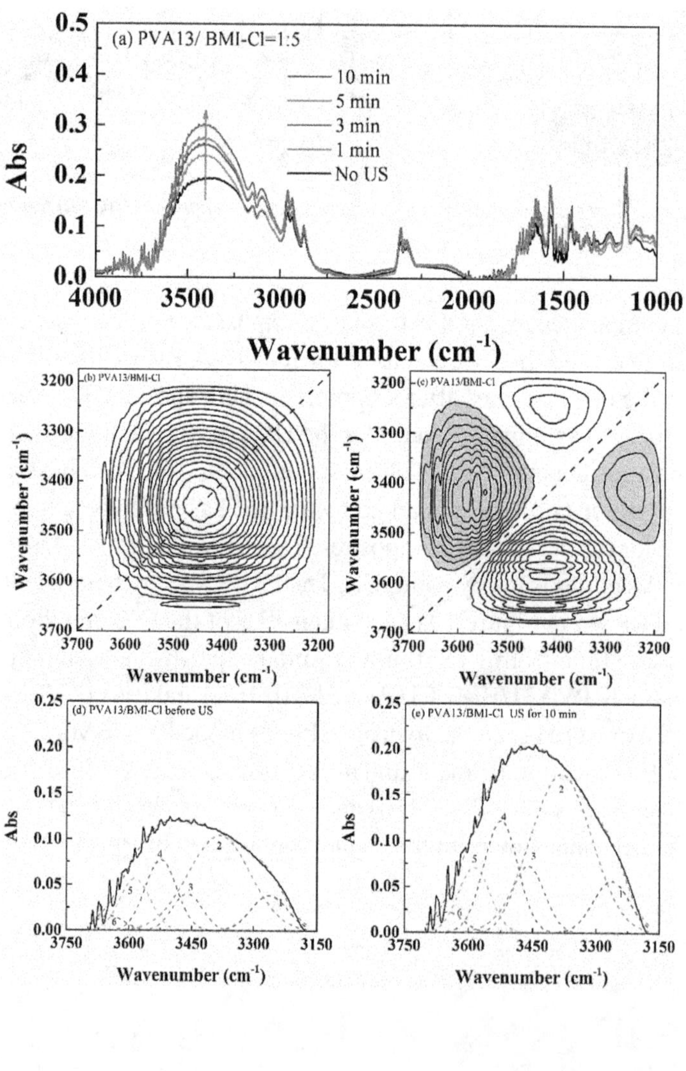

- (3580, 3450),
- (3529, 3410) and
- (3410, 3245), respectively.

Therefore, the OH band of PVA13/BMI-Cl was split into six bands in the OH stretching region, namely 3245, 3410, 3450, 3529, 3580 and 3648 cm^{-1}. In the deconvolution results for the FT-IR of the PVA/BMI-Cl composite before and after the US exposure (Figure 10d and e), the peak assignments are listed as followed. The Peak 1 belonged to the OH group hydrogen bonded with the neighbor OH group and the peak 2 for the OH group hydrogen bonded with intermolecular crosslink of PVA. Peak 3 referred to the free OH group and the peak 4 was the OH group hydrogen bonded with the Cl^- anion of BMI-Cl. Peak 5 and 6 belonged to the OH group hydrogen bonded with the C_2-H and $C_{4(5)}$-H in the imidazolium ring, respectively. As shown in Figure 10d and e, the FT-IR intensities of the peaks of 1, 2, 3, 4 and 5 were intensified after the US exposure. Especially, the observation of the peaks of 3, 4 and 5 implied that the US exposure broke hydrogen bonds between PVA13 and BMI-Cl. As assigned above, these peaks belonged to the OH group hydrogen bonded with the neighbor OH group, intermolecular OH group, Cl^- and C_2-H of imidazolium ring. After the US exposure, the free OH groups were released from the breakage of hydrogen bonds, reflecting from the enhancement of the peak 3, which belonged to the free OH groups of the PVA. For other PVA/BMI-Cl composites, similar enhancement of the fitting peak after US exposure was observed. Therefore, it could be seen that the hydrogen bonds formed the neighbor OH groups. The OH groups in different PVA chains, and the OH group with the Cl^- anion were easily influenced by the US exposure.

CONCLUSION

In the present chapter, use of green solvent for ILs and PILs were reviewed. In addition, US effect on three ionic liquid (IL) based systems, namely aqueous ionic liquids, aqucous poly(ionic liquid) and poly(vinyl alcohol)/IL composites, was introduced. Especially, hydrogen bonding in ILs, polymeric ILs and PVA/IL composites responded to US exposure and thus the exposure could strongly influence their hydrogen bonding networks. FT-IR spectroscopy was applied for analysis of their hydrogen bonds in the imidazole based ILs with different counter anions, indicating water solvation under US exposure played an importance role. In the US effects on the PIL solution, it was concluded that US caused the breakage and reformation of hydrogen bonds in the PIL and water. Moreover, the responsive behavior of the PVA/BMI-Cl gelatinous composites to US were due to influence of hydrogen bonding networks in the gelatinous materials. Therefore, ILs systems could have the potential to be applied as smart materials responded to US.

REFERENCES

Anderson, J. L., Ding, J., Welton, T., & Armstrong, D. W. (2002). Characterizing ionic liquids on the basis of multiple solvation interactions. *Journal of the American Chemical Society*, *124*(47), 14247–14254. doi:10.1021/ja028156h PMID:12440924

Angell, C. A., Liu, C., & Sanchez, E. (1993). Rubbery solid electrolytes with dominant cationic transport and high ambient conductivity. *Nature*, *362*(6416), 137–139. doi:10.1038/362137a0

Bester-Rogac, M., Stoppa, A., Hunger, J., Hefter, G., & Buchner, R. (2011). Association of ionic liquids in solution: A combined dielectric and conductivity study of [bmim][Cl] in water and in acetonitrile. *Physical Chemistry Chemical Physics*, *13*(39), 17588–17598. doi:10.1039/c1cp21371g PMID:21892477

Bi, W., Tang, B., & Row, K. (2013). Adsorption of alcohol from water by poly(ionic liquid)s. *Bioprocess and Biosystems Engineering*, *36*(6), 651–658. doi:10.1007/s00449-012-0828-8 PMID:23010726

Cammarata, L., Kazarian, S. G., Salter, P. A., & Welton, T. (2001). Molecular states of water in room temperature ionic liquids. *Physical Chemistry Chemical Physics*, *3*(23), 5192–5200. doi:10.1039/b106900d

Crowhurst, L., Mawdsley, P. R., Perez-Arlandis, J. M., Salter, P. A., & Welton, T. (2003). Solvent-solute interactions in ionic liquids. *Physical Chemistry Chemical Physics*, *5*(13), 2790–2794. doi:10.1039/B303095D

Deguchi, Y., Kohno, Y., & Ohno, H. (2015). A fine tuning of LCST-type phase transition of poly(ionic liquid)s in water. *Chemistry Letters*, *44*(3), 238–240. doi:10.1246/cl.141016

Ding, J., Zhou, D., Spinks, G., Wallace, G., Forsyth, S., Forsyth, M., & MacFarlane, D. (2003). Use of ionic liquids as electrolytes in electromechanical actuator systems based on inherently conducting polymers. *Chemistry of Materials*, *15*(12), 2392–2398. doi:10.1021/cm020918k

Dong, K., Zhang, S., Wang, D., & Yao, X. (2006). Hydrogen bonds in imidazolium ionic liquids. *The Journal of Physical Chemistry A*, *110*(31), 9775–9782. doi:10.1021/jp054054c PMID:16884211

Dupont, J., Consorti, C. S., Suarez, P. A. Z., & Souza, R. F. (2002). Preparation of 1-butyl-3-methylimidazolium based room temperature ionic liquids. *Organic Syntheses*, *79*, 236–239. doi:10.15227/orgsyn.079.0236

Dymek, C. J. Jr, Grossie, D. A., Fratini, A. V., & Wade Adams, W. (1989). Evidence for the presence of hydrogen-bonded ion-ion interactions in the molten salt precursor, 1-methyl-3-ethylimidazolium chloride. *Journal of Molecular Structure*, *213*, 25–34. doi:10.1016/0022-2860(89)85103-8

Gemmei-Ide, M., Motonaga, T., & Kitano, H. (2006). State of Irremovable Water in Solid Polymer Films Examined by Fourier Transform Infrared Spectroscopy I: Poly(Ethylene Glycol) Dimethyl Ether. *Langmuir*, *22*(6), 2422–2425. doi:10.1021/la053117f PMID:16519432

Gruzdev, M. S., Ramenskaya, L. M., Chervonova, U. V., & Kumeev, R. S. (2009). Preparation of 1-butyl-3-methylimidazolium salts and study of their phase behavior and intramolecular intractions. *Russian Journal of General Chemistry*, *79*(8), 1720–1727. doi:10.1134/S1070363209080246

Hallett, J. P., & Welton, T. (2011). Room-temperature ionic liquids: Solvents for synthesis and catalysis. 2. *Chemical Reviews*, *111*(5), 3508–3576. doi:10.1021/cr1003248 PMID:21469639

Hayashi, S., & Hamaguchi, H. (2004). Discovery of a magnetic ionic liquid [Bmim]$FeCl_4$. *Chemistry Letters*, *33*(12), 1590–1591. doi:10.1246/cl.2004.1590

He, Y., Boswell, P. G., Bühlmann, P., & Lodge, T. P. (2007). Ion gels by self-assembly of a triblock copolymer in an ionic liquid. *The Journal of Physical Chemistry B*, *111*(18), 4645–4652. doi:10.1021/jp064574n PMID:17474692

He, Y., & Lodge, T. P. (2007). A thermoreversible ion gel by triblock copolymer self-assembly in an ionic liquid. *Chemical Communications*, (26): 2732–2734. doi:10.1039/b704490a PMID:17594036

Heimer, N. E., Del Sesto, R. E., Meng, Z., Wilkes, J. S., & Carper, W. R. (2006). Vibrational spectra of imidazolium tetrafluoroborate ionic liquids. *Journal of Molecular Liquids*, *124*(1–3), 84–95. doi:10.1016/j.molliq.2005.08.004

Holbrey, D., & Seddon, K. R. (1999). The phase behaviour of 1-alkyl-3-methylimidazolium tetrafluoroborates; ionic liquids and ionic liquid crystals. *Journal of the Chemical Society, Dalton Transactions: Inorganic Chemistry*, (13): 2133–2140. doi:10.1039/a902818h

Horton, J. M., Bao, C., Bai, Z., Lodge, T. P., & Zhao, B. (2011). Temperature- and pH-triggered reversible transfer of doubly responsive hairy particles between water and a hydrophobic ionic liquid. *Langmuir*, *27*(21), 13324–13334. doi:10.1021/la2031818 PMID:21919469

Hunt, P. A., Kirchner, B., & Welton, T. (2006). Characterising the electronic structure of ionic liquids: An examination of the 1-butyl-3-methylimidazolium chloride ion pair. *Chemistry (Weinheim an der Bergstrasse, Germany)*, *12*(26), 6762–6775. doi:10.1002/chem.200600103 PMID:16807969

Huo, D., & Kobayashi, T. (2006). Sonoresponsive volume phase-transition behavior of crosslinked poly (N-isopropylacrylamide-co-acrylic acid) microgel. *Chemistry Letters*, *35*(7), 776–777. doi:10.1246/cl.2006.776

Jacquemin, J., Husson, P., Padua, A. A. H., & Majer, V. (2006). Density and viscosity of several pure and water-saturated ionic liquids. *Green Chemistry*, *8*(2), 172–180. doi:10.1039/B513231B

Jiang, J.-C., Li, S.-C., Shih, P.-M., Hung, T.-C., Chang, S.-C., Lin, S. H., & Chang, H.-C. (2010). A high-pressure infrared spectroscopic study on the interaction of ionic liquids with PEO-PPO-PEO block copolymers and 1,4-Dioxane. *The Journal of Physical Chemistry B*, *115*(5), 883–888. doi:10.1021/jp109600c PMID:21190359

Kakiuchi, T., & Nishi, N. (2006). Ionic liquid|water interface: A new electrified system for electrochemistry. *Electrochemistry*, *74*(12), 942–948. doi:10.5796/electrochemistry.74.942

Koddermann, T., Wertz, C., Heintz, A., & Ludwig, R. (2006). The association of water in ionic liquids: A reliable measure of polarity. *Angewandte Chemie International Edition*, *45*(22), 3697–3702. doi:10.1002/anie.200504471 PMID:16637092

Kohno, Y., & Ohno, H. (2012). Temperature-responsive ionic liquid/water interfaces: Relation between hydrophilicity of ions and dynamic phase change. *Physical Chemistry Chemical Physics*, *14*(15), 5063–5070. doi:10.1039/c2cp24026b PMID:22334119

Kotsuchibashi, Y., Yamamoto, K., & Aoyagi, T. (2009). Assembly behavior of double thermo-responsive block copolymers with controlled response temperature in aqueous solution. *Journal of Colloid and Interface Science*, *336*(1), 67–72. doi:10.1016/j.jcis.2009.03.093 PMID:19442985

Lee, J., Panzer, M. J., He, Y., Lodge, T. P., & Frisbie, C. D. (2007). Ion gel gated polymer thin-film transistors. *Journal of the American Chemical Society*, *129*(15), 4532–4533. doi:10.1021/ja070875e PMID:17381097

Lentacker, I., De Geest, B. G., Vandenbroucke, R. E., Peeters, L., Demeester, J., De Smedt, S. C., & Sanders, N. N. (2006). Ultrasound-responsive polymer-coated microbubbles that bind and protect DNA. *Langmuir*, *22*(17), 7273–7278. doi:10.1021/la0603828 PMID:16893226

Li, K., & Kobayashi, T. (2016a). A FT-IR spectroscopic study of ultrasound effect on aqueous imidazole based ionic liquids having different counter ions. *Ultrasonics Sonochemistry*, *28*(0), 39–46. doi:10.1016/j.ultsonch.2015.06.030 PMID:26384881

Li, K., & Kobayashi, T. (2016b). Ultrasound response of aqueous poly(ionic liquid) solution. *Ultrasonics Sonochemistry*, *30*, 52–60. doi:10.1016/j.ultsonch.2015.10.021 PMID:26597539

Li, K., Noguchi, S., & Kobayashi, T. (2016). Ultrasound-responsive behavior of gelatinous ionic liquid/poly(vinyl alcohol) composites. *Ind. Eng. Chem. Res., 55(37), 9915-9924.* doi: 10.1021/acs.iecr.6b02264

Li, K., Yuan, C., Zheng, S., & Fang, Q. (2012). A facile and economical procedure for the synthesis of maleimide derivatives using an acidic ionic liquid as a catalyst. *Tetrahedron Letters*, *53*(32), 4245–4247. doi:10.1016/j.tetlet.2012.06.025

Liew, C.-W., Ramesh, S., & Arof, A. K. (2014). A novel approach on ionic liquid-based poly(vinyl alcohol) proton conductive polymer electrolytes for fuel cell applications. *International Journal of Hydrogen Energy*, *39*(6), 2917–2928. doi:10.1016/j.ijhydene.2013.07.092

Liu, Z., Yang, Z., Zhao, Y., Ji, G., Zhang, H., Yu, B., & Gao, X. (2014). Fluoro-functionalized polymeric ionic liquids: Highly efficient catalysts for CO_2 cycloaddition to cyclic carbonates under mild conditions. *Green Chemistry*, *16*(8), 3724–3728. doi:10.1039/C4GC00730A

Lodge, T. P. (2008). A unique platform for materials design. *Science*, *321*(5885), 50–51. doi:10.1126/science.1159652 PMID:18599764

López-Pastor, M., Ayora-Cañada, M. J., Valcárcel, M., & Lendl, B. (2006). Association of methanol and water in ionic liquids elucidated by infrared spectroscopy using two-dimensional correlation and multivariate curve resolution. *The Journal of Physical Chemistry B*, *110*(22), 10896–10902. doi:10.1021/jp057398b PMID:16771342

Lu, W., Fadeev, A. G., Qi, B., Smela, E., Mattes, B. R., Ding, J., & Forsyth, M. et al. (2002). Use of ionic liquids for π-conjugated polymer electrochemical devices. *Science*, *297*(5583), 983–987. doi:10.1126/science.1072651 PMID:12098704

Matsumoto, K., & Hagiwara, R. (2007). Structural characteristics of alkylimidazolium-based salts containing fluoroanions. *Journal of Fluorine Chemistry*, *128*(4), 317–331. doi:10.1016/j.jfluchem.2006.10.016

Meng, H., & Jinlian Hu, . (2010). A brief review of stimulus-active polymers responsive to thermal, light, magnetic, electric, and water/solvent stimuli. *Journal of Intelligent Material Systems and Structures*, *21*(9), 859–885. doi:10.1177/1045389X10369718

Mori, H., Yahagi, M., & Endo, T. (2009). Raft polymerization of N-vinylimidazolium salts and synthesis of thermoresponsive ionic liquid block copolymers. *Macromolecules*, *42*(21), 8082–8092. doi:10.1021/ma901180j

Ngoc, N. L., & Kobayashi, T. (2010). Ultrasound stimulus effect on hydrogen bonding in networked alumina and polyacrylic acid slurry. *Ultrasonics Sonochemistry*, *17*(1), 186–192. doi:10.1016/j.ultsonch.2009.04.007 PMID:19464939

Noda, I. (1993). Generalized two-dimensional correlation method applicable to infrared, raman, and other types of spectroscopy. *Applied Spectroscopy*, *47*(9), 1329–1336. doi:10.1366/0003702934067694

Noda, I., Dowrey, A. E., Marcoli, C., Story, G. M., & Ozaki, Y. (2000). Generalized two-dimensional correlation spectroscopy. *Applied Spectroscopy*, *54*(7), 236A–248A. doi:10.1366/0003702001950454

Osada, I., de Vries, H., Scrosati, B., & Passerini, S. (2015). Ionic-liquid-based polymer electrolytes for battery applications. *Angewandte Chemie International Edition*, *55*(2), 500–513. doi:10.1002/anie.201504971 PMID:26783056

Reichert, W. M., Holbrey, J. D., Swatloski, R. P., Gutowski, K. E., Visser, A. E., Nieuwenhuyzen, M., & Rogers, R. D. et al. (2007). Solid-state analysis of low-melting 1, 3-dialkylimidazolium hexafluorophosphate salts (ionic liquids) by combined X-ray crystallographic and computational analyses. *Crystal Growth & Design*, *7*(6), 1106–1114. doi:10.1021/cg060746u

Rogers, R. D. (2007). Materials science: Reflections on ionic liquids. *Nature*, *447*(7147), 917–918. doi:10.1038/447917a PMID:17581570

Shaplov, A. S., Lozinskaya, E. I., & Vygodskii, Y. S. (2010). Polymer ionic liquids: synthesis, design and application in electrochemistry as ion conducting materials. In A. A. J. Torriero & M. J. A. Shiddiky (Eds.), *Electrochemical Properties and Applications of Ionic Liquids* (pp. 203–298). New York: Nova Publishers.

Sun, X., Luo, H., & Dai, S. (2011). Ionic liquids-based extraction: A promising strategy for the advanced nuclear fuel cycle. *Chemical Reviews*, *112*(4), 2100–2128. doi:10.1021/cr200193x PMID:22136437

Susan, M. A. B. H., Kaneko, T., Noda, A., & Watanabe, M. (2005). Ion gels prepared by in situ radical polymerization of vinyl monomers in an ionic liquid and their characterization as polymer electrolytes. *Journal of the American Chemical Society*, *127*(13), 4976–4983. doi:10.1021/ja045155b PMID:15796564

Tang, J., Tang, H., Sun, W., Plancher, H., Radosz, M., & Shen, Y. (2005). Poly(ionic liquid)s: A new material with enhanced and fast CO_2 absorption. *Chemical Communications*, (26): 3325–3327. PMID:15983662

Ueki, T., Usui, R., Kitazawa, Y., Lodge, T. P., & Watanabe, M. (2015). Thermally reversible ion gels with photohealing properties based on triblock copolymer self-assembly. *Macromolecules*, *48*(16), 5928–5933. doi:10.1021/acs.macromol.5b01366

Ueki, T., & Watanabe, M. (2006). Upper critical solution temperature behavior of poly(*N*-isopropylacrylamide) in an ionic liquid and preparation of thermo-sensitive nonvolatile gels. *Chemistry Letters*, *35*(8), 964–965. doi:10.1246/cl.2006.964

Venegas-Sanchez, J. A., Kusunoki, T., Yamamoto, M., & Kobayashi, T. (2013). Sono-respond on thermosensitive polymer microgels based on cross-linked poly(N-isopropylacrylamide-co-acrylic acid). *Ultrasonics Sonochemistry*, *20*(5), 1271–1275. doi:10.1016/j.ultsonch.2013.02.003 PMID:23528945

Venegas-Sanchez, J. A., Tagaya, M., & Kobayashi, T. (2013a). Effect of ultrasound on the aqueous viscosity of several water-soluble polymers. *Polymer Journal*, *45*(12), 1224–1233. doi:10.1038/pj.2013.47

Venegas-Sanchez, J. A., Tagaya, M., & Kobayashi, T. (2013b). Ultrasound effect used as external stimulus for viscosity change of aqueous carrageenans. *Ultrasonics Sonochemistry*, *20*(4), 1081–1091. doi:10.1016/j.ultsonch.2013.01.002 PMID:23395152

Venegas-Sanchez, J. A., Tagaya, M., & Kobayashi, T. (2014). Ultrasound stimulus inducing change in hydrogen bonded crosslinking of aqueous polyvinyl alcohols. *Ultrasonics Sonochemistry*, *21*(1), 295–309. doi:10.1016/j.ultsonch.2013.06.011 PMID:23835399

Vijayakrishna, K., Jewrajka, S. K., Ruiz, A., Marcilla, R., Pomposo, J. A., Mecerreyes, D., & Gnanou, Y. et al. (2008). Synthesis by RAFT and ionic responsiveness of double hydrophilic block copolymers based on ionic liquid monomer units. *Macromolecules*, *41*(17), 6299–6308. doi:10.1021/ma800677h

Walden, P. (1914). Article. Bull. Acad. Imper. Sci.(St. Petersburg), 1800.

Wang, G., Tong, X., & Zhao, Y. (2004). Preparation of azobenzene-containing amphiphilic diblock copolymers for light-responsive micellar aggregates. *Macromolecules*, *37*(24), 8911–8917. doi:10.1021/ma048416a

Wang, G., & Wu, P. (2015). In-depth study of the phase separation behaviour of a thermoresponsive ionic liquid and poly(ionic liquid) in concentrated aqueous solution. *Soft Matter*, *11*(26), 5253–5264. doi:10.1039/C5SM00603A PMID:26052832

Wang, Y., Li, H., & Han, S. (2006). A theoretical investigation of the interactions between water molecules and ionic liquids. *The Journal of Physical Chemistry B*, *110*(48), 24646–24651. doi:10.1021/jp064134w PMID:17134227

Welton, T. (1999). Room-temperature ionic liquids. Solvents for synthesis and catalysis. *Chemical Reviews*, *99*(8), 2071–2084. doi:10.1021/cr980032t PMID:11849019

Widegren, J. A., Laesecke, A., & Magee, J. W. (2005). The effect of dissolved water on the viscosities of hydrophobic room temperature ionic liquids. *Chemical Communications*, (12): 1610–1612. doi:10.1039/b417348a PMID:15770276

Widegren, J. A., Saurer, E. M., Marsh, K. N., & Magee, J. W. (2005). Electrolytic conductivity of four imidazolium-based room-temperature ionic liquids and the effect of a water impurity. *The Journal of Chemical Thermodynamics*, *37*(6), 569–575. doi:10.1016/j.jct.2005.04.009

Yan, F., & Texter, J. (2007). Solvent-reversible poration in ionic liquid copolymers. *Angewandte Chemie International Edition*, *119*(14), 2492–2495. doi:10.1002/ange.200604127 PMID:17328017

Yu, Y., Hu, T., Chen, X., Xu, K., Zhang, J., & Huang, J. (2011). Pd nanoparticles on a porous ionic copolymer: A highly active and recyclable catalyst for Suzuki-Miyaura reaction under air in water. *Chemical Communications*, *47*(12), 3592–3594. doi:10.1039/c0cc04498a PMID:21298188

Yuan, J., Mecerreyes, D., & Antonietti, M. (2013). Poly(ionic liquid)s: An update. *Progress in Polymer Science*, *38*(7), 1009–1036. doi:10.1016/j.progpolymsci.2013.04.002

Zhang, Q., Wu, S., Zhang, L., Lu, J., Verproot, F., Liu, Y., & Song, X. M. et al. (2011). Fabrication of polymeric ionic liquid/graphene nanocomposite for glucose oxidase immobilization and direct electrochemistry. *Biosensors & Bioelectronics*, *26*(5), 2632–2637. doi:10.1016/j.bios.2010.11.024 PMID:21159504

Zhao, Q., Heyda, J., Dzubiella, J., Täuber, K., Dunlop, J. W. C., & Yuan, J. (2015). Sensing solvents with ultrasensitive porous poly(ionic liquid) actuators. *Advanced Materials*, *27*(18), 2913–2917. doi:10.1002/adma.201500533 PMID:25828569

Zheng, Y. Z., Wang, N. N., Luo, J. J., Zhou, Y., & Yu, Z. W. (2013). Hydrogen-bonding interactions between [BMIM][BF_4] and acetonitrile. *Physical Chemistry Chemical Physics*, *15*(41), 18055–18064. doi:10.1039/c3cp53356e PMID:24061707

Zhou, X., Wu, T., Hu, B., Yang, G., & Han, B. (2010). Synthesis of graphene/polyaniline composite nanosheets mediated by polymerized ionic liquid. *Chemical Communications*, *46*(21), 3663–3665. doi:10.1039/c0cc00049c PMID:20396832

Zhu, W., Wu, P., Yang, L., Chang, Y., Chao, Y., Li, H., & Xun, S. et al. (2013). Pyridinium-based temperature-responsive magnetic ionic liquid for oxidative desulfurization of fuels. *Chemical Engineering Journal*, *229*, 250–256. doi:10.1016/j.cej.2013.05.115

KEY TERMS AND DEFINITIONS

Green Chemistry: An area of chemistry and chemical engineering focused on the designing of products and processes that minimize the use and generation of hazardous substances.

Hydrogen Bond: The electrostatic attraction between two polar groups that occurs when a hydrogen atom covalently bound to a highly electronegative atom such as nitrogen or oxygen experiences the electrostatic field of another highly electronegative atom nearby.

Infrared Spectroscopy: The spectroscopy that deals with the infrared region of the electromagnetic spectrum, that is light with a longer wavelength and lower frequency than visible light.

Ionic Liquids: A salt in the liquid state below 100 ◦ C.

Poly(Ionic Liquid): A kind of polymer that contains at least one ionic center, similar by composition to the structure of commonly used ILs, as a covalently bonded part of their constitutional repeating (monomer) unit.

Ultrasound: Sound waves with frequencies higher than the upper audible limit of human hearing.

Viscosity: A fluid is a measure of its resistance to gradual deformation by shear stress or tensile stress.

Compilation of References

Abdulkareem, O. A., Mustafa Al Bakri, A. M., Kamarudin, H., Khairul Nizar, I., & Saif, A. A. (2014). Effects of elevated temperatures on the thermal behavior and mechanical performance of fly ash geopolymer paste, mortar and lightweight concrete. *Construction & Building Materials*, *50*, 377–387. doi:10.1016/j.conbuildmat.2013.09.047

Abdullah, M. M. A. B., Hussin, K., Bnhussain, M., Ismail, K. N., Yahya, Z., & Abdul Razak, R. (2012). Fly Ash-based Geopolymer Lightweight Concrete Using Foaming Agent. *International Journal of Molecular Sciences*, *13*(6), 7186–7198. doi:10.3390/ijms13067186 PMID:22837687

Abdullah, M. M. A. B., Jamaludin, L., Hussin, K., Bnhussain, M., Ghazali, C. M. R., & Ahmad, M. I. (2012). Fly Ash Porous Material using Geopolymerization Process for High Temperature Exposure. *International Journal of Molecular Sciences*, *13*(4), 4388–4395. doi:10.3390/ijms13044388 PMID:22605984

Abdullah, M. O., Tan, I. A. W., & Limb, L. S. (2011). Automobile adsorption air-conditioning system using oil palm biomass-based activated carbon: A review. *Renewable & Sustainable Energy Reviews*, *15*(4), 2061–2072. doi:10.1016/j.rser.2011.01.012

Acasigua, G. A. X., de Olyveira, G. M., Costa, L. M. M., Braghirolli, D. I., Fossati, A. C. M., Guastaldi, A. C., & Basmaji, P. et al. (2014). Novel chemically modified bacterial cellulose nanocomposite as potential biomaterial for stem cell therapy applications. *Current Stem Cell Research & Therapy*, *9*(2), 117–123. doi:10.2174/1574888X08666131124135654 PMID:24266631

Achma, R. B., Ghorbel, A., Dafinov, A., & Medina, F. (2008). Copper-supported pillared clay catalysts for the wet hydrogen peroxide catalytic oxidation of model pollutant tyrosol. *Applied Catalysis A, General*, *349*(1-2), 20–28. doi:10.1016/j.apcata.2008.07.021

Agency for Natural Resources and Energy. (2014). *Green Power*. Retrieved from http://www.enecho.meti.go.jp/en/category/others/basic_plan/pdf/4th_strategic_energy_plan.pdf

Ahamat, M. A., & Tierney, M. J. (2012). Calorimetric assessment of adsorbents bonded to metal surfaces: Application to type A silica gel bonded to aluminium. *Applied Thermal Engineering*, *40*, 258–266. doi:10.1016/j.applthermaleng.2012.02.021

Ahmaruzzaman, M. (2010). A review on the utilization of fly ash. *Progress in Energy and Combustion Science*, *36*(3), 327–363. doi:10.1016/j.pecs.2009.11.003

Ahrem, H., Pretzel, D., Endres, M., Conrad, D., Courseau, J., Müller, H., & Kinne, R. W. et al. (2014). Laser-structured bacterial nanocellulose hydrogels support ingrowth and differentiation of chondrocytes and show potential as cartilage implants. *Acta Biomaterialia*, *10*(3), 1341–1353. doi:10.1016/j.actbio.2013.12.004 PMID:24334147

Akahira, A., Alam, K. C. A., & Hamamoto, Y. (2005). Mass recovery four-bed adsorption refrigeration cycle with energy cascading. *Applied Thermal Engineering*, *25*(11-12), 1764–1778. doi:10.1016/j.applthermaleng.2004.10.006

Akahira, A., Alam, K. C. A., Hamamoto, Y., Akisawa, A., & Kashiwagi, T. (2004). Mass recovery adsorption refrigeration cycle-improving cooling capacity. *International Journal of Refrigeration*, *27*(3), 225–234. doi:10.1016/j.ijrefrig.2003.10.004

Akbari, A., Majumder, M., & Tehrani, A. (2015). Polylactic acid (PLA) carbon nanotube Nanocomposites. In Handbook of Polymer Nanocomposites. Processing, Performance and Application – Volume B: Carbon Nanotube Based Polymer Composites (pp. 283-298). Springer-Verlag.

Akbarzadeh, A., Samiei, M., & Davaran, S. (2012). Magnetic nanoparticles: Preparation, physical properties, and applications in biomedicine. *Nanoscale Research Letters*, *7*(1), 1–13. doi:10.1186/1556-276X-7-144 PMID:22348683

Akolekar, D., Chaffee, A., & Howe, R. F. (1997). The transformation of kaolin to low-silica X zeolite. *Zeolites*, *19*(5-6), 359–365. doi:10.1016/S0144-2449(97)00132-2

Alaa, M., & Rashad, S. R. Z. (2011). The effect of activator concentration on the residual strength of alkali-activated fly ash pastes subjected to thermal load. *Construction & Building Materials*, *25*(7), 3098–3107. doi:10.1016/j.conbuildmat.2010.12.044

Alaba, P. A., Sani, Y. M., & Daud, W. M. A. W. (2015). Synthesis and characterization of hierarchical nanoporous HY zeolites from acid-activated kaolin. *Chinese Journal of Catalysis*, *36*(11), 1846–1851. doi:10.1016/S1872-2067(15)60962-7

Alexandre, M., & Dubois, P. (2000). Polymer-layered silicate nanocomposites: Preparation, properties and uses of a new class of materials. *Materials Science and Engineering*, *28*(1-2), 1–63. doi:10.1016/S0927-796X(00)00012-7

Alghoula, M. A., Sulaiman, M. Y., Sopian, K., & Azmi, B. Z. (2009). Performance of a dual-purpose solar continuous adsorption system. *Renewable Energy*, *34*(3), 920–927. doi:10.1016/j.renene.2008.05.037

Ali Shah, A., Hasan, F., Shah, Z., Kanwal, N., & Zeb, S. (2013). Biodegradation of natural and synthetic rubbers: A review. *International Biodeterioration & Biodegradation*, *83*, 145–157. doi:10.1016/j.ibiod.2013.05.004

Al-Sabagh, A. M., Kabel, K. I., Noor El-Din, M. R., & Elsharaky, E. A. (2012). Synthesis of polyalkylacrylate nanolatexes by microemulsion polymerization method. *Egyptian Journal of Petroleum*, *21*(2), 81–87. doi:10.1016/j.ejpe.2012.11.001

Álvarez-Ayuso, E. (2009). Approaches for the treatment of waste streams of the aluminium anodising industry. *Journal of Hazardous Materials*, *164*(2-3), 409–414. doi:10.1016/j.jhazmat.2008.08.054 PMID:18823703

Al-Zboon, K., Al-Harahsheh, M. S., & Hani, F. B. (2011). Fly ash-based geopolymer for Pb removal from aqueous solution. *Journal of Hazardous Materials*, *188*(1–3), 414–421. doi:10.1016/j.jhazmat.2011.01.133 PMID:21349635

Amari, A., Chlendi, M., Gannouni, A., & Bellagi, A. (2010). Optimised activation of bentonite for toluene adsorption. *Applied Clay Science*, *47*(3-4), 457–461. doi:10.1016/j.clay.2009.11.035

Amnuaikit, T., Chusuit, T., Raknam, P., & Boonme, P. (2011). Effects of a cellulose mask synthesized by a bacterium on facial skin characteristics and user satisfaction. *Medical Devices: Evidence and Research*, *4*(1), 77–81. PMID:22915933

Anakabe, J., Huici, A. M. Z., Eceiza, A., & Arbelaiz, A. (2016). The effect of the addition of poly(styrene-co-glycidyl methacrylate) copolymer on the properties of polylactide/ poly(methyl methacrylate) blend. *Journal of Applied Polymer Science*, *133*(37), 43935–43945. doi:10.1002/app.43935

Anand, A., Rani, N., Saxena, P., Bhandari, H., & Dhawan, S. K. (2015). Development of polyaniline/zinc oxide nanocomposite impregnated fabric as an electrostatic charge dissipative material. *Polymer International*, *64*(9), 1096–1103. doi:10.1002/pi.4870

Anderson, C. W. N., Brooks, R. R., Stewart, R. B., & Simcock, R. (1998). Harvesting a crop of gold in plants. *Nature*, *395*(6702), 553–554. doi:10.1038/26875

Anderson, J. L., Ding, J., Welton, T., & Armstrong, D. W. (2002). Characterizing ionic liquids on the basis of multiple solvation interactions. *Journal of the American Chemical Society*, *124*(47), 14247–14254. doi:10.1021/ja028156h PMID:12440924

Andersson, A. D. (2006). A study of the nanostructure of the cell wall of the tracheids of conifer xylem by C-ray scattering. *University of Helsinki, Reports Series in Physics. HU-P-D*, *135*, 18–25.

Andini, S., Cioffi, R., Colangelo, F., Grieco, T., Montagnaro, F., & Santoro, L. (2008). Coal fly ash as raw material for the manufacture of geopolymer-based products. *Waste Management (New York, N.Y.)*, *28*(2), 416–423. doi:10.1016/j.wasman.2007.02.001 PMID:17382528

Andrews, J. E., Brecombe, P. B., Jikel, T. D., & Brianreid, P. L. (2004). *An introduction to environmental chemistry*. Blackwell Sieme, LTD.

Angell, C. A., Liu, C., & Sanchez, E. (1993). Rubbery solid electrolytes with dominant cationic transport and high ambient conductivity. *Nature*, *362*(6416), 137–139. doi:10.1038/362137a0

Anonymous, . (2000). Valuable raw materials extracted from egg shells. *Food Eng. Int.*, *25*(2), 45–48.

Anoop-Anand, K. (2010). Natural rubber-carbon nanotube composites through latex compounding. *International Journal of Polymeric Materials*, *59*(1), 33–44. doi:10.1080/00914030903172916

Ansarifar, A., Lim, H. P., & Nijhawan, R. (2004). Assessment of the effect of a bifunctional organosilane on the bound rubber and properties of some natural rubber compounds. *International Journal of Adhesion and Adhesives*, *24*(1), 9–22. doi:10.1016/S0143-7496(03)00095-2

Antonio, P. F. L., & Michel, D. (2000). Performance of a new solid adsorption ice maker with solar energy regeneration. *Energy Conversion and Management*, *4*, 1625–1647.

Anupam, K., Chatterjee, A., Halder, G. N., & Sarkar, S. C. (2011a). Experimental investigation of a single-bed pressure swing adsorption refrigeration system towards replacement of halogenated refrigerants. *Chemical Engineering Journal*, *171*(2), 541–548. doi:10.1016/j.cej.2011.04.025

Anwar, J., Shafique, U., Waheed-uz-Zaman, , Rehman, R., Salman, M., Dar, A., & Ashraf, S. et al. (2015). Microwave chemistry: Effect of ions on dielectric heating in microwave ovens. *Arabian Journal of Chemistry*, *8*(1), 100–104. doi:10.1016/j.arabjc.2011.01.014

Anwer, M. A. S., & Naguib, H. E. (2016). Study on the morphological, dynamic mechanical and thermal properties of PLA carbon nanofibre composites. *Composites. Part B, Engineering*, *91*, 631–639. doi:10.1016/j.compositesb.2016.01.039

Anyanwu, E., & Ogueke, N. (2005). Thermodynamic design procedure for solid adsorption solar refrigerator. *Renewable Energy*, *30*(1), 81–96. doi:10.1016/j.renene.2004.05.005

Apilux, A., Siangroph, W., Praphairaksit, N., & Chailapakul, O. (2012). Simple and rapid colorimetric detection of Hg(II) by a paper-based device using silver nanoplates. *Talanta*, *97*, 388–394. doi:10.1016/j.talanta.2012.04.050 PMID:22841097

Aranda, P., & Ruiz-Hitzky, E. (1992). Poly (ethylene oxide)-silicate intercalation materials. *Chemistry of Materials*, *4*(6), 1395–1403. doi:10.1021/cm00024a048

Arfaoui, S., Frini-Srasra, N., & Srasra, E. (2008). Modelling of the adsorption of the chromium ion by modified clays. *Desalination*, *222*(1-3), 474–481. doi:10.1016/j.desal.2007.03.014

Aristov, Y. (2007). Novel Materials for adsorptive heat pumping and storage: Screening and nanotailoring of sorption properties. *Journal of Chemical Engineering of Japan*, *40*(13), 1242–1251. doi:10.1252/jcej.07WE228

Aristov, Yu. I., Dawoud, B., Glaznev, I. S., & Elyas, A. (2008). A new methodology of studying the dynamics of water sorption/desorption under real operating conditions of adsorption heat pumps: Experiment. *International Journal of Heat and Mass Transfer*, *51*(19–20), 4966–4972. doi:10.1016/j.ijheatmasstransfer.2007.10.042

Arroyo, M., Manchado, M. A., & Herero, B. (2003). Organo-montmorillonite as substitute of carbon black in natural rubber compounds. *Polymer*, *44*(8), 2447–2453. doi:10.1016/S0032-3861(03)00090-9

Arruda, L. C., Magaton, M., Bretas, R. E. S., & Ueki, M. M. (2015). Influence of chain extender on mechanical, thermal and morphological properties of blown films of PLA/PBAT blends. *Polymer Testing*, *43*, 27–37. doi:10.1016/j.polymertesting.2015.02.005

Ashori, A., & Nourbakhsh, A. (2010). Bio-based composites from waste agricultural residues. *Waste Management (New York, N.Y.)*, *30*(4), 680–684. doi:10.1016/j.wasman.2009.08.003 PMID:19740643

Askalany, A. A., & Saha, B. B. (2015). Experimental and theoretical study of adsorption kinetics of Difluoromethane onto activated carbons. *International Journal of Refrigeration*, *4*(9), 160–168. doi:10.1016/j.ijrefrig.2014.10.009

Askalany, A. A., Saha, B. B., & Ismail, I. M. (2014). Adsorption isotherms and kinetics of HFC410A onto activated carbons. *Applied Thermal Engineering*, *72*(2), 237–243. doi:10.1016/j.applthermaleng.2014.04.075

Askalany, A., Salem, M., Ismail, I. M., Ali, A. H. H., & Morsy, M. G. (2012). Experimental study on adsorptionedesorption characteristics of granular activated carbon/R134a pair. *International Journal of Refrigeration*, *35*(3), 494–498. doi:10.1016/j.ijrefrig.2011.04.002

ASTM. (2009). *ASTM D6400-04 Standard Specification of Compostable Plastics*. West Conshohocken, PA: ASTM International.

Attalla, M., & Sadek, S. (2014). Experimental Investigation of Granular Activated Carbon/R-134a Pair for Adsorption Cooling System Applications. *Journal of Power and Energy Engineering*, *2*(02), 11–20. doi:10.4236/jpee.2014.22002

Ayana, B., Suin, S., & Khatua, B. B. (2014). Highly exfoliated eco-friendly thermoplastic starch (TPS)/poly (lactic acid)(PLA)/clay nanocomposites using unmodified nanoclay. *Carbohydrate Polymers*, *110*, 430–439. doi:10.1016/j.carbpol.2014.04.024 PMID:24906776

Ayele, L., & Pariente, J. P. (2015). Synthesis of zeolite A from Ethiopian kaolin. *Microporous and Mesoporous Materials*, *215*, 29–36. doi:10.1016/j.micromeso.2015.05.022

Babaki, H., Salem, A., & Jafarizad, A. (2008). Kinetic model for the isothermal activation of bentonite by sulfuric acid. *Materials Chemistry and Physics*, *108*(2-3), 263–268. doi:10.1016/j.matchemphys.2007.09.034

Baba, Y., Kubota, F., Kamiya, N., & Goto, M. (2011). Selective recovery of dysprosium and neodymium ions by a supported liquid membrane based on ionic liquids. *Solv. Extr. Res. Dev. Jpn.*, *18*(0), 193–198. doi:10.15261/serdj.18.193

Bachir, C., Lan, Y., Mereacre, V., Powell, A. K., Koch, C. B., & Weidler, P. G. (2009). Magnetic titanium-pillared clays (Ti-M-PILC): Magnetic studies and Mossbauer spectroscopy. *Clays and Clay Minerals*, *57*(4), 433–443. doi:10.1346/CCMN.2009.0570404

Bäckdahl, H., Esguerra, M., Delbro, D., Risberg, B., & Gatenholm, P. (2008). Engineering microporosity in bacterial cellulose scaffolds. *Journal of Tissue Engineering and Regenerative Medicine*, *2*(6), 320–330. doi:10.1002/term.97 PMID:18615821

Bajiai, P. (2013). *Advances in bioethanol*. New York: Springer. doi:10.1007/978-81-322-1584-4

Bajpai, P. K., Singh, I., & Madaan, J. (2014). Development and characterization of PLA-based green composites: A review. *Journal of Thermoplastic Composite Materials*, *227*(1), 52–81. doi:10.1177/0892705712439571

Baker, A. J. M., & Brooks, R. R. (1989). Terrestrial higher plants which hyperaccumulate metallic elements—a review of their distribution, ecology and phytochemistry. *Biorecovery*, *1*, 81–126.

Baker, A. J. M., McGrath, S. P., Reeves, R. D., & Smith, J. A. C. (2000). Metal hyperaccumulator plants: A review of the ecology and physiology of a biological resource for phytoremediation of metal-polluted soils. In N. Terry & G. Bañuelos (Eds.), *Phytoremediation of Contaminated Soil and Water* (pp. 85–108). Boca Raton, FL: CRC Press.

Bakharev, T. (2005). Geopolymeric materials prepared using Class F fly ash and elevated temperature curing. *Cement and Concrete Research*, *35*(6), 1224–1232. doi:10.1016/j.cemconres.2004.06.031

Banker, N. D., Prasad, M., Dutta, P., & Srinivasan, K. (2010). Development and transient performance results of a single stage activated carbon – HFC 134a closed cycle adsorption cooling system. *Applied Thermal Engineering*, *30*(10), 1126–1132. doi:10.1016/j.applthermaleng.2010.01.027

Bansal, R. C., & Goyal, M. (2005). *Activated Carbon Adsorption*. Taylor & Francis Group. doi:10.1201/9781420028812

Barbier, F., Duc, G., & Petit-Ramel, M. (2000). Adsorption of lead and cadmium ions from aqueous solution to the montmorillonite: Water interface. *Colloids and Surfaces. A, Physicochemical and Engineering Aspects*, *166*(1-3), 153–159. doi:10.1016/S0927-7757(99)00501-4

Barman, S. C., Sahu, R. K., Bhargava, S. K., & Chatterjee, C. (2000). Distribution of heavy metals in wheat, mustard, and weed grown in fields irrigated with industrial effluents. *Bulletin of Environmental Contamination and Toxicology*, *64*(4), 489–496. doi:10.1007/s001280000030 PMID:10754044

Bartell, F.E. & Bower, J.E. (1952). *Adsorption of vapors by silica gels of different structures*. DOI: 10.1016/0095-8522(52)90022-6

Basci, N., Kocadagistan, E., & Kocadagistan, B. (2004). Biosorption of copper (II) from aqueous solutions by wheat shell. *Desalination*, *164*(2), 1, 135–1400. doi:10.1016/S0011-9164(04)00172-9

Bashaiwoldu, A. B., Podczeck, F., & Newton, J. M. (2004). A study on the effect of drying techniques on the mechanical properties of pellets and compacted pellets. *European Journal of Pharmaceutical Sciences*, *21*(2-3), 119–129. doi:10.1016/j.ejps.2003.09.013 PMID:14757483

Bebon, C., Colson, D., Marrot, B., Klein, J. P., & Renzo, F. D. (2002). Synthesis of zeolites: Study and application of a new process of homogeneous shaking out of the medium to minimize the shear rate during the crystallization. *Microporous and Mesoporous Materials*, *53*(1-3), 13–20. doi:10.1016/S1387-1811(02)00318-9

Bendou, S., & Amrani, M. (2014). Effect of Hydrochloric Acid on the Structural of Sodic-Bentonite Clay. *Journal of Minerals and Materials Characterization and Engineering*, *2*(05), 404–413. doi:10.4236/jmmce.2014.25045

Berber, S., Kwon, Y. K., & Tománek, D. (2000). Unusually high thermal conductivity of carbon nanotubes. *Physical Review Letters*, *84*(20), 4613–4616. doi:10.1103/PhysRevLett.84.4613 PMID:10990753

Bernardo, E. B., Neilan, B. A., & Couperwhite, I. (1998). Characterization, Differentiation and Identification of Wild-type Cellulose-synthesizing Acetobacter strains Involved in Nata de Coco Production. *Systematic and Applied Microbiology*, *21*(4), 599–608. doi:10.1016/S0723-2020(98)80073-8

Bester-Rogac, M., Stoppa, A., Hunger, J., Hefter, G., & Buchner, R. (2011). Association of ionic liquids in solution: A combined dielectric and conductivity study of [bmim][Cl] in water and in acetonitrile. *Physical Chemistry Chemical Physics, 13*(39), 17588–17598. doi:10.1039/c1cp21371g PMID:21892477

Betancourt-Galindo, R., Cabrera Miranda, C., & Puente Urbina, B. A. (2012). Casta, #xf1, eda-Facio, A., . . . ndez, O. S. (2012). Encapsulation of Silver Nanoparticles in a Polystyrene Matrix by Miniemulsion Polymerization and Its Antimicrobial Activity. *ISRN Nanotechnology, 5*. doi:10.5402/2012/186851

Bhattacharyya, K. G., & Gupta, S. S. (2008). Adsorption of a few heavy metals on natural and modified kaolinite and montmorillonite: A review. *Advances in Colloid and Interface Science, 140*(2), 114–131. doi:10.1016/j.cis.2007.12.008 PMID:18319190

Bijarimi, M., Ahmad, S., Rasid, R., Khushairi, M. A., & Zakir, M. (2016). Poly(lactic acid) / Poly(ethylene glycol) blends: Mechanical, thermal and morphological properties. AIP Conference Proceedings, 1727, 020002-1-02002-9.

Biller, P., & Ross, A. B. (2011). Potential yields and properties of oil from the hydrothermal liquefaction of microalgae with different biochemical content. *Bioresource Technology, 102*(1), 215–225. doi:10.1016/j.biortech.2010.06.028 PMID:20599375

Bineesh, K. V., Kim, D.-K., Kim, M.-I., & Park, D.-W. (2011). Selective catalytic oxidation of H_2S over V_2O_5 supported on TiO_2-pillared clay catalysts in the presence of water and ammonia. *Applied Clay Science, 53*(2), 204–211. doi:10.1016/j.clay.2010.12.022

Bi, W., Tang, B., & Row, K. (2013). Adsorption of alcohol from water by poly(ionic liquid)s. *Bioprocess and Biosystems Engineering, 36*(6), 651–658. doi:10.1007/s00449-012-0828-8 PMID:23010726

Blackstock, J. M., Neill, J., & Mcintosh, J. A. (2012). *Geopolymeric structural building units and methods of manufacture thereof.* Google Patents.

Böke, N., Birch, G. D., Nyale, S. M., & Petrik, L. F. (2015). New synthesis method for the production of coal fly ash-based foamed geopolymers. *Construction & Building Materials, 75*, 189–199. doi:10.1016/j.conbuildmat.2014.07.041

Bokobza, L. (2007). Multiwall carbon nanotube elastomeric composites: A review. *Polymer, 48*(17), 4907–4920. doi:10.1016/j.polymer.2007.06.046

Bokobza, L. (2012). Enhanced electrical and mechanical properties of multiwall carbon nanotube rubber composites. *Polymers for Advanced Technologies, 23*(12), 1543–1549. doi:10.1002/pat.3027

Bolto, B., & Gregory, J. (2007). Organic polyelectrolytes in water treatment. *Water Research, 41*(11), 2301–2324. doi:10.1016/j.watres.2007.03.012 PMID:17462699

Boonmahitthisud A., & Chuayjuljit, S. (2011). Rubber blend of 80/20 NR/SBR reinforced with nanosilica and PS-encapsulated nanosilica. *Materials Science Forum, 695*, 332-335.

Boonmahitthisud, A., & Chuayjuljit, S. (2012). NR/XSBR Nanocomposites with Carbon Black and Carbon Nanotube Prepared by Latex Compounding. *Journal of Metals, Materials and Minerals, 22*, 77–85.

Bouakaz, B. S., Pillin, I., Habi, A., & Grohens, Y. (2015). Synergy between fillers in organomontmorillonite/graphene-PLA nanocomposites. *Applied Clay Science, 116-117*, 69–77. doi:10.1016/j.clay.2015.08.017

Boubakri, A., Guilleminot, J. J., & Meunier, F. (2000). Adsorptive solar powered ice maker: Experiments and model. *Solar Energy, 69*(3), 249–263. doi:10.1016/S0038-092X(00)00063-3

Bouchenafa-Saïb, N., Khouli, K., & Mohammedi, O. (2007). Preparation and characterization of pillared montmorillonite: Application in adsorption of cadmium. *Desalination*, *217*(1-3), 282–290. doi:10.1016/j.desal.2007.03.007

BP Statistical Review of World Energy. (2015). Retrieved from bp.com/statisticalreview.

Bradaric, C. J., Downard, A., Kennedy, C., Robertson, A. J., & Zhou, Y. (2003a). Phosphonium Ionic Liquids. *The Strem Chemiker*, *XX*(1), 2–11.

Bradaric, C. J., Downard, A., Kennedy, C., Robertson, A. J., & Zhou, Y. (2003b). Industrial Preparation of Phosphonium Ionic Liquids. *Green Chemistry*, *5*(2), 143–152. doi:10.1039/b209734f

Bradley, M., & Grieser, F. (2002). Emulsion Polymerization Synthesis of Cationic Polymer Latex in an Ultrasonic Field. *Journal of Colloid and Interface Science*, *251*(1), 78–84. doi:10.1006/jcis.2002.8383 PMID:16290704

Branco, A., Santos, J. D., Pimente, M., Osuno, J. T., Lima, L. S., & David, J. M. (2010). D-Mannitol from Agave sisalana biomass waste. *Industrial Crops and Products*, *32*(3), 507–512. doi:10.1016/j.indcrop.2010.06.025

Brett, C., & Waldron, K. (1990). *Physiology and biochemistry of plant cell walls*. London, UK: Unwin Hyman. doi:10.1007/978-94-010-9641-6

Broadbent, J. (1998). *Environmental Politics in Japan*. Cambric, UK: Cambridge University press. doi:10.1017/CBO9780511571060

Brooks, R. R. (1998). *Plants That Hyperaccumulate Heavy Metals*. Oxford, UK: CAB International.

Brooks, R. R., Chambers, M. F., Nicks, L. J., & Robinson, B. H. (1998). Phytomining. *Trends in Plant Science*, *3*(9), 359–362. doi:10.1016/S1360-1385(98)01283-7

Brown, A. J. (1886). XLIII.-On an acetic ferment which forms cellulose. *Journal of the Chemical Society Transactions*, *49*(0), 432–439. doi:10.1039/CT8864900432

Brown, M. R. Jr. (1991). *Advances in Cellulose Biosynthesis Polymers from Biobased Materials*. Prentice hall.

Brundtland, G. H. (1987). *Towards Sustainable Development*. A/42/427. Our Common Future: Report of the World Commission on Environment and Development.

Buchsteiner, A., Lerf, A., & Pieper, A. Jr. (2006). Water dynamics in graphite oxide investigated with neutron scattering. *The Journal of Physical Chemistry B*, *110*(45), 22328–223238. doi:10.1021/jp0641132 PMID:17091972

Budhiono, A., Rosidi, B., Taher, H., & Iguchi, M. (1999). Kinetic aspects of bacterial cellulose formation in nata-de-coco culture system. *Carbohydrate Polymers*, *40*(2), 137–143. doi:10.1016/S0144-8617(99)00050-8

Bulgariu, L., Rǎţoi, M., Bulgariu, D., & Macoveanu, M. (2008). Equilibrium Study of Pb(II) and Hg(II) Sorption from Aqueous Solutions by Moss Peat. *Environmental Engineering and Management Journal.*, *7*(5), 511–516.

Burchell, T. D. (1999). *Adsorption refrigerators and heat pumps*. Ltd Carbon Materials for Advanced Technologies.

Cacicedo, , M., León, I., Gonzalez, J., Porto, L., Alvarez, V., & Castro, G. R. (. (2016). Modified bacterial cellulose scaffolds for localized doxorubicin release in human colorectal HT-29 cells. *Colloids and Surfaces. B, Biointerfaces*, *140*, 421–429. doi:10.1016/j.colsurfb.2016.01.007 PMID:26784658

Çağlar, A. (2012). *Design and experimental testing of an adsorbent bed for a thermal wave adsorption cooling cycle* (Unpublished doctoral dissertation). Middle East Technical University, Ankara, Turkey.

Cai, H. L., Sharma, S., Liu, W. Y., Mu, W., Liu, W., Zhang, X. D., & Deng, Y. L. (2014). Aerogel microspheres from natural cellulose nanofibrils and their applications as cell culture scaffolds. *Biomacromolecules, 15*(7), 2540–2547. doi:10.1021/bm5003976 PMID:24894125

Cailloux, J., Hakim, R. N., Santana, O. O., Bou, J., Abt, T., Sánchez-Soto, M., & Maspoch, M. L. et al. (2016, September). Reactive extrusion: A useful process to manufacture structurally modified PLA/o-MMT composites. *Composites. Part A, Applied Science and Manufacturing, 88*, 106–115. doi:10.1016/j.compositesa.2016.05.024

Cai, Z., & Kim, J. (2010). Preparation and Characterization of Novel Bacterial Cellulose/Gelatin Scaffold for Tissue Regeneration Using Bacterial Cellulose Hydrogel. *Journal of Nanotechnology in Engineering and Medicine, 1*(2), 1–6. doi:10.1115/1.4000858

Camargo, P. H. C., Satyanarayana, K. G., & Wypych, F. (2009). Nanocomposites: Synthesis, structure, properties and new application opportunities. *Materials Research, 12*(1), 1–39. doi:10.1590/S1516-14392009000100002

Cammarata, L., Kazarian, S. G., Salter, P. A., & Welton, T. (2001). Molecular states of water in room temperature ionic liquids. *Physical Chemistry Chemical Physics, 3*(23), 5192–5200. doi:10.1039/b106900d

Caramalău, C., Bulgariu, L., & Macoveanu, M. (2009). Cobalt (II) Removal from Aqueous Solutions by Adsorption on. Modified Peat Moss. *Chem. Bull. "POLITEHNICA" Univ. (Timisoara), 54*(68), 1.

Carson, J. (2002). *Prediction of the thermal conductivity of porous foods* (PhD Thesis). Messay University.

Chandel, A. K., daSilva, S. S., Carvalho, W., & Singh, O. V. (2012). Sugarcane Bagasse and Leaves:Foreseeable Biomass of Biofuel and Bio-products. *Journal of Chemical Technology and Biotechnology (Oxford, Oxfordshire), 87*(1), 11–20. doi:10.1002/jctb.2742

Chandrasekhar, S., & Pramada, P. N. (2008). Microwave assisted synthesis of zeolite A from metakaolin. *Microporous and Mesoporous Materials, 108*(1-3), 152–161. doi:10.1016/j.micromeso.2007.04.003

Chandy, T., & Sharma, C. P. (1998). Activated Charcoal Microcapsules and their Applications. *Journal of Biomaterials Applications, 13*, 128–157. PMID:9777464

Chaney, R. L., Angle, R. S., Baker, A. J. M., & Li, Y. M. (1998). *Method for phytomining of nickel cobalt and other metals from soil.* U.S. Patent No. 5711784.

Chang, C. Y., Duan, B., Cai, J., & Zhang, L. N. (2010). Superabsorbent hydrogels based on cellulose for smart swelling and controllable delivery. *European Polymer Journal, 46*(1), 92–100. doi:10.1016/j.eurpolymj.2009.04.033

Chan, K. C., Chao, C. Y. H., Sze-To, G. N., & Hui, K. S. (2012). Performance predictions for a new zeolite 13X/CaCl2 composite adsorbent for adsorption cooling systems. *International Journal of Heat and Mass Transfer, 55*(11-12), 3214–3224. doi:10.1016/j.ijheatmasstransfer.2012.02.054

Charpentier, P. A., Maguire, A., & Wan, W. (2006). Surface modification of polyester to produce a bacterial cellulose-based vascular prosthetic device. *Applied Surface Science, 252*(18), 6360–6367. doi:10.1016/j.apsusc.2005.09.064

Charreau, H., Foresti, M. L., & Vázquez, A. (2013). Nanocellulose patents trends: A comprehensive review on patents on cellulose nanocrystals, microfibrillated and bacterial cellulose. *Recent Patents on Nanotechnology, 7*(1), 56–80. doi:10.2174/187221013804484854 PMID:22747719

Chavez-Guerrero, L., & Hinojosa, M. (2010). Bagasse from the mescal industry. *Fuel, 89*, 4049–4052.

Chawla, P. R., Bajaj, I. B., Survase, S. A., & Singhal, R. S. (2009). Microbial cellulose: Fermentative production and applications. *Food Technology and Biotechnology, 47*(2), 107–124.

Chekirou, W., Boukheit, N., & Karaali, A. (2013). Optimal Design Study of Tubular Adsorber for Solid Adsorption Solar Cooling Machine Working with Activated Carbon–Methanol Pair. *16èmes Journées Internationales de Thermique (JITH 2013).*

Cheknane, B., Zermane, F., Baudu, M., Bouras, O., & Basly, J. P. (2012). Sorption of basic dyes onto granulated pillared clays: Thermodynamic and kinetic studies. *Journal of Colloid and Interface Science*, *381*(1), 158–163. doi:10.1016/j.jcis.2012.05.045 PMID:22721789

Chellaram, C., Murugaboopathi, G., John, A. A., Sivakumar, R., Ganesan, S., Krithika, S., & Priya, G. (2014). Significance of Nanotechnology in Food Industry. *APCBEE Procedia*, *8*, 109–113. doi:10.1016/j.apcbee.2014.03.010

Chen, C. J., Wang, R. z., Oliveira, R. G., & Hu, J. (2009). Adsorption Models and Structural Characterization for Activated Carbon Fibers. J. Shanghai Jiaotong Univ. (Sci.), 14(1), 35-39.

Cheng, K.-C., Catchmark, J., & Demirci, A. (2009). Enhanced production of bacterial cellulose by using a biofilm reactor and its material property analysis. *Journal of Biological Engineering*, *3*(1), 1–12. doi:10.1186/1754-1611-3-12 PMID:19630969

Cheng, T. W., Lee, M. L., Ko, M. S., Ueng, T. H., & Yang, S. F. (2012). The Heavy Metal Adsorption Characteristics on Metakaolin-Based Geopolymer. *Applied Clay Science*, *56*, 90–96. doi:10.1016/j.clay.2011.11.027

Chen, H., Yuan, L., Song, W., Wu, Z., & Li, D. (2008). Biocompatible polymer materials. Role of protein-surface interactions. *Progress in Polymer Science*, *33*(11), 1059–1065. doi:10.1016/j.progpolymsci.2008.07.006

Chiaoprakobkij, N., Sanchavanakit, N., Subbalekha, K., Pavasant, P., & Phisalaphong, M. (2011). Characterization and biocompatibility of bacterial cellulose/alginate composite sponges with human keraticytes and gingival fibroblast. *Carbohydrate Polymers*, *85*(3), 548–553. doi:10.1016/j.carbpol.2011.03.011

Chindaprasirt, P., Chareerat, T., & Sirivivatnanon, V. (2007). Workability and strength of coarse high calcium fly ash geopolymer. *Cement and Concrete Composites*, *29*(3), 224–229. doi:10.1016/j.cemconcomp.2006.11.002

Chindaprasirt, P., Jaturapitakkul, C., Chalee, W., & Rattanasak, U. (2009). Comparative study on the characteristics of fly ash and bottom ash geopolymers. *Waste Management (New York, N.Y.)*, *29*(2), 539–543. doi:10.1016/j.wasman.2008.06.023 PMID:18715775

Chindaprasirt, P., Rattanasak, U., & Taebuanhuad, S. (2013). Resistance to acid and sulfate solutions of microwave-assisted high calcium fly ash geopolymer. *Materials and Structures*, *46*(3), 375–381. doi:10.1617/s11527-012-9907-1

Chiulan, I., Mihaela Panaitescu, D., Nicoleta Frone, A., Teodorescu, M., Andi Nicolae, C., Căşărică, A., & Sălăgeanu, A. et al. (2016). Biocompatible polyhydroxyalkanoates/bacterial cellulose composites: Preparation, characterization, and in vitro evaluation. *Journal of Biomedical Materials Research. Part A*, *104*(10), 2576–2584. doi:10.1002/jbm.a.35800 PMID:27242044

Choi, S.-S. (2001). Improvement of properties of silica-filled styrene–butadiene rubber compounds using acrylonitrile–butadiene rubber. *Journal of Applied Polymer Science*, *79*(6), 1127–1133. doi:10.1002/1097-4628(20010207)79:6<1127::AID-APP170>3.0.CO;2-8

Chojnacka, K. (2005). Biosorption of Cr(III) ions by eggshellsOriginal Research Article. *Journal of Hazardous Materials*, *121*(1-3), 167–173. doi:10.1016/j.jhazmat.2005.02.004 PMID:15885418

Chruściel, J. J., & Leśniak, E. (2012). Thermoplastic Elastomers. *InTech Journal.*

Chuayjuljit, S., & Luecha, W. (2011). XSBR/NR Rubber Blends Filled with Polystyrene-Encapsulated Nanosilica Prepared by in situ Differential Microemulsion Polymerization. *Journal of Elastomers and Plastics*. doi: 10.1177/0095244311405001

Chuayjuljit, S., & Boonmahitthisud, A. (2010). Natural rubber nanocomposites using polystyrene-encapsulated nanosilica prepared by differential microemulsion polymerization. *Applied Surface Science*, *256*(23), 7211–7216. doi:10.1016/j.apsusc.2010.05.053

Chuayjuljit, S., Sukasem, N., & Boonmahitthisud, A. (2014). Effects of Silica, Poly(methyl methacrylate) and Poly(methyl methacrylate)-Grafted-Silica Nanoparticles on the Physical Properties of Plasticized-Poly(vinyl chloride). *Polymer-Plastics Technology and Engineering*, *53*(2), 116–122. doi:10.1080/03602559.2013.843695

Chumeka, W., Pasetto, P., Pilard, J.-F., & Tanrattanakul, V. (2014). Bio-based triblock copolymers from natural rubber and poly(lactic acid): Synthesis and application in polymer blending. *Polymer*, *55*(17), 4478–4487. doi:10.1016/j.polymer.2014.06.091

Chung, K.-H., Jung, H.-Y., Lee, Y.-W., & Lee, K.-Y. (2010). Preparation of TiO2-loaded nanocapsules and their sun protection behaviors. *Journal of Industrial and Engineering Chemistry*, *16*(2), 261–266. doi:10.1016/j.jiec.2010.01.039

Cioffi, R., Maffucci, L., & Santoro, L. (2003). Optimization of geopolymer synthesis by calcination and polycondensation of a kaolinitic residue. *Resources, Conservation and Recycling*, *40*(1), 27–38. doi:10.1016/S0921-3449(03)00023-5

Ciolacu, D., Oprea, A. M., Anghel, N., Cazacu, G., & Cazacu, M. (2012). New cellulose-lignin hydrogels and their application in controlled release of polyphenols. *Materials Science and Engineering C*, *32*(3), 452–463. doi:10.1016/j.msec.2011.11.018

Clarey, M., Edwards, J., Tsipursky, S. J., Beall, G. W., & Eisenhour, D. D. (2000), Method of manufacturing polymer-grade clay for use in nanocomposites, US patent 6,050,509.

Cocalia, V. A., Jensen, M. P., Holbrey, J. D., Spear, S. K., Stepinski, D. C., & Rogers, R. D. (2005). Identical extraction behavior and coordination of trivalent or hexavalent f-element cations using ionic liquid and molecular solvents. *Dalton Transactions (Cambridge, England)*, (11): 1966–1971. doi:10.1039/b502016f PMID:15909044

Colonetti, E., Hobold Kammer, E., & De Noni Junior, A. (2014). Chemically-bonded phosphate ceramics obtained from aluminum anodizing waste for use as coatings. *Ceramics International*, *40*(99, Part A), 14431–14438. doi:10.1016/j.ceramint.2014.06.039

Comini, E., Baratto, C., Concina, I., Faglia, G., Falasconi, M., Ferroni, M., & Sberveglieri, G. et al. (2013). Metal oxide nanoscience and nanotechnology for chemical sensors. *Sensors and Actuators. B, Chemical*, *179*, 3–20. doi:10.1016/j.snb.2012.10.027

Comrie, D. C., & Kriven, W. M. (2006). *Composite Cold Ceramic Geopolymer in a Refractory Application. In Advances in Ceramic Matrix Composites IX* (pp. 211–225). John Wiley & Sons, Inc.

Correa-Pacheco, Z. N., Jiménez-Pérez, J. L., Sabino, M. A., Cruz-Orea, A., & Loaiza, M. (2015). Photothermal and morphological characterization of PLA/PCL polymer blends. *Applied Physics. A, Materials Science & Processing*, *120*(4), 1323–1329. doi:10.1007/s00339-015-9309-7

Cotton, J., Pometto, A., & Gvozdenovic-Jeremic, J. (2001). Continuous lactic acid fermentation using a plastic composite support biofilm reactor. *Applied Microbiology and Biotechnology*, *57*(5-6), 626–630. doi:10.1007/s002530100820 PMID:11778870

Crini, G. (2005). Recent developments in polysaccharide-based materials used as adsorbents in wastewater treatment. *Progress in Polymer Science*, *30*(1), 38–70. doi:10.1016/j.progpolymsci.2004.11.002

Crini, G., & Badot, P. M. (2008). Application of chitosan, a natural aminopolysaccharide, for dye removal from aqueous solutions by adsorption processes using batch studies: A review of recent literature. *Progress in Polymer Science*, *33*(4), 399–447. doi:10.1016/j.progpolymsci.2007.11.001

Cristóbal, A. G. S., Castelló, R., Luengo, M. A. M., & Vizcayno, C. (2010). Zeolites prepared from calcined and mechanically modified kaolins A comparative study. *Applied Clay Science*, *49*(3), 239–246. doi:10.1016/j.clay.2010.05.012

Critoph, R. E. (1996). Evaluation of alternative refrigerant-adsorbent pairs for refrigeration cycles. *Applied Thermal Engineering*, *16*(11), 891–900. doi:10.1016/1359-4311(96)00008-7

Critoph, R. E. (1998). Forced Convection Adsorption Cycles. *Applied Thermal Engineering*, *18*(9-10), 799–807. doi:10.1016/S1359-4311(97)00110-5

Critoph, R. E. (1999). Forced Convection Adsorption Cycle with Packed Bed Heat Regeneration. *International Journal of Refrigeration*, *22*(1), 38–46. doi:10.1016/S0140-7007(97)00036-4

Crowhurst, L., Mawdsley, P. R., Perez-Arlandis, J. M., Salter, P. A., & Welton, T. (2003). Solvent-solute interactions in ionic liquids. *Physical Chemistry Chemical Physics*, *5*(13), 2790–2794. doi:10.1039/B303095D

Cui, Q., Tao, G., Chen, H., Guo, X., & Yao, H. (2005). Environmentally benign working pairs for adsorption refrigeration. *Energy*, *30*(2-4), 261–271. doi:10.1016/j.energy.2004.05.005

Czaja, W., Krystynowicz, A., Bielecki, S., & Brown, R. M. Jr. (2006). Microbial cellulose - The natural power to heal wounds. *Biomaterials*, *27*(2), 145–151. doi:10.1016/j.biomaterials.2005.07.035 PMID:16099034

Daengprok, W., Garnjanagoonchorn, W., & Mine, Y. (2002). Fermented pork sausage fortified with commercial or hen eggshell calcium lactate. *Meat Science*, *62*(2), 199–204. doi:10.1016/S0309-1740(01)00247-9 PMID:22061412

Dahman, Y. (2009). Nanostructured biomaterials and biocomposites from bacterial Cellulose nanofibers. *Journal of Nanoscience and Nanotechnology*, *9*(9), 5105–5122. doi:10.1166/jnn.2009.1466 PMID:19928189

Dai, L., Soundarrajan, P., & Kim, T. (2002). Sensors and sensor arrays based on conjugated polymers and carbon nanotubes. *Pure and Applied Chemistry*, *74*(9), 1753–1772. doi:10.1351/pac200274091753

Dai, S., Ju, Y. H., & Barnes, C. E. (1999). Solvent extraction of strontium nitrate by a crown ether using room-temperature ionic liquids. *Journal of the Chemical Society, Dalton Transactions: Inorganic Chemistry*, (8): 1201–1202. doi:10.1039/a809672d

Daou, K., Wang, R. Z., & Xia, Z. Z. (2006). Development of a new synthesized adsorbent for refrigeration and air conditioning applications. *Applied Thermal Engineering*, *26*(1), 56–65. doi:10.1016/j.applthermaleng.2005.04.024

Daou, K., Wang, R. Z., Yang, G. Z., & Xia, Z. Z. (2008). Theoretical comparison of the refrigerating performances of a CaCl2 impregnated composite adsorbent to those of the host silica gel. *International Journal of Thermal Sciences*, *47*(1), 68–75. doi:10.1016/j.ijthermalsci.2007.01.003

Davidovits, J. (1982). *Mineral polymers and methods of making them.* Google Patents.

Davidovits, J. (1984). *Synthetic mineral polymer compound of the silicoaluminates family and preparation process.* Google Patents.

Davidovits, J. (1994). Properties of geopolymer cements. *First international conference on alkaline cements and concretes, 1*, 131-149.

Davidovits, J. (2008). *Geopolymer Chemistry & Applications.* Institut Géopolymère 16 rue Galilée F-02100 Saint-Quentin France.

Davidovits, J. (1991). Geopolymers - Inorganic polymeric new materials. *Journal of Thermal Analysis*, *37*(8), 1633–1656. doi:10.1007/BF01912193

Dawoud, B., Sohel, M. I., Freni, A., Vasta, S., & Restuccia, G. (2011). On the effective thermal conductivity of wetted zeolite under the working conditions of an adsorption chiller. *Applied Thermal Engineering*, *31*(14-15), 2241–2246. doi:10.1016/j.applthermaleng.2011.03.016

De Mestrado, D. (2010). *Technical and Economic Assessment of Medium Sized Solar-Assisted Air-Conditioning in Brazil*. Academic Press. http://www.urb.puc-rio.br/dissertacao/dissertacao_till.pdf

de Oliveira, A. G., da Silva, A. L. N., de Sousa, A. M. F., Leite, M. C. A. M., Jandorno, J. C., & Escócio, V. A. (2016). Composites based on green high-density polyethylene, polylactide and nanosized calcium carbonate: Effect of the processing parameter and blend composition. *Materials Chemistry and Physics*, *181*, 344–351. doi:10.1016/j.matchemphys.2016.06.068

de Oliveira, M. A., da Silva, P. M. L. C., Alves, M. G., de Oliveira, C. P., & dos Santos, M. A. (2005). Encapsulation of TiO2 by emulsion polymerization with methyl metacrylate (MMA). *Polymer Bulletin*, *55*(6), 477–484. doi:10.1007/s00289-005-0457-3

Deguchi, Y., Kohno, Y., & Ohno, H. (2015). A fine tuning of LCST-type phase transition of poly(ionic liquid)s in water. *Chemistry Letters*, *44*(3), 238–240. doi:10.1246/cl.141016

Del Sesto, R. E., Corley, C., Robertson, A., & Wilkes, J. S. (2005). Tetraalkylphosphonium-Based Ionic Liquids. *Journal of Organometallic Chemistry*, *690*(10), 2536–2542. doi:10.1016/j.jorganchem.2004.09.060

Demir, H., Mobedi, M., & Ulku, S. (2011). Microcalorimetric investigation of water vapor adsorption on silica gel. *Journal of Thermal Analysis and Calorimetry*, *105*(1), 375–382. doi:10.1007/s10973-011-1395-y

Deng, Y., & Thomas, N. L. (2015). Blending poly(butylene succinate) with poly(lactic acid): Ductility and phase inversion effects. *European Polymer Journal*, *71*, 534–546. doi:10.1016/j.eurpolymj.2015.08.029

Deng, Y., Wang, L., Yang, W., Fu, S., & Elaassari, A. (2003). Preparation of magnetic polymeric particles via inverse microemulsion polymerization process. *Journal of Magnetism and Magnetic Materials*, *257*(1), 69–78. doi:10.1016/S0304-8853(02)00987-3

Detphan, S., & Chindaprasirt, P. (2009). Preparation of fly ash and rice husk ash geopolymer. *International Journal of Minerals. Metals and Materials*, *16*(6), 720–726. doi:10.1016/S1674-4799(10)60019-2

Diallo, M.S., Johnson, J., & Balogh, L. (2000). *Removal of Toxic Metal Ions from Contaminated Water by Dendrimer Enhanced Ultrafiltration.* National Science Foundation.

Diallo, M. S., Christie, S., Swaminathan, P., Johnson, J. H. Jr, & Goddard, W. A. III. (2005). Dendrimer Enhanced Ultrafiltration. 1. Recovery of Cu(II) from Aqueous Solutions Using PAMAM Dendrimers with Ethylene Diamine Core and Terminal NH_2 Groups. *Environmental Science & Technology*, *39*(5), 1366–1377. doi:10.1021/es048961r PMID:15787379

Dietz, M. L., & Stepinski, D. C. (2005). A Ternary Mechanism for the Facilitated Transfer of Metal Ions into Room-Temperature Ionic Liquids (RTILs): Implications for the Greenness of RTILs as Extraction Solvents. *Green Chemistry*, *7*(10), 747–750. doi:10.1039/b508604c

Dil, E. J., Carreau, P. J., & Favis, B. D. (2015). Morphology, miscibility and continuity development in poly(lactic acid)/poly(butylene adipate-co-terephthalate) blends. *Polymer*, *68*, 202–212. doi:10.1016/j.polymer.2015.05.012

Ding, J., Zhou, D., Spinks, G., Wallace, G., Forsyth, S., Forsyth, M., & MacFarlane, D. (2003). Use of ionic liquids as electrolytes in electromechanical actuator systems based on inherently conducting polymers. *Chemistry of Materials, 15*(12), 2392–2398. doi:10.1021/cm020918k

Ding, X., Zhao, J., Liu, Y., Zhang, H., & Wang, Z. (2004). Silica nanoparticles encapsulated by polystyrene via surface grafting and in situ emulsion polymerization. *Materials Letters, 58*(25), 3126–3130. doi:10.1016/j.matlet.2004.06.003

Don T. N., Huyen P.T, & Hong N.K.D. (2013). Synthesis, characterization and application of nanozeolite NaX from Vietnamese kaolin. *Adv. Nat. Sci.: Nanosci. Nanotechnol., 4.*

Dong, H., Chen, Y., Sheng, G., Li, J., Cao, J., Li, Z., & Li, Y. (2016). The roles of a pillared bentonite on enhancing Se(VI) removal by ZVI and the influence of co-existing solutes in groundwater. *Journal of Hazardous Materials, 304*, 306–312. doi:10.1016/j.jhazmat.2015.10.072 PMID:26561754

Dong, K., Zhang, S., Wang, D., & Yao, X. (2006). Hydrogen bonds in imidazolium ionic liquids. *The Journal of Physical Chemistry A, 110*(31), 9775–9782. doi:10.1021/jp054054c PMID:16884211

Douss, N., & Meunier, F. (1989). Experimental study of cascading adsorption cycles. *Chemical Engineering Science, 44*(2), 225–235. doi:10.1016/0009-2509(89)85060-2

Dresselhaus, M. (1996). Carbon connections promise nanoelectronics. *Physics World, 9*(5), 18–19. doi:10.1088/2058-7058/9/5/18

Dreyer, D. R., Park, S., Bielawski, C. W., & Ruoff, R. S. (2010). The chemistry of graphene oxide. *Chemical Society Reviews, 39*(1), 228–240. doi:10.1039/B917103G PMID:20023850

Duan, P., Yan, C., & Zhou, W. (2016). Influence of partial replacement of fly ash by metakaolin on mechanical properties and microstructure of fly ash geopolymer paste exposed to sulfate attack. *Ceramics International, 42*(2, Part B), 3504–3517. doi:10.1016/j.ceramint.2015.10.154

Duan, P., Yan, C., Zhou, W., & Luo, W. (2016). Fresh properties, mechanical strength and microstructure of fly ash geopolymer paste reinforced with sawdust. *Construction & Building Materials, 111*, 600–610. doi:10.1016/j.conbuildmat.2016.02.091

Du, C., Yeh, J., & Pan, N. (2005). High power density supercapacitors using locally aligned carbon nanotube electrodes. *Nanotechnology, 16*(4), 350–353. doi:10.1088/0957-4484/16/4/003

Du, F., Fischer, J. E., & Winey, K. I. (2003). Coagulation method for preparing single-walled carbon nanotube/poly(methyl methacrylate) composites and their modulus, electrical conductivity, and thermal stability. *Journal of Polymer Science. Part B, Polymer Physics, 41*(17), 3333–3338. doi:10.1002/polb.10701

Du, J., & Sun, H. (2014). Polymer/TiO2 Hybrid Vesicles for Excellent UV Screening and Effective Encapsulation of Antioxidant Agents. *ACS Applied Materials & Interfaces, 6*(16), 13535–13541. doi:10.1021/am502663j PMID:25059274

Duncan, T. V. (2011). Applications of nanotechnology in food packaging and food safety: Barrier materials, antimicrobials and sensors. *Journal of Colloid and Interface Science, 363*(1), 1–24. doi:10.1016/j.jcis.2011.07.017 PMID:21824625

Dupont, A. L., Egasse, C., Morin, A., & Vasseur, F. (2007). Comprehensive characterization of cellulose and lignocellulose degradation products in aged papers: Capillary zone electrophoresis of low molar mass organic acids, carbohydrates, and aromatic lignin derivatives. *Carbohydrate Polymers, 68*(1), 1–8. doi:10.1016/j.carbpol.2006.07.005

Dupont, J., Consorti, C. S., Suarez, P. A. Z., & Souza, R. F. (2002). Preparation of 1-butyl-3-methylimidazolium based room temperature ionic liquids. *Organic Syntheses, 79*, 236–239. doi:10.15227/orgsyn.079.0236

Duxson, P., Fernández-Jiménez, A., Lukey, G. C., Palomo, A., Provis, J. L., & Van Deventer, J. S. J. (2007). Geopolymer technology: The current state of the art. *Journal of Materials Science*, *42*(9), 2917–2933. doi:10.1007/s10853-006-0637-z

Dymek, C. J. Jr, Grossie, D. A., Fratini, A. V., & Wade Adams, W. (1989). Evidence for the presence of hydrogen-bonded ion-ion interactions in the molten salt precursor, 1-methyl-3-ethylimidazolium chloride. *Journal of Molecular Structure*, *213*, 25–34. doi:10.1016/0022-2860(89)85103-8

Ebbesen, T. W., & Ajayan, P. M. (1992). Large-scale synthesis of carbon nanotubes. *Nature*, *358*(6383), 220–222. doi:10.1038/358220a0

Ebelman, M. (1845). Untersuchungen uber die Verbindung der Borsaure und Kieselsaure mit Aether. *Ann. Chem*, *57*(3), 319–355. doi:10.1002/jlac.18460570303

Ebnesajjsd, S. (2013). *Handbook of Biopolymers and Biodegradable Plastics: Properties, Processing and Applications*. MA, USA: Elsevier Inc.

Edgar, K. J., Buchanan, C. M., Debenham, J. S., Rundquist, P. A., Seiler, B. D., Shelton, D., & Tindall, D. (2001). Advances in cellulose ester performance and application. *Progress in Polymer Science*, *26*(9), 1605–1611. doi:10.1016/S0079-6700(01)00027-2

Ehrlich, H., Hanke, T., Simon, P., Born, R., Fischer, C., Frolov, A., & Worch, H. et al. (2010). Carboxymethylation of the fibrillar collagen with respect to formation of hydroxyapatite. *Journal of Biomedical Materials Research. Part B, Applied Biomaterials*, *92B*(2), 542–551. PMID:19957363

EL-Mekkawi, D. M., Ibrahim, F. A., & Selim, M. M.. (2015). Stability and thermal transformation studies of Zn^{2+}- and Fe^{2+} - loaded zeolite Y prepared from Egyptian kaolin. *Solid State Sciences*, *48*, 294–299. doi:10.1016/j.solidstatesciences.2015.09.002

EL-Mekkawi, D. M., & Selim, M. M.. (2014). Removal of Pb^{2+} from water by using Na-Y zeolites prepared from Egyptian kaolins collected from different sources. *Journal of Environmental Chemical Engineering*, *2*(1), 723–730. doi:10.1016/j.jece.2013.11.014

Elsayed, A., Al-Dadah, R. K., Mahmoud, S., & Kaialy, W. (2014). Investigation of activated carbon /Ethanol for low temperature adsorption cooling. *International Journal of Green Energy*.

Elsayed, A., Al-Dadah, R., Mahmoud, S., Elshaer, A., & Bowen, J. (2014). Investigation of efficient adsorbents using transient analysis of adsorption chillers with Simulink. *Proceedings of the International Conference on Heat Transfer and Fluid Flow*.

Elsayed, A., & Elsayed, E. AL-Dadaha, R., Mahmoud, S., Elshaer,A., &Kaialy, W.(2016). Thermal energy storage using metal–organic framework materials. *Applied Energy*. doi:10.1016/j.apenergy.2016.03.113

El-Sharkawy, I. I., Hassan, M., Saha, B. B., Koyama, S., & Nasr, M. M. (2009). Study on adsorption of methanol onto carbon based adsorbents. *International Journal of Refrigeration*, *32*(7), 1579–1586. doi:10.1016/j.ijrefrig.2009.06.011

El-Sharkawy, I. I., Kuwahara, K., Saha, B. B., Koyama, S., & Ng, K. C. (2006). Experimental investigation of activated carbon fibers/ethanol pairs for adsorption cooling system application. *Applied Thermal Engineering*, *26*(8-9), 859–865. doi:10.1016/j.applthermaleng.2005.10.010

El-Sharkawy, I. I., Saha, B. B., Koyama, S., He, J., Yap, K. C., & Ng, C. (2008). Experimental investigation on activated carbon–ethanol pair for solar powered adsorption cooling applications. *International Journal of Refrigeration*, *31*(8), 1407–1413. doi:10.1016/j.ijrefrig.2008.03.012

El-Sharkawy, I. I., Uddin, K., Miyazaki, T., Saha, B. B., Koyama, S., Miyawaki, J., & Yoon, S. H. (2014). Adsorption of ethanol onto parent and surface treated activated carbon powders. *International Journal of Heat and Mass Transfer, 73*, 445–455. doi:10.1016/j.ijheatmasstransfer.2014.02.046

Emmenegger, Ch., Mauron, Ph., Sudan, P., Wenger, P., Hermann, V., Gallay, R., & Züttel, A. (2003). Investigation of electrochemical double-layer (ECDL) capacitors electrodes based on carbon nanotubes and activated carbon materials. *Journal of Power Sources, 124*(1), 321–329. doi:10.1016/S0378-7753(03)00590-1

Engler, A. J., Sen, S., Sweeney, H. L., & Discher, D. E. (2006). Matrix Elasticity Directs Stem Cell Lineage Specification. *Cell, 126*(4), 677–689. doi:10.1016/j.cell.2006.06.044 PMID:16923388

Environmental Protection Agency (EPA). (1990). *Green chemistry*. Retrieved from http://www.epa.gov/greenchemistry/basics-green-chemistry

EPA. (1981). *Summary Report: Control and Treatment Technology for the Metal Finishing Industry; Ion Exchange.* EPA 625/8-81-007.

Eun, T. H., Song, H. K., Han, J. H., Lee, K.-H., & Kim, J. N. (2000). Enhancement of heat and mass transfer in silica gel-expanded graphite composite blocks for adsorption heat pumps: Part I. characterization of the composite blocks. *International Journal of Refrigeration, 23*, 64–73. doi:10.1016/S0140-7007(99)00035-3

European Commission. (2008). *European Union Risk Assessment Report, Zinc metal.* DOI 10.2788/40041

Fakhrul-Razi, A., Atieh, M. A., Girun, N., Chuah, T. G., El-Sadig, M., & Biak, D. R. A. (2006). Effect of multi-wall carbon nanotubes on the mechanical properties of natural rubber. *Composite Structures, 75*(1-4), 496–500. doi:10.1016/j.compstruct.2006.04.035

Farber, L. (1935). Application of pervaporation. *Science, 82*(2120), 158–158. doi:10.1126/science.82.2120.158 PMID:17811950

Farooq, J., Kozinski, A., Khan, M. A., & Athar, M. (2010). Biosorption of heavy metal ions using wheat based biosorbents. A review of the recent literature. *Bioresource Technology, 101*(14), 5043–5053. doi:10.1016/j.biortech.2010.02.030 PMID:20223652

Faucheux, N., Schweiss, R., Lützow, K., Werner, C., & Groth, T. (2004). Self-assembled monolayers with different terminating groups as model substrates for cell adhesion studies. *Biomaterials, 25*(14), 2721–2730. doi:10.1016/j.biomaterials.2003.09.069 PMID:14962551

Favi, P. M., Benson, R. S., Neilsen, N. R., Hammonds, R. L., Bates, C. C., Stephens, C. P., & Dhar, M. S. (2013). Cell proliferation, viability, and in vitro differentiation of equine mesenchymal stem cells seeded on bacterial cellulose hydrogel scaffolds. *Materials Science and Engineering C, 33*(4), 1935–1944. doi:10.1016/j.msec.2012.12.100 PMID:23498215

Feng, H., Li, C., & Shan, H. (2009). Effect of calcination temperature of kaolin microspheres on the in situ synthesis of ZSM-5. *Catalysis Letters, 129*(1-2), 71–78. doi:10.1007/s10562-008-9794-9

Ferri, J. M., Fenollar, O., Jorda-Vilaplana, A., García-Sanoguera, D., & Balart, R. (2016a). Effect of miscibility on mechanical and thermal properties of poly(lactic acid)/polycaprolactone blends. *Polymer International, 65*(4), 453–463. doi:10.1002/pi.5079

Ferri, J. M., Garcia, D. G., Sánchez-Nacher, L., Fenollar, O., & Balart, R. (2016b). The effect of maleinized linseed oil (MLO) on mechanical performance of poly(lactic acid)-thermoplastic starch (PLA-TPS) blends. *Carbohydrate Polymers, 147*, 60–68. doi:10.1016/j.carbpol.2016.03.082 PMID:27178909

Feuser, P. E., Fernandes, A. C., Nele, M., Viegas, A. C., Ricci-Junior, E., Tedesco, A. C., & de Araújo, P. H. H. et al. (2015). Simultaneous encapsulation of magnetic nanoparticles and zinc phthalocyanine in poly(methyl methacrylate) nanoparticles by miniemulsion polymerization and in vitro studies. *Colloids and Surfaces. B, Biointerfaces*, *135*, 357–364. doi:10.1016/j.colsurfb.2015.07.067 PMID:26263221

Filho, G. R., Ribeiro, S. D., & Meireles, C. S. (2011). Release of doxycycline through cellulose acetate symmetric and asymmetric membrane produced from recycled agroindustrial residue: Sugarcane bagasse. *Industrial Crops and Products*, *33*(3), 566–576. doi:10.1016/j.indcrop.2010.10.037

Fink, J. K. (2013). *Silicones Reactive Polymers Fundamentals and Applications* (2nd ed.). Oxford, UK: William Andrew Publishing. doi:10.1016/B978-1-4557-3149-7.00008-5

Flower, D. J. M., & Sanjayan, J. G. (2007). Green house gas emissions due to concrete manufacture. *The International Journal of Life Cycle Assessment*, *12*(5), 282–288. doi:10.1007/s11367-007-0327-3

Foletto, E. L., Paz, D. S., & Gündel, A. (2013). Acid-activation assisted by microwave of a Brazilian bentonite and its activity in the bleaching of soybean oil. *Applied Clay Science*, *83–84*, 63–67. doi:10.1016/j.clay.2013.08.017

Fontana, J. D., Franco, V. C., De Souza, S. J., Lyra, I. N., & De Souza, A. M. (1991). Nature of plant stimulators in the production of Acetobacter xylinum (tea fungus) biofilm used in skin therapy. *Applied Biochemistry and Biotechnology*, *28-29*(1), 341–351. doi:10.1007/BF02922613 PMID:1929372

Fontana, J., De Souza, A., Fontana, C., Torriani, I., Moreschi, J., Gallotti, B., & Farah, L. et al. (1990). Acetobacter cellulose pellicle as a temporary skin substitute. *Applied Biochemistry and Biotechnology*, *24-25*(1), 253–264. doi:10.1007/BF02920250 PMID:2353811

Forouharshad, M., Gardella, L., Furfaro, D., Galimberti, M., & Monticelli, O. (2015). A low-environmental-impact approach for novel biocomposites based on PLLA/PCL blends and high surface area graphite. *European Polymer Journal*, *70*, 28–36. doi:10.1016/j.eurpolymj.2015.06.016

Franco, F., Benítez-Guerrero, M., Gonzalez-Triviño, I., Pérez-Recuerda, R., Assiego, C., Cifuentes-Melchor, J., & Pascual-Cosp, J. (2016). Low-cost aluminum and iron oxides supported on dioctahedral and trioctahedral smectites: A comparative study of the effectiveness on the heavy metal adsorption from water. *Applied Clay Science*, *119*, 321–332. doi:10.1016/j.clay.2015.10.035

Freni, A., Tokarev, M. M., Restuccia, G., Okunev, A. G., & Aristov, Y. I. (2002). Thermal conductivity of selective water sorbents under the working conditions of a sorption chiller. *Applied Thermal Engineering*, *22*(14), 1631–1642. doi:10.1016/S1359-4311(02)00076-5

Fujioka, K., & Suzuki, H. (2013). Thermophysical properties and reaction rate of composite reactant of calcium chloride and expanded graphite. *Applied Thermal Engineering*, *50*(2), 1627–1632. doi:10.1016/j.applthermaleng.2011.08.024

Füldner, G., & Schnable, L. (2008). Non-isothermal kinetics of water adsorption in compact adsorbent layers on a metal support. *Comsol conference proceeding*. Retrieved from http://www.comsol.com/papers/5118/

Gaballah, I., & Kilbertus, G. (1998). Recovery of heavy metal ions through decontamination of synthetic solutions and industrial effluents using modified barks. *Journal of Geochemical Exploration*, *62*(1-3), 241–286. doi:10.1016/S0375-6742(97)00068-X

Galarneau, A., Barodawalla, A., & Pinnavaia, T. J. (1994). Porous clay heterostructures formed by gallery-templated synthesis. *Nature*, *374*(6522), 529–531. doi:10.1038/374529a0

Galimberti, M. (2001). *Rubber-clay nanocomposites*. Hoboken, NJ: John Wiley & Sons, Inc.

Gallucci, D. (2015). *Southern Thailand hit by weak rubber prices*. Retrieved 4 February, 2016, from http://www.ft.com/intl/cms/s/3/ded417fe-2ec1-11e5-8873-775ba7c2ea3d.html#axzz3zrcxpTeq

Gandia, L. M., Vicente, M. A., & Gil, A. (2000). Preparation and characterization of manganese oxide catalysts supported on alumina and zirconia-pillared clays. *Applied Catalysis A, General*, *196*(2), 281–292. doi:10.1016/S0926-860X(99)00479-2

Ganim, S. & Tran, L. (2016). *How tap water became toxic in Flint, Michigan*. CNN.

Ganter, M., Gronski, W., Reichart, P., & Mulhaupt, R. (2001). Rubber Nanocomposites: Morphology and mechanical properties of BR and SBR vulcanizates reinforced by organophilic layered silicates. *Rubber Chemistry and Technology*, *74*(2), 221–235. doi:10.5254/1.3544946

Gao, W. H., Chen, K. F., Yang, R. D., Yang, F., & Han, W. J. (2011). Properties of bacterial cellulose and its influence on the physical properties of paper. *BioResources*, *6*(1), 144–153.

Gardea-Torresdey, J. L., Tang, L., & Salvador, J. M. (1995). Copper Adosoption by Sphagnum Peat Moss and Its Different Humic Fractions.*Proceedings of the 10th Annual Conference on Hazardous Waste Research*, (pp. 249-260).

Gavillon, R., & Budtova, T. (2008). Aerocellulose: New highly porous cellulose prepared from cellulose-NaOH aqueous solutions. *Biomacromolecules*, *9*(1), 269–277. doi:10.1021/bm700972k PMID:18085745

Geckeler, K. E., & Volchek, K. (1996). Removal of Hazardous Substances from Water Using Ultrafiltration in Conjunction with Soluble Polymers. *Environmental Science & Technology*, *30*(3), 725–734. doi:10.1021/es950326l

Gelin, K., Bodin, A., Gatenholm, P., Mihranyan, A., Edwards, K., & Stromme, M. (2007). Characterization of water in bacterial cellulose using dielectric spectroscopy and electron microscopy. *Polymer*, *48*(26), 7623–7631. doi:10.1016/j.polymer.2007.10.039

Gemmei-Ide, M., Motonaga, T., & Kitano, H. (2006). State of Irremovable Water in Solid Polymer Films Examined by Fourier Transform Infrared Spectroscopy I: Poly(Ethylene Glycol) Dimethyl Ether. *Langmuir*, *22*(6), 2422–2425. doi:10.1021/la053117f PMID:16519432

George, J., Ramana, K. V., Sabapathy, S. N., & Bawa, A. S. (2005). Physico-Mechanical Properties of Chemically Treated Bacterial (Acetobacter xylinum) Cellulose Membrane. *World Journal of Microbiology & Biotechnology*, *21*(8), 1323–1327. doi:10.1007/s11274-005-3574-0

Gerente, C., Lee, V. K. C., Cloirec, P. L., & McKay, G. (2007). Application of Chitosan for the Removal of Metals From Wastewaters by Adsorption—Mechanisms and Models Review. *Critical Reviews in Environmental Science and Technology*, *37*(1), 41–127. doi:10.1080/10643380600729089

Gericke, M., Trygg, J., & Fardim, P. (2013). Functional cellulose beads: Preparation, characterization, and applications. *Chemical Reviews*, *113*(7), 4812–4836. doi:10.1021/cr300242j PMID:23540980

Gibson, L. J. (2005). Biomechanics of cellular solids. *Journal of Biomechanics*, *38*(3), 377–399. doi:10.1016/j.jbiomech.2004.09.027 PMID:15652536

Gil, A., Korili, S. A., Trujillano, R., & Vicente, M. A. (2011). A review on characterization of pillared clays by specific techniques. *Applied Clay Science*, *53*(2), 97–105. doi:10.1016/j.clay.2010.09.018

Gil, A., Korili, S. A., Trujillano, R., & Vicente, M. A. (Eds.). (2010). *Pillared Clays and Related Catalysts*. Springer. doi:10.1007/978-1-4419-6670-4

Giridhar, P., Venkatesan, K. A., Srinivasan, T. G., & Vasudewa Rao, P. R. (2005). Extraction of Uranium(VI) from Nitric Acid Medium by 1.1 M Tri-n-butylphosphonate in Ionic Liquid Diluent. *Journal of Radioanalytical and Nuclear Chemistry*, *265*(1), 31–38. doi:10.1007/s10967-005-0785-7

Goldemberg, J. (2007). Ethanol for sustainable energy future. *Science*, *315*(5813), 808–810. doi:10.1126/science.1137013 PMID:17289989

Gong, X., Liu, J., Baskaran, S., Voise, R. D., & Young, J. S. (2000). Surfactant-assisted processing of carbon nanotube/polymer composites. *Chemistry of Materials*, *12*(4), 1049–1052. doi:10.1021/cm9906396

Gonzalez, M. I., & Rodriguez, L. R. (2007). Solar powered adsorption refrigerator with CPC collection system: Collector design and experimental test. *Energy Conversion and Management*, *48*(9), 2587–2594. doi:10.1016/j.enconman.2007.03.016

González, N., Custal, M. À., Lalaouna, S., Riba, J.-R., & Armelin, E. (2016). Improvement of dielectric properties of natural rubber by adding perovskite nanoparticles. *European Polymer Journal*, *75*, 210–222. doi:10.1016/j.eurpolymj.2015.12.023

Görhan, G., & Kürklü, G. (2014). The influence of the NaOH solution on the properties of the fly ash-based geopolymer mortar cured at different temperatures. *Composites. Part B, Engineering*, *58*(0), 371–377. doi:10.1016/j.compositesb.2013.10.082

Grothe, J., Kaskel, S., & Leuteritz, A. (2012). *8.08 - Nanocomposites and Hybrid Materials A2 - Möller, Krzysztof MatyjaszewskiMartin. In Polymer Science: A Comprehensive Reference* (pp. 177–209). Amsterdam: Elsevier.

Grundke, K., Boerner, M., & Jacobasch, H. J. (1991). Characterization of fillers and fibres by wetting and electrokinetic measurements. *Colloids and Surfaces*, *58*(1), 47–59. doi:10.1016/0166-6622(91)80197-V

Gruzdev, M. S., Ramenskaya, L. M., Chervonova, U. V., & Kumeev, R. S. (2009). Preparation of 1-butyl-3-methylimidazolium salts and study of their phase behavior and intramolecular intractions. *Russian Journal of General Chemistry*, *79*(8), 1720–1727. doi:10.1134/S1070363209080246

Gubb, T. A., Baranova, I., Allan, S. M., Fall, M. L., Shulman, H. S., & Kriven, W. M. (2011). *Microwave Enhanced Drying and Firing of Geopolymers. In Developments in Strategic Materials and Computational Design II* (pp. 35–43). John Wiley & Sons, Inc. doi:10.1002/9781118095393.ch4

Guibal, E. (2004). Interactions of metal ions with chitosan-based sorbents: A review. *Separation and Purification Technology*, *38*(1), 43–74. doi:10.1016/j.seppur.2003.10.004

Guilbert, S., Cuq, B., & Gontard, N. (1997). Recent innovations in edible and/or biodegradable packaging materials. *Food Additives and Contaminants*, *14*(6-7), 741–751. doi:10.1080/02652039709374585 PMID:9373537

Guimaraes, J. F., Frollini, E., da Silva, C. G., Wypych, F. K. G., & Satyanarayana, K. G. (2009). Characterization of banana, sugarcane bagasse and sponge gourd fibers of Brazil. *Industrial Crops and Products*, *30*(3), 407–415. doi:10.1016/j.indcrop.2009.07.013

Guo, B., Finne-Wistrand, A., & Albertsson, A.-C. (2010). Molecular Architecture of Electroactive and Biodegradable Copolymers Composed of Polylactide and Carboxyl-Capped Aniline Trimer. *Biomacromolecules*, *11*(4), 855–863. doi:10.1021/bm9011248 PMID:20201489

Gupta, S., Nayek, S., Saha, R. N., & Satpati, S. (2008). Assessment of heavy metal accumulation in macrophyte, agricultural soil, and crop plants adjacent to discharge zone of sponge iron factory. *Environmental Geology*, *55*(4), 731–739. doi:10.1007/s00254-007-1025-y

Gurgel, J. M., Oliveira, P. A., Marques Ailson, S., Oliveira, L. G., & Tavares, F. R. M. (2005). Experimental results for thermal conductivity of adsorbed natural gas on activated carbon., *17th European conference on thermophysical properties.*

Gwadera, M., Kupiec, K. (2011). *Adsorption cooling as an effective method of waste heat utilization*. Academic Press. https://suw.biblos.pk.edu.pl/resources/i4/i5/i4/i8/r4548/GwaderaM_AdsorptionCooling.pdf

Gyftopoulou, M. E., Millan, M., Bridgwater, A. V., Dugwell, D., Kandiyoti, R., & Hriljac, J. A. (2005). Pillared clays as catalysts for hydrocracking of heavy liquid fuels. *Applied Catalysis A, General*, *282*(1-2), 205–214. doi:10.1016/j.apcata.2004.12.012

Habib, K., Majid, M. A. B. A., & Sulaiman, S. A. B. (2013). Study on two stage activated carbon/HFC-134a based adsorption chiller.*4th International Conference on Energy and Environment*. doi:10.1088/1755-1315/16/1/012084

Habib, K., & Saha, B. (2013). Performance evaluation of solar driven activated carbon fiber-ethanol based adsorption cooling system in Malaysia. *Asian Journal of Scientific Research*, *6*, 146–156. doi:10.3923/ajsr.2013.146.156

Habib, K., Saha, B. B., & Koyama, S. (2014). Study of various adsorbent-refrigerant pairs for the application of solar driven adsorption cooling in tropical climates. *Applied Thermal Engineering*, *72*(2), 266–274. doi:10.1016/j.applthermaleng.2014.05.102

Habib, K., Saha, B. B., Rahman, K. A., Chakraborty, A., Koyama, S., & Ng, K. C. (2010). Experimental study on adsorption kinetics of activated carbon/R134a and activated carbon/R507A pairs. *International Journal of Refrigeration*, *33*(4), 706–713. doi:10.1016/j.ijrefrig.2010.01.006

Haimer, E., Wendland, M., Schlufter, K., Frankenfeld, K., Miethe, P., Potthast, A., & Liebner, F. et al. (2010). Loading of Bacterial Cellulose Aerogels with Bioactive Compounds by Antisolvent Precipitation with Supercritical Carbon Dioxide. *Macromolecular Symposia*, *294*(2), 64–74. doi:10.1002/masy.201000008

Hakim, R. N., & Ismail, H. (2009). Comparison of the effects of organoclay loadings on the curing and mechanical properties of organoclay-filled epoxidixed natural rubber nanocomposites and organoclay-filled natural rubber nanocomposites. *Journal of Applied Polymer Science*, *20*(2), 37–59.

Hallett, J. P., & Welton, T. (2011). Room-temperature ionic liquids: Solvents for synthesisand catalysis. 2. *Chemical Reviews*, *111*(5), 3508–3576. doi:10.1021/cr1003248 PMID:21469639

Hamad, K., Kaseem, M., Deri, F., & Ko, Y. G. (2016). Mechanical properties and compatibility of polylactic acid/polystyrene polymer blend. *Materials Letters*, *164*, 409–412. doi:10.1016/j.matlet.2015.11.029

Hamamoto, Y., Alam, K. C. A., Saha, B. B., Koyama, S., Akisawa, A., & Kashiwagi, T. (2006). Study on adsorption refrigeration cycle utilizing activated carbon fibers. Part 2. Cycle Performance evaluation. *International Journal of Refrigeration*, *29*(2), 315–327. doi:10.1016/j.ijrefrig.2005.06.001

Hamdy, M., Askalany, A., Harby, K., & Kora, N. (2015). *An overview on adsorption cooling systems powered by waste heat from internal combustion engine*. Academic Press. Renewable Sustainable Energy Rev., 51,1223–1234.

Hamisso, A. M., Lodi, A., Seffen, M., Finocchio, E., Botter, R., & Converti, A. (2010). Asorption of Cd (II) and Pb (II) from aqueous solution onto Agave Americana fibers. *Chemical Engineering Journal*, *159*(1-3), 67–76. doi:10.1016/j.cej.2010.02.036

Hanjitsuwan, S., Hunpratub, S., Thongbai, P., Maensiri, S., Sata, V., & Chindaprasirt, P. (2014). Effects of NaOH concentrations on physical and electrical properties of high calcium fly ash geopolymer paste. *Cement and Concrete Composites*, *45*, 9–14. doi:10.1016/j.cemconcomp.2013.09.012

Hao, X., Kaschta, J., Pan, Y., Liu, X., & Schubert, D. W. (2016). Intermolecular cooperativity and entanglement network in a miscible PLA/PMMA blend in the presence of nanosilica. *Polymer*, *82*, 57–65. doi:10.1016/j.polymer.2015.11.029

Hao, Y., Yan, L., Yu, H., Yang, K., Yu, S., Shan, R., & Du, B. (2014). Comparative study on adsorption of basic and acid dyes by hydroxy-aluminum pillared bentonite. *Journal of Molecular Liquids*, *199*, 202–207. doi:10.1016/j.molliq.2014.09.005

Harada, K. (2008, January 11). *Press Release*. National Institute for Materials Science.

Harada, M. (1995). Minamata disease: Methylmercury poisoning in Japan caused by environmental pollution. *Critical Reviews in Toxicology*, *25*(1), 1–24. doi:10.3109/10408449509089885 PMID:7734058

Hasanbeigi, A., Price, L., & Lin, E. (2012). Emerging energy-efficiency and CO2 emission-reduction technologies for cement and concrete production: A technical review. *Renewable & Sustainable Energy Reviews*, *16*(8), 6220–6238. doi:10.1016/j.rser.2012.07.019

Hashim, A. S., Azahari, B., Ikeda, Y., & Kohjiya, S. (1998). The effect of Bis(3-triethoxysilylpropyl) tetrasulfide on silica reinforcement of styrene-butadiene rubber. *Rubber Chemistry and Technology*, *71*(2), 289–299. doi:10.5254/1.3538485

Hashim, A. S., Kawabata, N., & Kohjiya, S. (1995). Silica reinforcement of epoxidized natural rubber by the sol-gel method. *Journal of Sol-Gel Science and Technology*, *5*(3), 211–218. doi:10.1007/BF00487018

Hassan, H. Z. (2013). Energy Analysis and Performance Evaluation of the Adsorption refrigeration System. *ISRN Mechanical Engineering*, *2013*, 14. doi:10.1155/2013/704340

Hayashi, S., & Hamaguchi, H. (2004). Discovery of a magnetic ionic liquid [Bmim]$FeCl_4$. *Chemistry Letters*, *33*(12), 1590–1591. doi:10.1246/cl.2004.1590

He, Z. H., Huang, H.Y., Deng, L.S., Yuan, H.R., Kobayashi, N., Kubota, M., … Zhao, D.D. (2014). Development of novel type of two-stage adsorption chiller with different adsorbents. *The 6th International Conference on Applied Energy – ICAE2014, Energy Procedia, 61*, 1996 – 1999.

Heath, A., Paine, K., & McManus, M. (2014). Minimising the global warming potential of clay based geopolymers. *Journal of Cleaner Production*, *78*, 75–83. doi:10.1016/j.jclepro.2014.04.046

He, G., Pan, Q., & Rempel, G. L. (2003). Synthesis of Poly(methyl methacrylate) Nanosize Particles by Differential Microemulsion Polymerization. *Macromolecular Rapid Communications*, *24*(9), 585–588. doi:10.1002/marc.200390089

Heimer, N. E., Del Sesto, R. E., Meng, Z., Wilkes, J. S., & Carper, W. R. (2006). Vibrational spectra of imidazolium tetrafluoroborate ionic liquids. *Journal of Molecular Liquids*, *124*(1–3), 84–95. doi:10.1016/j.molliq.2005.08.004

Henninger, S. K., Schicktanz, M., Hugenell, P. P. C., Sievers, H., & Henning, H. M. (2012). Evaluation of methanol adsorption on activated carbon for thermally driven chillers part I: Thermophysical characterisation. *International Journal of Refrigeration*, *35*(3), 543–553. doi:10.1016/j.ijrefrig.2011.10.004

Hernandez, M., Bernal, M., Verdejo, R., Ezquerra, T., & Lopez-Manchado, M. (2012). Overall performance of natural rubber/graphene nanocomposites. *Composites Science and Technology*, *73*, 40–46. doi:10.1016/j.compscitech.2012.08.012

Hestrin, S., & Schramm, M. (1954). Synthesis of cellulose by Acetobacter xylinum. 2. Preparation of freeze-dried cells capable of polymerizing glucose to cellulose. *The Biochemical Journal*, *58*(2), 345–352. doi:10.1042/bj0580345 PMID:13208601

He, W., Li, G., Ma, X., Wang, H., Huang, J., Xu, M., & Huang, C. (2006). Recovery strategies and the WEEE treatment status in China. *Journal of Hazardous Materials*, *136*(3), 502–512. doi:10.1016/j.jhazmat.2006.04.060 PMID:16820262

Hewitt, N. (2007). *Compounding Precipitated Silica in Elastomers*. Plastics Design Library: William Andrew.

He, Y., Boswell, P. G., Bühlmann, P., & Lodge, T. P. (2007). Ion gels by self-assembly of a triblock copolymer in an ionic liquid. *The Journal of Physical Chemistry B*, *111*(18), 4645–4652. doi:10.1021/jp064574n PMID:17474692

He, Y., & Lodge, T. P. (2007). A thermoreversible ion gel by triblock copolymer self-assembly in an ionic liquid. *Chemical Communications*, (26): 2732–2734. doi:10.1039/b704490a PMID:17594036

Hioki, N., Hori, Y., Watanabe, K., Morinaga, Y., Yoshinaga, F., Hibino, Y., & Ogura, T. (1995). Bacterial Cellulose; as a New Material for Papermaking. *Japan Tappi Journal*, *49*(4), 718–723. doi:10.2524/jtappij.49.718

Hirayama, N., Okamura, H., Kidani, K., & Imura, H. (2008). Ionic Liquid Synergistic Cation-Exchange System for the Selective Extraction of Lanthanum(III) Using 2-Thenoyltrifluoroacetone and 18-Crown-6. *Analytical Sciences*, *24*(6), 697–699. doi:10.2116/analsci.24.697 PMID:18544855

Hlaváček, P., Šmilauer, V., Škvára, F., Kopecký, L., & Šulc, R. (2015). Inorganic foams made from alkali-activated fly ash: Mechanical, chemical and physical properties. *Journal of the European Ceramic Society*, *35*(2), 703–709. doi:10.1016/j.jeurceramsoc.2014.08.024

Hoenich, N. (2006). Cellulose for medical applications: Past, present, and future. *BioResources*, *2*, 270–280.

Hofsetz, K., & Silva, M. A. (2012). Brazilian sugarcane bagasse:Energy and non-energy consumption. *Biomass and Bioenergy*, *46*, 564–573. doi:10.1016/j.biombioe.2012.06.038

Holbrey, D., & Seddon, K. R. (1999). The phase behaviour of 1-alkyl-3-methylimidazolium tetrafluoroborates; ionic liquids and ionic liquid crystals. *Journal of the Chemical Society, Dalton Transactions: Inorganic Chemistry*, (13): 2133–2140. doi:10.1039/a902818h

Hollander, A. P., & Hatton, P. V. (2004). *Biopolymer methods in tissue engineering*. Human press.

Hon, D. N., (1997). *Chemical modification of lignocelluosic materials*. New York: Marcel Dekker, Inc.

Hong, D., & Chang, W. V. (2007). *Preparation of nano polystyrene beads by microemulsion polymerization*. Retrieved 12 February, 2016, from http://chems.usc.edu/assets/004/56944.pdf

Hong, F., & Qiu, K. (2008). An alternative carbon source from konjac powder for enhancing production of bacterial cellulose in static cultures by a model strain Acetobacter aceti subsp. xylinus ATCC 23770. *Carbohydrate Polymers*, *72*(3), 545–549. doi:10.1016/j.carbpol.2007.09.015

Hong, S., Kim, M., Hong, C. K., Jung, D., & Shim, S. E. (2008). Encapsulation of multi-walled carbon nanotubes by poly(4-vinylpyridine) and its dispersion stability in various solvent media. *Synthetic Metals*, *158*(21–24), 900–907. doi:10.1016/j.synthmet.2008.06.023

Hongyu, H., Zhaohong, H., Haoran, Y., Noriyuki, K., Dandan, Z., Mitsuhiro, K., & Ze-Shao, C. et al. (2009). Analysis for composite zeolite/foam aluminum–water mass recovery adsorption refrigeration system driven by engine exhaust heat. *Energy Conversion and Management*, *50*(2), 255–261. doi:10.1016/j.enconman.2008.09.022

Hoogerstraete, T. V., Onghena, B., & Binnemans, K. (2013). Homogeneous Liquid–Liquid Extraction of Metal Ions with a Functionalized Ionic Liquid. *The Journal of Physical Chemistry Letters*, *4*(10), 1659–1663. doi:10.1021/jz4005366 PMID:26282975

Hornung, M., Ludwig, M., & Schmauder, H. P. (2007). Optimizing the Production of Bacterial Cellulose in Surface Culture: A Novel Aerosol Bioreactor Working on a Fed Batch Principle (Part 3). *Engineering in Life Sciences*, *7*(1), 35–41. doi:10.1002/elsc.200620164

Horton, J. M., Bao, C., Bai, Z., Lodge, T. P., & Zhao, B. (2011). Temperature- and pH-triggered reversible transfer of doubly responsive hairy particles between water and a hydrophobic ionic liquid. *Langmuir*, *27*(21), 13324–13334. doi:10.1021/la2031818 PMID:21919469

Hossner, L. R., Loeppert, R. H., Newton, R. J., & Szaniszlo, P. J. (1998). Literature review: Phytoaccumulation of chromium, uranium, and plutonium in plant systems. Amarillo National Resource Center for Plutonium, TX (United States) Technical Report.

Hou, M.-F., Ma, C.-X., Zhang, W.-D., Tang, X.-Y., Fan, Y.-N., & Wan, H.-F. (2011). Removal of rhodamine B using iron-pillared bentonite. *Journal of Hazardous Materials*, *186*(2-3), 1118–1123. doi:10.1016/j.jhazmat.2010.11.110 PMID:21168960

Howell, J. A. (2002). Future research and developments in the membrane field. *Desalination*, *144*(1), 127–131. doi:10.1016/S0011-9164(02)00300-4

Hrachova, J., Komadel, P., Jochec-Moskova, D., Krajcl, J., Janigova, I., Slouf, M., & Chodak, I. (2013). Properties of organo-clay/natural rubber nanocomposites: Effect of orgonophilic modifiers. *Journal of Applied Polymer Science*, *127*(5), 3447–3455. doi:10.1002/app.37602

Hsueh, C. L., Kuo, J. F., Huang, Y. H., Wang, C. C., & Chen, C. Y. (2005). Separation of ethanol–water solution by poly(acrylonitrile-co-acrylic acid) membranes. *Separation and Purification Technology*, *41*(1), 39–47. doi:10.1016/j.seppur.2004.04.002

Huafang, G. (2014). Effect of Adsorbent Diameter on the Performance of Adsorption Refrigeration. *Chinese Journal of Chemical Engineering*, *22*(5), 602–606. doi:10.1016/S1004-9541(14)60074-4

Huang, S. J. (1995). Polymer Waste Management–Biodegradation, Incineration, and Recycling. *Journal of Macromolecular Science, Part A*, *32*(4), 593–597. doi:10.1080/10601329508010272

Hubbard, C. D., Illner, P., & van Eldik, R. (2011). Understanding Chemical Reaction Mechanisms in Ionic Liquids: Successes and Challenges. *Chemical Society Reviews*, *40*(1), 272–290. doi:10.1039/C0CS00043D PMID:21079861

Huneau, B. (2011). Strain-induced crystallization of natural rubber: a review of x-ray diffraction investigations. *Rubber Chemistry and Technology*, *84*(3), 425–452. doi:10.5254/1.3601131

Hunt, P. A., Kirchner, B., & Welton, T. (2006). Characterising the electronic structure of ionic liquids: An examination of the 1-butyl-3-methylimidazolium chloride ion pair. *Chemistry (Weinheim an der Bergstrasse, Germany)*, *12*(26), 6762–6775. doi:10.1002/chem.200600103 PMID:16807969

Huo, D., & Kobayashi, T. (2006). Sonoresponsive volume phase-transition behavior of crosslinked poly (N-isopropylacrylamide-co-acrylic acid) microgel. *Chemistry Letters*, *35*(7), 776–777. doi:10.1246/cl.2006.776

Hu, Y., Daoud, W. A., Cheuk, K. K. L., & Lin, C. S. K. (2016). Newly developed techniques on polycondensation, ring-opening polymerization and polymer modification: Focus on poly(lactic acid). *Materials (Basel)*, *9*(3), 133–147. doi:10.3390/ma9030133

Hu, Y., Zhou, S., & Wu, L. (2009). Surface mechanical properties of transparent poly(methyl methacrylate)/zirconia nanocomposites prepared by in situ bulk polymerization. *Polymer*, *50*(15), 3609–3616. doi:10.1016/j.polymer.2009.03.028

Iguchi, M., Mitsuhashi, S., Ichimura, K., Nishi, Y., Uryu, M., Yamanaka, S., & Watanabe, K. (1988). *Bacterial cellulose-containing molding material having high dynamic strength*. Google Patents.

Iguchi, M., Yamanaka, S., & Budhiono, A. (2000a). Bacterial cellulose - a masterpiece of natures arts. *Journal of Materials Science*, *35*(2), 261–270. doi:10.1023/A:1004775229149

Ikenoue, T., Trindade, M. C. D., Lee, M. S., Lin, E. Y., Schurman, D. J., Goodman, S. B., & Smith, R. L. (2003). Mechanoregulation of human articular chondrocyte aggrecan and type II collagen expression by intermittent hydrostatic pressure in vitro. *Journal of Orthopaedic Research*, *21*(1), 110–116. doi:10.1016/S0736-0266(02)00091-8 PMID:12507587

Ikingura, J. R., Akagi, H., Mujumba, J., & Messo, C. (2006). Environmental assessment of mercury dispersion, transformation and bioavailability in the Lake Victoria Goldfields, Tanzania. *Journal of Environmental Management*, *81*(2), 167–173. doi:10.1016/j.jenvman.2005.09.026 PMID:16782263

International Environmental Technology Center. (2013, May 13). Policy Brief on E-waste. International Environmental Technology Center.

Interstate Technology & Regulatory Council (ITRC). Mining Waste Team. (2010). *Technology overview Re-use and reprocess (r2) technologies.* Retrieved August, 2010, from http://www.itrcweb.org/miningwaste-guidance/to_reuse.htm

Iñuguez-Covarrubias G., Diaz-Terez R., Sanjuan-Duenas R., Anzaldo-Hernandez J., & Rowel R. M. (n.d.). Utilization of by-products from the tequila industry. Part 2: potential value of agave tequilana weber azul leaves. *Bioresources Technology, 77*, 101-108.

Iñuguez-Covarrubias, G., Lang, S. E., & Rowel, R. M. (2001). Utilization of by-products from the tequila industry. Part 1: Potential value of agave tequilana weber azul leaves. *Bioresource Technology*, *77*, 25–40. doi:10.1016/S0960-8524(00)00137-1 PMID:11211072

Ip, A. W. M., Barford, J. P., & McKay, G. (2009). Reactive Black dye adsorption/desorption onto different adsorbents: Effect of salt, surface chemistry, pore size and surface area. *Journal of Colloid and Interface Science*, 33732–33738. PMID:19523648

Irena. (2015). *Solar Heating and Cooling for Residential Applications Technology Brief.* Retrieved from http://www.irena.org/documentdownloads/publications/irena_etsap_tech_brief_r12_solar_thermal_residential_2015.pdf

Ishaq, M., Sultan, S., Ahmad, I., Ullah, H., Yaseen, M., & Amir, A. (2015). (in press). Adsorptive desulfurization of model oil using untreated, acid activated and magnetite nanoparticle loaded bentonite as adsorbent. *Journal of Saudi Chemical Society*. doi:10.1016/j.jscs.2015.02.003

Ismail, A. B., Li, A., Thu, K., Ng, K. C., & Chun, W. (2014). Pressurized adsorption cooling cycles driven by solar/waste heat. *Applied Thermal Engineering*, *67*(1-2), 106–113. doi:10.1016/j.applthermaleng.2014.02.063

Ismail, H., Ramly, F., & Othman, N. (2010). Multiwall carbon nanotube-filled natural rubber: The effects of filler loading and mixing method. *Polymer-Plastic Technology and Engineering*, *49*(3), 260–266. doi:10.1080/03602550903413888

Israelachvili, J. (1991). *Intermolecular and surface forces.* San Diego, CA: Academic Press.

Ivanova, E. P., Kostova, M. A., & Koumanova, B. K. (2010). Kinetics of water and alcohol vapors adsorption on natural zeolite. *Asia-Pacific Journal of Chemical Engineering*, *5*(6), 869–881. doi:10.1002/apj.412

Jacquemin, J., Husson, P., Padua, A. A. H., & Majer, V. (2006). Density and viscosity of several pure and water-saturated ionic liquids. *Green Chemistry*, *8*(2), 172–180. doi:10.1039/B513231B

Jamil, T. S., Ibrahim, H. S., El-Maksoud, I. H. A., & El-Wakeel, S. T. (2010). Application of zeolite prepared from Egyptian kaolin for removal of heavy metals: I. Optimum conditions. *Desalination*, *258*(1-3), 34–40. doi:10.1016/j.desal.2010.03.052

Jänchen, J., Schumann, K., Thrun, E., Brandt,A., Unger B., & Hellwig, U. (2012). Preparation, hydrothermal stability and thermal adsorption storage properties of binderless zeolite beads. *International Journal of Low-Carbon Technologies.* doi: 10.1093/ijlct/cts037

Jang, J. G., & Lee, H. K. (2016). Effect of fly ash characteristics on delayed high-strength development of geopolymers. *Construction & Building Materials*, *102*(Part 1), 260–269. doi:10.1016/j.conbuildmat.2015.10.172

Jang, J., Bae, J., Choi, M., & Yoon, S.-H. (2005). Fabrication and characterization of polyaniline coated carbon nanofiber for supercapacitor. *Carbon*, *43*(13), 2730–2736. doi:10.1016/j.carbon.2005.05.039

Jansen, S., Broadley, M. R., Robbrecht, E., & Smets, E. (2003). Aluminium hyperaccumulation in angiosperms: A review of its phylogenetic significance. *Botanical Review*, *68*(2), 235–269. doi:10.1663/0006-8101(2002)068[0235:AHIAAR]2.0.CO;2

Järup, L., & Åkesson, A. (2009). Current status of cadmium as an environmental health problem. *Toxicology and Applied Pharmacology*, *238*(3), 201–208. doi:10.1016/j.taap.2009.04.020 PMID:19409405

Jašo, V., Glenn, G., Klamczynski, A., & Petrović, Z. S. (2015). Biodegradability study of polylactic acid/ thermoplastic polyurethane blends. *Polymer Testing*, *47*, 1–3. doi:10.1016/j.polymertesting.2015.07.011

Jeenpadiphat, S., & Tungasmita, D. N. (2013). Acid-activated pillar bentonite as a novel catalyst for the esterification of high FFA oil. *Powder Technology*, *237*, 634–640. doi:10.1016/j.powtec.2013.02.001

Jensen, M. P., Neuefeind, J., Beitz, J. V., Skanthakumar, S., & Soderholm, L. (2003). Mechanisms of metal ion transfer into room-temperature ionic liquids: The role of anion exchange. *Journal of the American Chemical Society*, *125*(50), 15466–15473. doi:10.1021/ja037577b PMID:14664592

Jerai, F., Miyazaki, T., Saha, B., & Koyama, S. (2015). Overview of adsorption cooling system based on activated carbon –Alcohol Pair. *Joint Journal of Novel Carbon Resources Sciences & Green Asia Strategy*, 30-40.

Jiang, H., Tovar-Carrillo, K., & Kobayashi, T. (2016). Ultrasound Release of Mimosa Medicine Included in Cellulose Hydrogel Films. *Ultrasonics Sonochemistry*, *32*, 398–406.

Jiang, J.-C., Li, S.-C., Shih, P.-M., Hung, T.-C., Chang, S.-C., Lin, S. H., & Chang, H.-C. (2010). A high-pressure infrared spectroscopic study on the interaction of ionic liquids with PEO-PPO-PEO block copolymers and 1,4-Dioxane. *The Journal of Physical Chemistry B*, *115*(5), 883–888. doi:10.1021/jp109600c PMID:21190359

Jiang, L., Wang, L. W., & Wang, R. Z. (2014). Investigation on thermal conductive consolidated composite CaCl2 for adsorption refrigeration. *International Journal of Thermal Sciences*, *81*, 68–75. doi:10.1016/j.ijthermalsci.2014.03.003

Jiayou, Q. (2003). *Characterization of silica gel-water vapor adsorption and its measuring facility* (Unpublished master dissertation). National University of Singapore, Singapore.

Jinrong, W., & Huang, G. (2013). Enhanced mechanical and gas barrier properties of rubber nanocomposite with surface functionalized graphene oxide at low content. *Polymer*, *54*(7), 1930–1937. doi:10.1016/j.polymer.2013.01.049

Ji, X., Li, M., Fan, J., Zhang, P., Luo, B., & Wang, L. (2014). Structure optimization and performance experiments of a solar-powered finned-tube adsorption refrigeration system. *Applied Energy*, *113*, 1293–1300. doi:10.1016/j.apenergy.2013.08.088

Johnson, E. B. G., & Arshad, S. E. (2014). Hydrothermally synthesized zeolites based on kaolinite: A review. *Applied Clay Science*, *97–98*, 215–221. doi:10.1016/j.clay.2014.06.005

Jo, J. O., Saha, P., Kim, N. G., Chang Ho, C., & Kim, J. K. (2015). Development of nanocomposite with epoxidized natural rubber and functionalized multiwalled carbon nanotubes for enhanced thermal conductivity and gas barrier property. *Materials & Design*, *83*, 777–785. doi:10.1016/j.matdes.2015.06.045

Jonas, R., & Farah, L. F. (1998). Production and application of microbial cellulose. *Polymer Degradation & Stability, 59*(1–3), 101–106. doi:10.1016/S0141-3910(97)00197-3

Jorapur, R., & Rajvanshi, A. K. (1997). Sugarcan leaf-bagasse gasifiers for industrial heating applications. *Biomass and Bioenergy, 13*(3), 141–146. doi:10.1016/S0961-9534(97)00014-7

Jozala, A. F., de Lencastre-Novaes, L. C., Lopes, A. M., de Carvalho Santos-Ebinuma, V., Mazzola, P. G., Pessoa-Jr, A., & Chaud, M. V. et al. (2016). Bacterial nanocellulose production and application: A 10-year overview. *Applied Microbiology and Biotechnology, 100*(5), 2063–2072. doi:10.1007/s00253-015-7243-4 PMID:26743657

Jribi, S., Chakraborty, A., El-Sharkawy, I., Saha, B. B., & Koyama, S. (2009). Thermodynamic Analysis of Activated Carbon CO2 based Adsorption Cooling Cycles. *International Journal of Engineering and Natural Sciences, 3*(3), 142–145.

Jribi, S., El-Sharkawy, I. I., Koyama, S., & Saha, B. B. (2010b). Study on activated carbon-co2 pair: adsorption characteristics and cycle performance. *International Symposium on Next-generation Air Conditioning and Refrigeration Technology.*

Jribi, S., Koyama, S., & Saha, B. B. (2010a). Performance Investigation of a Novel CO2 Compression-Adsorption Based Hybrid Cooling Cycle. *Engineering Sciences Reports. Kyushu University, 32*(3), 12–18.

Jribi, S., Saha, B. B., Koyama, S., & Bentaher, H. (2014). Modeling and simulation of an activated carbon–CO2 four bed based adsorption cooling system. *Energy Conversion and Management, 78*, 985–991. doi:10.1016/j.enconman.2013.06.061

Jribi, S., Saha, B. B., Koyama, S., Chakraborty, A., & Ng, K. C. (2013). Study on activated carbon/HFO-1234ze(E) based adsorption cooling cycle. *Applied Thermal Engineering, 50*(2), 1570–1575. doi:10.1016/j.applthermaleng.2011.11.066

Jumrat, S., Chatveera, B., & Rattanadecho, P. (2011). Dielectric properties and temperature profile of fly ash-based geopolymer mortar. *International Communications in Heat and Mass Transfer, 38*(2), 242–248. doi:10.1016/j.icheatmasstransfer.2010.11.020

Kaikake, K., Hoaki, K., Sunada, H., Dhakal, R. P., & Baba, Y. (2007). Removal characteristics of metal ions using degreased coffee beans: Adsorption equilibrium of cadmium(II). *Bioresource Technology, 98*(15), 2787–2791. doi:10.1016/j.biortech.2006.02.040 PMID:17400448

Kaiyoorawong, S., & Yangdee, B. (2006). *Rights of rubber farmers in Thailand under free trade.* Retrieved 02 February, 2016, from http://wrm.org.uy/oldsite/countries/Thailand/Rights_of_rubber_farmers_in_Thailand.pdf

Kakiuchi, T., & Nishi, N. (2006). Ionic liquid|water interface: A new electrified system for electrochemistry. *Electrochemistry, 74*(12), 942–948. doi:10.5796/electrochemistry.74.942

Kalia, S., & Averous, L. (2011). Biopolymers: Biomedical and environmental applications. Scrivener Publishing. doi:10.1002/9781118164792

Kalia, S., Dufresne, A., Cherian, B. M., Kaith, B. S., Averous, L., Njuguna, J., & Nassiopoulos, E. (2011). Cellulose-based bio and nanocomposites: A review. *International Journal of Polymer Science, 2011*, 1–35. doi:10.1155/2011/837875

Kalia, S., Kaith, B. S., & Kaur, I. (2009). Pretreatments of natural fibers and their applications as reinforcing materials in polymers composites- a review. *Polymer Engineering and Science, 49*(7), 1253–1272. doi:10.1002/pen.21328

Kamarudin, S., Mohd Sahaid, K., Mohd Sobri, T., Wan Mohtar, W. Y., Dayang Radiah, A. B., & Norhasliza, H. (2013). Different media formulation on biocellulose production by Acetobacter xylinum (0416). *Pertanika Journal of Science and Technology, 21*(1), 29–36.

Kanjanamosit, N., Muangnapoh, C., & Phisalaphong, M. (2010). Biosynthesis and characterization of bacteria cellulose–alginate film. *Journal of Applied Polymer Science, 115*(3), 1581–1588. doi:10.1002/app.31138

Karakoç, M. B., Türkmen, İ., Maraş, M. M., Kantarci, F., Demirboğa, R., & Uğur Toprak, M. (2014). Mechanical properties and setting time of ferrochrome slag based geopolymer paste and mortar. *Construction & Building Materials*, *72*(0), 283–292. doi:10.1016/j.conbuildmat.2014.09.021

Kashi, S., Gupta, R. K., Baum, T., Kao, N., & Bhattacharya, S. N. (2016). Morphology, electromagnetic properties and electromagnetic interference shielding performance of poly lactide/graphene nanoplatelet nanocomposites. *Materials & Design*, *95*, 119–126. doi:10.1016/j.matdes.2016.01.086

Katepetch, C., Rujiravanit, R., & Tamura, H. (2013). Formation of nanocrystalline ZnO particles into bacterial cellulose pellicle by ultrasonic-assisted in situ synthesis. *Cellulose (London, England)*, *20*(3), 1275–1292. doi:10.1007/s10570-013-9892-8

Kato, S. (1982). Laboratory culture and morphology of Colacium vesiculosum Ehrb. (Euglenophyceae). *Japanese Journal of Phycology*, *30*, 63–67.

Khalifa, A. H. N., Hussein, F. M., & Hadi, F. M. (2011). Experimental Study on Two Beds Adsorption Chiller with Regeneration. *Modern Applied Science*, *5*(4), 43–52. doi:10.5539/mas.v5n4p43

Khattab, N. M. (2004). A novel solar-powered adsorption refrigeration module. *Applied Thermal Engineering*, *24*(17-18), 2747–2760. doi:10.1016/j.applthermaleng.2004.04.001

Killmann, W. (2001). *Non-forest tree plantations* (D. J. Mead, Ed.). Forestry Department.

Kim, M. K., Kim, C. A., Ahn, S. D., Kang, S.-R., & Suh, K. S. (2004). Density compatibility of encapsulation of white inorganic TiO2 particles using dispersion polymerization technique for electrophoretic display. *Synthetic Metals*, *146*(2), 197–199. doi:10.1016/j.synthmet.2004.07.002

Kim, S. H., & Kim, Y. H. (1999). Direct condensation polymerization of lactic acid. *Macromolecular Symposia*, *144*(1), 277–287. doi:10.1002/masy.19991440125

Kirali, E. G., & Lac, O. (2006). Statistical modelling of acid activation on cotton oil bleaching by Turkish bentonite. *Journal of Food Engineering*, *75*(1), 137–141. doi:10.1016/j.jfoodeng.2005.06.010

Kirdponpattara, S., Khamkeaw, A., Sanchavanakit, N., Pavasant, P., & Phisalaphong, M. (2015). Structural modification and characterization of bacterial cellulose–alginate composite scaffolds for tissue engineering. *Carbohydrate Polymers*, *132*, 146–155. doi:10.1016/j.carbpol.2015.06.059 PMID:26256335

Klechkovskaya, V. V., Volkov, V. V., Shtykova, E. V., Arkharova, N. A., Baklagina, Y. G., Khripunov, A. K., & Tkachenko, A. A. (2008). Network Model of Acetobacter Xylinum Cellulose Intercalated by Drug Nanoparticles. In M. Giersig & G. Khomutov (Eds.), *Nanomaterials for Application in Medicine and Biology* (pp. 165–177). Springer Netherlands. doi:10.1007/978-1-4020-6829-4_14

Klemm, Fink, & Bohn. (2004). Cellulose: Fascinating Biopolymerand Sustainable Raw Material. *Ang. Chemie (Intl. Edn.)*, *44*, 3358.

Klemm, D., Heublein, B., Fink, H. P., & Bohn, A. (2005). Cellulose: Fascinating biopolymer and sustainable raw material. *Angewandte Chemie International Edition*, *44*(22), 3358–3393. doi:10.1002/anie.200460587 PMID:15861454

Klocking, R., & Helbing, B. (2005). *Medical aspects and human applications of humic substances*. Wiley-VCH.

Kobayashi, K., Tovar-Carrillo, K. L., & Tagaya, M. (2015). Bagasse sustainable polymers for cellulose hydrogel sheets showing tissue regeneration. In Handbook of Sustainable Polymers: Structure and Chemistry. Pan Stanford Publishing Pte Ltd.

Kobayashi, T. (2015). *Fabrication of cellulose hydrogels and characterization of their biocompatible films. In Studies in Natural Products Chemistry, 45* (pp. 1–15). Amsterdam, Netherlands: Elsevier B.V.

Kobayashi, T., Ohshiro, M., Nakamoto, K., & Syuji Uchida, S. (2016). Decontamination of Extra-Diluted Radioactive Cesium in Fukushima Water Using Zeolite–Polymer Composite Fibers. *Industrial & Engineering Chemistry Research, 55*(25), 6996–7002. doi:10.1021/acs.iecr.6b00903

Kobayashi, T., Ohshiro, M., & Uchida, S. (2014). *Zeolite fukugouka seni niyoru Cesium zyosen gizyutu (Japanese).* Tokyo, Japan: NTS.

Kobayashi, T., & Tovar-Carrillo, K. (2015). Fibroblast Cell Cultivation on Wooden Pulp Cellulose hydrogels for cytocompatibility scaffold method. *Pharmaceutic Analylical Acta, 6*, 10. doi:10.4172/2153-2435.1000423

Koddermann, T., Wertz, C., Heintz, A., & Ludwig, R. (2006). The association of water in ionic liquids: A reliable measure of polarity. *Angewandte Chemie International Edition, 45*(22), 3697–3702. doi:10.1002/anie.200504471 PMID:16637092

Kohler, N., Sun, C., Fichtenholtz, A., Gunn, J., Fang, C., & Zhang, M. (2006). Methotrexate-Immobilized Poly(ethylene glycol) Magnetic Nanoparticles for MR Imaging and Drug Delivery. *Small, 2*(6), 785–792. doi:10.1002/smll.200600009 PMID:17193123

Kohno, Y., & Ohno, H. (2012). Temperature-responsive ionic liquid/water interfaces: Relation between hydrophilicity of ions and dynamic phase change. *Physical Chemistry Chemical Physics, 14*(15), 5063–5070. doi:10.1039/c2cp24026b PMID:22334119

Komnitsas, K., & Zaharaki, D. (2007). Geopolymerisation: A Review and Prospects for the Minerals Industry. *Minerals Engineering, 20*(14), 1261–1277. doi:10.1016/j.mineng.2007.07.011

Kongsinlark, A., Rempel, G. L., & Prasassarakich, P. (2012). Synthesis of monodispersed polyisoprene–silica nanoparticles via differential microemulsion polymerization and mechanical properties of polyisoprene nanocomposite. *Chemical Engineering Journal, 193–194*, 215–226. doi:10.1016/j.cej.2012.04.008

Korichi, S., Elias, A., Mefti, A., & Bensmaili, A. (2012). The effect of microwave irradiation and conventional acid activation on the textural properties of smectite: Comparative study. *Applied Clay Science, 59-60*, 76–83. doi:10.1016/j.clay.2012.01.020

Kotsuchibashi, Y., Yamamoto, K., & Aoyagi, T. (2009). Assembly behavior of double thermo-responsive block copolymers with controlled response temperature in aqueous solution. *Journal of Colloid and Interface Science, 336*(1), 67–72. doi:10.1016/j.jcis.2009.03.093 PMID:19442985

Kou, M. R. S., Mendioroz, S., Salerno, P., & Muñoz, V. (2003). Catalytic activity of pillared clays in methanol conversion. *Applied Catalysis A, General, 240*(1-2), 273–285. doi:10.1016/S0926-860X(02)00466-0

Kozonoi, N., & Ikeda, Y. (2007). Extraction Mechanism of Metal Ion from Aqueous Solution to the Hydrophobic Ionic Liquid, 1-Butyl-3-methylimidazolium Nonafluorobutanesulfonate. *Monatshefte fur Chemie, 138*(11), 1145–1151. doi:10.1007/s00706-007-0727-x

Krajewska, B. (2001). Diffusion of metal ions through gel chitosan membranes. *Reactive & Functional Polymers, 47*(1), 37–47. doi:10.1016/S1381-5148(00)00068-7

Krause, U. (2009). *Fires in Silos: Hazards, Prevention, and Fire Fighting*. Wiley Publisher, VCH. doi:10.1002/9783527623822

Królicki, Z., Wilk, T., Zajączkowski, B., & Kaczmarczyk, J. (2006). Research on adsorption characteristics of active carbon and silica gel and application in solar refrigeration and heat pump.*61st ATI National Congress –International Session "Solar Heating and Cooling".*

Kubota, F., Koyanagi, Y., Nakashima, K., Shimojo, K., Kamiya, N., & Goto, M. (2008). Extraction of Lanthanide Ions with an Organophosphorous Extractant into Ionic Liquids. *Solv. Extr. Res. Dev. Jpn.*, *15*, 81–87.

Kubota, F., Shimobori, Y., Baba, Y., Koyanagi, Y., Shimojo, K., Kamiya, N., & Goto, M. (2011). Application of ionic liquids to extraction separation of rare earth metals with an effective diglycol amic acid extractant. *Journal of Chemical Engineering of Japan*, *44*(5), 307–312. doi:10.1252/jcej.11we005

Kubota, F., Shimobori, Y., Koyanagi, Y., Shimojo, K., Kamiya, N., & Goto, M. (2010). Uphill transport of rare-earth metals through a highly stable supported liquid membrane based on an ionic liquid. *Analytical Sciences*, *26*(3), 289–290. doi:10.2116/analsci.26.289 PMID:20215675

Kucińska-Lipka, J., Gubanska, I., & Janik, H. (2015). Bacterial cellulose in the field of wound healing and regenerative medicine of skin: Recent trends and future prospectives. *Polymer Bulletin (Berlin)*, *72*(9), 2399–2419. doi:10.1007/s00289-015-1407-3

Kumar, M. (2011). *Adsorption of methane on activated carbonby volumetric method* (Unpublished master dissertation). National Institute of Technology, Rourkela, India.

Kumar, A., Srivastava, A., Galaev, I. Y., & Mattiasson, B. (2007). Smart polymers: Physical forms and bioengineering applications. *Progress in Polymer Science*, *32*(10), 1205–1237. doi:10.1016/j.progpolymsci.2007.05.003

Kumar, S., & Kumar, R. (2011). Mechanical activation of fly ash: Effect on reaction, structure and properties of resulting geopolymer. *Ceramics International*, *37*(2), 533–541. doi:10.1016/j.ceramint.2010.09.038

Kurita, K. (2001). Controlled functionalization of the polysaccharide chitin. *Progress in Polymer Science*, *26*(9), 1921–1971. doi:10.1016/S0079-6700(01)00007-7

Kurniawan, H., Ye, Y.-S., Kuo, W.-H., Lai, J.-T., Wang, M.-J., & Liu, H.-S. (2013). Improvement of biofouling resistance on bacterial cellulose membranes. *Biochemical Engineering Journal*, *78*, 138–145. doi:10.1016/j.bej.2013.03.021

Kurniawan, T. A., Chan, G. Y. S., Lo, W., & Babel, S. (2006). Physico-chemical treatment techniques for wastewater laden with heavy metals. *Chemical Engineering Journal*, *118*(1-2), 83–98. doi:10.1016/j.cej.2006.01.015

Kurosumi, A., Sasaki, C., Yamashita, Y., & Nakamura, Y. (2009). Utilization of various fruit juices as carbon source for production of bacterial cellulose by Acetobacter xylinum NBRC 13693. *Carbohydrate Polymers*, *76*(2), 333–335. doi:10.1016/j.carbpol.2008.11.009

Laachachi, A., Cochez, M., Leroy, E., Ferriol, M., & Lopez-Cuesta, J. M. (2007, January). Fire retardant systems in poly(methyl methacrylate): Interactions between metal oxide nanoparticles and phosphinates. *Polymer Degradation & Stability*, *92*(1), 61–69. doi:10.1016/j.polymdegradstab.2006.09.011

Laachachi, A., Leroy, E., Cochez, M., Ferriol, M., & Lopez Cuesta, J. M. (2005, August). Use of oxide nanoparticles and organoclays to improve thermal stability and fire retardancy of poly(methyl methacrylate). *Polymer Degradation & Stability*, *89*(2), 344–352. doi:10.1016/j.polymdegradstab.2005.01.019

Lai, S.-M., Wu, W.-L., & Wang, Y.-J. (2016). Annealing effect on the shape memory properties of polylactic acid (PLA)/thermoplastic polyurethane (TPU) bio-based blends. *Journal of Polymer Research*, *23*(5), 99–111. doi:10.1007/s10965-016-0993-6

Langer, R., & Vacanti, J. P. (1993). Tissue engineering. *Science*, *260*(5110), 920–926. doi:10.1126/science.8493529 PMID:8493529

Lartey, R. B., & Francis, A. (1999). *Developing national capability for manufacture of activated carbon from agricultural wastes.* Institute of Industrial Research, CSIR, Accra, Ghana. Retrieved from http://www.africantechnologyforum.org/GhIE/ActCarbon.htm

Latt, K. K., & Takahashi, Y. (2011). Fabrication and characterization of a α,β,γ,δ-Tetrakis(1-methylpyridinium-4-yl) porphine/silica nanocomposite thin-layer membrane for detection of ppb-level heavy metal ion. *Analytica Chimica Acta, 689*(1), 103–109. doi:10.1016/j.aca.2011.01.032 PMID:21338764

Lecomte, I., Henrist, C., Liégeois, M., Maseri, F., Rulmont, A., & Cloots, R. (2006). (Micro)-structural comparison between geopolymers, alkali-activated slag cement and Portland cement. *Journal of the European Ceramic Society, 26*(16), 3789–3797. doi:10.1016/j.jeurceramsoc.2005.12.021

Lee, D.J., & Logan, B. E. (2011). Preface. *Bioresource Technology, 102*(1), 1.

Lee, S-J., Lee, M-E., Chung, J. W., Park J. H., Huh, K. Y., & Jun, G-I. (2013). Immobilization of Lead from Pb-Contaminated Soil Amended with Peat Moss. *Journal of Chemistry*, 1-6.

Lee, J. W., & Kjeang, E. (2013). Nanofluidic fuel cell. *Journal of Power Sources, 242*, 472–477. doi:10.1016/j.jpowsour.2013.05.129

Lee, J., Panzer, M. J., He, Y., Lodge, T. P., & Frisbie, C. D. (2007). Ion gel gated polymer thin-film transistors. *Journal of the American Chemical Society, 129*(15), 4532–4533. doi:10.1021/ja070875e PMID:17381097

Lee, S. H., Lim, Y. M., Jeong, S. I., An, S. J., Kang, S. S., Jeong, C. M., & Huh, J. B. (2015). The effect of bacterial cellulose membrane compared with collagen membrane on guided bone regeneration. *Journal of Advanced Prosthodontics, 7*(6), 484–495. doi:10.4047/jap.2015.7.6.484 PMID:26816579

Lee, S.-J., Park, J. H., Ahn, Y.-T., & Chung, J. W. (2015, February). Comparison of Heavy Metal Adsorption by Peat Moss and Peat Moss-Derived Biochar Produced Under Different Carbonization Conditions. *Water, Air, and Soil Pollution, 226*(2), 9. doi:10.1007/s11270-014-2275-4

Lee, S., van Riessen, A., Chon, C.-M., Kang, N.-H., Jou, H.-T., & Kim, Y.-J. (2016). Impact of activator type on the immobilisation of lead in fly ash-based geopolymer. *Journal of Hazardous Materials, 305*, 59–66. doi:10.1016/j.jhazmat.2015.11.023 PMID:26642447

Leitch, M. E., Li, C., Ikkala, O., Mauter, M. S., & Lowry, G. V. (2016). Bacterial Nanocellulose Aerogel Membranes: Novel High-Porosity Materials for Membrane Distillation. *Environmental Science and Technology Letters, 3*(3), 85–91. doi:10.1021/acs.estlett.6b00030

Lenarda, M., Ganzerla, R., Storaro, L., Enzo, S., & Zanoni, R. (1994). Bifunctional catalysts from pillared clays: Vapour phase conversion of propene to acetone catalyzed by iron and ruthenium containing alurninum pillared bentonites. *Journal of Molecular Catalysis, 92*(2), 201–215. doi:10.1016/0304-5102(94)00064-6

Lentacker, I., De Geest, B. G., Vandenbroucke, R. E., Peeters, L., Demeester, J., De Smedt, S. C., & Sanders, N. N. (2006). Ultrasound-responsive polymer-coated microbubbles that bind and protect DNA. *Langmuir, 22*(17), 7273–7278. doi:10.1021/la0603828 PMID:16893226

Lewtas, P., Fujikawa, Y., & Wajrak, M. (2015). Stripping Voltammetry for On-Site Analysis of Heavy Metals: Working Electrode and Automation Options. *Kankyo Gijyutsu, 144*, 537–544.

Li, C., Wang, R. Z., Wang, L. W., Li, T. X., & Chen, Y. (2013). Experimental study on an adsorption icemaker driven by parabolic trough solar collector. *Renewable Energy, 57*, 223–233. doi:10.1016/j.renene.2013.01.040

Liebner, F., Haimer, E., Wendland, M., Neouze, M. A., Schlufter, K., Miethe, P., & Rosenau, T. et al. (2010). Aerogels from unaltered bacterial cellulose: Application of scCO <inf>2</inf> drying for the preparation of shaped, ultra-lightweight cellulosic aerogels. *Macromolecular Bioscience*, *10*(4), 349–352. doi:10.1002/mabi.200900371 PMID:20166232

Liew, C.-W., Ramesh, S., & Arof, A. K. (2014). A novel approach on ionic liquid-based poly(vinyl alcohol) proton conductive polymer electrolytes for fuel cell applications. *International Journal of Hydrogen Energy*, *39*(6), 2917–2928. doi:10.1016/j.ijhydene.2013.07.092

Li, F.-J., Liang, J.-Z., Zhang, S.-D., & Zhu, B. (2015). Tensile properties of polylactide/poly(ethylene glycol) blends. *Journal of Polymers and the Environment*, *23*(3), 407–415. doi:10.1007/s10924-015-0718-7

Li, F., Yan, N., Zhan, Y., Fei, G., & Xia, H. (2013). Probing the reinforcing mechanism of graphene and graphene Oxide in natural Rubber. *Journal of Applied Physics*, *129*(4), 2342–2351.

Li, K., & Kobayashi, T. (2016a). A FT-IR spectroscopic study of ultrasound effect on aqueous imidazole based ionic liquids having different counter ions. *Ultrasonics Sonochemistry*, *28*(0), 39–46. doi:10.1016/j.ultsonch.2015.06.030 PMID:26384881

Li, K., & Kobayashi, T. (2016b). Ultrasound response of aqueous poly(ionic liquid) solution. *Ultrasonics Sonochemistry*, *30*, 52–60. doi:10.1016/j.ultsonch.2015.10.021 PMID:26597539

Li, K., Noguchi, S., & Kobayashi, T. (2016). Ultrasound-responsive behavior of gelatinous ionic liquid/poly(vinyl alcohol) composites. *Ind. Eng. Chem. Res., 55(37), 9915-9924.* doi: 10.1021/acs.iecr.6b02264

Li, K., Satoh, M., & Kobayashi, T. (2016). A Facile Method to Determine the Cu^{2+} Concentration Using Ionic Liquids as Concentrated Reagent. *Journal of Chemical Engineering of Japan*, *49*(1), 16–21. doi:10.1252/jcej.15we137

Li, K., Yuan, C., Zheng, S., & Fang, Q. (2012). A facile and economical procedure for the synthesis of maleimide derivatives using an acidic ionic liquid as a catalyst. *Tetrahedron Letters*, *53*(32), 4245–4247. doi:10.1016/j.tetlet.2012.06.025

Likos, W. J., & Lu, N. (2002). Water vapor sorption behavior of smectite-kaolinite mixtures. *Clays and Clay Minerals*, *50*(5), 553–561. doi:10.1346/000986002320679297

Li, L., Wang, S. B., & Zhu, Z. H. (2006). Geopolymeric adsorbents from fly ash for dye removal from aqueous solution. *Journal of Colloid and Interface Science*, *300*(1), 52–59. doi:10.1016/j.jcis.2006.03.062 PMID:16626729

Li, M., Wang, R. Z., Xu, Y. X., Wu, J. Y., & Dieng, A. O. (2002). Experimental study on dynamic performance analysis of a flat-plate solar solid-adsorption refrigeration for ice maker. *Renewable Energy*, *27*(2), 211–221. doi:10.1016/S0960-1481(01)00188-4

Lim, S.-W., Choi, M.-C., Jeong, J.-H., Park, E.-Y., & Ha, C.-S. (2016). Toughening poly(lactic acid) (PLA) through *in situ* reactive blending with liquid polybutadiene rubber (LPB). *Composite Interfaces*, *23*(8), 807–818. doi:10.1080/09276440.2016.1175168

Li, N., Li, Y., & Liu, S. (2016). Rapid prototyping of continuous carbon fiber reinforced polylacticacid composites by 3D printing. *Journal of Materials Processing Technology*, *238*, 218–225. doi:10.1016/j.jmatprotec.2016.07.025

Lin, C. W., Liang, S. S., Chen, S. W., & Lai, J. T. (2013). Sorption and transport properties of 2-acrylamido-2-methyl-1-propanesulfonic acid-grafted bacterial cellulose membranes for fuel cell application. *Journal of Power Sources*, *232*, 297–305. doi:10.1016/j.jpowsour.2013.01.047

Lindgren, G. (2003). *Impending world oil shortage*. Retrieved 02 February, 2016, from http://www.imc-cim.org/mmap/pdf/prod-lindgren-oil-e.pdf

Lindma, B., Karlstrom, G., & Stigsson, L. (2010). On the mechanism of dissolution of cellulose. *Journal of Molecular Liquids*, *156*(1), 76–81. doi:10.1016/j.molliq.2010.04.016

Lin, S. B., Chen, L. C., & Chen, H. H. (2011). Physical characteristics of surimi and bacterial cellulose composite gel. *Journal of Food Process Engineering*, *34*(4), 1363–1379. doi:10.1111/j.1745-4530.2009.00533.x

Lin, S. P., Loira Calvar, I., Catchmark, J. M., Liu, J. R., Demirci, A., & Cheng, K. C. (2013). Biosynthesis, production and applications of bacterial cellulose. *Cellulose (London, England)*, *20*(5), 2191–2219. doi:10.1007/s10570-013-9994-3

Li, T. X., Wang, R. Z., Wang, L. W., Lu, Z. S., & Wu, J. Y. (2008). Influence of mass recovery on the performance of a heat pipe type ammonia sorption refrigeration system using CaCl2 /activated carbon as compound adsorbent. *Applied Thermal Engineering*, *28*(13), 1638–1646. doi:10.1016/j.applthermaleng.2007.10.027

Liu, C., Hsu, P., Lee, H., Ye, M., Zheng, G., Liu, N., & Cui, Y. et al. (2015). Transparent air filter for high-efficiency PM2.5 capture. *Nature Communications*, *6*, 1–9. doi:10.1038/ncomms7205 PMID:25683688

Liu, H., Lv, M., Deng, B., Li, J., Yu, M., Huang, Q., & Fan, C. (2014). Laundering durable antibacterial cotton fabrics grafted with pomegranate-shaped polymer wrapped in silver nanoparticle aggregations. *Scientific Reports*, *4*, 5920. doi:10.1038/srep05920 PMID:25082297

Liu, H., & Zhang, J. (2011). Research progress in toughening modification of poly(lactic acid). *Journal of Polymer Science. Part B, Polymer Physics*, *49*(15), 1051–1083. doi:10.1002/polb.22283

Liu, J., & Ueda, M. (2009). High refractive index polymers: Fundamental research and practical applications. *Journal of Materials Chemistry*, *19*(47), 8907. doi:10.1039/b909690f

Liu, X., Ma, L., Mao, Z., & Gao, C. (2011). Chitosan-based biomaterials for tissue repair and regeneration. *Advances in Polymer Science*, *244*, 81–87. doi:10.1007/12_2011_118

Liu, Y.-L., Hsu, C.-Y., Su, Y.-H., & Lai, J.-Y. (2005). Chitosan−Silica Complex Membranes from Sulfonic Acid Functionalized Silica Nanoparticles for Pervaporation Dehydration of Ethanol−Water Solutions. *Biomacromolecules*, *6*(1), 368–373. doi:10.1021/bm049531w PMID:15638541

Liu, Y., Zhu, L., Sun, X., Chen, J., & Luo, F. (2009). Silica materials doped with bifunctional ionic liquid extractant for yttrium extraction. *Industrial & Engineering Chemistry Research*, *48*(15), 7308–7313. doi:10.1021/ie900468c

Liu, Z., Yang, Z., Zhao, Y., Ji, G., Zhang, H., Yu, B., & Gao, X. (2014). Fluoro-functionalized polymeric ionic liquids: Highly efficient catalysts for CO_2 cycloaddition to cyclic carbonates under mild conditions. *Green Chemistry*, *16*(8), 3724–3728. doi:10.1039/C4GC00730A

Livage, J. (1997). Sol-gel processes. *Current Opinion in Solid State and Materials Science*, *2*(2), 132–138. doi:10.1016/S1359-0286(97)80057-5

Livage, J., Henry, M., & Sanchez, C. (1988). Sol-gel chemistry of transition metal oxides. *Progress in Solid State Chemistry*, *18*(4), 259–341. doi:10.1016/0079-6786(88)90005-2

Livingston, W. (2016). *New Liquid Cooled Servers with Adsorption Chillers Improve Efficiency, Reduce Footprint*. Retrieved from http://www.ppiway.com/new-liquid-cooled-servers-with-adsorption-chillers-improve-efficiency-reduce-footprint/

Li, Y., Mai, Y.-W., & Ye, L. (2000). Sisal fibre and its composites: A review of recent developments. *Composites Science and Technology*, *60*(11), 2037–2048. doi:10.1016/S0266-3538(00)00101-9

Lloyd, L. L., Kennedy, J. F., Methacanon, P., Paterson, M., & Knill, C. J. (1998). Carbohydrates polymers as wound management aids. *Carbohydrate Polymers*, *37*(3), 315–319. doi:10.1016/S0144-8617(98)00077-0

Lodge, T. P. (2008). A unique platform for materials design. *Science*, *321*(5885), 50–51. doi:10.1126/science.1159652 PMID:18599764

Loh, W., Shah, B., Chakraborty, A., Ng, K., & Chun, W. (2010). Performance analysis of waste heat driven pressurized adsorption chiller. *Journal of Thermal Science and Technology*, *5*(2), 252-265.

Loh, W. S., El-Sharkawy, I. I., Ng, K. C., & Saha, B. B. (2009). Adsorption cooling cycles for alternative adsorbent/adsorbate pairs working at partial vacuum and pressurized conditions. *Applied Thermal Engineering*, *29*(4), 793–798. doi:10.1016/j.applthermaleng.2008.04.014

Loiola, A. R., Andrade, J. C. R. A., Sasaki, J. M., & da Silva, L. R. D. (2012). Structural analysis of zeolite NaA synthesized by a cost-effective hydrothermal method using kaolin and its use as water softener. *Journal of Colloid and Interface Science*, *367*(1), 34–39. doi:10.1016/j.jcis.2010.11.026 PMID:22074689

Lopes, M. S., Jardini, A. L., & Filho, R. M. (2014). Synthesis and characterizations of poly (lactic acid) by ring-opening polymerization for biomedical applications. *Chemical Engineering Transactions*, *38*, 331–336.

Lopes, T. D., Riegel-Vidotti, I. C., Grein, A., Tischer, C. A., & Faria-Tischer, P. C. S. (2014). Bacterial cellulose and hyaluronic acid hybrid membranes: Production and characterization. *International Journal of Biological Macromolecules*, *67*, 401–408. doi:10.1016/j.ijbiomac.2014.03.047 PMID:24704166

Lopez, F. J., Sugita, S., & Kobayashi, T. (2014). Cesium-adsorbent geopolymer foams based on silica from rice husk and metakaolin. *Chemistry Letters*, *43*(1), 128–130. doi:10.1246/cl.130851

Lopez, F. J., Sugita, S., Tagaya, M., & Kobayashi, T. (2014). Geopolymers using rice husk silica and metakaolin derivatives; Preparation and their characteristics. *J. Mater. Sci. Chem. Eng.*, *2*, 35–43.

López, F. J., Sugita, S., Takaya, M., & Kobayashi, T. (2014). Geopolymers Using Rice Husk Silica and Metakaolin Derivatives; Preparation and Their Characteristics. *Journal of Materials Science and Chemical Engineering*, *2*(05), 35–43. doi:10.4236/msce.2014.25006

López, F. J., Sugita, S., Takaya, M., & Kobayashi, T. (2014). Metakaolin-Based Geopolymers for Targeted Adsorbents to Heavy Metal Ion Separation. *Journal of Materials Science and Chemical Engineering*, *2*(07), 16–27. doi:10.4236/msce.2014.27002

López-Pastor, M., Ayora-Cañada, M. J., Valcárcel, M., & Lendl, B. (2006). Association of methanol and water in ionic liquids elucidated by infrared spectroscopy using two-dimensional correlation and multivariate curve resolution. *The Journal of Physical Chemistry B*, *110*(22), 10896–10902. doi:10.1021/jp057398b PMID:16771342

Lü, C., & Yang, B. (2009). High refractive index organic–inorganic nanocomposites: Design, synthesis and application. *Journal of Materials Chemistry*, *19*(19), 2884. doi:10.1039/b816254a

Luo, X. G., Liu, S. L., Zhou, J. P., & Zhang, L. N. (2009). In situ synthesis of. Fe3O4/cellulose microspheres with magnetic-induced protein delivery. *Journal of Materials Chemistry*, *19*(21), 3538–3545. doi:10.1039/b900103d

Lu, W., Fadeev, A. G., Qi, B., Smela, E., Mattes, B. R., Ding, J., & Forsyth, M. et al. (2002). Use of ionic liquids for π-conjugated polymer electrochemical devices. *Science*, *297*(5583), 983–987. doi:10.1126/science.1072651 PMID:12098704

Lu, Z. S., & Wang, R. Z. (2013). Performance improvement and comparison of mass recovery in CaCl2/activated carbon adsorption refrigerator and silica gel/LiCl adsorption chiller driven by low grade waste heat. *International Journal of Refrigeration*, *36*(5), 1504–1511. doi:10.1016/j.ijrefrig.2013.03.008

Lu, Z. S., Wang, R. Z., & Wang, L. W. (2011). Dynamic characteristics of a novel adsorption refrigerator with compound mass-heat recovery. *International Journal of Energy Research*, *37*(1), 59–68. doi:10.1002/er.1911

Lv, X., Yang, J., Feng, C., Li, Z., Chen, S., Xie, M., & Xu, Y. et al. (2016). Bacterial Cellulose-Based Biomimetic Nanofibrous Scaffold with Muscle Cells for Hollow Organ Tissue Engineering. *ACS Biomaterial Science and Engineering*, *2*(1), 19–29. doi:10.1021/acsbiomaterials.5b00259

Lychnos, G., & Tamainot-Telto, Z. (2014). Performance of hybrid refrigeration system using ammonia. *Applied Thermal Engineering*, *62*(2), 560–565. doi:10.1016/j.applthermaleng.2013.10.013

Madhu, G., Bhunia, H., Bajpai, P. K., & Nando, G. B. (2016). Physico-Mechanical Properties and Biodegradation of Oxo-degradable HDPE/PLA Blends. *Polymer Science, Series A*, *58*(1), 57–75. doi:10.1134/S0965545X16010077

Madhusoodana, K. N., & Varghese, S. (2006). Technological and processing properties of natural rubber layered silicate-nanocomposites by melt intercalation Process. *Journal of Applied Polymer Science*, *102*(3), 2537–2543. doi:10.1002/app.24680

Mae, T.-M. E. (2007). Development of Functional Foods in the Philippines. *Food Science and Technology Research*, *13*(3), 179–186. doi:10.3136/fstr.13.179

Mahmood, W. A. K., Rahman Khan, M. M., & Azarian, M. H. (2013). Sol-gel synthesis and morphology, thermal and optical properties of epoxidized natural rubber/zirconia hybrid films. *Journal of Non-Crystalline Solids*, *378*, 152–157. doi:10.1016/j.jnoncrysol.2013.07.002

Maia, A. A. B., Neves, R. F., Angélica, R. S., & Pöllmann, H. (2015). Synthesis, optimisation and characterisation of the zeolite NaA using kaolin waste from the Amazon Region. Production of Zeolites KA, MgA and CaA. *Applied Clay Science*, *108*, 55–60. doi:10.1016/j.clay.2015.02.017

Ma, J., Zhang, L.-Q., & Geng, L. (2010). *Manufacturing Techniques of Rubber Nanocomposites. In Rubber Nanocomposites* (pp. 21–62). John Wiley & Sons, Ltd.

Mallah, M.H., Shemirani, F., & Maragheh, M. G. (2009). Ionic liquids for simultaneous preconcentration of some lanthanoids using dispersive liquid-liquid microextraction technique in uranium dioxide powder. *Environ Sci Technol.*, *43*(6), 1947-1951.

Mallah, M. H., Shemirani, F., Maragheh, M. G., & Jamali, M. R. (2010). Evaluation of synergism in dispersive liquid–liquid microextraction for simultaneous preconcentration of some lanthanoids. *Journal of Molecular Liquids*, *151*(2-3), 122–124. doi:10.1016/j.molliq.2009.12.003

Mamun, A., & Bledzki, A. K. (2013). Micro fibre reinforced PLA and PP composites: Enzyme modification, mechanical and thermal properties. *Composites Science and Technology*, *78*, 10–17. doi:10.1016/j.compscitech.2013.01.013

Maneerung, T., Tokura, S., & Rujiravanit, R. (2008). Impregnation of silver nanoparticles into bacterial cellulose for antimicrobial wound dressing. *Carbohydrate Polymers*, *72*(1), 43–51. doi:10.1016/j.carbpol.2007.07.025

Mao, H., Zhu, K., Li, B., Yao, C., & Kong, Y. (2014). Synthesis of titania modified silica-pillared clay (SPC) with highly ordered interlayered mesoporous structure for removing toxic metal ion Cr(VI) from aqueous state. *Applied Surface Science*, *292*, 1009–1019. doi:10.1016/j.apsusc.2013.12.113

Margui, E., Hidalgo, M., Queralt, I., Van Meel, K., & Fontas, C. (2012). Analytical capabilities of laboratory, bench-top and hand-held X-ray fluorescence systems for detection of metals in aqueous samples pre-concentrated with solid phase extraction disks, *Spectrochimica Acta. B*, *67*, 17–23.

Margui, E., Zawisza, B., & Sitko, R. (2014). Trace and ultratrace analysis of liquid samples by X-ray fluorescence spectrometry. *Trends in Analytical Chemistry*, *53*, 73–83. doi:10.1016/j.trac.2013.09.009

Market and trade of coconut products. (2013). Retrieved March 2015 from http://www.fao.org/fileadmin/templates/rap/files/meetings/2013/131030-market.pdf

Market, B. (2003). *Element concentration in ecosystems*. Zittau, Germany: International Institute of Advanced Ecological and Economic Studies.

Martin, A. R., Martins, M. A., da Silva, O. R., & Motto, L. H. (2010). Studies on the thermal properties of sisal fiber and its constituents. *Thermochimica Acta*, *506*(1-2), 14–22. doi:10.1016/j.tca.2010.04.008

Maruszewski, K. (2004). Synthesis and properties of an inorganic-organic hybrid prepared by the sol-gel method. *Opt. Mater. (Amst)*, *26*(2), 207–211. doi:10.1016/j.optmat.2003.11.022

Masi, G., Rickard, W. D. A., Vickers, L., Bignozzi, M. C., & van Riessen, A. (2014). A comparison between different foaming methods for the synthesis of light weight geopolymers. *Ceramics International*, *40*(9, Part A), 13891–13902. doi:10.1016/j.ceramint.2014.05.108

Masoud S.N., & G., D. (2011). Polymeric Nanocomposite Materials. *InTech Journal*.

Matis, K. A., Zouboulis, A. I., & Lazaridis, N. K. (1998). Removal and recovery of Metals from Dilute Solutions: Applications of Flotation Techniques. *Mineral Processing and the Environment*, *43*, 165–196. doi:10.1007/978-94-017-2284-1_9

Matsui, S. (2003). Nanostructure fabrication using electron and ion beams A2 - Iwamoto, M. In K. Kaneto & S. Mashiko (Eds.), *Nanotechnology and Nano-Interface Controlled Electronic Devices* (pp. 3–20). Amsterdam: Elsevier. doi:10.1016/B978-044451091-4/50002-9

Matsumoto, K., & Hagiwara, R. (2007). Structural characteristics of alkylimidazolium-based salts containing fluoroanions. *Journal of Fluorine Chemistry*, *128*(4), 317–331. doi:10.1016/j.jfluchem.2006.10.016

Ma, Y., Yan, C., Alshameri, A., Qiu, X., Zhou, C., & Li, D. (2014). Synthesis and characterization of 13X zeolite from low-grade natural kaolin. *Advanced Powder Technology*, *25*(2), 495–499. doi:10.1016/j.apt.2013.08.002

Mazur, P. (1970). Cryobiology: The freezing of biological systems. *Science*, *168*(3934), 939–949. doi:10.1126/science.168.3934.939 PMID:5462399

McBride, M. B. (1994). *Environmental Chemistry of Soils*. New York: Oxford University Press.

McGlinchey, D. (2005). *Characterisation of Bulk Solids*. Blackwell Publishing Ltd. doi:10.1002/9781444305456

Meadows, D. H., Meadows, D. L., Randers, J., & Behrens, W. W. (1972). *The limits to glow*. New York: Universe Books.

Meena, A. K., Kadirvelu, K., Mishra, G. K., Rajagopal, C., & Nagar, P. N. (2008). Adsorptive Removal of Heavy Metals from Aqueous Solution by Treated Sawdust (Acacia arabica). *Journal of Hazardous Materials*, *150*(3), 604–611. doi:10.1016/j.jhazmat.2007.05.030 PMID:17600619

Mekki, S., Wai, C. M., Billard, I., Moutiers, G., Burt, J., Yen, C. H., & Hesemann, P. et al. (2006). Extraction of Lanthanides from Aqueous Solution by Using Room-Temperature Ionic Liquid and Supercritical Carbon Dioxide in Conjunction. *Chemistry (Weinheim an der Bergstrasse, Germany)*, *12*(6), 1760–1766. doi:10.1002/chem.200500559 PMID:16311989

Melo, C. R., Riella, H. G., Kuhnen, N. C., Angioletto, E., Melo, A. R., Bernardin, A. M., & Silva, L. et al. (2012). Review: Synthesis of 4A zeolites from kaolin for obtaining 5A zeolites through ionic exchange for adsorption of arsenic. *Materials Science and Engineering B*, *177*(4), 345–349. doi:10.1016/j.mseb.2012.01.015

Meng, H., & Jinlian Hu, . (2010). A brief review of stimulus-active polymers responsive to thermal, light, magnetic, electric, and water/solvent stimuli. *Journal of Intelligent Material Systems and Structures*, *21*(9), 859–885. doi:10.1177/1045389X10369718

Messenger, B. (Ed.). (2013). *Analysis: Bioplastics sector set for rapid growth up to 2017. In Waste Management World.* Vienna, Austria: Industriemagazin Verlag.

Metcalf, S.J., Tamainot-Telto, Z., & Critoph, R.E. (2011). Application of a compact sorption generator to solar refrigeration: Case study of Dakar (Senegal). *Applied Thermal Engineering, 31*, 2197-2204.

Meunier, F. (1985). Second law analysis of a solid adsorption heat pump operating on reversible cascade cycles: Application to the zeolite-water pair. *Heat Recovery Systems*, *5*(2), 133–141. doi:10.1016/0198-7593(85)90045-1

Meunier, F. (1986). Theoretical performances of solid adsorbent cascading cycles using the zeolite-water and active carbon-methanol pairs: Four case studies. *Heat Recovery Systems*, *6*(6), 491–498. doi:10.1016/0198-7593(86)90042-1

Meunier, F. (1993). Solid sorption: An alternative to CFCs. *Heat Recovery Systems & CHP*, *13*(4), 289–295. doi:10.1016/0890-4332(93)90051-V

Meyers, M. A., Chen, P.-Y., Lin, A. Y.-M., & Seki, Y. (2008). Biological materials: Structure and mechanical properties. *Progress in Materials Science*, *53*(1), 1–206. doi:10.1016/j.pmatsci.2007.05.002

Miles, D. J., & Shelton, S. V. (1996). Design and Testing of a Solid-Sorption Heat-Pump System. *Applied Thermal Engineering*, *16*(5), 389–394. doi:10.1016/1359-4311(95)00021-6

Ming, H. J. (2009). *Adsorption evaporative emission control system for vehicles* (Unpublished doctoral dissertation). National University of Singapore, Singapore.

Ministry of Agriculture, Forestry and Fisheries, Japan. (2014). *FY2013 Annual Report on Food, Agriculture and Rural Areas in Japan.* Author.

Ministry of Economy. Japan. (2010). *Trade and Industry technology strategy map 2010.* Retrieved from http://www.meti.go.jp/policy/economy/gijutsu_kakushin/kenkyu_kaihatu/str2010.html

Ministry of the Environment, Government of Japan. (2003). *Environmental Quality Standards for Water Pollution.* Retrieved from https://www.env.go.jp/en/water/wq/wp.pdf

Ministry of the Environment, Government of Japan. (2015). *National Effluent Standards.* Retrieved from http://www.env.go.jp/en/water/wq/nes.html

Miretzky, P., & Cirelli, A. F. (2009). Hg(II) removal from water by chitosan and chitosan derivatives: A review. *Journal of Hazardous Materials*, *167*(1–3), 10–23. doi:10.1016/j.jhazmat.2009.01.060 PMID:19232467

Miyazaki, T., El-Sharkawy, I. I., Saha, B. B., & Koyama, S. (2014). Optimized Performance of One-Bed Adsorption Cooling System.*International Refrigeration and Air Conditioning Conference.*

Mizunoa, T., Emorib, K., & Itoa, S. (2013). Manganese hyperaccumulation from non-contaminated soil in Chengiopanax sciadophylloides Franch. et Sav. and its correlation with calcium accumulation. *Soil Science and Plant Nutrition*, *59*(4), 591–602. doi:10.1080/00380768.2013.807213

Mohamad, N., Mohd Amin, M. C. I., Pandey, M., Ahmad, N., & Rajab, N. F. (2014). Bacterial cellulose/acrylic acid hydrogel synthesized via electron beam irradiation: Accelerated burn wound healing in an animal model. *Carbohydrate Polymers*, *114*, 312–320. doi:10.1016/j.carbpol.2014.08.025 PMID:25263896

Mohamed, A., Anas, A. K., Bakar, S. A., Ardyani, T., Zin, W. M. W., Ibrahim, S., & Eastoe, J. et al. (2015). Enhanced dispersion of multiwall carbon nanotubes in natural rubber latex nanocomposites by surfactants bearing phenyl. *Journal of Colloid and Interface Science*, *455*, 179–187. doi:10.1016/j.jcis.2015.05.054 PMID:26070188

Mohanty. (2005). *Natural Fibers, Biopolymers and Biocomposites*. CRC Press.

Mohanty, A. K., Depan, D., Tomer, N. S., & Singh, R. P. (2007). Nanoscale particles for polymer degradation and stabilization—Trends and future perspectives. *Progress in Polymer Science*, *34*, 479–515.

Mohiuddin, E., Isa, Y. M., Mdleleni, M. M., Sincadu, N., Key, D., & Tshabalala, T. (2016). Synthesis of ZSM-5 from impure and beneficiated Grahamstown kaolin: Effect of kaolinite content, crystallisation temperatures and time. *Applied Clay Science*, *119*, 213–221. doi:10.1016/j.clay.2015.10.008

Mokhothu, T. H., Luyt, A. S., Morselli, D., Bondioli, F., & Messori, M. (2015). Influence of in situ-generated silica nanoparticles on EPDM morphology, thermal, thermomechanical, and mechanical properties. *Polymer Composites*, *36*(5), 825–833. doi:10.1002/pc.23002

Mori, H., Yahagi, M., & Endo, T. (2009). Raft polymerization of N-vinylimidazolium salts and synthesis of thermoresponsive ionic liquid block copolymers. *Macromolecules*, *42*(21), 8082–8092. doi:10.1021/ma901180j

Morino, Y., Ohara, T., Watanabe, M., Hayashi, S., & Nishizawa, M. (2013). Episode analysis of deposition of radiocesium from Fukushima daiich neuclear power plant accident. *Environmental Science & Technology*, *47*, 2314–2322.

Mori, R., Nakai, T., Enomoto, K., Uchio, Y., & Yoshino, K. (2011). Increased antibiotic release from a bone cement containing bacterial cellulose. *Clinical Orthopaedics and Related Research*, *469*(2), 600–606. doi:10.1007/s11999-010-1626-8 PMID:20945120

Morishita, T., & Boratynski, J. K. (1992). Accumulation of cadmium and other metals in organs of plants growing around metal smelters in Japan. *Soil Science and Plant Nutrition*, *38*(4), 781–785. doi:10.1080/00380768.1992.10416712

Moritz, S., Wiegand, C., Wesarg, F., Hessler, N., Müller, F. A., Kralisch, D., & Fischer, D. et al. (2014). Active wound dressings based on bacterial nanocellulose as drug delivery system for octenidine. *International Journal of Pharmaceutics*, *471*(1–2), 45–55. doi:10.1016/j.ijpharm.2014.04.062 PMID:24792978

MREPC. (2016). *Important Facts about Latex and Latex Gloves*. Retrieved 04 February, 2016, from http://www.latex-gloves.info/manufacturing.php

Muenduen, P., & Suchata, K. (2015). *Synthesis and Characterization of Bacterial Cellulose-Based Composites and Their Applications. In Handbook of Sustainable Polymers* (pp. 495–532). Pan Stanford.

Mugnier, D., & Goetz, V. (2001). Energy storage comparison of sorption systems for cooling and refrigeration. *Solar Energy*, *71*(1), 47–55. doi:10.1016/S0038-092X(01)00013-5

Mühlethaler, K. (1949). The structure of bacterial cellulose. *Biochimica et Biophysica Acta*, *3*(0), 527–535. doi:10.1016/0006-3002(49)90126-2

Mukoma, P., Jooste, B. R., & Vosloo, H. C. M. (2004). Synthesis and characterization of cross-linked chitosan membranes for application as alternative proton exchange membrane materials in fuel cells. *Journal of Power Sources*, *136*(1), 16–23. doi:10.1016/j.jpowsour.2004.05.027

Mulder, M. (1991). *Basic Principles of Membrane Technology*. Dordrecht, The Netherland: Kluwer. doi:10.1007/978-94-017-0835-7

Muñiz-Villarreal, M. S., Manzano-Ramírez, A., Sampieri-Bulbarela, S., Gasca-Tirado, J. R., Reyes-Araiza, J. L., Rubio-Ávalos, J. C., & Amigó-Borrás, V. et al. (2011). The effect of temperature on the geopolymerization process of a metakaolin-based geopolymer. *Materials Letters*, *65*(6), 995–998. doi:10.1016/j.matlet.2010.12.049

Murphy, C. M., & OBrien, F. J. (2010). Understanding the effect of mean pore size on cell activity in collagen-glycosaminoglycan scaffolds. *Cell Adhesion & Migration*, *4*(3), 377–381. doi:10.4161/cam.4.3.11747 PMID:20421733

Murray, H. H. (1991). Overview - clay mineral applications. *Applied Clay Science*, *5*(5-6), 379–395. doi:10.1016/0169-1317(91)90014-Z

Mutanda, T., Ramesh, D., Karthikeyan, S., Kumari, S., Anandraj, A., & Bux, F. (2011). Bioprospecting for hyper-lipid producing microalgal strains for sustainable biofuel production. *Bioresource Technology*, *102*(1), 57–70. doi:10.1016/j.biortech.2010.06.077 PMID:20624676

Myers, R. S., & Clever, H. L. (1969). Surface tension of octamethylcyclotetrasiloxane and hexamethyldisilazane and their solutions with carbon tetrachloride and n-hexadecane. *Journal of Chemical & Engineering Data*, *14*(2), 161–164. doi:10.1021/je60041a014

Naef, A. A. (2013). Improving ice productivity and performance for an activated carbon/methanol solar adsorption ice-maker. *Solar Energy*, *98*, 523–542. doi:10.1016/j.solener.2013.10.018

Nair, A. B., & Joseph, R. (2014). 9 - Eco-friendly bio-composites using natural rubber (NR) matrices and natural fiber reinforcements A2 - Kohjiya, Shinzo. In Y. Ikeda (Ed.), *Chemistry, Manufacture and Applications of Natural Rubber* (pp. 249–283). Woodhead Publishing. doi:10.1533/9780857096913.2.249

Nair, L. S., & Laurencin, C. T. (2007). Biodegradable polymers as biomaterials. *Progress in Polymer Science*, *32*(8-9), 762–767. doi:10.1016/j.progpolymsci.2007.05.017

Nakamura, K., Saiwaki, T., & Fukao, K. (2010). Dielectric Relaxation Behavior of Polymerized Ionic Liquid. *Macromolecules*, *43*(14), 6092–6098. doi:10.1021/ma100918e

Nakarai e-kit. (2015). Retrieved from http://www.nacalai.co.jp/products/recommend/fatty_acid_methylation_kit.html

Nakaramontri, Y., Nakasan, C., & Kummerlowe, C. (2016). Enhancement of electrical conductivity and filler dispersion of carbon nanotube filled natural rubber composites latex mixing and in situ silanization. *Rubber Chemistry and Technology*, *89*(2), 272–291. doi:10.5254/rct.15.84848

Nakashima, K., Kubota, F., Maruyama, T., & Goto, M. (2003). Ionic liquids as a novel solvent for lanthanide extraction. *Analytical Sciences*, *19*(8), 1097–1098. doi:10.2116/analsci.19.1097 PMID:12945658

Nakashima, K., Kubota, F., Maruyama, T., & Goto, M. (2005). Feasibility of ionic liquidsas alternative separation mediaforindustrial solvent extraction processes. *Industrial & Engineering Chemistry Research*, *44*(12), 4368–4372. doi:10.1021/ie049050t

Nakashima, T., & Kawai, T. (2005). Quantum Dots-Ionic Liquid Hybrids: Efficient Extraction of Cationic Cdte Nanocrystals into An Ionic Liquid. *Chemical Communications (Cambridge)*, (12): 1643–1645. doi:10.1039/B418001A PMID:15770287

Nakasone, K., Ikematsu, S., & Kobayashi, T. (2016). Biocompatibility evaluation of cellulose hydrogel film regenerated from sugarcane bagasse waste and its in vivo behavior in mice. *Industrial & Engineering Chemistry Research*, *55*(1), 30–37. doi:10.1021/acs.iecr.5b03926

Nakasone, K., & Kobayashi, T. (2016). Effect of pre-treatment of sugarcane bagasse on the cellulose solution and application for the cellulose hydrogel films. *Polymers for Advanced Technologies*, *27*(7), 973–980. doi:10.1002/pat.3757

Nakasone, K., & Kobayashi, T. (2016, July). Cytocompatible cellulose hydrogels containing trace lignin. *Materials Science and Engineering C*, *64*, 269–277. doi:10.1016/j.msec.2016.03.108 PMID:27127053

Nasti, G., Gentile, G., Cerruti, P., Carfagna, C., & Ambrogi, V. (2016). Double percolation of multiwalled carbon nanotubes in polystyrene/polylactic acid blends. *Polymer*, *99*, 193–203. doi:10.1016/j.polymer.2016.06.058

National Environmental Methods Index. (n.d.). Retrieved from https://www.nemi.gov/home/

National Institute for Environmental Studies. (2004). *Collection LIST of STRAINS*. Microalgae and Protozoa.

NatureWorks. (2007). *Technology Focus Report: Polylactic Acid Containing Fillers and Fibers*. Minnetonka, MN: NatureWorks.

Nazari, A., Bagheri, A., & Riahi, S. (2011). Properties of geopolymer with seeded fly ash and rice husk bark ash. *Materials Science and Engineering A*, *528*(24), 7395–7401. doi:10.1016/j.msea.2011.06.027

Nazari, A., & Sanjayan, J. G. (2015). Synthesis of geopolymer from industrial wastes. *Journal of Cleaner Production*, *99*, 297–304. doi:10.1016/j.jclepro.2015.03.003

Nema, A. H. K., & Fadiel, H. J. (2009). Experimental Study on A Hybrid Adsorption *Refrigeration System. Eng. & Tech. Journal*, *27*(9), 1790–1800.

Nery, W. E., & Kubota, L. T. (2013). Sensing approaches on paper-based devices: A review. *Analytical and Bioanalytical Chemistry*, *405*(24), 7573–7595. doi:10.1007/s00216-013-6911-4 PMID:23604524

Ng, C. C., Rahman, M. M., Boyce, A. N., & Abas, M. R. (2016). Heavy metals phyto-assessment in commonly grown vegetables: Water spinach (I. aquatica) and okra (A. esculentus). *SpringerPlus*, *5*(1), 469. doi:10.1186/s40064-016-2125-5 PMID:27119073

Ngoc, N. L., & Kobayashi, T. (2010). Ultrasound stimulus effect on hydrogen bonding in networked alumina and polyacrylic acid slurry. *Ultrasonics Sonochemistry*, *17*(1), 186–192. doi:10.1016/j.ultsonch.2009.04.007 PMID:19464939

Nguetnkam, J. P., Kamga, R., Villiéras, F., Ekodeck, G. E., Razafitianamaharavo, A., & Yvon, J. (2005). Assessment of the surface areas of silica and clay in acid-leached clay materials using concepts of adsorption on heterogeneous surfaces. *Journal of Colloid and Interface Science*, *289*(1), 104–115. doi:10.1016/j.jcis.2005.03.053 PMID:16009222

Nguyen, V. T., Gidley, M. J., & Dykes, G. A. (2008). Potential of a nisin-containing bacterial cellulose film to inhibit Listeria monocytogenes on processed meats. *Food Microbiology*, *25*(3), 471–478. doi:10.1016/j.fm.2008.01.004 PMID:18355672

Ni, B., Liu, M., Lü, S., Xie, L., & Wang, Y. (2010). Multifunctional Slow-Release Organic–Inorganic Compound Fertilizer. *Journal of Agricultural and Food Chemistry*, *58*(23), 12373–12378. doi:10.1021/jf1029306 PMID:21058723

Nishi, Y., Uryu, M., Yamanaka, S., Watanabe, K., Kitamura, N., Iguchi, M., & Mitsuhashi, S. (1990). The structure and mechanical properties of sheets prepared from bacterial cellulose. *Journal of Materials Science*, *25*(6), 2997–3001. doi:10.1007/BF00584917

Noble R. D., & Stern, S. A. (Eds.). (1995). *Membrane Separation Technology, Principles and Applications*. Elsevier.

Noda, I. (1993). Generalized two-dimensional correlation method applicable to infrared, raman, and other types of spectroscopy. *Applied Spectroscopy*, *47*(9), 1329–1336. doi:10.1366/0003702934067694

Noda, I., Dowrey, A. E., Marcoli, C., Story, G. M., & Ozaki, Y. (2000). Generalized two-dimensional correlation spectroscopy. *Applied Spectroscopy*, *54*(7), 236A–248A. doi:10.1366/0003702001950454

Nofar, M., Heuzey, M. C., Carreau, P. J., & Kamal, M. R. (2016). Effects of nanoclay and its localization on the morphology stabilization of PLA/PBAT blends under shear flow. *Polymer*, *98*, 353–364. doi:10.1016/j.polymer.2016.06.044

No, H. K., & Meyers, S. P. (1995). Preparation and Characterization of Chitin and Chitosan – A review. *Journal of Aquatic Food Product Technology*, *4*(2), 27–52. doi:10.1300/J030v04n02_03

Nomura, T., Itouga, M., Kojima, M., Kato, Y., Sakakibara, H., & Hasezawa, S. (2015). Copper mediates auxin signalling to control cell differentiation in the copper moss Scopelophila cataractae. *Journal of Experimental Botany*, *66*(5), 1205–1213. doi:10.1093/jxb/eru470 PMID:25428998

Norazlina, H., & Kamal, Y. (2015). Graphene modifications in polylactic acid nanocomposites: A review. *Polymer Bulletin*, *72*(4), 931–961. doi:10.1007/s00289-015-1308-5

Noreen, A., Zia, K. M., Zuber, M., Ali, M., & Mujahid, M. A critical review of algal biomass: A versatile platform of bio-based polyesters from renewable resources. *International Journal of Biological Macromolecules*. doi:10.1016/j.ijbiomac.2016.01.067 PMID:26808018

Novoselov, K. S., Geim, A. K., Morozov, S. V., Jiang, D., Dubonos, S. V., Grigorieva, I. V., & Firsoc, A. (2004). Electric Field Effect in Atomically Thin Carbon Films. *Science*, *306*(5696), 666–669. doi:10.1126/science.1102896 PMID:15499015

Noyan, H., Onal, M., & Sarikaya, Y. (2007). The effect of sulphuric acid activation on the crystallinity, surface area, porosity, surface acidity, and bleaching power of a bentonite. *Food Chemistry*, *105*(1), 156–163. doi:10.1016/j.foodchem.2007.03.060

Oberdisse, J. (2006). Aggregation of colloidal nanoparticles in polymer matrices. *Soft Matter*, *2*(1), 29–36. doi:10.1039/B511959F PMID:16733636

Ochi, M., Nii, D., & Harada, M. (2011, September). Preparation of epoxy/zirconia hybrid materials via in situ polymerization using zirconium alkoxide coordinated with acid anhydride. *Materials Chemistry and Physics*, *129*(1–2), 424–432. doi:10.1016/j.matchemphys.2011.04.034

Odinets, I. L., Sharova, E. V., Artyshin, O. I., Lyssenko, K. A., Nelyubina, Y. V., Myasoedova, G. V., & Zakharchenro, E. A. et al. (2010). Novel class of functionalized ionic liquids with grafted CMPO-moieties for actinides and rare-earth elements recovery. *Dalton Transactions (Cambridge, England)*, *39*(17), 4170–4178. doi:10.1039/b926377b PMID:20390180

Oike, T., Huang, H., Watanabe, F., Hayashi, N., Oshima, T., Hasatani, M., & Kobayashi, N. (2007). Development of novel composite adsorbents used for adsorption heat pump.*Proceedings of International Symposium on Eco Topia Science, ISETS07.*

Okada, A., Kawasumi, M., Kurauchi, T., & Kamigaito, O. (1987). Synthesis and characterization of a nylon 6-clay hybrid. *Papers Presented at the ... Meeting. American Chemical Society. Division of Polymer Chemistry*, *28*, 447.

Okamura, H., Hirayama, N., Morita, K., Shimojo, K., Naganawa, H., & Imura, H. (2010). Synergistic effect of 18-crown-6 derivatives on chelate extraction of lanthanoids(III) into an ionic liquid with 2-thenoyltrifluoroacetone. *Analytical Sciences*, *26*(5), 607–611. doi:10.2116/analsci.26.607 PMID:20467139

Okoye, F. N., Durgaprasad, J., & Singh, N. B. (2016). Effect of silica fume on the mechanical properties of fly ash based-geopolymer concrete. *Ceramics International*, *42*(2, Part B), 3000–3006. doi:10.1016/j.ceramint.2015.10.084

Oliveira Barud, H. G., Barud, H. D. S., Cavicchioli, M., Do Amaral, T. S., De Oliveira Junior, O. B., Santos, D. M., & Ribeiro, S. J. L. et al. (2015). Preparation and characterization of a bacterial cellulose/silk fibroin sponge scaffold for tissue regeneration. *Carbohydrate Polymers*, *128*, 41–51. doi:10.1016/j.carbpol.2015.04.007 PMID:26005138

Oliveira, R. G., Silveira, V. Jr, & Wang, R. Z. (2006). Experimental study of mass recovery adsorption cycles for ice making at low generation temperature. *Applied Thermal Engineering*, *26*(2-3), 303–311. doi:10.1016/j.applthermaleng.2005.04.021

Olivier, J. H., Camerel, F., & Ziessel, R. (2011). Lanthanide ion extraction by trifluoromethyl-1,3diketonate-functionalised ionic liquids adsorbed on silica. *Chemistry (Weinheim an der Bergstrasse, Germany)*, *17*(33), 9113–9122. doi:10.1002/chem.201100296 PMID:21739493

Önal, M. (2007). Swelling and cation exchange capacity relationship for the samples obtained from a bentonite by acid activations and heat treatments. *Applied Clay Science*, *37*(1-2), 74–80. doi:10.1016/j.clay.2006.12.004

Önal, M., & Sarikaya, Y. (2007). Preparation and characterization of acid-activated bentonite powders. *Powder Technology*, *172*(1), 14–18. doi:10.1016/j.powtec.2006.10.034

Onghena, B., & Binnemans, K. (2015). Recovery of Scandium(III) from Aqueous Solutions by Solvent Extraction with the Functionalized Ionic Liquid Betainium Bis(trifluoromethylsulfonyl)imide. *Industrial & Engineering Chemistry Research*, *54*(6), 1887–1898. doi:10.1021/ie504765v

Ono, Y. (2014). *Seven countries allying to halt rubber price slide*. Retrieved 03 February, 2016, from http://asia.nikkei.com/Politics-Economy/International-Relations/Seven-countries-allying-to-halt-rubber-price-slide

Onutai, S., Jiemsirilers, S., Thavorniti, P., & Kobayashi, T. (2016). Fast microwave syntheses of fly ash based porous geopolymers in the presence of high alkali concentration. *Ceramics International*, *42*(8), 9866–9874. doi:10.1016/j.ceramint.2016.03.086

Onutai, S., Sirithan, J., Thavorniti, P., & Kobayashi, T. (2015). Aluminium hydroxide waste based geopolymer composed of fly ash for sustainable cement materials. *Construction & Building Materials*, *101*, 298–308. doi:10.1016/j.conbuildmat.2015.10.097

O-Rak, K., Ummartyotin, S., Sain, M., & Manuspiya, H. (2013). Covalently grafted carbon nanotube on bacterial cellulose composite for flexible touch screen application. *Materials Letters*, *107*, 247–250. doi:10.1016/j.matlet.2013.05.126

Osada, I., de Vries, H., Scrosati, B., & Passerini, S. (2015). Ionic-liquid-based polymer electrolytes for battery applications. *Angewandte Chemie International Edition*, *55*(2), 500–513. doi:10.1002/anie.201504971 PMID:26783056

Ougiya, H., Watanabe, K., Morinaga, Y., & Yoshinaga, F. (1997). Emulsion-stabilizing Effect of Bacterial Cellulose. *Bioscience, Biotechnology, and Biochemistry*, *61*(9), 1541–1545. doi:10.1271/bbb.61.1541

Oyarzún, J. (1999). Pigment Processing: Physico-chemical Principles. Vincentz Verlag.

Oya, S., Kanno, D., Watanabe, H., Tamura, M., Nakagawa, Y., & Tomishige, K. (2015). Catalytic Production on Branched Small Alkanes from Biohydrocarbons. *ChemSusChem*, *8*(15), 2472–2475. doi:10.1002/cssc.201500375 PMID:26097221

Palkovits, R., Althues, H., Rumplecker, A., Tesche, B., Dreier, A., Holle, U., & Kaskel, S. et al. (2005). Polymerization of w/o Microemulsions for the Preparation of Transparent SiO2/PMMA Nanocomposites. *Langmuir*, *21*(13), 6048–6053. doi:10.1021/la050630k PMID:15952859

Pandey, M., Amin, M. C. I. M., Ahmad, N., & Abeer, M. M. (2013). Rapid synthesis of superabsorbent smart-swelling bacterial cellulose/acrylamide-based hydrogels for drug delivery. *International Journal of Polymer Science*.

Pandey, A., Soccol, C. R., Nigam, P., & Soccol, V. T. (2000). Biotechnological potential of agro-industrial residues I, Sugar Bagasse. *Bioresource Technology*, *74*(1), 69–80. doi:10.1016/S0960-8524(99)00142-X

Pan, F., Lu, X., Wang, Y., Chen, S., Wang, T., & Yan, Y. (2014). Synthesis and crystallization kinetics of ZSM-5 without organic template from coal-series kaolinite. *Microporous and Mesoporous Materials*, *184*, 134–140. doi:10.1016/j.micromeso.2013.10.013

Pangdaeng, S., Phoo-ngernkham, T., Sata, V., & Chindaprasirt, P. (2014). Influence of curing conditions on properties of high calcium fly ash geopolymer containing Portland cement as additive. *Materials & Design*, *53*, 269–274. doi:10.1016/j.matdes.2013.07.018

Panias, D., Giannopoulou, I. P., & Perraki, T. (2007). Effect of synthesis parameters on the mechanical properties of fly ash-based geopolymers. *Colloids and Surfaces. A, Physicochemical and Engineering Aspects*, *301*(1–3), 246–254. doi:10.1016/j.colsurfa.2006.12.064

Park, E. J., Kim, W. S., Hwang, H. S., Park, C., & Lim, K. T. (2007). PMMA Encapsulated Silica Particles by Dispersion Polymerization in Supercritical Carbon Dioxide. *Macromolecular Symposia*, *249-250*(1), 196–201. doi:10.1002/masy.200750332

Park, J.-G., Ye, Q., Topp, E. M., Misra, A., & Spencer, P. (2009). Water sorption and dynamic mechanical properties of dentin adhesives with a urethane-based multifunctional methacrylate monomer. *Dental Materials*, *25*(12), 1569–1575. doi:10.1016/j.dental.2009.07.010 PMID:19709724

Park, S., An, J. H., Jung, I. W., Piner, R. D., An, S. J., Li, X. S., & Ruoff, R. S. et al. (2009). Colloidal suspensions of highly reduced graphene oxide in a wide variety of organic solvents. *Nano Letters*, *9*(4), 1593–1597. doi:10.1021/nl803798y PMID:19265429

Park, S., & Ruoff, R. S. (2009). Chemical methods for the production of graphenes. *Nature Nanotechnology*, *5*(4), 217–224. doi:10.1038/nnano.2009.58 PMID:19350030

Park, Y., Abolmaali, A., Kim, Y. H., & Ghahremannejad, M. (2016). Compressive strength of fly ash-based geopolymer concrete with crumb rubber partially replacing sand. *Construction & Building Materials*, *118*, 43–51. doi:10.1016/j.conbuildmat.2016.05.001

Patole, A. S., Patole, S. P., Kang, H., Yoo, J.-B., Kim, T.-H., & Ahn, J.-H. (2010). A facile approach to the fabrication of graphene/polystyrene nanocomposite by in situ microemulsion polymerization. *Journal of Colloid and Interface Science*, *350*(2), 530–537. doi:10.1016/j.jcis.2010.01.035 PMID:20673907

Patterson, J. W., & Minear, R. A. (1975). Physical-Chemical Methods of Heavy Metal Removal. In P. A. Kenkel (Ed.), *Heavy Metals in the Aquatic Environment* (pp. 261–276). Oxford, UK: Pergamon Press. doi:10.1016/B978-0-08-018068-7.50040-X

Pavaloiu, R.-D., Stoica, A., Stroescu, M., & Dobre, T. (2014). Controlled release of amoxicillin from bacterial cellulose membranes. *Central European Journal of Chemistry*, *12*(9), 962–967. doi:10.2478/s11532-014-0541-3

Pavithra, P., Srinivasula Reddy, M., Dinakar, P., Hanumantha Rao, B., Satpathy, B. K., & Mohanty, A. N. (2016). A mix design procedure for geopolymer concrete with fly ash. *Journal of Cleaner Production*, *133*, 117–125. doi:10.1016/j.jclepro.2016.05.041

Pawar, R. R., Lalhmunsiama, , Bajaj, H. C., & Lee, S.-M. (2016). Activated bentonite as a low-cost adsorbent for the removal of Cu(II) and Pb(II) from aqueous solutions: Batch and column studies. *Journal of Industrial and Engineering Chemistry*, *34*, 213–223. doi:10.1016/j.jiec.2015.11.014

Paydayesh, A., Arefazar, A., & Jalaliarani, A. (2016). A morphological study on the migration and selective localization of graphene in the PLA/PMMA blends. *Journal of Applied Polymer Science*, *133*(34), 43799–43810. doi:10.1002/app.43799

Perelomov, L., Sarkar, B., Rahman, M. M., Goryacheva, A., & Naidu, R. (2016). Uptake of lead by Na-exchanged and Al-pillared bentonite in the presence of organic acids with different functional groups. *Applied Clay Science*, *119*, 417–423. doi:10.1016/j.clay.2015.11.004

Perez de los Rios, A., Hernandez-Fernandez, F. J., Sanchez-Segado, S., Lozano, L. J., Moreno, J. I., & Godinez, C. (2010). Selective Separation of Zn(II) and Fe(III) from Acidic Media Using Ionic Liquids as Sole Extraction Agents. *Chem. Eng. Transactions*, *21*, 625–630.

Performance of Philippines' Top Non-Traditional Coco Products Export. (n.d.). Retrieved March 2015 from http://ucap.org.ph/news-and-events/category/industry-news

Pernak, J., Stefaniak, F., & Weglewski, J. (2005). Phosphonium Acesulfamate Based Ionic Liquids. *European Journal of Organic Chemistry*, (4): 650–652. doi:10.1002/ejoc.200400658

Pertile, R. A. N., Andrade, F. K., Alves, C. Jr, & Gama, M. (2010). Surface modification of bacterial cellulose by nitrogen-containing plasma for improved interaction with cells. *Carbohydrate Polymers*, *82*(3), 692–698. doi:10.1016/j.carbpol.2010.05.037

Petersen, N., & Gatenholm, P. (2011). Bacterial cellulose-based materials and medical devices: Current state and perspectives. *Applied Microbiology and Biotechnology*, *91*(5), 1277–1286. doi:10.1007/s00253-011-3432-y PMID:21744133

Peterson, P. J. (1983). CHAPTER 4 Adaptation to Toxic Metals. In D. A. Robb (Ed.), *Metals and Micronutrients* (pp. 51–69). doi:10.1016/B978-0-12-589580-4.50010-9

Petkovic, M., Seddon, K. R., Rebelo, L. P., & Pereira, C. S. (2011). Ionic Liquids: A Pathway to Environmental Acceptability. *Chemical Society Reviews*, *40*(3), 1383–1403. doi:10.1039/C004968A PMID:21116514

Pham, J. Q., Mitchell, C. A., Bahr, J. L., Tour, J. M., Krishanamoorti, R., & Green, P. F. (2003). Glass transition of polymer/single-walled carbon nanotube composite films. *Journal of Polymer Science. Part B, Polymer Physics*, *41*(17), 3339–4920. doi:10.1002/polb.10702

Phetwarotai, W., Potiyaraj, P., & Aht-Ong, D. (2012). Characteristics of biodegradable polylactide/gelatinized starch films: Effects of starch, plasticizer, and compatibilizer. *Journal of Applied Polymer Science*, *126*(S1), E162–E172. doi:10.1002/app.36736

Phetwarotai, W., Potiyaraj, P., & Aht-Ong, D. (2013). Biodegradation of polylactide and gelatinized starch blend films under controlled soil burial conditions. *Journal of Polymers and the Environment*, *21*(1), 95–107. doi:10.1007/s10924-012-0530-6

Pilla, S. (2011). Handbook of bioplastics and biocomposites Engineering applications. Scrivener Publishing. doi:10.1002/9781118203699

Pivsa-Art, W., Fujii, K., Nomura, K., Aso, Y., Ohara, H., & Yamane, Y. (2015). The effect of poly(ethylene glycol) as plasticizer in blends of poly(lactic acid) and poly(butylene succinate). *Journal of Applied Polymer Science*, *133*(8), 43044–43053.

Plechkova, N. V., & Seddon, K. R. (2008). Applications of Ionic Liquids in The Chemical Industry. *Chemical Society Reviews*, *37*(1), 123–150. doi:10.1039/B006677J PMID:18197338

Pokhrel, D. T., & Viroraghavan, T. (2004). Treatment of pulp and paper mill wastewater a review. *The Science of the Total Environment*, *333*(1-3), 37–43. doi:10.1016/j.scitotenv.2004.05.017 PMID:15364518

Pollution Control Department, Thailand. (n.d.a). *Industrial Effluent Standards*. Retrieved from http://www.pcd.go.th/info_serv/en_reg_std_water04.html

Pollution Control Department, Thailand. (n.d.b). *Surface water quality standards*. Retrieved from http://www.pcd.go.th/info_serv/en_reg_std_water05.html

Poompiew, N., & Potiyaraj, P. (2016, January) *Preparation and properties of poly (lactic acid) reinforced with graphene nanoparticle*. Paper presented at International Conference on Applied Sciences, Fukuoka, Japan.

Potts, J., Lynch, M., Wilkings, A., Huppé, G., Cunningham, M., & Voora, V. (2014). Chaper 8 Coffee Market. In R. Ilnyckyj, D. Holmes, & E. Rickert (Eds.), *SSI Review* (pp. 155–186). London: International Institute for Environment and Development.

Prasertsri, S., & Rattanasom, N. (2011). Mechanical and damping properties of silica/natural rubber composites prepared from latex system. *Polymer Testing*, *30*(5), 515–526. doi:10.1016/j.polymertesting.2011.04.001

Puertas, F., Martínez-Ramírez, S., Alonso, S., & Vázquez, T. (2000). Alkali-activated fly ash/slag cements: Strength behaviour and hydration products. *Cement and Concrete Research*, *30*(10), 1625–1632. doi:10.1016/S0008-8846(00)00298-2

Qi, D., Bao, Y., Weng, Z., & Huang, Z. (2006). Preparation of acrylate polymer/silica nanocomposite particles with high silica encapsulation efficiency via miniemulsion polymerization. *Polymer*, *47*(13), 4622–4629. doi:10.1016/j.polymer.2006.04.024

Qi, Z. (2005). Study on hybrid system of solar powered water heater and adsorption ice maker. *International Journal on Architectural Science.*, *6*, 168–172.

Qu, T. F., Wang, R. Z., & Wang, W. (2001). Study on heat and mass recovery in adsorption refrigeration cycles. *Applied Thermal Engineering*, *21*(4), 439–452. doi:10.1016/S1359-4311(00)00050-8

Rahman, A. F., Ueda, Y., Akisawa, A., Miyazaki, T., & Saha, B. B. (2013). Design and Performance of an Innovative Four-Bed, Three-Stage Adsorption Cycle. *Energies*, *6*(3), 1365–1384. doi:10.3390/en6031365

Rahman, M. H., Akhanda, A. R., & Islam, A. K. M. S. (2006). Study, Design and Fabrication of a Solar Powered Adsorption Refrigeration System. *ISESCO*, *2*(1), 10–13.

Raja, M., Ryu, S. H., & Shanmugharaj, A. M. (2013). Thermal, mechanical and electroactive shape memory properties of polyurethane (PU)/poly (lactic acid) (PLA)/CNT nanocomposites. *European Polymer Journal*, *49*(11), 3492–3500. doi:10.1016/j.eurpolymj.2013.08.009

Raj, S., Jose, S., Sumod, U. S., & Sabitha, M. (2012). Nanotechnology in cosmetics: Opportunities and challenges. *Journal of Pharmacy & Bioallied Sciences*, *4*(3), 186–193. doi:10.4103/0975-7406.99016 PMID:22923959

Ramanathan, T., Liu, H., & Brinson, L. C. (2005). Functionalized SWNT/polymer nanocomposites for dramatic property improvement. *Journal of Polymer Science. Part B, Polymer Physics*, *43*(17), 2269–2279. doi:10.1002/polb.20510

Ramesan, M. T. (2004). Thermogravimetric analysis, flammability and oil resistance properties in natural rubber and dichlorocarbene modified styrene butadiene rubber blends. *Reactive & Functional Polymers*, *59*(3), 267–274. doi:10.1016/j.reactfunctpolym.2004.02.005

Ramesh, A., Hasegawa, H., Sugimoto, W., Maki, T., & Ueda, K. (2008). Adsorption of gold(III), platinum(IV) and palladium(II) onto glycine modified crosslinked chitosan resin. *Bioresource Technology*, *99*(9), 3801–3809. doi:10.1016/j.biortech.2007.07.008 PMID:17698356

Ramirez-Salgado, J. (2007). Study of basic biopolymer as proton membrane for fuel cell systems. *Electrochimica Acta*, *52*(11), 3766–3778. doi:10.1016/j.electacta.2006.10.051

Ramji, H. R., Leo, S. L., Tan, I. A. W., & Abdullah, M. O. (2014). Comparative study of three different adsorbent-adsorbate working pairs for a waste heat driven adsorption air conditioning system based on simulation. *IJRRAS*, *18*(2), 109–121.

Ramos, L. P. (2003). The chemistry involved in the steam treatment of lignocellulosic materials. *Quimica Nova*, *26*(6), 23–31. doi:10.1590/S0100-40422003000600015

Rascio, N., & Navari-Izzo, F. (2011). Heavy metal hyperaccumulating plants: How and why do they do it? And what makes them so interesting? *Plant Science*, *180*(2), 169–181. doi:10.1016/j.plantsci.2010.08.016 PMID:21421358

Rattanasom, N., Saowapark, T., & Deeprasertkul, C. (2007). Reinforcement of natural rubber with silica/carbon black hybrid filler. *Polymer Testing*, *26*(3), 369–377. doi:10.1016/j.polymertesting.2006.12.003

Ravi Kumar, M. N. V. (2000). A review of chitin and chitosan applications. *Reactive & Functional Polymers*, *46*(1), 1–27. doi:10.1016/S1381-5148(00)00038-9

Raymond, A. W. (2010). *Investigation of microparticle to system level phenomena in thermally activated adsorption heat pumps* (Unpublished master dissertation). Georgia Institute of Technology.

Ray, S. S., & Okamoto, M. (2003). Polymer/layered silicate nanocomposites: A review from the preparation to processing. *Progress in Polymer Science*, *28*(11), 1539–1641. doi:10.1016/j.progpolymsci.2003.08.002

Reddy, K. R., Sin, B. C., Yoo, C. H., Sohn, D., & Lee, Y. (2009). Coating of multiwalled carbon nanotubes with polymer nanospheres through microemulsion polymerization. *Journal of Colloid and Interface Science*, *340*(2), 160–165. doi:10.1016/j.jcis.2009.08.044 PMID:19800632

Reichert, W. M., Holbrey, J. D., Swatloski, R. P., Gutowski, K. E., Visser, A. E., Nieuwenhuyzen, M., & Rogers, R. D. et al. (2007). Solid-state analysis of low-melting 1, 3-dialkylimidazolium hexafluorophosphate salts (ionic liquids) by combined X-ray crystallographic and computational analyses. *Crystal Growth & Design*, *7*(6), 1106–1114. doi:10.1021/cg060746u

Reid, J. S. G. (1997). Carbohydrate methbilism: Structural carbohydrates. In Plant Biochemistry. Academic Press.

Retegi, A., Algar, I., Martin, L., Altuna, F., Stefani, P., Zuluaga, R., & Mondragon, I. et al. (2012). Sustainable optically transparent composites based on epoxidized soy-bean oil (ESO) matrix and high contents of bacterial cellulose (BC). *Cellulose (London, England)*, *19*(1), 103–109. doi:10.1007/s10570-011-9598-8

Retegi, A., Gabilondo, N., Peña, C., Zuluaga, R., Castro, C., Gañan, P., & Mondragon, I. et al. (2010). Bacterial cellulose films with controlled microstructure-mechanical property relationships. *Cellulose (London, England)*, *17*(3), 661–669. doi:10.1007/s10570-009-9389-7

Rezende, C. A., Bragança, F. C., Doi, T. R., Lee, L.-T., Galembeck, F., & Boué, F. (2010). Natural rubber-clay nanocomposites: Mechanical and structural properties. *Polymer*, *51*(16), 3644–3652. doi:10.1016/j.polymer.2010.06.026

Ribeiro, M. J., & Labrincha, J. A. (2008). Properties of sintered mullite and cordierite pressed bodies manufactured using Al-rich anodising sludge. *Ceramics International*, *34*(3), 593–597. doi:10.1016/j.ceramint.2006.12.005

Rickard, W. D. A., Vickers, L., & van Riessen, A. (2013). Performance of fibre reinforced, low density metakaolin geopolymers under simulated fire conditions. *Applied Clay Science*, *73*, 71–77. doi:10.1016/j.clay.2012.10.006

Riffat, S. B., Williams, M. D., & Stuart, C. (1997). Adsorption heat pump using HFC refrigerants. *International Journal of Energy Research*, *21*(6), 481–494. doi:10.1002/(SICI)1099-114X(199705)21:6<481::AID-ER277>3.0.CO;2-7

Riffel, D. B., Belo, F. A., & Leite, A. P. F. (2009). Heat and mass recovery in an activated carbon-methanol adsorptive chiller.*20th International Congress of Mechanical Engineering*, Gramado, RS, Brazil.

Ring, D. F., Nashed, W., & Dow, T. (1987). *Pressing, releasing polysaccharide pellicles to rearrange into lamellas*. Google Patents.

Rivas, B. L., & Geckeler, K. E. (1992). Synthesis and metal complexation of poly(ethylenimine) and derivatives. *Advances in Polymer Science*, *102*, 171–188. doi:10.1007/3-540-55090-9_6

Rivas, B. L., Pereira, E. D., & Moreno-Villoslada, I. (2003). Water-soluble polymer–metal ion interactions Review Article. *Progress in Polymer Science*, *28*(2), 173–208. doi:10.1016/S0079-6700(02)00028-X

Robbins, C. (n.d.). *Solar Cooling Technology Review*. Retrieved from https://www.dri.edu/images/stories/editors/receditor/Solar_Cooling_Review.pdf

Roberts, A. D. (1998). *Natural Rubber Science and Technology*. Oxford, UK: Oxford University Press.

Roberts, J. C. (1997). *The chemistry of paper*. Cambridge, UK: RSC Paperbacks.

Robert, W. P., & Young, K. H. (1985). Evaluation of recent treatment techniques for removal of heavy metals from industrial wastewaters.*American Industrial Chemical Engineering Symposium Series, No. 243,*Vol. 81, 165-203.

Rogers, R. D. (2007). Materials science: Reflections on ionic liquids. *Nature*, *447*(7147), 917–918. doi:10.1038/447917a PMID:17581570

Romio, A. P., Rodrigues, H. H., Peres, A., Da Cas Viegas, A., Kobitskaya, E., Ziener, U., & Araújo, P. H. H. et al. (2013). Encapsulation of magnetic nickel nanoparticles via inverse miniemulsion polymerization. *Journal of Applied Polymer Science*, *129*(3), 1426–1433. doi:10.1002/app.38840

Rose, K., & Steinbüchel, A. (2005). Biodegradation of natural rubber and related compounds: Recent insights into a hardly understood catabolic capability of microorganisms. *Applied and Environmental Microbiology*, *71*(6), 2803–2812. doi:10.1128/AEM.71.6.2803-2812.2005 PMID:15932971

Rosenberg, A. A., Fogarty, M. J., Sissenwine, M. P., Beddington, J. R., & Shepherd, J. G. (1993). Achieving sustainable use of renewable resources. *Science*, *262*(5), 828–829. doi:10.1126/science.262.5135.828 PMID:17757341

Ross, P., Mayer, R., & Benziman, M. (1991). Cellulose biosynthesis and function in bacteria. *Microbiological Reviews*, *55*(1), 35–58. PMID:2030672

Rouabhia, M., Asselin, J., Tazi, N., Messaddeq, Y., Levinson, D., & Zhang, Z. (2014). Production of biocompatible and antimicrobial bacterial cellulose polymers functionalized by RGDC grafting groups and gentamicin. *ACS Applied Materials & Interfaces*, *6*(3), 1439–1446. doi:10.1021/am4027983 PMID:24422537

Rout, A., & Binnemans, K. (2014). Liquid–liquid extraction of europium(III) and other trivalent rare-earth ions using a non-fluorinated functionalized ionic liquid. *Dalton Transactions (Cambridge, England)*, *43*(4), 1862–1872. doi:10.1039/C3DT52285G PMID:24257814

Rout, A., Karmakar, S., Venkatesan, K. A., Srinivasan, T. G., & Rao, P. R. V. (2011). Room temperature ionic liquid diluent for the mutual separation of europium (III) from americium (III). *Separation and Purification Technology*, *81*(2), 109–115. doi:10.1016/j.seppur.2011.04.033

Rout, A., Kotlarska, J., Dehaen, W., & Binnemans, K. (2013). Liquid–liquid extraction of neodymium(III) by dialkylphosphate ionic liquids from acidic medium: The importance of the ionic liquid cation. *Physical Chemistry Chemical Physics*, *15*(39), 16533–16541. doi:10.1039/c3cp52218k PMID:23949284

Rout, A., Wellens, S., & Binnemans, S. (2014). Separation of rare earths and nickel by solvent extraction with two mutually immiscible ionic liquids. *RSC Advances*, *4*(11), 5753–5758. doi:10.1039/c3ra46261g

Roviello, G., Ricciotti, L., Ferone, C., Colangelo, F., & Tarallo, O. (2015). Fire resistant melamine based organic-geopolymer hybrid composites. *Cement and Concrete Composites*, *59*, 89–99. doi:10.1016/j.cemconcomp.2015.03.007

Rovnaník, P. (2010). Effect of curing temperature on the development of hard structure of metakaolin-based geopolymer. *Construction & Building Materials*, *24*(7), 1176–1183. doi:10.1016/j.conbuildmat.2009.12.023

Rowe, K. C. (2014). Numerical analysis of multi-bed activated carbon/methanol adsorption refrigeration system with allowance for dynamic pressure variation.*3rd IIR Conference on Cold Chain and Sustainability*, London

Rowlands, A. S., George, P. A., & Cooper-White, J. J. (2008). Directing osteogenic and myogenic differentiation of MSCs: Interplay of stiffness and adhesive ligand presentation. *American Journal of Physiology. Cell Physiology*, *295*(4), C1037–C1044. doi:10.1152/ajpcell.67.2008 PMID:18753317

Rozploch, F., Patyk, J., & Stankowski, J. (2007). Graphenes bonding forces in vulcanization kinetics of graphene/natural rubber nanocomposites. *Acta Phisica Polanica A*, *112*(3), 557.

Ruess, V. G., & Vogt, F. (1948). Hochstlamellarer Kohlenstoff aus Graphitoxyhydroxyd. *Monatshefte fur Chemie*, *78*(3-4), 222–242. doi:10.1007/BF01141527

Ryu, G. S., Lee, Y. B., Koh, K. T., & Chung, Y. S. (2013). The mechanical properties of fly ash-based geopolymer concrete with alkaline activators. *Construction & Building Materials*, *47*(0), 409–418. doi:10.1016/j.conbuildmat.2013.05.069

Saeed, A., Akhter, M. E., & Iqbal, M. (2005). Removal and recovery of heavy metals from aqueous solution using papaya wood as a new biosorbent. *Separation and Purification Technology*, *45*(1), 25–31. doi:10.1016/j.seppur.2005.02.004

Saeed, A., Iqbal, M., & Akhtar, M. W. (2005). Removal and recovery of lead (II) from single and multimetal (Cd, Cu, Ni, Zn) solutions by crop milling waste (black gram husk). *Journal of Hazardous Materials*, *117*(1), 65–73. doi:10.1016/j.jhazmat.2004.09.008 PMID:15621354

Saha, B. B., El-Sharkawy, I. I., Chakraborty, A., & Koyama, S. (2007). Study on an activated carbon fiber-ethanol adsorption chiller: Part I-system description and modelling. *International Journal of Refrigeration*, *30*(1), 86–95. doi:10.1016/j.ijrefrig.2006.08.004

Saha, B. B., Habib, K., El-Sharkawy, I. I., & Koyama, S. (2009). Adsorption characteristics and heat of adsorption measurements of R-134a on activated carbon. *International Journal of Refrigeration*, *32*(7), 1563–1569. doi:10.1016/j.ijrefrig.2009.03.010

Sahoo, P. K., John, M., Newalkar, B. L., Choudhary, N. V., & Ayappa, K. G. (2011). Filling Characteristics for an Activated Carbon Based Adsorbed Natural Gas Storage System. *Industrial & Engineering Chemistry Research*, *50*(23), 13000–13011. doi:10.1021/ie200241x

Sahu, O. P., & Chaudhari, P. K. (2013). Review on Chemical treatment of Industrial Waste Water. *J. Appl. Sci. Environ. Manage.*, *17*(2), 241–257.

Saibuatong, O., & Phisalaphong, M. (2010). Novo aloe vera–bacterial cellulose composite film from biosynthesis. *Carbohydrate Polymers*, *79*(2), 455–460. doi:10.1016/j.carbpol.2009.08.039

Sai, H., Fu, R., Xing, L., Xiang, J., Li, Z., Li, F., & Zhang, T. (2015). Surface Modification of Bacterial Cellulose Aerogels Web-like Skeleton for Oil/Water Separation. *ACS Applied Materials & Interfaces*, *7*(13), 7373–7381. doi:10.1021/acsami.5b00846 PMID:25799389

Sakkas, K., Panias, D., Nomikos, P. P., & Sofianos, A. I. (2014). Potassium based geopolymer for passive fire protection of concrete tunnels linings. *Tunnelling and Underground Space Technology*, *43*, 148–156. doi:10.1016/j.tust.2014.05.003

Salem, A. K., Tendler, S. J., & Roberts, C. J. (2002). Interactions of 3T3 fibroblasts and endothelial cells with defined pore features. *Journal of Biomedical Materials Research*, *61*(2), 212–218. doi:10.1002/jbm.10195 PMID:12007201

Salt, D. E., Smith, R. D., & Raskin, I. (1998). Phytoremediation. *Annual Review of Plant Physiology and Plant Molecular Biology*, *49*(1), 643–668. doi:10.1146/annurev.arplant.49.1.643 PMID:15012249

Sanchavanakit, N., Sangrungraungroj, W., Kaomongkolgit, R., Banaprasert, T., Pavasant, P., & Phisalaphong, M. (2006). Growth of human keratinocytes and fibroblasts on bacterial cellulose film. *Biotechnology Progress*, *22*(4), 1194–1199. doi:10.1021/bp060035o PMID:16889398

Sanchez, C., Julián, B., Belleville, P., & Popall, M. (2005). Applications of hybrid organic-inorganic nanocomposites. *Journal of Materials Chemistry*, *15*(35–36), 3559–3592. doi:10.1039/b509097k

San, J. Y., & Lin, W.-M. (2008). Comparison among three adsorption pairs for using as the working substances in a multi-bed adsorption heat pump. *Applied Thermal Engineering*, *28*(8-9), 988–997. doi:10.1016/j.applthermaleng.2007.06.026

Sanjayan, J. G., Nazari, A., Chen, L., & Nguyen, G. H. (2015). Physical and mechanical properties of lightweight aerated geopolymer. *Construction & Building Materials*, *79*, 236–244. doi:10.1016/j.conbuildmat.2015.01.043

Sanna, R., Fortunati, E., Alzari, V., Nuvoli, D., Terenzi, A., Casula, M. F., & Mariani, A. et al. (2013). Poly (N-vinylcaprolactam) nanocomposites containing nanocrystalline cellulose: A green approach to thermoresponsive hydrogels. *Cellulose (London, England)*, *20*(5), 2393–2402. doi:10.1007/s10570-013-9988-1

Sapienza, A., Santamaria, S., Frazzica, A., Freni, A., & Aristov, Y. I. (2014). Dynamic study of adsorbers by a new gravimetric version of the Large Temperature Jump method. *Applied Energy*, *113*, 1244–1251. doi:10.1016/j.apenergy.2013.09.005

Saputro, S. S., Kai, S., Satoda, M., Matsuoka, S., & Yoshimura, K. (2013). Solid-phase Spectrophotometry for on-site analysis of trace elements in natural water. *Analytical Sciences*, *29*, 677–680. doi:10.2116/analsci.29.677 PMID:23749138

Sareeladdanon, S., & Potiyaraj, P. (2014). Mechanical properties of PLA/LLDPE films reinforced with silica from rice husk. *Advanced Materials Research*, *1025-1026*, 221–226. doi:10.4028/www.scientific.net/AMR.1025-1026.221

Sarma, H. (2011). Metal hyperaccumuulation in plants: A Review focusing on phytoremediation technology. *Journal of Environmental Science and Technology.*, *4*(2), 118–138. doi:10.3923/jest.2011.118.138

Saska, S., Teixeira, L. N., de Oliveira, P. T., Gaspar, A. M. M., Ribeiro, S. J. L., Messaddeq, Y., & Marchetto, R. (2012). Bacterial Cellulose-collagen nanocomposite for bone tissue engineering. *Journal of Materials Chemistry*, *22*(41), 22102–22112. doi:10.1039/c2jm33762b

Satake, K., Shibata, K., Nishikawa, M., & Fuwa, K. (1988). Copper accumulation and location in the moss Scopelophila cataractae. *Journal of Bryology*, *15*(2), 353–376. doi:10.1179/jbr.1988.15.2.353

Sato, K., Kishimoto, T., Morimoto, M., Saimoto, H., & Shigemasa, Y. (2003). Hydrolysis of acetals in water under hydrothermal conditions. *Tetrahedron Letters*, *44*(47), 8623–8625. doi:10.1016/j.tetlet.2003.09.087

Saucedo-luna, J., Castro-Montoya, A. J., Campos-Garcia, J., & Rico, J. L. (2010). Optimization of acid hydrolysis bagasse from agave tequilana weber. *Revista Mexicana de Ingenieia Quimica*, *9*(1), 91–97.

Schaafsma, A., Pakan, I., Hofstede, G. J. H., Muskiet, F. A. J., Van Der Veer, E., & De Vries, P. J. F. (2000). Mineral, amino acid, and hormonal composition of chicken eggshell powder and the evaluation of its use in human nutrition. *Poultry Science*, *79*(12), 1833–1838. doi:10.1093/ps/79.12.1833 PMID:11194049

Schneider, J., Manjure, S., & Narayan, R. (2016). Reactive modification and compatibilization of poly(lactide) and poly(butylene adipate-co-terephthalate) blends with epoxy functionalized-poly(lactide) for blown film applications. *Journal of Applied Polymer Science*, *133*(16), 43310–43318. doi:10.1002/app.43310

Schurz, J. (1999). A bright future for cellulose. *Progress in Polymer Science*, *24*(4), 481–483. doi:10.1016/S0079-6700(99)00011-8

Seentrakoon, B., Junhasavasdikul, B., & Chavasiri, W. (2013). Enhanced UV-protection and antibacterial properties of natural rubber/rutile-TiO2 nanocomposites. *Polymer Degradation & Stability*, *98*(2), 566–578. doi:10.1016/j.polymdegradstab.2012.11.018

Sekhar, M., & Chakravarthy, C. (2001). Coal and clay minerals removal of zinc by adsorption. In Advances in Environmental Materials, Volume II Environmentally preferred materials. MRS (Singapore).

Serafica, G., Mormino, R., Oster, G. A., Lentz, K. E., & Koehler, K. P. (2010). *Microbial cellulose wound dressing for treating chronic wounds*. Google Patents.

Serafica, G., Mormino, R., & Bungay, H. (2002). Inclusion of solid particles in bacterial cellulose. *Applied Microbiology and Biotechnology*, *58*(6), 756–760. doi:10.1007/s00253-002-0978-8 PMID:12021795

Serive, B., Kaas, R., Bérard, J. B., Pasquet, V., Picot, L., & Cadoret, J. P. (2012). Selection and optimisation of a method for efficient metabolites extraction from microalgae. *Bioresource Technology*, *124*, 311–320. doi:10.1016/j.biortech.2012.07.105 PMID:22989659

Seville, J., & Wu, C. (2016). *Particle Technology and Engineering: An Engineer's Guide to Particles and Powders: Fundamentals and Computational Approaches*. Butterworth Heinemann.

Shahidi, F., Arachchi, J. K. V., & Jeon, Y.-J. (1999). Food applications of chitin and chitosans. *Trends in Food Science & Technology*, *10*(2), 37–51. doi:10.1016/S0924-2244(99)00017-5

Shahzadi, K., Wu, L., Ge, X., Zhao, F., Li, H., Pang, S., & Mu, X. et al. (2016). Preparation and characterization of bio-based hybrid film containing chitosan and silver nanowires. *Carbohydrate Polymers*, *137*, 732–738. doi:10.1016/j.carbpol.2015.11.012 PMID:26686186

Shanmughara, A. M., Bae, J. H., Lee, K. Y., Noh, W. H., Lee, S. H., & Ryu, S. H. (2007). Physical and chemical characteristics of multiwalled carbon nanotubes functionalized with aminosilane and its influence on the properties of natural rubber composites. *Composites Science and Technology*, *67*(9), 1813–1822. doi:10.1016/j.compscitech.2006.10.021

Shaplov, A. S., Lozinskaya, E. I., & Vygodskii, Y. S. (2010). Polymer ionic liquids: synthesis, design and application in electrochemistry as ion conducting materials. In A. A. J. Torriero & M. J. A. Shiddiky (Eds.), *Electrochemical Properties and Applications of Ionic Liquids* (pp. 203–298). New York: Nova Publishers.

Shaw, A. J. (1994). Adaptation to metals in widespread and endemic plants. *Environmental Health Perspectives*, *102*(Suppl 12), 105–108. doi:10.1289/ehp.94102s12105 PMID:7713025

Sheehan, J., Dunahay, T., Benemann, J., & Roessler, P. (1998). *A Look Back at the U.S. Department of Energy's Aquatic Species Program-Biodiesel from Algae*. National Renewable Energy Laboratory. doi:10.2172/15003040

She, H., Malotky, D., & Chaudhury, M. K. (1998). Estimation of Adhesion Hysteresis at Polymer/Oxide Interfaces Using Rolling Contact Mechanics. *Langmuir*, *14*(11), 3090–3100. doi:10.1021/la971061m

Shelton, S. V., Wepfer, W. J., & Miles, D. J. (1990). Ramp Wave Analysis of the Solid/ Vapor Heat Pump. *Journal of Energy Resources Technology*, *112*(1), 69–78. doi:10.1115/1.2905715

Shen, W., Chen, S., Shi, S., Li, X., Zhang, X., Hu, W., & Wang, H. (2009). Adsorption of Cu(II) and Pb(II) onto diethylenetriamine-bacterial cellulose. *Carbohydrate Polymers*, *75*(1), 110–114. doi:10.1016/j.carbpol.2008.07.006

Shen, X., Shamshina, J. L., Berton, P., Gurrau, G., & Rogers, R. D. (2016). Hydrogels based on cellulose and chitin: Fabrication, properties, and applications. *Green Chemistry*, *18*(1), 53–75. doi:10.1039/C5GC02396C

Shi, B. (2015). *Development of an MOF based Adsorption air conditioning System for automotive Application* (PhD thesis). University of Birmingham.

Shiklomanov, I. A. (2000). Appraisal and Assessment of World Water Resources. *Water International*, *25*(1), 11–32. doi:10.1080/02508060008686794

Shimojo, K., & Goto, M. (2004). Solvent Extraction and Stripping of Silver Ions in Room-Temperature Ionic Liquids Containing Calixarenes. *Analytical Chemistry*, *76*(17), 5039–5044. doi:10.1021/ac049549x PMID:15373439

Shimojo, K., Kurahashi, K., & Naganawa, H. (2008). Extraction behavior of lanthanides using a diglycolamide derivative TODGA in ionic liquids. *Dalton Transactions (Cambridge, England)*, (37): 5083–5088. doi:10.1039/b810277p PMID:18802624

Shimojo, K., Naganawa, H., Kubota, F., & Goto, M. (2006). Solvent Extraction of Lanthanides into an Ionic Liquid Containing N, N, N, N-tetrakis(2-yridylmethyl)ethylenediamine. *Chemistry Letters*, *35*(5), 484–485. doi:10.1246/cl.2006.484

Shin-Etsu, C. (2011). *Silane coupling agent*. Retrieved 10 February 2016, from http://www.shinetsusilicone-global.com/catalog/pdf/SilaneCouplingAgents_e.pdf

Shmroukh, A.N., Ali, A.H.H., Abel-Rahman, A.K., & Ookwara, S. (2015). Experimental Investigation on Adsorption Capacity of a Variety of Activated Carbon/Refrigerant Pairs. *Int. Journal of Engineering Research and Applications*, *5*(4), 66-76.

Shmroukh, A. N., Ali, A. H. H., & Abel-Rahman, A. K. (2013). Adsorption Refrigeration Working Pairs: The state-of-the-art in the application. *International Journal of Chemical Materials Science and Engineering*, *7*(11b), 793–805.

Shmroukh, A. N., Ali, A. H. H., & Ookawara, S. (2015). Adsorption working pairs for adsorption cooling chillers: A review based on adsorption capacity and environmental impact. *Renewable & Sustainable Energy Reviews*, *50*, 445–456. doi:10.1016/j.rser.2015.05.035

Shoda, M., & Sugano, Y. (2005). Recent advances in bacterial cellulose production. *Biotechnology and Bioprocess Engineering*, *10*(1), 1–8. doi:10.1007/BF02931175

Shroeder, H. (1962). Properties and Applications of Oxide Layers Deposited on Glass from Organic Solutions, Opt. *Acta Int. J. Opt.*, *9*(3), 249–254. doi:10.1080/713826428

Siegfried, K. D. B. (1999). Experiments on solar adsorption refrigeration using zeolite and water. University of Applied Sciences Luebeck.

Sigmund, S., Akgün, E., Meyer, J., Hubbuch, J., Wörner, M., & Kasper, G. (2014). Defined polymer shells on nanoparticles via a continuous aerosol-based process. *Journal of Nanoparticle Research*, *16*(8), 1–9. doi:10.1007/s11051-014-2533-9

Singha, A. S., & Guleria, A. (2015). Utility of chemically modified agricultural waste okra biomass for removal of toxic heavy metal ions from aqueous solution. *Engineering in Agriculture. Environment Folio*, *8*(1), 52–60.

Sionkowske, A. (2011). Current research on the blends of natural and synthetic polymers as new biomaterials. *Progress in Polymer Science*, *36*(9), 1254–1261. doi:10.1016/j.progpolymsci.2011.05.003

Smook, G. A. (1992). Handbook for pulp and paper technologists. Agus Wilde Publications.

Solair. (n.d.). *Solar Cooling; an Intelligent Option.* Retrieved from http://www.solair-project.eu/fileadmin/SOLAIR_uploads/DOCS/Guidelines/SOLAIR_Brochure_EN.pdf

Solway, D. R., Clark, W. A., & Levinson, D. J. (2011). A parallel open-label trial to evaluate microbial cellulose wound dressing in the treatment of diabetic foot ulcers. *International Wound Journal, 8*(1), 69–73. doi:10.1111/j.1742-481X.2010.00750.x PMID:21159127

Somaratna, J., Ravikumar, D., & Neithalath, N. (2010). Response of alkali activated fly ash mortars to microwave curing. *Cement and Concrete Research, 40*(12), 1688–1696. doi:10.1016/j.cemconres.2010.08.010

Somna, K., Jaturapitakkul, C., Kajitvichyanukul, P., & Chindaprasirt, P. (2011). NaOH-activated ground fly ash geopolymer cured at ambient temperature. *Fuel, 90*(6), 2118–2124. doi:10.1016/j.fuel.2011.01.018

Songpiriyakij, S., Kubprasit, T., Jaturapitakkul, C., & Chindaprasirt, P. (2010). Compressive strength and degree of reaction of biomass- and fly ash-based geopolymer. *Construction & Building Materials, 24*(3), 236–240. doi:10.1016/j.conbuildmat.2009.09.002

Son, H.-J., Kim, H.-G., Kim, K.-K., Kim, H.-S., Kim, Y.-G., & Lee, S.-J. (2003). Increased production of bacterial cellulose by Acetobacter sp. V6 in synthetic media under shaking culture conditions. *Bioresource Technology, 86*(3), 215–219. doi:10.1016/S0960-8524(02)00176-1 PMID:12688462

Spahis, N., Addoun, A., & Mahmoudi, H. (2007). Study on solar adsorption refrigeration cycle utilizing activated carbon prepared from olive stones. *Revue des Energies Renouvelables, 10*(3), 415–420.

SPI Bioplastics Council. (2012). *Bioplastics Industry Overview Guide.* Washington, DC: The Society of the Plastics Industry.

Steavens, E. S. (2002). *Green Plastics: An Introduction to the New Science of Biodegradable Plastics.* United Kingdom: Princeton University Press.

Stephens, R. S., Westland, J. A., & Neogi, A. N. (1990). *Method of using bacterial cellulose as a dietary fiber component.* Google Patents.

Sthiannopkaoa, S., & Wongb, M. H. (2013). Handling e-waste in developed and developing countries: Initiatives, practices, and consequences. *The Science of the Total Environment, 463–464*, 1147–1153. doi:10.1016/j.scitotenv.2012.06.088 PMID:22858354

Stoica-Guzun, A., Stroescu, M., Tache, F., Zaharescu, T., & Grosu, E. (2007). Effect of electron beam irradiation on bacterial cellulose membranes used as transdermal drug delivery systems. *Nuclear Instruments & Methods in Physics Research. Section B, Beam Interactions with Materials and Atoms, 265*(1), 434–438. doi:10.1016/j.nimb.2007.09.036

Striegel, A. M. (1997). Theory and applications of DMAc/LiCl in the analysis of polysaccharides. *Carbohydrate Polymers, 34*(4), 267–273. doi:10.1016/S0144-8617(97)00101-X

Strnad, Z. (1986). *Glass-ceramic materials.* Amsterdam: Elsevier.

Strozi Cilla, M., Colombo, P., & Raymundo Morelli, M. (2014). Geopolymer foams by gelcasting. *Ceramics International, 40*(4), 5723–5730. doi:10.1016/j.ceramint.2013.11.011

Strozi Cilla, M., Raymundo Morelli, M., & Colombo, P. (2014). Effect of process parameters on the physical properties of porous geopolymers obtained by gelcasting. *Ceramics International, 40*(88, Part B), 13585–13590. doi:10.1016/j.ceramint.2014.05.074

Studart, A. R., Gonzenbach, U. T., Tervoort, E., & Gauckler, L. J. (2006). Processing routes to macroporous ceramics: A review. *Journal of the American Ceramic Society*, *89*(6), 1771–1789. doi:10.1111/j.1551-2916.2006.01044.x

Suchiai, H. T., Shizaka, T. I., Nakamura, H., Ueno, T., & Hattori, H. (1996). Study of flue gas desulfurizeation absorbent prepared from coal fly ash. *Industrial & Engineering Chemistry Research*, *35*, 322–2326.

Sui, G., Zhong, W. H., Yang, X. P., & Yu, Y. H. (2008). Curing kinetics and mechanical behavior of natural rubber reinforced with pretreated carbon nanotubes. *Materials Science and Engineering A*, *485*(1-2), 524–531. doi:10.1016/j.msea.2007.09.007

Sui, G., Zhong, W., & Yang, X. (2008). Preparation and properties of natural rubber composites reinforced with pretreated carbon nanotubes. *Polymers for Advanced Technologies*, *19*(August), 1543–1549.

Sumio, I. (1991). Helical microtubules of graphitic carbon. *Nature*, *354*(6348), 56–58. doi:10.1038/354056a0

Sun, X., Bell, J. R., Luo, H., & Dai, S. (2011). Extraction separation of rare-earth ions via competitive ligand complexations between aqueous and ionic-liquid phases. *Dalton Transactions (Cambridge, England)*, *40*(31), 8019–8023. doi:10.1039/c1dt10873e PMID:21735029

Sun, X., Ji, Y., Guo, L., Chen, J., & Li, D. (2011). A novel ammonium ionic liquid based extraction strategy for separating scandium from yttrium and lanthanides. *Separation and Purification Technology*, *81*(1), 25–30.

Sun, X., Ji, Y., Hu, F., He, B., Chen, J., & Li, D. (2010). The inner synergistic effect of bifunctional ionic liquid extractant for solvent extraction. *Talanta*, *81*(4-5), 1877–1883. doi:10.1016/j.talanta.2010.03.041 PMID:20441990

Sun, X., Luo, H., & Dai, S. (2011). Ionic liquids-based extraction: A promising strategy for the advanced nuclear fuel cycle. *Chemical Reviews*, *112*(4), 2100–2128. doi:10.1021/cr200193x PMID:22136437

Sun, X., Wu, D., Chen, J., & Li, D. (2007). Separation of scandium(III) from lanthanides(III) with room temperature ionic liquid based extraction containing. Cyanex 925. *Journal of Chemical Technology and Biotechnology (Oxford, Oxfordshire)*, *82*(3), 267–272. doi:10.1002/jctb.1664

Suratago, T., Taokaew, S., Kanjanamosit, N., Kanjanaprapakul, K., Burapatana, V., & Phisalaphong, M. (2015). Development of bacterial cellulose/alginate nanocomposite membrane for separation of ethanol–water mixtures. *Journal of Industrial and Engineering Chemistry*, *32*, 305–312. doi:10.1016/j.jiec.2015.09.004

Susan, M. A. B. H., Kaneko, T., Noda, A., & Watanabe, M. (2005). Ion gels prepared by in situ radical polymerization of vinyl monomers in an ionic liquid and their characterization as polymer electrolytes. *Journal of the American Chemical Society*, *127*(13), 4976–4983. doi:10.1021/ja045155b PMID:15796564

Svensson, A., Nicklasson, E., Harrah, T., Panilaitis, B., Kaplan, D. L., Brittberg, M., & Gatenholm, P. (2005). Bacterial cellulose as a potential scaffold for tissue engineering of cartilage. *Biomaterials*, *26*(4), 419–425. doi:10.1016/j.biomaterials.2004.02.049 PMID:15275816

Swanepoel, J. C., & Strydom, C. A. (2002). Utilisation of fly ash in a geopolymeric material. *Applied Geochemistry*, *17*(8), 1143–1148. doi:10.1016/S0883-2927(02)00005-7

Taguchi, S., Asaoka, M., Hirokami, E., Hata, N., Kuramitz, H., Kawakami, T., & Miyake, R. (2015). A Simple and Rapid Method for Simultaneous Preconcentration of Eight Trace-heavy-metals in Water Using 1-(2-Pyridylazo)-2-naphthol and Yttrium for X-ray Fluorescence Spectrometry. *Analytical Methods*, *7*(16), 6545–6551. doi:10.1039/C5AY00657K

Taguchi, S., Hui-Feng, S., Hata, N., & Kasahara, I. (2000). Design of simple and sensitive methods for trace analysis by mean of concentration with a membrane filter. *Bunseki Kagaku*, *49*, 941–952. doi:10.2116/bunsekikagaku.49.941

Taguchi, S., Nakayama, K., Hata, N., Kasahara, I., & Goto, K. (1994). Membrane filters for concentration of trace elements in water: Distribution of 2-(2-pyridylazo)-5-diethylaminophenol between the membrane filter and the aqueous phases. *Analyst (London)*, *119*(1), 135–139. doi:10.1039/an9941900135

Takahashi, Y. (2012). Elimination of interferences in the Mercury(II) determination with a dithizone nanofiber-coated test strip based on filtration-enrichment. *Bunseki Kagaku*, *61*(2), 123–126. doi:10.2116/bunsekikagaku.61.123

Takahashi, Y. (2014). Dye nanoparticle-coated test strips for detection of ppb-level ions in water. In A. Street, R. Sustich, J. Duncan, & N. Savage (Eds.), *Nanotechnology Applications for Clean Water* (2nd ed.; pp. 63–72). Oxford, UK: Elsevier. doi:10.1016/B978-1-4557-3116-9.00004-4

Takahashi, Y. (2015a). Dye Nanoparticle or Nanocomposite-Coated Test Papers for Detection at Ppb Levels of Harmful Ions. In M. Aliofkhazraei (Ed.), *Handbook of Nanoparticles* (pp. 1–13). New York: Springer. doi:10.1007/978-3-319-13188-7_38-1

Takahashi, Y. (2015b). Preparation and characterization of dye nanoparticle-coated test strips composed of anionic dyes and basic nano-adsorbents. *Bunseki Kagaku*, *63*, 525–531. doi:10.2116/bunsekikagaku.63.525

Takahashi, Y., Danwittayakul, S., & Suzuki, T. M. (2009). Dithizone nanofiber-coated membrane for filtration-enrichment and colorimetric detection of trace Hg(II) ion. *Analyst (London)*, *134*(7), 1380–1385. doi:10.1039/b816461d PMID:19562205

Takahashi, Y., Kasai, H., Nakanishi, H., & Suzuki, T. M. (2006). Test strips for heavy-metal ions fabricated from nanosized dye compounds. *Angewandte Chemie International Edition*, *45*(6), 913–916. doi:10.1002/anie.200503015 PMID:16382478

Takahashi, Y., Kondo, K., Miyaji, A., Umeo, M., Honma, T., & Asaoka, S. (2012). Recovery and separation of rare earth elements using columns loaded with DNA-filter hybrid. *Analytical Sciences*, *28*(10), 985–992. doi:10.2116/analsci.28.985 PMID:23059995

Takahashi, Y., Kondo, K., Miyaji, A., Watanabe, Y., Fan, Q., Honma, T., & Tanaka, K. (2014). Recovery and Separation of Rare *Earth Elements Using Salmon Milt. PLoS ONE*, *9*(12), e114858. doi:10.1371/journal.pone.0114858 PMID:25490035

Takahashi, Y., Souma, S., & Wakui, Y. (2012). Preparation of 2-(5-bromo-2-pyridylazo)-5-diethylaminophenol nanoparticle-coated test strip and its detection of trace Cd ^{II} Ion by immersion test. *Bunseki Kagaku*, *61*, 229–234. doi:10.2116/bunsekikagaku.61.229

Tamada, Y., & Ikada, Q. (1993). Effect of preadsorbed proteins on cell adhesion to polymer surface. *Journal of Colloid and Interface Science*, *155*(2), 334–341. doi:10.1006/jcis.1993.1044

Tamainot-Telto, Z., & Critoph, R.E. (1996). Adsorption refrigerator using monolithic carbon-ammonia pair. *International Journal of Refrigeration*, *20*(2), 146–155.

Tamainot-Telto, Z., Metcalf, S. J., & Critoph, R. E. (2009). Novel compact sorption generators for car air conditioning. *International Journal of Refrigeration*, *32*(4), 727–733. doi:10.1016/j.ijrefrig.2008.11.010

Tancharernrat, T., Rempel, G. L., & Prasassarakich, P. (2014). Preparation of styrene butadiene copolymer–silica nanocomposites via differential microemulsion polymerization and NR/SBR–SiO2 membranes for pervaporation of water–ethanol mixtures. *Chemical Engineering Journal*, *258*, 290–300. doi:10.1016/j.cej.2014.05.151

Tancharernrat, T., Rempel, G. L., & Prasassarakich, P. (2015). Synthesis of polybutadiene-silica nanoparticles via differential microemulsion polymerization and their hydrogenated nanoparticles by diimide reduction. *Polymer Degradation & Stability*, *118*, 69–81. doi:10.1016/j.polymdegradstab.2015.04.008

Tang, J., Tang, H., Sun, W., Plancher, H., Radosz, M., & Shen, Y. (2005). Poly(ionic liquid)s: A new material with enhanced and fast CO_2 absorption. *Chemical Communications*, (26): 3325–3327. PMID:15983662

Tanimura, S., Nakao, S.-i., & Kimura, S. (1990). Ethanol-selective membrane for reverse osmosis of ethanol/water mixture. *AIChE Journal. American Institute of Chemical Engineers*, *36*(7), 1118–1120. doi:10.1002/aic.690360719

Tanrattanakul, V., Sungthong, N., & Raksa, P. (2008). Rubber toughening of nylon 6 with epoxidized natural rubber. *Polymer Testing*, *27*(7), 794–800. doi:10.1016/j.polymertesting.2008.05.013

Taokaew, S., Phisalaphong, M., Yin, L. Y., & Newby, B. M. Z. (2013a). Effects of drying methods of bacterial cellulose membrane on differentiation and proliferation of induced pluripotent stem cells. *Abstracts of Papers of the American Chemical Society*, *245*.

Taokaew, S., Nunkaew, N., Siripong, P., & Phisalaphong, M. (2014a). Characteristics and anticancer properties of bacterial cellulose films containing ethanolic extract of mangosteen peel. *Journal of Biomaterials Science. Polymer Edition*, *25*(9), 907–922. doi:10.1080/09205063.2014.913464 PMID:24802115

Taokaew, S., Phisalaphong, M., & Newby, B. M. Z. (2015). Modification of bacterial cellulose with organosilanes to improve attachment and spreading of human fibroblasts. *Cellulose (London, England)*, *22*(4), 2311–2324. doi:10.1007/s10570-015-0651-x PMID:26478661

Taokaew, S., Phisalaphong, M., & Zhang Newby, B. M. (2014b). In vitro behaviors of rat mesenchymal stem cells on bacterial celluloses with different moduli. *Materials Science and Engineering C*, *38*(1), 263–271. doi:10.1016/j.msec.2014.02.005 PMID:24656377

Taokaew, S., Seetabhawang, S., Siripong, P., & Phisalaphong, M. (2013b). Biosynthesis and characterization of nanocellulose-gelatin films. *Materials (Basel)*, *6*(3), 782–794. doi:10.3390/ma6030782

The World Bank. (2005). *Towards a More Effective Operational Response: Arsenic Contamination of Groundwater in South and East Asian Countries*. No 71518. United States Environmental Protection Agency. Effluent Guidelines. Retrieved from http://www.epa.gov/eg

Theng, B. K. G. (1974). *The chemistry of clay-organic reactions*. Academic Press.

Theng, B. K. G. (1979). *Formation and properties of clay-polymer complexes*. Amsterdam: Elsevier Scientific Publishing Company.

Theppradit, T., Prasassarakich, P., & Poompradub, S. (2014). Surface modification of silica particles and its effects on cure and mechanical properties of the natural rubber composites. *Materials Chemistry and Physics*, *148*(3), 940–948. doi:10.1016/j.matchemphys.2014.09.003

Ting, G. (1995). Electrical conductors comprising single-wall carbon nanotubes. *Chemical Physics Letters*, *243*, 49.

Ting, G. (1995). Self-assembly of tubular fullerenes. *Journal of Physical Chemistry*, *99*(27), 10694–10697. doi:10.1021/j100027a002

Tomalia, D. A. (2004). Birth of a New Macromolecular Architecture: Dendrimers as Quantized Building Blocks for Nanoscale Synthetic Organic Chemistry. *Aldrichimica Acta*, *37*(2), 39.

Toor, M., Jin, B., Dai, S., & Vimonses, V. (2015). Activating natural bentonite as a cost-effective adsorbent for removal of Congo-red in wastewater. *Journal of Industrial and Engineering Chemistry*, *21*, 653–661. doi:10.1016/j.jiec.2014.03.033

Tovar-Carrillo, K. L. (2014). *Study on Cellulose Hydrogel Films Regenerated From Natural Plant Bagasse and Their Bio and Cytocompatible Properties for Tissue Engineering* (Thesis). Nagaoka University of Technology.

Tovar-Carrillo, K. L., & Kobayashi, T. (2015). Fibroblast Cell Cultivation on Wooden Pulp Cellulose Hydrogels for Cytocompatibility Scaffold Method. Pharm Anal Acta, 6(10), 423.

Tovar-Carrillo, K. L., Tagaya, M., & Kobayashi, T. (2013). Bamboo fibers elaborating cellulose hydrogel films for medical applications. *J. Mater. Sci. Chemical Engineering (Albany, N.Y.)*, *1*(7), 7–12.

Tovar-Carrillo, K. L., Tagaya, M., & Kobayashi, T. (2013). Fibroblast compatibility on scaffold hydrogels prepared from Agave tequilana weber bagasse for tissue regeneration. *Industrial & Engineering Chemistry Research*, *60*(33), 11607–11673. doi:10.1021/ie401793w

Tovar-Carrillo, K., Nakasone, K., Sugita, S., Tagaya, M., & Kobayashi, T. (2014). Effects of sodium hypochlorite on Agave tequilana Weber bagasse fibers used to elaborate cyto and biocompatible hydrogel films. *Materials Science and Engineering C*, *42*, 808–815. doi:10.1016/j.msec.2014.06.023 PMID:25063183

Tovar-Carrillo, K., Tagaya, M., & Kobayashi, T. (2013). Bamboo Fibers Elaborating Cellulose Hydrogel Films for Medical Applications. *Journal of Material Science Chemical Engineering*, *1*(07), 7–12. doi:10.4236/msce.2013.17002

Toworfe, G. K., Bhattacharyya, S., Composto, R. J., Adams, C. S., Shapiro, I. M., & Ducheyne, P. (2009). Effect of functional end groups of silane self-assembled monolayer surfaces on apatite formation, fibronectin adsorption and osteoblast cell function. *Journal of Tissue Engineering and Regenerative Medicine*, *3*(1), 26–36. doi:10.1002/term.131 PMID:19012271

TRA. (2015a). *NR situation*. Retrieved 4 February, 2016, from http://www.thainr.com/en/index.php?detail=situation&selectmonth=01

TRA. (2015b). *Thai rubber statistics*. Retrieved 4 February, 2016, from http://www.thainr.com/uploadfile/20151103111354.pdf

Trang, T. C., & Kobayashi, T. (2011). Vulcanized Paper for Separation of Alcohol Aqueous Solutions by Pervaporation. *Journal of Applied Polymer Science*, *121*(2), 639–647. doi:10.1002/app.33258

Trang, T. C., Nhung, N. T., & Kobayashi, T. (2011). Fabrication and Characterization of Pulp/Chitosan Composite Membranes Crosslinked with 3-Methylglutaric Anhydride for Pervaporation of Ethanol/Water Mixture. *Engineering*, *3*, 110–118.

Tripathi, B. P., & Shahi, V. K. (2008). Functionalized Organic–Inorganic Nanostructured N-p-Carboxy Benzyl Chitosan–Silica–PVA Hybrid Polyelectrolyte Complex as Proton Exchange Membrane for DMFC Applications. *The Journal of Physical Chemistry B*, *112*(49), 15678–15690. doi:10.1021/jp806337b PMID:19368033

Trovatti, E., Freire, C. S. R., Pinto, P. C., Almeida, I. F., Costa, P., Silvestre, A. J. D., & Rosado, C. et al. (2012). Bacterial cellulose membranes applied in topical and transdermal delivery of lidocaine hydrochloride and ibuprofen: In vitro diffusion studies. *International Journal of Pharmaceutics*, *435*(1), 83–87. doi:10.1016/j.ijpharm.2012.01.002 PMID:22266531

Trovatti, E., Silva, N. H. C. S., Duarte, I. F., Rosado, C. F., Almeida, I. F., Costa, P., & Neto, C. P. et al. (2011). Biocellulose Membranes as Supports for Dermal Release of Lidocaine. *Biomacromolecules*, *12*(11), 4162–4168. doi:10.1021/bm201303r PMID:21999108

Tseng, R. L., & Wu, F. C. (2009). Analyzing a liquid–solid phase counter current two- and three-stage adsorption process with the Freundlich equation. *Journal of Hazardous Materials*, *162*(1), 237–248. doi:10.1016/j.jhazmat.2008.05.031 PMID:18556116

Tso, C. Y., & Chao, C. Y. H. (2012). Activated carbon, silica-gel and calcium chloride composite adsorbents for energy efficient solar adsorption cooling and dehumidification systems. *International Journal of Refrigeration*, *35*(6), 1626–1638. doi:10.1016/j.ijrefrig.2012.05.007

Tso, C. Y., Chao, C. Y. H., & Fu, S. C. (2012). Performance analysis of a waste heat driven activated carbon based composite adsorbent – Water adsorption chiller using simulation model. *International Journal of Heat and Mass Transfer*, *55*(25-26), 7596–7610. doi:10.1016/j.ijheatmasstransfer.2012.07.064

Tsutsumi, Y., Koga, H., Qi, Z. D., Saito, T., & Isogai, A. (2014). Nanofibrillar chitin aerogels as renewable nased catalysts. *Biomacromolecules*, *15*(11), 4314–4319. doi:10.1021/bm501320b PMID:25285573

Turanov, A. N., Karandashev, V. K., & Baulin, V. E. (2008). Effect of ionic liquids on the extraction of rare-earth elements by bidentate neutral organophosphorus compounds from chloride solutions. *Russian Journal of Inorganic Chemistry*, *53*(6), 970–975. doi:10.1134/S0036023608060272

Uddin, K., El-Sharkawy, I. I., Miyazaki, T., Saha, B. B., Koyama, S., Kil, H. S., & Yoon, S. H. et al. (2014). Adsorption characteristics of ethanol onto functional activated carbons with controlled oxygen content. *Applied Thermal Engineering*, *72*(2), 211–218. doi:10.1016/j.applthermaleng.2014.03.062

Ueki, T., Usui, R., Kitazawa, Y., Lodge, T. P., & Watanabe, M. (2015). Thermally reversible ion gels with photohealing properties based on triblock copolymer self-assembly. *Macromolecules*, *48*(16), 5928–5933. doi:10.1021/acs.macromol.5b01366

Ueki, T., & Watanabe, M. (2006). Upper critical solution temperature behavior of poly(*N*-isopropylacrylamide) in an ionic liquid and preparation of thermo-sensitive nonvolatile gels. *Chemistry Letters*, *35*(8), 964–965. doi:10.1246/cl.2006.964

Ulku, S. (1986). Adsorption heat pumps. *Heat Recovery Systems*, *6*(4), 277–284. doi:10.1016/0198-7593(86)90113-X

Ullaha, Z., Hussain, S., Gul, S., Khan, S., & Bangasha, F. K. (2015). Use of HCl-modified bentonite clay for the adsorption of Acid Blue 129 from aqueous solutions. *Desalination and Water Treatment*, 1–10.

Ullah, K. R., Saidur, R., Ping, H. W., Akikur, R. K., & Shuvo, N. H. (2013). A review of solar thermal refrigeration and cooling methods. *Renewable & Sustainable Energy Reviews*, *24*, 499–513. doi:10.1016/j.rser.2013.03.024

Umair, M., Akisawa, A., & Ueda, Y. (2014). Simulation Study of Continuous Solar Adsorption Refrigeration System Driven by Compound Parabolic Concentrator. *The Open Renewable Energy Journal*, *7*(1), 1–12. doi:10.2174/1876387101407010001

United Nations Environment Programme (UNEP). (2014). *Climate change Minimizing the scale and impact of climate change,* Retrieved 2014, from http://www.unep.org/annualreport/2014/en/pdf/climate_change.pdf

United States Environmental Protection Agency. (n.d.). *Table of Regulated Drinking Water Contaminants*. Retrieved from https://www.epa.gov/your-drinking-water/table-regulated-drinking-water-contaminants

Uragami, T. (2014). *Pervaporation Membranes. In Encyclopedia of Polymeric Nanomaterials*. Berlin: Springer Berlin Heidelberg. doi:10.1007/978-3-642-36199-9_133-1

Uragami, T., & Takigawa, K. (1990). Permeation and separation characteristics of ethanol-water mixtures through chitosan derivative membranes by pervaporation and evapomeation. *Polymer*, *31*(4), 668–672. doi:10.1016/0032-3861(90)90287-9

Uraki, Y., Nemoto, J., Otsuka, H., Tamai, Y., Sugiyama, J., Kishimoto, T., & Shimomura, M. et al. (2007). Honeycomb-like architecture produced by living bacteria, Gluconacetobacter xylinus. *Carbohydrate Polymers*, *69*(1), 1–6. doi:10.1016/j.carbpol.2006.08.021

Urquijo, J., Dagréou, S., Guerrica-Echevarría, G., & Eguiazábal, J. I. (2016). Structure and properties of poly(lactic acid)/poly(ε-caprolactone) nanocomposites with kinetically induced nanoclay location. *Journal of Applied Polymer Science*, *133*(33), 43815–43826. doi:10.1002/app.43815

Vaia, R. A., Ishii, H., & Giannelis, E. P. (1993). Synthesis and properties of two-dimensional nanostructures by direct intercalation of polymer melts in layered silicates. *Chemistry of Materials*, *5*(12), 1694–1696. doi:10.1021/cm00036a004

Valadares, L. F., Letle, C. A. P., & Galembeck, F. (2006). Preparation of natural rubber-montmorillonite nanocomposites in aqueous medium: Evidence for polymer-platelet adhesion. *Polymer*, *47*(2), 672–678. doi:10.1016/j.polymer.2005.11.062

Van Jaarsveld, J. G. S., Van Deventer, J. S. J., & Lorenzen, L. (1997). The potential use of geopolymeric materials to immobilise toxic metals: Part I. Theory and applications. *Minerals Engineering*, *10*(7), 659–669. doi:10.1016/S0892-6875(97)00046-0

Vaou, V., & Panias, D. (2010). Thermal insulating foamy geopolymers from perlite. *Minerals Engineering*, *23*(14), 1146–1151. doi:10.1016/j.mineng.2010.07.015

Varghese, S., & Karger-Kocsis, J. (2003, August). (2203). Natural rubber-based nanocomposites by latex compounding with layered silicates. *Polymer*, *44*(17), 4921–4927. doi:10.1016/S0032-3861(03)00480-4

Varghese, S., & Karger-Kocsis, J. (2004). Melt-compounded natural rubber nanocomposites with pristine and organophilic layered silicates of natural and synthetic origin. *Journal of Applied Polymer Science*, *91*(2), 813–5819. doi:10.1002/app.13173

Varma, A. J., Deshpande, S. V., & Kennedy, J. F. (2004). Metal complexation by chitosan and its derivatives: A review. *Carbohydrate Polymers*, *55*(1), 77–93. doi:10.1016/j.carbpol.2003.08.005

Vazquez, M. I., De Lara, R., & Benavente, J. (2008). Chemical surface diffusional, electrical and elastic characterization of two different dense regenerated cellulose membranes. *Journal of Colloid and Interface Science*, *328*(2), 331–338. doi:10.1016/j.jcis.2008.09.019 PMID:18842272

Venegas-Sanchez, J. A., Kusunoki, T., Yamamoto, M., & Kobayashi, T. (2013). Sono-respond on thermosensitive polymer microgels based on cross-linked poly(N-isopropylacrylamide-co-acrylic acid). *Ultrasonics Sonochemistry*, *20*(5), 1271–1275. doi:10.1016/j.ultsonch.2013.02.003 PMID:23528945

Venegas-Sanchez, J. A., Tagaya, M., & Kobayashi, T. (2013a). Effect of ultrasound on the aqueous viscosity of several water-soluble polymers. *Polymer Journal*, *45*(12), 1224–1233. doi:10.1038/pj.2013.47

Venegas-Sanchez, J. A., Tagaya, M., & Kobayashi, T. (2013b). Ultrasound effect used as external stimulus for viscosity change of aqueous carrageenans. *Ultrasonics Sonochemistry*, *20*(4), 1081–1091. doi:10.1016/j.ultsonch.2013.01.002 PMID:23395152

Venegas-Sanchez, J. A., Tagaya, M., & Kobayashi, T. (2014). Ultrasound stimulus inducing change in hydrogen bonded crosslinking of aqueous polyvinyl alcohols. *Ultrasonics Sonochemistry*, *21*(1), 295–309. doi:10.1016/j.ultsonch.2013.06.011 PMID:23835399

Vicente, M. A., Gil, A., & Bergaya, F. (2013). Pillared Clays and Clay Minerals. In E. Bergaya & L. Lagaly (Eds.), *Handbook of Clay Science* (pp. 525–557). Elsevier Ltd. doi:10.1016/B978-0-08-098258-8.00017-1

Vijayakrishna, K., Jewrajka, S. K., Ruiz, A., Marcilla, R., Pomposo, J. A., Mecerreyes, D., & Gnanou, Y. et al. (2008). Synthesis by RAFT and ionic responsiveness of double hydrophilic block copolymers based on ionic liquid monomer units. *Macromolecules*, *41*(17), 6299–6308. doi:10.1021/ma800677h

Vijayakumar, B., Nagendrappa, G., & Prakash, B. S. J. (2009). Acid Activated Indian Bentonite, an Efficient Catalyst for Esterification of Carboxylic Acids. *Catalysis Letters*, *128*(1-2), 183–189. doi:10.1007/s10562-008-9729-5

Vijayaraghavan, K., Palanivelu, K., & Velan, M. (2006). Biosorption of copper(II) and cobalt(II) from aqueous solutions by crab shell particles. *Bioresource Technology*, *97*(12), 1411–1419. doi:10.1016/j.biortech.2005.07.001 PMID:16112568

Vilgis, T. A. (2005). Time scales in the reinforcement of elastomers. *Polymer*, *46*(12), 4223–4229. doi:10.1016/j.polymer.2005.02.060

Visser, A. E., Swatloski, R. P., Griffin, S. T., Hartman, D. H., & Rogers, R. D. (2001). Liquid/liquid extraction of metal ions in room temperature ionic liquids. *Separation Science and Technology*, *36*(5-6), 785–804. doi:10.1081/SS-100103620

Visser, A. E., Swatloski, R. P., Reichert, W. M., Griffin, S. T., & Rogers, R. D. (2000). Traditional Extractants in Non-traditional Solvents: Groups 1 and 2 Extraction by Crown Ethers in Room-Temperature Ionic Liquids. *Industrial & Engineering Chemistry Research*, *39*(10), 3596–3604. doi:10.1021/ie000426m

Visser, A. E., Swatloski, R. P., Reichert, W. M., Mayton, R., Sheff, S., Wierzbicki, A., & Rogers, R. D. et al. (2002). Task Specific Ionic Liquids Incorporating Novel Cations for the Coordination and Extraction of Hg $^{2+}$ and Cd $^{2+}$: Synthesis, Characterization and Extraction Studies. *Environmental Science & Technology*, *36*(11), 2523–2529. doi:10.1021/es0158004 PMID:12075816

Vollmer, M. (2004). Physics of the microwave oven. *Physics Education*, *39*(1), 74–81. doi:10.1088/0031-9120/39/1/006

Vrsaljko, D., Macut, D., & Kovačević, V. (2014). Potential role of nanofillers as compatibilizers in immiscible PLA/LDPE blends. *Journal of Applied Polymer Science*, *132*(6), 41414–41428.

Walden, P. (1914). Article. Bull. Acad. Imper. Sci.(St. Petersburg), 1800.

Wang, R. (2000). Performance improvement of adsorption heat pump by heat and mass recovery operations. *IV Minsk International Seminar "Heat Pipes, Heat Pumps, Refrigerators"*.

Wang, G., Tong, X., & Zhao, Y. (2004). Preparation of azobenzene-containing amphiphilic diblock copolymers for light-responsive micellar aggregates. *Macromolecules*, *37*(24), 8911–8917. doi:10.1021/ma048416a

Wang, G., & Wu, P. (2015). In-depth study of the phase separation behaviour of a thermoresponsive ionic liquid and poly(ionic liquid) in concentrated aqueous solution. *Soft Matter*, *11*(26), 5253–5264. doi:10.1039/C5SM00603A PMID:26052832

Wang, H., Li, H., Wang, Y., & Yan, F. (2015). Preparation of macroporous ceramic from metakaolinite-based geopolymer by calcination. *Ceramics International*, *41*(99, Part A), 11177–11183. doi:10.1016/j.ceramint.2015.05.067

Wang, J.-Q., Huang, Y.-X., Pan, Y., & Mi, J.-X. (2014). Hydrothermal synthesis of high purity zeolite A from natural kaolin without calcination. *Microporous and Mesoporous Materials*, *199*, 50–56. doi:10.1016/j.micromeso.2014.08.002

Wang, J., Zhou, X., & Xiao, H. (2013). Structure and properties of cellulose/poly (N-isopropylacrylamide) hydrogels prepared by SIPN strategy. *Carbohydrate Polymers*, *94*(2), 749–754. doi:10.1016/j.carbpol.2013.01.036 PMID:23544629

Wang, K., Wu, J. Y., Wang, R. Z., & Wang, L. W. (2006a). Composite adsorbent of CaCl2 and expanded graphite for adsorption ice maker on fishing boats. *International Journal of Refrigeration*, *29*(2), 199–210. doi:10.1016/j.ijrefrig.2005.06.004

Wang, K., Wu, J. Y., Wang, R. Z., & Wang, L. W. (2006b). Effective thermal conductivity of expanded graphite–CaCl2 composite adsorbent for chemical adsorption chillers. *Energy Conversion and Management*, *47*(13–14), 1902–1912. doi:10.1016/j.enconman.2005.09.005

Wang, L. F., Rhim, J. W., & Hong, S. I. (2016). Preparation of poly(lactide)/poly(butylene adipate-co-terephthalate) blend films using a solvent casting method and their food packaging application. *LWT-. Food Science and Technology (Campinas.)*, *68*, 454–461.

Wang, L. W., Metcalf, S. J., Critoph, R. E., Thorpe, R., & Tamainot-Telto, Z.. (2012). Development of thermal conductive consolidated activated carbon for adsorption refrigeration. *Carbon*, *50*(3), 977–986. doi:10.1016/j.carbon.2011.09.061

Wang, L. W., Metcalf, S. J., Wang, R., Wang, L., & Wu, J. (2014). *Adsorption Refrigeration Technology*. Wiley. doi:10.1002/9781118197448

Wang, L. W., Tamainot-Telto, Z., Thrope, R., Critoph, R. E., & Metcalf, S. J. (2011). Study of thermal conductivity permeability and adsorption performance of solidified composite activated carbon adsorbent for refrigeration. *Renewable Energy*, *36*(8), 2062–2066. doi:10.1016/j.renene.2011.01.005

Wang, L. W., Wang, R. Z., Lu, Z. S., Chen, C. J., Wang, K., & Wu, J. Y. (2006). The performance of two adsorption ice making test units using activated carbon and a carbon composite as adsorbents. *Carbon*, *44*(13), 2671–2680. doi:10.1016/j.carbon.2006.04.013

Wang, L. W., Wang, R. Z., & Oliveira, R. G. (2009a). A review on adsorption working pairs for refrigeration. *Renewable & Sustainable Energy Reviews*, *13*(3), 518–534. doi:10.1016/j.rser.2007.12.002

Wang, L. W., Wu, J. Y., Wang, R. Z., Xu, Y. X., Wang, S. G., & Li, X. R. (2003). Study of the performance of activated carbon–methanol adsorption systems concerning heat and mass transfer. *Applied Thermal Engineering*, *23*(13), 1605–1617. doi:10.1016/S1359-4311(03)00104-2

Wang, L., & Dang, Z. M. (2005). Carbon nanotube composites with high dielectric constant at low percolation threshold. *Applied Physics Letters*, *87*, 0429031–0429033.

Wang, L., Dormer, N. H., Bonewald, L. F., & Detamore, M. S. (2010). Osteogenic differentiation of human umbilical cord mesenchymal stromal cells in polyglycolic acid scaffolds. *Tissue Engineering. Part A*, *16*(6), 1937–1948. doi:10.1089/ten.tea.2009.0706 PMID:20070186

Wang, R. (2001). Performance improvement of adsorption cooling by heat and mass recovery operation. *International Journal of Refrigeration*, *24*(7), 602–611. doi:10.1016/S0140-7007(01)00004-4

Wang, R. Z., & Oliveira, R. G. (2005). Adsorption refrigeration- an efficient way to make good use of waste heat and solar energy.*International Sorption Heat Pump Conference*.

Wang, R. Z., Wu, J. Y., Xu, Y. X., & Wang, W. (2001). Performance researches and improvements on heat regenerative adsorption refrigerator and heat pump. *Energy Conversion and Management*, *42*(2), 233–249. doi:10.1016/S0196-8904(99)00189-2

Wang, S., Li, L., & Zhu, Z. H. (2007). Solid-state conversion of fly ash to effective adsorbents for Cu removal from wastewater. *Journal of Hazardous Materials*, *139*(2), 254–259. doi:10.1016/j.jhazmat.2006.06.018 PMID:16839666

Wang, S., Li, Y., Xiang, H., Zhou, Z., Chang, T., & Zhu, M. (2015). Low cost carbon fibers from bio-renewable Lignin/poly(lactic acid) (PLA) blends. *Composites Science and Technology*, *119*, 20–25. doi:10.1016/j.compscitech.2015.09.021

Wang, W., Yang, H., Cui, H., Zhang, D., Liu, Y., & Chen, J. (2011). Application of Bifunctional Ionic Liquid Extractants [A336][CA-12] and [A336][CA-100] to the Lanthanum Extraction and Separation from Rare Earths in the Chloride Medium. *Industrial & Engineering Chemistry Research*, *50*(12), 7534–7541. doi:10.1021/ie2001633

Wang, Y., Li, H., & Han, S. (2006). A theoretical investigation of the interactions between water molecules and ionic liquids. *The Journal of Physical Chemistry B*, *110*(48), 24646–24651. doi:10.1021/jp064134w PMID:17134227

Wang, Y., Zhang, L., Tang, C., & Yu, D. (2000). Preparation and characterization of rubber-clay nanocomposites. *Journal of Applied Polymer Science*, *78*(11), 1879–1883. doi:10.1002/1097-4628(20001209)78:11<1879::AID-APP50>3.0.CO;2-1

Watanabe, K., Tabuchi, M., Morinaga, Y., & Yoshinaga, F. (1998). Structural Features and Properties of Bacterial Cellulose Produced in Agitated Culture. *Cellulose (London, England)*, *5*(3), 187–200. doi:10.1023/A:1009272904582

Webb, K., Hlady, V., & Tresco, P. A. (1998). Relative importance of surface wettability and charged functional groups on NIH 3T3 fibroblast attachment, spreading, and cytoskeletal organization. *Journal of Biomedical Materials Research*, *41*(3), 422–430. doi:10.1002/(SICI)1097-4636(19980905)41:3<422::AID-JBM12>3.0.CO;2-K PMID:9659612

Wei, G. T., Yang, Z., & Chen, C. J. (2003). Room Temperature Ionic Liquid as a Novel Medium for Liquid/Liquid Extraction of Metal Ions. *Analytica Chimica Acta*, *488*(2), 183–192. doi:10.1016/S0003-2670(03)00660-3

Weigel, T., Schinkel, G., & Lendlein, A. (2006). Design and preparation of polymeric scaffolds for tissue engineering. *Expert Review of Medical Devices*, *3*(6), 835–851. doi:10.1586/17434440.3.6.835 PMID:17280547

Wei, Q., Nagi, R., Sadeghi, K., Feng, S., Yan, E., Ki, S. J., & Ozcan, A. et al. (2014). Detection and spatial mapping of mercury contamination in water samples using a smart-phone. *ACS Nano*, *8*(2), 1121–1129. doi:10.1021/nn406571t PMID:24437470

Welton, T. (1999). Room-Temperature Ionic Liquids. Solvents for Synthesis and Catalysis. *Chemical Reviews*, *99*(8), 2071–2084. doi:10.1021/cr980032t PMID:11849019

Weng, P., Wei, Q., Tang, Z., Lin, T., & Guo, B. (2015). The influence of molybdenum disulfide nanoplatelets on the dispersion of nano silica in natural rubber composites. *Applied Surface Science*, *359*, 782–789. doi:10.1016/j.apsusc.2015.10.172

West, N. K. (1990). Properties and Applications of Oxide Layers Deposited on Glass from Organic Solutions. *Chemical Reviews*, *90*, 33–72.

White, T., & Stegemann, J. A. (2001). Advances in Environmental Materials, Volume II Environmentally preferred materials. MRS (Singapore).

Widegren, J. A., Laesecke, A., & Magee, J. W. (2005). The effect of dissolved water on the viscosities of hydrophobic room temperature ionic liquids. *Chemical Communications*, (12): 1610–1612. doi:10.1039/b417348a PMID:15770276

Widegren, J. A., Saurer, E. M., Marsh, K. N., & Magee, J. W. (2005). Electrolytic conductivity of four imidazolium-based room-temperature ionic liquids and the effect of a water impurity. *The Journal of Chemical Thermodynamics*, *37*(6), 569–575. doi:10.1016/j.jct.2005.04.009

Wiegand, C., Elsner, P., Hipler, U.-C., & Klemm, D. (2006). Protease and ROS activities influenced by a composite of bacterial cellulose and collagen type I in vitro. *Cellulose (London, England)*, *13*(6), 689–696. doi:10.1007/s10570-006-9073-0

Williams, L. (2001). Environmental chemistry a modular approach. John Wiley & Sons, LTD.

Williams, D. F. (2004). *The Biomaterials Silver Jubilee Compendium*. Amsterdam: Elsevier B. V.

Wnuk, R., & Chmiel, H. (1992). Direct heating of composite membranes in pervaporation and gas separation processes. *Journal of Membrane Science*, *68*(3), 293–300. doi:10.1016/0376-7388(92)85030-M

Wong Y. W. H., Leung, M. Y. S., Ku, S. K. A., & Lam, H. L. I. (2006). Selected applications of nanotechnology in textiles. *AUTEX Research Journal, 6*, 1-8.

Wongsampanwech, S., & Potiyaraj, P. (2014). Toughening of poly(lactic acid) by blending with poly (ethylene-co-octene). *Advanced Materials Research*, *1025-1026*, 461–465. doi:10.4028/www.scientific.net/AMR.1025-1026.461

Wooding, C. (2001). *Regenerated cellulose fibers*. Cambridge, UK: Woodhead Publishing Limited. doi:10.1533/9781855737587

World Health Organization. (2011). Guidelines for drinking water quality (4th ed.). Author.

Wu, H., Wan, S., & Zhu, D. (2007). Effects of impregnating variables on dynamic sorption characteristics and storage properties of composite sorbnet for solar heat storage. *Solar Energy*, *81*(7), 864–871. doi:10.1016/j.solener.2006.11.013

Wu, J., Fukuzawa, K., & Ohtani, J. (1990). Lignin analysis in some tropical hardwoods using ultraviolet microscopy, Res. Bull. Of College Exp. For. *Hokkaido University*, *47*, 353–361.

Wu, Y. P., Wang, Y. Q., Zhang, H. F., Wang, Y. Z., Yu, D. S., Zhang, L. Q., & Yang, J. (2005). Rubber–pristine clay nanocomposites prepared by co-coagulating rubber latex and clay aqueous suspension. *Composites Science and Technology*, *65*, 1195–1202.

Xiao, L., Wang, B., Yang, G., & Gauthier, M. (2012). Poly(lactic acid)-based biomaterials: Synthesis, modification and applications. In D. N. Ghista (Ed.), *Biomedical Science, Engineering and Technology* (pp. 247–282). Rijeka, Croatia: Intech Europe. doi:10.5772/23927

Xie, H., Liu, B., Yuan, Z., Shen, J., & Cheng, R. (2004). Cure kinetics of carbon nanotube/tetrafunctional epoxy nanocomposites by isothermal differential scanning calorimetry. *Journal of Polymer Science. Part B, Polymer Physics*, *42*(20), 3701–3712. doi:10.1002/polb.20220

Xing, S., Tan, L. H., Yang, M., Pan, M., Lv, Y., Tang, Q., & Chen, H. et al. (2009). Highly controlled core/shell structures: Tunable conductive polymer shells on gold nanoparticles and nanochains. *Journal of Materials Chemistry*, *19*(20), 3286–3291. doi:10.1039/b900993k

Xiong, W., Chen, L., Zhao, B., Wang, D.-y., & Wang, Y. (2012). Polyamide 6 with a flame retardant encapsulated by polyamide 66: Flame retardation, thermo-decomposition and the potential mechanism. *Chinese Journal of Polymer Science*, *30*(2), 297–307. doi:10.1007/s10118-012-1126-2

Xiu, H. J., Wang, Z. J., & Li, J. B. (2005). Effect of bacterial cellulose on paper property. *Chung-kuo Tsao Chih/China Pulp and Paper*, *24*(3), 14-17.

Xu, C., Chen, J., Wu, D., Chen, Y., Lv, Q., & Wang, M. (2016). Polylactide/acetylated nanocrystalline cellulose composites preparedby a continuous route: A phase interface-property relation study. *Carbohydrate Polymers*, *146*, 58–66. doi:10.1016/j.carbpol.2016.03.058 PMID:27112851

Xue, R., Li, J. Y.-S., Yeh, Y., Yang, L., & Chien, S. (2013). Effects of matrix elasticity and cell density on human mesenchymal stem cells differentiation. *Journal of Orthopaedic Research*, *31*(9), 1360–1365. doi:10.1002/jor.22374 PMID:23606500

Xu, G. C., Li, A. Y., Zhang, L. D., Wu, G. S., Yuan, X. Y., & Xie, T. (2003). Synthesis and characterization of silica nanocomposite in situ photopolymerization. *Journal of Applied Polymer Science*, *90*(8), 837–840. doi:10.1002/app.12650

Xu, H., & Van Deventer, J. S. J. (2002). Geopolymerisation of multiple minerals. *Minerals Engineering*, *15*(12), 1131–1139. doi:10.1016/S0892-6875(02)00255-8

Xu, J., & Song, J. (2015). Polylactic acid (PLA)-based shape-memory materials for biomedical applications. In L. Yahia (Ed.), *Shape Memory Polymers for Biomedical Applications* (pp. 197–217). London, UK: Woodhead Publishing. doi:10.1016/B978-0-85709-698-2.00010-6

Xu, T., Jia, Z., Luo, Y., Jia, D., & Peng, Z. (2015). Interfacial interaction between the epoxidized natural rubber and silica in natural rubber/silica composites. *Applied Surface Science*, *328*, 306–313. doi:10.1016/j.apsusc.2014.12.029

Xu, Y., Zhang, Y., & Suhir, E. (2006). Thermal properties of carbon nanotube array used for integrated circuit cooling. *Journal of Applied Physics*, *100*, 074306.

Yamak, H. B. (2013). *Emulsion Polymerization: Effects of Polymerization Variables on the Properties of Vinyl Acetate Based Emulsion Polymers. InTech Journal.*

Yaman, S. (2004). Pyrolisis of biomass to produce fuels and chemical feedstock. *Energy Conversion and Management, 45*(5), 651–671. doi:10.1016/S0196-8904(03)00177-8

Yamaoka, H., Ogasawara, T., Nishizawa, H., Takahashi, T., Nakatsuka, T., Koshima, I., & Hoshi, K. et al. (2006). Cartilage tissue engineering using human auricular chondrocytes embedded in different hydrogel materials. *Journal of Biomedical Materials Research. Part A, 78*(1), 1–8. doi:10.1002/jbm.a.30655 PMID:16596585

Yamashita, A., Yoshida, K., Oshima, T., & Baba, Y. (2015). Synthesis of an Aminophenylthiomethyl Chitosan Derivative and Its Adsorption of Precious Metals. *Journal of Chemical Engineering of Japan, 48*(11), 897–902. doi:10.1252/jcej.15we025

Yamasue, E., Minamino, R., Numata, T., Nakajima, K., Murakami, S., Daigo, I., & Ishihara, K. N. et al. (2009). Novel Evaluation Method of Elemental Recyclability from Urban Mine. *Materials Transactions, 50*(6), 1536–1540. doi:10.2320/matertrans.MBW200816

Yan, F., & Texter, J. (2007). Solvent-reversible poration in ionic liquid copolymers. *Angewandte Chemie International Edition, 119*(14), 2492–2495. doi:10.1002/ange.200604127 PMID:17328017

Yang, F., Baba, Y., Kubota, F., Kamiya, N., & Goto, M. (2012). Extraction and separation of rare earth metal ions with DODGAA in ionic liquids. *Solvent Extr. Res. Dev., 19*(0), 69–76. doi:10.15261/serdj.19.69

Yang, F., & Nelson, G. L. (2004). PMMA/silica nanocomposite studies: Synthesis and properties. *Journal of Applied Polymer Science, 91*(6), 3844–3850. doi:10.1002/app.13573

Yang, H., Wang, W., Cui, H., Zhang, D., Liu, Y., & Chen, J. (2012). Recovery of rare earth elements from simulated fluorescent powder using bifunctional ionic liquid extractants (Bif-ILEs). *Journal of Chemical Technology and Biotechnology (Oxford, Oxfordshire), 87*(2), 198–205. doi:10.1002/jctb.2696

Yang, J., Kubota, F., Baba, Y., Kamiya, N., & Goto, M. (2014). One Step Effective Separation of Platinum and Palladium in an Acidic Chloride Solution by Using Undiluted Ionic Liquids. *Solvent Extr. Res. Dev., 21*(2), 129–135. doi:10.15261/serdj.21.129

Yang, J., Kubota, F., Baba, Y., Kamiya, N., & Goto, M. (2014). Separation of Precious Metals by Using Undiluted Ionic Liquids. *Solvent Extr. Res. Dev., 21*(1), 89–94. doi:10.15261/serdj.21.89

Yang, R. T. (2003). *Adsorbents: Fundamentals and applications*. John Wiley & Sons. doi:10.1002/047144409X

Yang, T. C., & Zall, R. R. (1984). Absorption of metals by natural polymers generated from seafood processing wastes. *Industrial & Engineering Chemistry Product Research and Development, 23*(1), 168–172. doi:10.1021/i300013a033

Yan, L., Shan, X., Wen, B., & Owens, G. (2008). Adsorption of cadmium onto Al_{13}-pillared acid-activated montmorillonite. *Journal of Hazardous Materials, 156*(1-3), 499–508. doi:10.1016/j.jhazmat.2007.12.045 PMID:18243531

Yano, H., Sugiyama, J., Nakagaito, A. N., Nogi, M., Matsuura, T., Hikita, M., & Handa, K. (2005). Optically transparent composites reinforced with networks of bacterial nanofibers. *Advanced Materials, 17*(2), 153–155. doi:10.1002/adma.200400597

Yao, X., Zhang, Z., Zhu, H., & Chen, Y. (2009). Geopolymerization process of alkali–metakaolinite characterized by isothermal calorimetry. *Thermochimica Acta, 493*(1–2), 49–54. doi:10.1016/j.tca.2009.04.002

Yaragallaa, S., Meera, A. P., Kalarikkal, N., & Thomas, S. (2015). Chemistry associated with natural rubber–graphene nanocomposites and its effect on physical and structural properties. *Industrial Crops and Products*, *74*, 792–802. doi:10.1016/j.indcrop.2015.05.079

Yeo, T. H. C., Tan, I. A. W., & Abdullah, M. O. (2012). Development of adsorption air-conditioning technology using modified activated carbon – A review. *Renewable & Sustainable Energy Reviews*, *16*(5), 3355–3363. doi:10.1016/j.rser.2012.02.073

Yin, N., Chen, S., Li, Z., Ouyang, Y., Hu, W., Tang, L., & Wang, H. et al. (2012). Porous bacterial cellulose prepared by a facile surfactant-assisted foaming method in azodicarbonamide-NaOH aqueous solution. *Materials Letters*, *81*, 131–134. doi:10.1016/j.matlet.2012.04.133

Yoon, S. J., Lee, J. G., Tajima, H., Yamasaki, A., Kiyono, F., Nakazato, T., & Tao, H. (2010). Extraction of lanthanide ions from aqueous solution by bis(2-ethylhexyl)phosphoric acid with room-temperature ionic liquids. *Journal of Industrial and Engineering Chemistry*, *16*(3), 350–354. doi:10.1016/j.jiec.2009.09.063

Yoshimoto, Y., Kaneko, K., & Onodera, M. (2009). Influence of Fatty Acid Methy Ester Composition on the Combustion Characteristics of Biodiesel Fuels. The Japan Society of Mechanical Engineers, 75(752), 847-854.

Yoshinaga, F., Tonouchi, N., & Watanabe, K. (1997). Research progress in production of bacterial cellulose by aeration and agitation culture and its application as a new industrial material. *Bioscience, Biotechnology, and Biochemistry*, *61*(2), 219–224. doi:10.1271/bbb.61.219

Yoshino, T., Asakura, T., & Toda, K. (1996). Cellulose production by Acetobacter pasteurianus on silicone membrane. *Journal of Fermentation and Bioengineering*, *81*(1), 32–36. doi:10.1016/0922-338X(96)83116-3

Yuan, J. J., Armes, S. P., Takabayashi, Y., Prassides, K., Leite, C. A. P., Galembeck, F., & Lewis, A. L. (2006). Synthesis of Biocompatible Poly[2-(methacryloyloxy)ethyl phosphorylcholine]-Coated Magnetite Nanoparticles. *Langmuir*, *22*(26), 10989–10993. doi:10.1021/la061834j PMID:17154575

Yuan, J., Mecerreyes, D., & Antonietti, M. (2013). Poly(ionic liquid)s: An update. *Progress in Polymer Science*, *38*(7), 1009–1036. doi:10.1016/j.progpolymsci.2013.04.002

Yun, J., Im, J. S., Lee, Y.-S., Bae, T.-S., Lim, Y.-M., & Kim, H.-I. (2010). pH and electro-responsive release behavior of MWCNT/PVA/PAAc composite microcapsules. *Colloids and Surfaces. A, Physicochemical and Engineering Aspects*, *368*(1–3), 23–30. doi:10.1016/j.colsurfa.2010.07.010

Yu, Y., Hu, T., Chen, X., Xu, K., Zhang, J., & Huang, J. (2011). Pd nanoparticles on a porous ionic copolymer: A highly active and recyclable catalyst for Suzuki-Miyaura reaction under air in water. *Chemical Communications*, *47*(12), 3592–3594. doi:10.1039/c0cc04498a PMID:21298188

Yu, Y., Ouyang, C., Gao, Y., Si, Z., Chen, W., Wang, Z., & Xue, G. (2005). Synthesis and characterization of carbon nanotube/polypyrrole core-shell nanocomposites via in situ inverse micro emulsion. *Journal of Polymer Science. Part A, Polymer Chemistry*, *43*(23), 6105–6115. doi:10.1002/pola.21114

Zaborowska, M., Bodin, A., Bäckdahl, H., Popp, J., Goldstein, A., & Gatenholm, P. (2010). Microporous bacterial cellulose as a potential scaffold for bone regeneration. *Acta Biomaterialia*, *6*(7), 2540–2547. doi:10.1016/j.actbio.2010.01.004 PMID:20060935

Zadegan, S., Hosainalipour, M., Rezaie, H. R., Ghassai, H., & Shokrgozar, M. A. (2011). Synthesis and biocompatibility evaluation of cellulose/ hydroxyapatite nanocomposite scaffold in 1-n-allyl-3-methylimidazolium chloride. *Materials Science and Engineering C*, *31*(5), 954–961. doi:10.1016/j.msec.2011.02.021

Zeng, J.-B., Lia, K.-A., & Du, A.-K. (2015). Compatibilization strategies in poly(lactic acid)-based blends. *RSC Advances*, *5*(41), 32546–32565. doi:10.1039/C5RA01655J

Zeng, Q. H., Wang, D. Z., Yu, A. B., & Lu, G. Q. (2002). Synthesis of polymer montmorillonite nanocomposites by in situ intercalative polymerization. *Nanotechnology*, *13*(13), 549–553. doi:10.1088/0957-4484/13/5/301

Zeng, Z., Yu, J., & Guo, Z.-X. (2004). Preparation of Carboxyl-Functionalized Polystyrene/Silica Composite Nanoparticles. *Macromolecular Chemistry and Physics*, *205*(16), 2197–2204. doi:10.1002/macp.200400201

Zermane, F., Bouras, O., Baudu, M., & Basly, J.-P. (2010). Cooperative coadsorption of 4-nitrophenol and basic yellow 28 dye onto an iron organo–inorgano pillared montmorillonite clay. *Journal of Colloid and Interface Science*, *350*(1), 315–319. doi:10.1016/j.jcis.2010.06.040 PMID:20638666

Zhang, K., Fu, Q., Huang, Y., & Zhou, D. Encapsulation of inorganic particles by dispersion polymerization through ultrasonic irradiation. *Science in China Series B: Chemistry*, *48*(6), 545-552. doi: 10.1360/042004-100

Zhang, L., Wang, Y., Wang, Y., Sui, Y., & Yu, D. (2000). Morphology and mechanical properties of clay/styrene-butadiene rubber nanocomposites. *Journal of Applied Polymer Science*, *78*(11), 1873–1878. doi:10.1002/1097-4628(20001209)78:11<1873::AID-APP40>3.0.CO;2-8

Zhang, Q., Wu, S., Zhang, L., Lu, J., Verproot, F., Liu, Y., & Song, X. M. et al. (2011). Fabrication of polymeric ionic liquid/graphene nanocomposite for glucose oxidase immobilization and direct electrochemistry. *Biosensors & Bioelectronics*, *26*(5), 2632–2637. doi:10.1016/j.bios.2010.11.024 PMID:21159504

Zhang, X., & Zhang, Y. (2016). Reinforcement effect of poly(butylene succinate) (PBS)-graftedcellulose nanocrystal on toughened PBS/polylactic acid blends. *Carbohydrate Polymers*, *140*, 374–382. doi:10.1016/j.carbpol.2015.12.073 PMID:26876864

Zhang, Z. H., Zhu, H. J., Zhou, C. H., & Wang, H. (2016). Geopolymer from kaolin in China: An overview. *Applied Clay Science*, *119*(Part 1), 31–41. doi:10.1016/j.clay.2015.04.023

Zhao, Y. (2011). *Study of activated carbon/methanol adsorption refrigeration tube and system integration* (Master Science thesis). Adelaide University.

Zhao, G. L., Bagayoko, D., & Yang, L. (2006). Optical properties of aligned carbon nanotube mats for photonic applications. *Journal of Applied Physics*, *99*(11), 114311–114315. doi:10.1063/1.2201738

Zhao, Q., Heyda, J., Dzubiella, J., Täuber, K., Dunlop, J. W. C., & Yuan, J. (2015). Sensing solvents with ultrasensitive porous poly(ionic liquid) actuators. *Advanced Materials*, *27*(18), 2913–2917. doi:10.1002/adma.201500533 PMID:25828569

Zhao, Y. J., Wang, L. W., Wan, R. Z., Ma, K. Q., & Jiang, L. (2013). Study on consolidated activated carbon: Choice of optimal adsorbent for refrigeration application. *International Journal of Heat and Mass Transfer*, *67*, 867–876. doi:10.1016/j.ijheatmasstransfer.2013.08.052

Zhao, Y. J., Wang, R. Z., Wang, L. W., & Yu, N. (2014). Development of highly conductive KNO3/NaNO3 composite for TES (thermal energy storage). *Energy*, *70*, 272–27. doi:10.1016/j.energy.2014.03.127

Zhao, Y., Ye, J., Lu, X., Liu, M., Lin, Y., Gong, W., & Ning, G. (2010). Preparation of sintered foam materials by alkali-activated coal fly ash. *Journal of Hazardous Materials*, *174*(1–3), 108–112. doi:10.1016/j.jhazmat.2009.09.023 PMID:19775812

Zheng, Y. Z., Wang, N. N., Luo, J. J., Zhou, Y., & Yu, Z. W. (2013). Hydrogen-bonding interactions between $[BMIM][BF_4]$ and acetonitrile. *Physical Chemistry Chemical Physics*, *15*(41), 18055–18064. doi:10.1039/c3cp53356e PMID:24061707

Zhijiang, C., & Guang, Y. (2011). Bacterial cellulose/collagen composite: Characterization and first evaluation of cytocompatibility. *Journal of Applied Polymer Science*, *120*(5), 2938–2944. doi:10.1002/app.33318

Zhong, Y., Critoph, R. E., & Thorpe, R. (2006). Evaluation of the performance of solid sorption refrigeration systems using carbon dioxide as refrigerant. *Applied Thermal Engineering*, *26*(16), 1807–1811. doi:10.1016/j.applthermaleng.2006.02.006

Zhong, Y., Critoph, R. E., Thorpe, R. N., & Tamainot-Telto, Z. (2009). Dynamics of BaCl2–NH3 adsorption pair. *Applied Thermal Engineering*, *29*(5-6), 1180–1186. doi:10.1016/j.applthermaleng.2008.06.015

Zhong, Z., Jian, Y., & Zhaoxia, G. (2006). Synthesis of composite nanoparticles bearing epoxy functional groups: Encapsulation of silica by emulsion polymerization of GMA. *Journal of Wuhan University of Technology-Mater. Science Editor*, *21*(2), 136–138. doi:10.1007/bf02840860

Zhou, W., Hu, X., Bai, X., Zhou, S., Sun, C., Yan, J., & Chen, P. (2011). Synthesis and Electromagnetic, Microwave Absorbing Properties of Core–Shell Fe3O4–Poly (3, 4-ethylenedioxythiophene) Microspheres. *ACS Applied Materials & Interfaces*, *3*(10), 3839–3845. doi:10.1021/am2004812 PMID:21913665

Zhou, X., Wu, T., Hu, B., Yang, G., & Han, B. (2010). Synthesis of graphene/polyaniline composite nanosheets mediated by polymerized ionic liquid. *Chemical Communications*, *46*(21), 3663–3665. doi:10.1039/c0cc00049c PMID:20396832

Zhu, W., Wu, P., Yang, L., Chang, Y., Chao, Y., Li, H., & Xun, S. et al. (2013). Pyridinium-based temperature-responsive magnetic ionic liquid for oxidative desulfurization of fuels. *Chemical Engineering Journal*, *229*, 250–256. doi:10.1016/j.cej.2013.05.115

Zimmermann, M. V. G., Borsoi, C., Lavoratti, A., Zanini, M., Zattera, A. J., & Santana, R. M. C. (2016). Drying techniques applied to cellulose nanofibers. *Journal of Reinforced Plastics and Composites*, *35*(8), 682–697. doi:10.1177/0731684415626286

Zuo, S., & Zhou, R. (2006). Al-pillared clays supported rare earths and palladium catalysts for deep oxidation of low concentration of benzene. *Applied Surface Science*, *253*(5), 2508–2514. doi:10.1016/j.apsusc.2006.05.119

Zuo, Y., Chen, J., & Li, D. (2008). Reversed micellar solubilization extraction and separation of thorium(IV) from rare earth(III) by primary amine N1923 in ionic liquid. *Separation and Purification Technology*, *63*(3), 684–690. doi:10.1016/j.seppur.2008.07.014

Zuo, Y., Liu, Y., Chen, J., & Li, D. Q. (2008). The separation of cerium(IV) from nitric acid solutions containing thorium(IV) and lanthanides(III) using pure [C8mim]PF6 as extracting phase. *Industrial & Engineering Chemistry Research*, *47*(7), 2349–2355. doi:10.1021/ie071486w

Zuo, Y., Liu, Y., Chen, J., & Li, D. Q. (2009). Extraction and recovery of cerium(IV) along with fluorine(I) from bastnasite leaching liquor by DEHEHP in [C_8mim]PF_6. *Journal of Chemical Technology and Biotechnology (Oxford, Oxfordshire)*, *84*(7), 949–956. doi:10.1002/jctb.2116

About the Contributors

Takaomi Kobayashi received his PhD in Osaka University, Faculty of Science. (1988). He worked in Nagaoka University of Technology (NUT) as research associate (1991). In 1995, he was visiting researcher in Emory University,USA, and then associate professor and professor (2006) in NUT at Department of Materials Science and Technology. On 2016, he also holds the post of professor of Science of Technology Innovation. Intelligent materials and their chemistry relating with polymer science and technology are his research interests. Recent researches are focused on sustainable technologies in waste biomass for the regeneration technologies and waste water treatments by using environmental materials.

* * *

Kanoktip Boonkerd received a Ph.D degree in Polymer Science from the University of Akron, Ohio, USA in 2006. Since 2006, she became a lecturer at Department of Materials Science, Faculty of Science, Chulalongkorn University, where she is now an assistant professor. During 2011-2013, she worked as a visiting researcher for 18 months at Gyeongsang National University, Republic of South Korea. Her current research interests include the development of novel fillers from the agricultural wastes, the rubber formulation for engineering construction, and chemical modification of rubber for advanced application.

Anyaporn Boonmahitthisud is a lecturer at Department of Materials Science, Faculty of Science, Chulalongkorn University, Thailand. Her bachelor, master and Phd were from Chulalonkorn University, Thailand. During 2015-2016, she worked as a post-doctoral fellow in Nagaoka University of Technology, Niigata, Japna. She has eleven publications in international journals and one book chapters to her credit. Her areas of interest include organic-inorganic nanohybrid materials, rubber and polymer nanocomposites, recycling of plastics, sustainable materials from natural resources and biodegradable plastics.

Karla Tovar Carrillo received her Bachelor Degree in Chemistry from the Universidad Autonoma de Cd. Juarez in 2003. In 2005, she was hired to work and coordinate the Chemistry Laboratories of all Enginnering Programs of the Universidad Autonoma de Cd. Juarez. In 2008, she received her Master Degree in Materials Science from Universidad Autonoma de Cd. Juarez. Between 2008 and 2010 she was invited to 3 Research Stays at Nagaoka University of Technology in Japan. She received her Doctoral Degree in Engineering from the Nagaoka University of Technology in 2014. In 2014 she was hired as Research Professor at the Universidad Autonoma de Cd. Juarez, in the Biomedical Institute to Development Research in Medical and Odontology fields developing medical materials based on Cellulose Hydrogels.

Ahmed Elsayed is a research associate at University of Bath - UK with a PhD in thermal engineering from university of Birmingham 2011 experiencing in modelling and testing of boiling and condensation heat transfer. During his career he joined several research projects including adsorption cooling systems for Weatherite Air Conditioning Limited 2012-2014 with KTP project, Stirling engines research program at university of Birmingham 2014-2015, and Computational fluid dynamics analysis for various exhaust systems using OpenFoam at Coventry University 2015-2016. He completed several consultation projects including development of thermosyphon performance software tool for Sterling Thermal Technology, enhanced coated condensers for Alkemy. Currently, he is investigating heating systems in energy efficient buildings. Dr Elsayed has more than 30 published articles in peer reviewed international journals, conferences and book chapters.

Pham Thanh Huyen received her PhD from the Hanoi University of Science and Technology (HUST) in 2004. She has been teaching in HUST for nearly 20 years and is currently the Deputy Head of Department of Organic and Petrochemical Technology in School of Chemical Engineering. She has been working on several projects related to the synthesis of microporous and mesoporous materials and their application as adsorbent and catalysis materials, especially on the conversion of clays to porous materials. She is the author of 1 invention application and 72 papers in international and domestic journals.

Kai Li completed his master degree in Polymer Chemistry and Physics from Zhengzhou University in 2013. During Sep. 2011 to Jun. 2013, he was a Joint Training Master between Zhengzhou University and Shanghai Institute of Organic Chemistry, Chinese Academy of Sciences. He received his Ph.D. degree in Energy and Environment Science at Nagaoka University of Technology in Aug. 2016 and then is working as PhD in McGill University, Canada. His research interests are ionic liquids, green chemistry, ultrasound and biomaterials.

Bi-min Zhang Newby received both her MS and PhD in Chemical Engineering from Lehigh University (Bethlehem, Pennsylvania, USA) in 1995 and 1999, respectively. Subsequent to her graduation, she worked as a post-doctoral researcher in the Laboratory of Research on the Structure of Matter (LRSM) at the University of Pennsylvania from August 1998 to June 2000. She joined the Department of Chemical and Biomolecular Engineering at The University of Akron in Fall of 2000. She is a full professor currently in the same department. Her research interests are in the areas of surface modification/patterning, and polymeric materials especially biomaterials.

Sujitra Onutai received her Bachelor's Degree in Materials Science from Chiangmai University in 2009. After graduation, she continued with her Master Degree in Materials Science, Faculty of Science, Chulalongkorn University. In 2013, she was doctoral students under Double Doctoral Degree Program from collaboration with Chulalongkorn University and Nagaoka University of Technology. She graduated the first of doctoral degree at Nagaoka University of Technology in 2016. She studied about geopolymer materials holds master and doctoral degrees. She is currently studying about geopolymer materials at Chulalongkorn University.

Muenduen Phisalaphong received her MS and PhD in Chemical Engineering from Lehigh University (Bethlehem, Pennsylvania, USA) and Colorado State University (Fort Collins, Colorado, USA) in 1996 and 1999, respectively. At present, she is an associate professor in the Department of Chemical Engineering, Faculty of Engineering, Chulalongkorn University. She had served as head of the biochemical engineering research group from 2006 to 2016. Her present research interests focus on biomaterials and bioprocess engineering such as bacterial cellulose composite as a yeast cell carrier for ethanol production. She has publications in international peer-reviewed journals and book chapters on the development of new bacterial cellulose-based materials.

Pranut Potiyaraj received his first degree in Materials Science from Chulalongkorn University. Since 1995, he became a lecturer at Department of Materials Science, Faculty of Science, Chulalongkorn University, where he is now an associate professor. In 2000, he received his Doctoral degree in Textiles from the University of Manchester (formerly UMIST). His current research interests include technical textiles, bioplastics as well as novel 2D materials.

Minoru Satoh received his Doctor of Science from Tohoku University in 1991. He worked at Yamagata Research Institute of Technology, as a Researcher, and Ibaraki National College of Technology, Department of Industrial Chemistry, as an assistant professor, an associate professor, and then a professor. He became a professor in the Department of Materials Science and Technology, Nagaoka University of Technology in 2014. Since 2015 he has been working as a professor in the Division of Department of Chemistry and Material Engineering, National Institute of Technology, Ibaraki College. He has been studying the removal of heavy metal ions by using some ligands.

Yukiko Takahashi received her Ph.D. in Engineering from Tohoku University in 1999. From 2000 she participated in some research projects as a postdoctoral fellow in National Institute of Advanced Industrial Science and Technology (AIST) and is currently associate professor for Department of Materials Science and Technology at Nagaoka University of Technology. She is carrying out basic and applied research for organic dye nanoparticles from the viewpoint of analytical chemistry and photochemistry.

Siriporn Taokaew received both her MS and PhD in Chemical Engineering from Chulalongkorn University (Bangkok, Thailand) in 2010 and 2014, respectively. After that, she was hired as a postdoctoral researcher by Chulalongkorn University. In 2015, she was hired as a research associate by the department of Chemical and Biomolecular Engineering, the University of Akron (Akron, Ohio, USA). Currently, she is an assistant professor in the Department of Materials Science and Technology, Graduate School of Engineering, Nagaoka University of Technology (Nagaoka, Japan). Her recent research interest is the development of biomaterials such as bacterial cellulose to be used in biomedical application. She has authored and coauthored peer-reviewed international journals and a book chapter on bacterial cellulose-based materials.

Index

F

G

H

I

K

L

M

N

O

P

R

S

T

U

V

W

X

Y

Z

www.ingramcontent.com/pod-product-compliance
Lightning Source LLC
LaVergne TN
LVHW060400300926
849162LV00006B/1313